Methods in Molecular Biology

Series Editor
John M. Walker
School of Life and Medical Sciences
University of Hertfordshire
Hatfield, Hertfordshire, AL10 9AB, UK

For further volumes:
http://www.springer.com/series/7651

G Protein-Coupled Receptor Signaling

Methods and Protocols

Edited by

Mario Tiberi

Ottawa Hospital Research Institute (Neurosciences), Ottawa, ON, Canada

Department of Cellular and Molecular Medicine, University of Ottawa Brain and Mind Research Institute, University of Ottawa, Ottawa, ON, Canada

Humana Press

Editor
Mario Tiberi
Ottawa Hospital Research Institute (Neurosciences)
Ottawa, ON, Canada

Department of Cellular and Molecular Medicine
University of Ottawa Brain and Mind Research Institute
University of Ottawa
Ottawa, ON, Canada

ISSN 1064-3745 ISSN 1940-6029 (electronic)
Methods in Molecular Biology
ISBN 978-1-4939-9120-4 ISBN 978-1-4939-9121-1 (eBook)
https://doi.org/10.1007/978-1-4939-9121-1

Library of Congress Control Number: 2019932813

Cover Art: Geneviève Laroche and Patrick Giguère

This Humana Press imprint is published by the registered company Springer Science+Business Media, LLC, part of Springer Nature.
The registered company address is: 233 Spring Street, New York, NY 10013, U.S.A.

Preface

G protein-coupled receptors (GPCRs) represent the largest transmembrane protein families in the proteome of most living organisms. They have evolved as key communication players by virtue of sensing extracellular cues such as light, odorants, tastants, and also a wide variety of chemically distinct molecules found in our body and environment. The sensing step triggers a series of intricate conformational changes in GPCRs which allow reshaping the receptor to recruit additional signaling proteins to ultimately relay the information inside cells via chemical reactions. This modus operandi permits GPCR systems to regulate most physiological functions in humans such as, for instance, vision, olfaction, smell, blood pressure, locomotion, memory, and thought process. In line with their widespread role in humans, deficiencies in GPCR systems have been linked to numerous diseases. In fact, approximately 40% of currently prescribed drugs target GPCRs. In spite of decades of GPCR research, a full knowledge of GPCR biology has yet to be achieved. This is of medical importance for improving current available therapeutics with improved efficacy and fewer undesirable side effects. In this volume, I have assembled chapters that comprehensively describe protocols to assist with the study of structural, molecular, cell biological, and in vivo facets of GPCRs and to enable the development of experimental tools for screening novel GPCR drugs. Protocols will be of interest to life scientists working in a variety of research fields including molecular pharmacology, cell and developmental biology, brain behavior and physiology, and drug development and screening.

This volume would not have been possible without the help of key people. First and foremost, I want to thank all the contributors for their efforts. I am enormously indebted to all of them for their willingness and generosity in providing their time and expertise to diligently write their expert protocols. I am also extremely grateful to Dr. John Walker, the editor in chief of Methods in Molecular Biology series, for his great guidance and support throughout the making of this volume. I want also to thank Patrick Marton, the executive editor of *Springer Protocols*, and Kursad Turksen, editor in chief of Springer's *Stem Cells and Regenerative Medicine*, for their enthusiastic support and valuable advice in the initial phase of making this volume. Lastly, my heartfelt thanks go to David C. Casey, the editor of *Springer Protocols*, for his efforts to assist me with the completion of this volume.

Ottawa, ON, Canada ***Mario Tiberi***

Preface

[illegible]

[illegible] assist me with the completion of this volume.

Ottawa, ON, Canada *Mario Tiberi*

Contents

Preface . . . . . *v*
Contributors . . . . . *xi*

PART I PROBING STRUCTURAL PROPERTIES OF GPCRs

1 Assessment of Conformational State Transitions of Class B GPCRs Using Molecular Dynamics 3
Chenyi Liao, Victor May, and Jianing Li

2 Molecular Dynamic Simulations to Probe Water Permeation Pathways of GPCRs 21
Katsufumi Tomobe, Eiji Yamamoto, and Kenji Yasuoka

3 Expression and Purification of a Functional *E. coli* $^{13}CH_3$-Methionine-Labeled Thermostable Neurotensin Receptor 1 Variant for Solution NMR Studies 31
Fabian Bumbak, Ross A. D. Bathgate, Daniel J. Scott, and Paul R. Gooley

4 A Combined Cell-Free Protein Synthesis and Fluorescence-Based Approach to Investigate GPCR Binding Properties 57
Anne Zemella, Theresa Richter, Lena Thoring, and Stefan Kubick

PART II TWEAKING LIGANDS FOR GPCR STUDIES

5 Furan Cross-Linking Technology for Investigating GPCR–Ligand Interactions 81
Marleen Van Troys, Willem Vannecke, Christophe Ampe, and Annemieke Madder

6 Optical Regulation of Class C GPCRs by Photoswitchable Orthogonal Remotely Tethered Ligands 103
Amanda Acosta-Ruiz, Johannes Broichhagen, and Joshua Levitz

7 Chemoselective Acylation of Hydrazinopeptides to Access Fluorescent Probes for Time-Resolved FRET Assays on GPCRs 137
Sridévi M. Ramanoudjame, Lucie Esteoulle, Stéphanie Riché, Jean-François Margathe, Thierry Durroux, Iuliia A. Karpenko, and Dominique Bonnet

PART III BIOLUMINESCENCE AND FLUORESCENCE RESONANCE TRANSFER ENERGY APPROACHES

8 Time-Resolved FRET-Based Assays to Characterize G Protein-Coupled Receptor Hetero-oligomer Pharmacology 151
Joyce Heuninck, Candide Hounsou, Elodie Dupuis, Eric Trinquet, Bernard Mouillac, Jean-Philippe Pin, Dominique Bonnet, and Thierry Durroux

9 Combining Conformational Profiling of GPCRs with CRISPR/Cas9 Gene Editing Approaches ... 169
Kyla Bourque, Dominic Devost, Asuka Inoue, and Terence E. Hébert

10 Measuring GPCR Stoichiometry Using Types-1, -2, and -3 Bioluminescence Resonance Energy Transfer-Based Assays ... 183
James H. Felce, John R. James, and Simon J. Davis

11 Combining SRET^2 and BiFC to Study GPCR Heteromerization and Protein–Protein Interactions ... 199
Amina M. Bagher, Melanie E. M. Kelly, and Eileen M. Denovan-Wright

12 Quantification and Comparison of Signals Generated by Different FRET-Based cAMP Reporters ... 217
Andreas Koschinski and Manuela Zaccolo

PART IV EXPLORING GPCR SIGNALING PROPERTIES

13 Measuring GPCR-Induced Activation of Protein Tyrosine Phosphatases (PTP) Using In-Gel and Colorimetric PTP Assays ... 241
Geneviève Hamel-Côté, Fanny Lapointe, and Jana Stankova

14 Measurement of β-Arrestin Recruitment at GPCRs Using the Tango Assay ... 257
Geneviève Laroche and Patrick M. Giguère

15 Probing the Interactome of Corticotropin-Releasing Factor Receptor Heteromers Using Mass Spectrometry ... 269
Burcu Hasdemir, Juan A. Oses-Prieto, Alma Burlingame, and Aditi Bhargava

PART V VISUALIZING SUBCELLULAR COMPARTMENTALIZATION OF GPCRS

16 Monitoring the Aggregation of GPCRs by Fluorescence Microscopy ... 289
Samuel Génier, Jade Degrandmaison, Christine L. Lavoie, Louis Gendron, and Jean-Luc Parent

17 Combining RNAi and Immunofluorescence Approaches to Investigate Post-endocytic Sorting of GPCRs into Multivesicular Bodies ... 303
Xuezhi Li, Stéphanie Rosciglione, Andréanne Laniel, and Christine Lavoie

18 Super-Resolution Imaging of G Protein-Coupled Receptors Using Ground State Depletion Microscopy ... 323
Fabiana A. Caetano Crowley, Bryan Heit, and Stephen S. G. Ferguson

19 Analysis of Spatial Assembly of GPCRs Using Photoactivatable Dyes and Localization Microscopy ... 337
Kim C. Jonas and Aylin C. Hanyaloglu

Part VI Investigating In Vivo Functions of GPCRs

20 Optical Modulation of Metabotropic Glutamate Receptor Type 5 In Vivo Using a Photoactive Drug 351
Marc López-Cano, Joan Font, Amadeu Llebaria, Víctor Fernández-Dueñas, and Francisco Ciruela

21 Imaging of Tissue-Specific and Temporal Activation of GPCR Signaling Using DREADD Knock-In Mice 261
Dmitry Akhmedov, Nicholas S. Kirkby, Jane A. Mitchell, and Rebecca Berdeaux

22 Laser Doppler Flowmetry to Study the Regulation of Cerebral Blood Flow by G Protein-Coupled Receptors in Rodents 377
Xavier Toussay, Mario Tiberi, and Baptiste Lacoste

23 Preparation of Pulmonary Artery Myocytes and Rings to Study Vasoactive GPCRs 389
Martha Hinton, Anurag Singh Sikarwar, and Shyamala Dakshinamurti

Index *403*

Contributors

AMANDA ACOSTA-RUIZ • *Department of Biochemistry, Weill Cornell Medicine, New York, NY, USA*

DMITRY AKHMEDOV • *Department of Integrative Biology and Pharmacology, McGovern Medical School, University of Texas Health Science Center at Houston, Houston, TX, USA*

CHRISTOPHE AMPE • *Department of Biochemistry, Faculty of Medicine and Health Sciences, Ghent University, Ghent, Belgium; Department of Biomolecular Medicine, Faculty of Medicine and Health Sciences, Ghent University, Ghent, Belgium*

AMINA M. BAGHER • *Department of Pharmacology, Dalhousie University, Halifax, NS, Canada; Department of Pharmacology and Toxicology, King AbdulAziz University, Jeddah, Saudi Arabia*

ROSS A. D. BATHGATE • *Department of Biochemistry and Molecular Biology, The University of Melbourne, Parkville, VIC, Australia; The Florey Institute of Neuroscience and Mental Health, The University of Melbourne, Parkville, VIC, Australia*

REBECCA BERDEAUX • *Department of Integrative Biology and Pharmacology, McGovern Medical School, University of Texas Health Science Center at Houston, Houston, TX, USA; Program in Biochemistry and Cell Biology, MD Anderson Cancer Center-UTHealth Graduate School of Biomedical Sciences, Houston, TX, USA*

ADITI BHARGAVA • *The Osher Center for Integrative Medicine, University of California, San Francisco, San Francisco, CA, USA; Department of Ob-Gyn, University of California, San Francisco, San Francisco, CA, USA*

DOMINIQUE BONNET • *Laboratoire d'Innovation Thérapeutique, Faculté de Pharmacie, UMR7200 CNRS/Université de Strasbourg, LabEx MEDALIS, Illkirch, France*

KYLA BOURQUE • *Department of Pharmacology and Therapeutics, McGill University, Montreal, QC, Canada*

JOHANNES BROICHHAGEN • *Department of Chemical Biology, Max Planck Institute for Medical Research, Heidelberg, Germany*

FABIAN BUMBAK • *Department of Biochemistry and Molecular Biology, The University of Melbourne, Parkville, VIC, Australia; Bio21 Molecular Science and Biotechnology Institute, The University of Melbourne, Parkville, VIC, Australia; The Florey Institute of Neuroscience and Mental Health, The University of Melbourne, Parkville, VIC, Australia*

ALMA BURLINGAME • *Department of Pharmaceutical Chemistry, University of California, San Francisco, San Francisco, CA, USA*

FABIANA A. CAETANO CROWLEY • *Department of Physiology and Pharmacology, Western University, London, ON, Canada*

FRANCISCO CIRUELA • *Unitat de Farmacologia, Departament Patologia i Terapèutica Experimental, Facultat de Medicina i Ciències de la Salut, IDIBELL, L'Hospitalet de Llobregat, Universitat de Barcelona, Barcelona, Spain; Institut de Neurociències, Universitat de Barcelona, Barcelona, Spain*

SHYAMALA DAKSHINAMURTI • *Biology of Breathing Group, Children's Hospital Research Institute of Manitoba, Winnipeg, MB, Canada; Department of Physiology, University of Manitoba, Winnipeg, MB, Canada; Departments of Pediatrics, University of Manitoba, Winnipeg, MB, Canada*

Simon J. Davis • *Medical Research Council Human Immunology Unit, Radcliffe Department of Clinical Medicine, Weatherall Institute of Molecular Medicine, University of Oxford, Oxford, UK*

Jade Degrandmaison • *Département de Médecine, Faculté de Médecine et des Sciences de la Santé, Institut de Pharmacologie de Sherbrooke, Université de Sherbrooke, Sherbrooke, QC, Canada*

Eileen M. Denovan-Wright • *Department of Pharmacology, Dalhousie University, Halifax, NS, Canada*

Dominic Devost • *Department of Pharmacology and Therapeutics, McGill University, Montreal, QC, Canada*

Elodie Dupuis • *Cisbio, Codolet, France*

Thierry Durroux • *CNRS, UMR 5203, Institut de Génomique Fonctionnelle, Montpellier, France; INSERM, U. 1191, Montpellier, France; Université de Montpellier, Montpellier, France*

Lucie Esteoulle • *Laboratoire d'Innovation Thérapeutique, Faculté de Pharmacie, UMR 7200 CNRS/Université de Strasbourg, LabEx MEDALIS, Illkirch, France*

James H. Felce • *Kennedy Institute of Rheumatology, University of Oxford, Oxford, UK*

Stephen S. G. Ferguson • *Department of Cellular and Molecular Medicine, University of Ottawa Brain and Mind Research Institute, University of Ottawa, Ottawa, ON, Canada*

Víctor Fernández-Dueñas • *Unitat de Farmacologia, Departament Patologia i Terapèutica Experimental, Facultat de Medicina i Ciències de la Salut, IDIBELL, L'Hospitalet de Llobregat, Universitat de Barcelona, Barcelona, Spain; Institut de Neurociències, Universitat de Barcelona, Barcelona, Spain*

Joan Font • *MCS, Laboratory of Medicinal Chemistry, Institute for Advanced Chemistry of Catalonia (IQAC-CSIC), Barcelona, Spain*

Louis Gendron • *Département de Pharmacologie-Physiologie, Faculté de Médecine et des Sciences de la Santé, Institut de Pharmacologie de Sherbrooke, Université de Sherbrooke, Sherbrooke, QC, Canada*

Samuel Génier • *Département de Médecine, Faculté de Médecine et des Sciences de la Santé, Institut de Pharmacologie de Sherbrooke, Université de Sherbrooke, Sherbrooke, QC, Canada*

Patrick M. Giguère • *Department of Biochemistry, Microbiology and Immunology, University of Ottawa Brain and Mind Research Institute, University of Ottawa, Ottawa, ON, Canada*

Paul R. Gooley • *Department of Biochemistry and Molecular Biology, The University of Melbourne, Parkville, VIC, Australia; Bio21 Molecular Science and Biotechnology Institute, The University of Melbourne, Parkville, VIC, Australia*

Geneviève Hamel-Côté • *Division of Immunology, Department of Pediatrics, Faculty of Medicine and Health Sciences, Université de Sherbrooke, Sherbrooke, QC, Canada*

Aylin C. Hanyaloglu • *Institute of Reproductive and Developmental Biology, Imperial College London, London, UK*

Burcu Hasdemir • *The Osher Center for Integrative Medicine, University of California, San Francisco, San Francisco, CA, USA; Department of Ob-Gyn, University of California, San Francisco, San Francisco, CA, USA*

Terence E. Hébert • *Department of Pharmacology and Therapeutics, McGill University, Montreal, QC, Canada*

Bryan Heit • *Department of Microbiology and Immunology, Western University, London, ON, Canada*

Joyce Heuninck • *CNRS, UMR 5203, Institut de Génomique Fonctionnelle, Montpellier, France; INSERM, U. 1191, Montpellier, France; Université de Montpellier, Montpellier, France*

Martha Hinton • *Biology of Breathing Group, Children's Hospital Research Institute of Manitoba, Winnipeg, MB, Canada*

Candide Hounsou • *CNRS, UMR 5203, Institut de Génomique Fonctionnelle, Montpellier, France; INSERM, U. 1191, Montpellier, France; Université de Montpellier, Montpellier, France*

Asuka Inoue • *Graduate School of Pharmaceutical Sciences, Tohoku University, Sendai, Miyagi, Japan; Japan Science and Technology Agency (JST), Precursory Research for Embryonic Science and Technology (PRESTO), Kawaguchi, Saitama, Japan*

John R. James • *Warwick Medical School, University of Warwick, Coventry, UK*

Kim C. Jonas • *Molecular and Clinical Sciences Research Institute, St George's University of London, London, UK; Institute of Medical and Biomedical Education, St George's University of London, London, UK; Institute of Reproductive and Developmental Biology, Imperial College London, London, UK*

Iuliia A. Karpenko • *Laboratoire d'Innovation Thérapeutique, Faculté de Pharmacie, UMR7200 CNRS/Université de Strasbourg, LabEx MEDALIS, Illkirch, France*

Melanie E. M. Kelly • *Department of Pharmacology, Dalhousie University, Halifax, NS, Canada; Department of Ophthalmology and Visual Sciences, Dalhousie University, Halifax, NS, Canada*

Nicholas S. Kirkby • *Cardiothoracic Pharmacology, National Heart and Lung Institute, Imperial College London, London, UK*

Andreas Koschinski • *Department of Physiology, Anatomy and Genetics, University of Oxford, Oxford, UK*

Stefan Kubick • *Cell-free and Cell-based Bioproduction, Fraunhofer Institute for Cell Therapy and Immunology (IZI), Branch Bioanalytics and Bioprocesses (IZI-BB), Potsdam, Germany*

Baptiste Lacoste • *Ottawa Hospital Research Institute (Neurosciences), Ottawa, ON, Canada; Department of Cellular and Molecular Medicine, University of Ottawa Brain and Mind Research Institute, University of Ottawa, Ottawa, ON, Canada*

Andréanne Laniel • *Département de Pharmacologie-Physiologie, Faculté de Médecine et des Sciences de la Santé, Institut de Pharmacologie de Sherbrooke, Université de Sherbrooke, Sherbrooke, QC, Canada*

Fanny Lapointe • *Division of Immunology, Department of Pediatrics, Faculty of Medicine and Health Sciences, Université de Sherbrooke, Sherbrooke, QC, Canada*

Geneviève Laroche • *Department of Biochemistry, Microbiology and Immunology, University of Ottawa Brain and Mind Research Institute, University of Ottawa, Ottawa, ON, Canada*

Christine L. Lavoie • *Département de Pharmacologie-Physiologie, Faculté de Médecine et des Sciences de la Santé, Institut de Pharmacologie de Sherbrooke, Université de Sherbrooke, Sherbrooke, QC, Canada*

Joshua Levitz • *Department of Biochemistry, Weill Cornell Medicine, New York, NY, USA*

Jianing Li • *Department of Chemistry, The University of Vermont, Burlington, VT, USA*

Xuezhi Li • *Département de Pharmacologie-Physiologie, Faculté de Médecine et des Sciences de la Santé, Institut de Pharmacologie de Sherbrooke, Université de Sherbrooke, Sherbrooke, QC, Canada*

Chenyi Liao • *Department of Chemistry, The University of Vermont, Burlington, VT, USA*

AMADEU LLEBARIA • *MCS, Laboratory of Medicinal Chemistry, Institute for Advanced Chemistry of Catalonia (IQAC-CSIC), Barcelona, Spain*
MARC LÓPEZ-CANO • *Unitat de Farmacologia, Departament Patologia i Terapèutica Experimental, Facultat de Medicina i Ciències de la Salut, IDIBELL, L'Hospitalet de Llobregat, Universitat de Barcelona, Barcelona, Spain; Institut de Neurociències, Universitat de Barcelona, Barcelona, Spain*
ANNEMIEKE MADDER • *Organic and Biomimetic Chemistry Research Group, Department of Organic and Macromolecular Chemistry, Ghent University, Ghent, Belgium*
JEAN-FRANÇOIS MARGATHE • *Laboratoire d'Innovation Thérapeutique, Faculté de Pharmacie, UMR 7200 CNRS/Université de Strasbourg, LabEx MEDALIS, Illkirch, France*
VICTOR MAY • *Department of Neurological Sciences, Larner College of Medicine, The University of Vermont, Burlington, VT, USA*
JANE A. MITCHELL • *Cardiothoracic Pharmacology, National Heart and Lung Institute, Imperial College London, London, UK*
BERNARD MOUILLAC • *CNRS, UMR 5203, Institut de Génomique Fonctionnelle, Montpellier, France; INSERM, U. 1191, Montpellier, France; Université de Montpellier, Montpellier, France*
JUAN A. OSES-PRIETO • *Pharmaceutical Chemistry, University of California San Francisco, San Francisco, CA, USA*
JEAN-LUC PARENT • *Département de Médecine, Faculté de Médecine et des Sciences de la Santé, Institut de Pharmacologie de Sherbrooke, Université de Sherbrooke, Sherbrooke, QC, Canada*
JEAN-PHILIPPE PIN • *CNRS, UMR 5203, Institut de Génomique Fonctionnelle, Montpellier, France; INSERM, U. 1191, Montpellier, France; Université de Montpellier, Montpellier, France*
SRIDÉVI M. RAMANOUDJAME • *Laboratoire d'Innovation Thérapeutique, Faculté de Pharmacie, UMR7200 CNRS/Université de Strasbourg, LabEx MEDALIS, Illkirch, France*
STÉPHANIE RICHÉ • *Laboratoire d'Innovation Thérapeutique, Faculté de Pharmacie, UMR7200 CNRS/Université de Strasbourg, LabEx MEDALIS, Illkirch, France*
THERESA RICHTER • *Cell-free and Cell-based Bioproduction, Fraunhofer Institute for Cell Therapy and Immunology (IZI), Branch Bioanalytics and Bioprocesses (IZI-BB), Potsdam, Germany; University of Potsdam, Potsdam, Germany*
STÉPHANIE ROSCIGLIONE • *Département de Pharmacologie-Physiologie, Faculté de Médecine et des Sciences de la Santé, Institut de Pharmacologie de Sherbrooke, Université de Sherbrooke, Sherbrooke, QC, Canada*
DANIEL J. SCOTT • *Department of Biochemistry and Molecular Biology, The University of Melbourne, Parkville, VIC, Australia; The Florey Institute of Neuroscience and Mental Health, The University of Melbourne, Parkville, VIC, Australia*
ANURAG SINGH SIKARWAR • *Biology of Breathing Group, Children's Hospital Research Institute of Manitoba, Winnipeg, MB, Canada; Department of Oral Biology, University of Manitoba, Winnipeg, MB, Canada*
JANA STANKOVA • *Division of Immunology, Department of Pediatrics, Faculty of Medicine and Health Sciences, Université de Sherbrooke, Sherbrooke, QC, Canada*
LENA THORING • *Cell-free and Cell-based Bioproduction, Fraunhofer Institute for Cell Therapy and Immunology (IZI), Branch Bioanalytics and Bioprocesses (IZI-BB), Potsdam, Germany*

MARIO TIBERI • *Ottawa Hospital Research Institute (Neurosciences), Ottawa, ON, Canada; Department of Cellular and Molecular Medicine, University of Ottawa Brain and Mind Research Institute, University of Ottawa, Ottawa, ON, Canada*

KATSUFUMI TOMOBE • *Department of Mechanical Engineering, Keio University, Yokohama, Japan*

XAVIER TOUSSAY • *Ottawa Hospital Research Institute (Neurosciences), Ottawa, ON, Canada; Department of Cellular and Molecular Medicine, University of Ottawa Brain and Mind Research Institute, University of Ottawa, Ottawa, ON, Canada*

ERIC TRINQUET • *Cisbio, Codolet, France*

MARLEEN VAN TROYS • *Department of Biochemistry, Faculty of Medicine and Health Sciences, Ghent University, Ghent, Belgium; Department of Biomolecular Medicine, Faculty of Medicine and Health Sciences, Ghent University, Ghent, Belgium*

WILLEM VANNECKE • *Organic and Biomimetic Chemistry Research Group, Department of Organic and Macromolecular Chemistry, Ghent University, Ghent, Belgium*

EIJI YAMAMOTO • *Department of System Design Engineering, Keio University, Yokohama, Japan*

KENJI YASUOKA • *Department of Mechanical Engineering, Keio University, Yokohama, Japan*

MANUELA ZACCOLO • *Department of Physiology, Anatomy and Genetics, University of Oxford, Oxford, UK*

ANNE ZEMELLA • *Cell-free and Cell-based Bioproduction, Fraunhofer Institute for Cell Therapy and Immunology (IZI), Branch Bioanalytics and Bioprocesses (IZI-BB), Potsdam, Germany*

Part I

Probing Structural Properties of GPCRs

Chapter 1

Assessment of Conformational State Transitions of Class B GPCRs Using Molecular Dynamics

Chenyi Liao, Victor May, and Jianing Li

Abstract

Class B G protein-coupled receptors (GPCRs) comprise a family of 15 peptide-binding members, which are crucial targets for endocrine, metabolic, and stress-related disorders. While their protein structures and dynamics remain largely unclear, computer modeling and simulations represent a promising means to help solve such puzzles. Herein, we present a basic introduction to the methodology of molecular dynamics (MD) simulations and two analytical methods to assess the conformational ensembles and transitions of Class B GPCRs, using our recent studies of the human pituitary adenylate cyclase activating polypeptide (PAC_1) receptor as an example. From long MD simulations, conformational ensembles with different roles in ligand binding and receptor activation are sampled to establish four states identified as either "open" or "closed" for the PAC_1 receptor. Next, the dynamical network can be applied to analyze the simulations and identify key features within each conformational ensemble, which help distinguish the ligand-bound states of the PAC_1 receptor from the ligand-free one. Further, the Markov State Model has emerged as a key approach to construct the transition network and connect the GPCR ensembles, providing detailed information for the transition pathways and kinetics. For the ligand-free PAC_1 receptor, the transitions within the closed states are near 10–30 times faster than the open-closed transitions, which is likely related to the activation mechanism of the receptor. Overall, long MD simulations and analyses are useful to assess conformational transitions for the Class B GPCRs and to gain mechanistic insight, which is difficult to obtain using other methods.

Key words Multiscale modeling, Molecular dynamics, Conformational ensemble, Markov state model, Communication networks

1 Introduction

G protein-coupled receptors (GPCRs), the largest family of transmembrane proteins in the human genome, transduce a variety of signals, including hormones, neurotransmitters, odorants, tastants, and light, to regulate virtually all physiological responses for homeostasis [1, 2]. The more than 800 canonical heptahelical receptors are divided into 5 major classes based on sequence and structural similarities: rhodopsin (Class A); secretin (Class B); glutamate (Class C); adhesion and frizzled/taste. Class A is the largest

Mario Tiberi (ed.), *G Protein-Coupled Receptor Signaling: Methods and Protocols*, Methods in Molecular Biology, vol. 1947,
https://doi.org/10.1007/978-1-4939-9121-1_1,

class of more than 700 members with ~40 unique structures available. With just 15 members, the Class B secretin/glucagon/VIP family of GPCRs is critically important with respect to neural development, body calcium homeostasis, glucose metabolism, circadian rhythm, thermoregulation, inflammation, feeding behavior, pain modulation, stress and related endocrine responses [3–5]. Accordingly, these receptors are pharmacological targets for a variety of disorders including osteoporosis, hypercalcemia, type 2 diabetes, obesity, migraine and related chronic pain disorders, anxiety and depression. Despite the great pharmacological interests, only two full-length Class B receptor structures [6–8] have been determined. The three-dimensional molecular structures remain entirely or partially elusive among most Class B members. As a result, the detailed knowledge of receptor activation and regulation, such as the transitions among various states, is still incomplete. Therefore, for insight into ligand selectivity, activation, and regulation mechanisms, it is crucial to apply computer modeling to study the conformational states, as well as the associated transition pathways. Such insight will also serve as the foundation to design structure- and mechanism-based strategies to modulate Class B GPCRs.

Class B GPCRs are activated by well-studied peptide hormones, such as corticotropin-releasing factor (CRF), calcitonin gene-related peptide (CGRP), glucagon, glucagon-like peptide (GLP), pituitary adenylate cyclase activating polypeptide (PACAP), parathyroid hormone (PTH), secretin, and vasoactive intestinal peptide (VIP). The functional states are associated with distinct protein conformations as well as the presence of various ligands. One of the major challenges in studying the conformational transitions of Class B receptors is capturing the dynamics of the full-length receptor structure to accurately simulate the two-domain binding model for receptor activation [9]. In contrast to Class A GPCRs, which contain only the heptahelical transmembrane domain (7TM), each Class B receptor possesses an additional extracellular domain (ECD) of ~120 amino acid residues that is crucial for high affinity peptide binding and dynamics that allow bound-ligand presentation to the 7TM for receptor activation. Given such structural complexity, the choreography of Class B receptor activation—induced by the neuropeptide—is likely different from that of Class A receptors.

Although only few are available in full-length structures in Class B family, *e.g.* the glucagon receptor (GCGR, PDBIDs: 5XEZ and 5YQZ) and glucagon-like peptide 1 receptor (GLP-1R, PDBID: 5NX2), the construction of full-length models of other Class B receptors is viable with current knowledge of the 7TM and ECD structures. The 7TM structures of four members had been determined between 2013 and mid-2018: GCGR (PDBIDs: 5XEZ, 5YQZ, 5EE7, and 4L6R), corticotropin-

releasing factor receptor (CRF_1R, PDBID: 4K5Y), GLP-1R (PDBIDs: 5NX2, 5VAI, 5VEW, and 5VEX), and calcitonin receptor (PDBID: 5UZ7). The ECD structures of nine members are known: glucose-dependent insulinotropic polypeptide (GIP) receptor (PDBID: 2QKH), GCGR (PDBIDs: 4ERS and 4LF3), GLP-1R (PDBIDs: 3IOL, 5E94, 3C59, 3C5T, and 4ZGM), PACAP receptor type 1 (PAC_1R, PDBIDs: 2JOD and 3N94), VIP/PACAP receptor type 2 ($VPAC_2R$, PDBID: 2X57), PTH_1 receptor (PDBIDs: 3CM4 and 3H3G), CRF_1R (PDBIDs: 2L27, 3EHS, and 3EHU), CRF_2R (PDBIDs: 1U34 and 3N96), and CGRP receptor (PDBIDs: 3N7P, 3N7R, and 3N7S). These experimental efforts allow the exploration of function-related conformational states and the assessment of transitions among these states via protein modeling and molecular dynamics (MD) simulations. Under physiological conditions, GPCRs are not static and often transition fluidly and fleetingly from one conformation to another, forming ensembles of receptor microstates that kinetically constitute macrostates. These constitutive macrostates may respond to binding of specific ligands and allosteric modulators or preferentially favor particular signaling events. Thus, long MD simulations (conventional or with enhanced sampling) followed by structural-based analysis are suitable to sample and distinguish these states, as well as to identify their transitions.

However, despite the large body of simulation studies of Class A GPCRs [10–16], the growth of MD simulations to investigate Class B GPCRs (with the full-length or 7TM models) has only emerged recently, due to the increasing amount of 7TM crystal structures. These simulation studies range from a few hundred nanoseconds to tens of microseconds and have provided valuable information regarding the molecular basis of receptor dynamics [17, 18], stabilizing effects from point mutations [19], hydrogen bonding network at an allosteric site [20], and inactive-to-active transitions focusing on TM displacement [21, 22]. Often in good agreement with experimental evidence, these recent studies provide invaluable mechanistic insight, which is typically costly and lengthy to study using experimental approaches alone. With advances in high-performance computing (HPC) [23, 24] and multiscale modeling technology [16, 17, 24–27], MD simulations have become useful tools to explore more complex systems (*i.e.*, Class B GPCRs in the oligomeric states, the ligand-bound states, or the G protein/arrestin-bound states) on biologically relevant timescales. In this chapter, we provide a basic introduction to the methodology of MD simulation and the analysis to study conformational transitions of Class B GPCRs involving large-scale domain motions and helical displacements, using our studies of the PAC_1 receptor as an example.

2 Methods

2.1 Description of the MD Protocol

2.1.1 Conventional MD Simulations

An MD simulation describes the motion of a collection of molecules over time in a system of interest according to the physical and chemical principles. Such a system often contains essential chemical/biological components (such as proteins, water, lipids, and ions) in a three-dimensional box at a condition to mimic real-world experiments (such as temperature and external pressure). Generally, in an all-atom model, each atom is represented by a particle at a specific position in the simulation box with periodic boundary conditions, while the covalent bonds are treated like springs. The interactions between particles are described by equations and parameters—the so-called force field. According to the Newton's laws, the particle motions are calculated in discrete time steps analogous to the film frames in a movie. The positions of atoms are updated from one step to the next; in a continuous fashion, as atoms move and time advances, a cinematic feature is constructed to show the conformational changes or transitions in the simulation box. In real practice, the time step for biological simulations is often chosen to be around 1 to 2 femtoseconds (fs). Therefore, for a typical simulation of 100 nanoseconds (ns), the number of steps approximates a million for over ten thousand atoms in the system; multiple simulation trajectories (or replicas) are needed for each model construct to ensure reliable data collection. With such high computational demands, MD simulations (*i.e.*, for GPCRs) often require supercomputing resources.

In one of our studies of the PAC_1 receptor, for example, each model system contains a PAC_1 receptor model, a lipid bilayer of ~219 1-palmitoyl-2-oleoyl-sn-glycero-3-phosphocholine (POPC) molecules, ~28,500 TIP3P water molecules, counter ions, and 0.15 M NaCl, totaling ~126,000 atoms in a periodic box of $95 \times 95 \times 134$ Å^3. All the simulations were performed with CHARMM36-cmap force field on the specialized Anton supercomputer using the Anton software 2.13.0 [23, 28]. Microsecond-long MD simulations (2–2.9 μs) for four conformational states were performed in an NPT ensemble (310 K, 1 bar, Berendesn thermostat and semi-isotropic barostat) with a time step of 2 fs (*see* **Note 1**).

2.1.2 Preparation of the Starting GPCR Models

The conformational transition between two states of biomolecules, initial state (A) and final state (B), is often under thermodynamic (downhill energy profile) and/or kinetic control (uphill energy profile) [29]. For systems where the probability of reaching B is very low, a strategy is to start with multiple intermediate points which may proceed to A or B while their early conformational evolutions can structurally overlap within a certain range (initial region). Our current class B GPCR study [17] indicated that in a

microsecond-long MD simulation, the dynamic free-swinging process of the ECD, exposed to the solution, occurred during the first few hundred nanoseconds, after which the ECD interacted with the extracellular loops (ECLs) of the 7TM to greatly restrict its mobility. The large-scale domain motions lead to different ECD orientations with slight positional variations at the ECD–7TM interface. Thus, it is appropriate to choose initial conformations that have ECD free of interactions with ECLs and of different orientations relative to 7TM to initiate the conformational sampling. Notably, these starting conformations are likely metastable, and thus require them to be well equilibrated in short MD simulations. In our preparation of the PAC_1 receptor models, four homology models, differing in the ECD orientations by dihedral rotation at the linker region, were generated from multiple short MD simulations for microsecond-long production simulations [17]. Additionally, the initial models can also be generated using enhanced sampling methods, such as replica exchange [30], adaptive tempering [31], steered MD [32], and accelerated MD [33]. In particular, these methods can accelerate a kinetic-control process, for example, the transition from inactive to active receptor states and vice versa involving the displacement of the intracellular structures related to transmembrane helix 6 (TM6) [21, 22].

2.2 Analytical Methods for Conformational Transitions

While MD simulations provide direct visualization of GPCR conformations and conformational changes, the rich detail of atom positions also allows qualitative and quantitative characterizations of each conformational state and the transition from one state to another. For example, from four microsecond-long MD simulations [17], we obtained two major conformations distinct in the ECD orientations as the closed (G1, G2, and G3) or open (G4) states (Fig. 1), which implicate differential roles in ligand binding and receptor activation. Herein, we introduce two typical methods to analyze the conformational ensembles of Class B GPCRs and their transitions.

2.2.1 Dynamical Network Analysis

Given the conformational ensembles of variant ECD orientations, analysis of the dynamical network has been applied to identify the shared features of interactions and communication within the ligand-free PAC_1 receptor (G1, G2, G3, and G4), in comparison with a trajectory in the ligand-bound state. Such analysis can be readily carried out with the *NetworkView* plugin [34–36] in the VMD program [37]. Dynamical networks were created from the last 150 ns of each trajectory. To define the dynamical network, each Cα atom of amino acid residues represents a node; two nodes are connected by an edge if any two of their heavy atoms are within 4.5 Å for more than 75% of the simulation time [36]. The edge distances d_{ij} are derived from the pairwise correlations (C_{ij}) calculated by the program Carma v1.3 [38]. The community

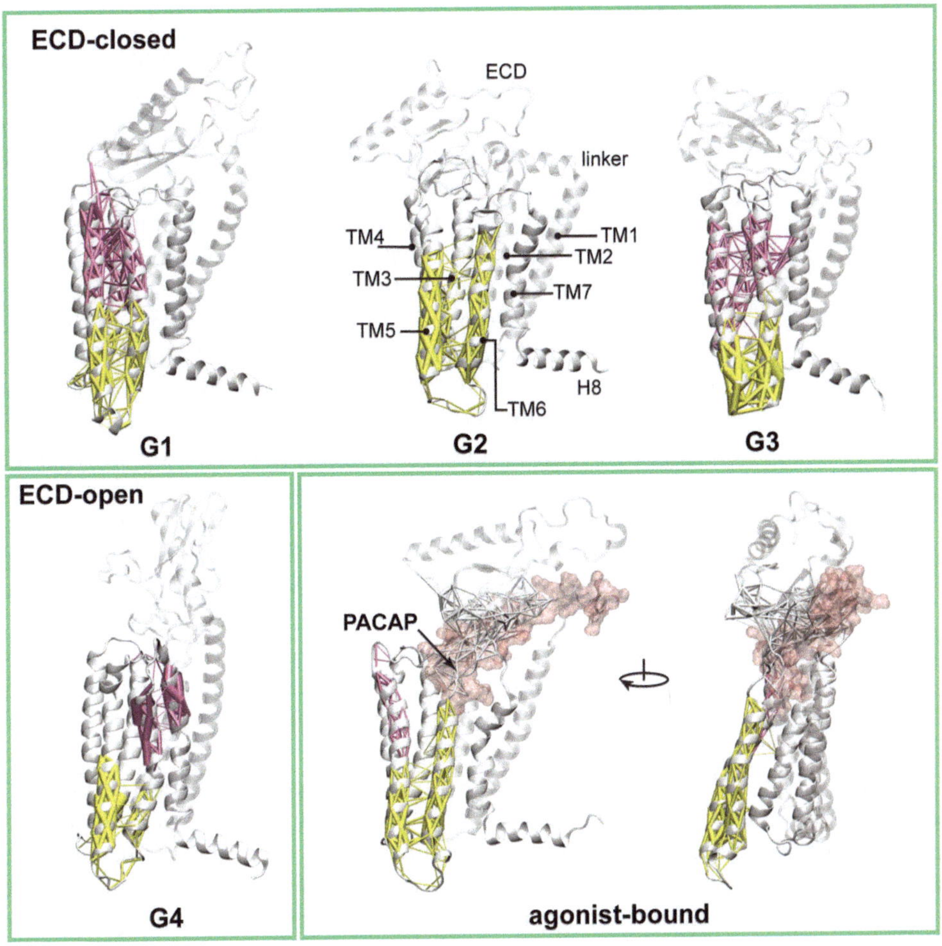

Fig. 1 Community analysis of the ligand-free ECD closed and open states, and the agonist-bound active state of the PAC_1 receptor. The community decompositions of TM6 are displayed with weighted edges (thicker edges show greater correlation). The dynamical networks were created from the last 150 ns of each trajectory. In the ligand-free states G1, G3, and G4, TM6 splits into two communities, its extracellular half joins communities containing TM3 or TM7. In G2, TM6 merges into the community with the intracellular half of TM3. In the agonist-bound conformation, the entire TM6 behaves as a single community with the intracellular half of TM5 and fewer correlations with TM3. Thus, in the agonist-bound PAC_1 receptor, residues within TM6 can propagate information relatively easily through multiple routes from the extracellular side to the intracellular face of the receptor without perturbations from other TM helices

substructure of the network is obtained by the Girvan–Newman algorithm [39]. With correlation-based weights, communities correspond to sets of residues that move in concert with each other. The connections between nodes (representing amino acid residues) within one community should be stronger than the connections between nodes across different communities.

TM6 and the third intracellular loop (ICL3) play an essential role in the signaling process of GPCRs—by facilitating the binding of G proteins or other effector proteins at the intracellular receptor

face after an outward conformational shift of the TM6/ICL3 segments (in the active state) [11]. The community decompositions of TM6 with weighted edges are displayed in Fig. 1 for four ligand-free states of the PAC_1 receptor, in comparison with a ligand-bound state. In the ligand-free G1, G3, and G4 states of the PAC_1 receptor, TM6 can be divided into two communities with its extracellular half (Fig. 1, purple) merging into communities containing TM3 or TM7. In G2, although the entire TM6 lies in the same community, it also joins the community containing the intracellular half of TM3. Thus, in all ECD open or closed ligand-free states, TM6 can be separated into two communities and exhibits topological dependency with adjacent TM helices (stronger correlations). By contrast, in the agonist-bound state, the entire length of TM6 behaves as a single community together with the intracellular half of TM5 and fewer correlations with TM3. The stronger communication within TM6 in the agonist-bound state may associate with signal propagation from the agonist-bound orthosteric pocket to the intracellular G protein-binding site. In direct terms, with fewer correlations with other neighboring TM helices, the single TM6 community permits an easier outward shift of the helix for the conformational changes necessary at the intracellular receptor face for G protein coupling and signaling.

2.2.2 Markov State Models

By applying the analysis based on the Markov State Model (MSM) to the extensive MD trajectory datasets, we have revealed the microsecond to millisecond-scale dynamics between open and closed PAC_1 receptor conformations, interconnected within an ensemble of transitional states—timescales toward GPCR activation. The open-to-closed transition, together with the intrinsic features of the receptor, can provide important insight into signaling mechanisms and potential druggable sites. In the following, we describe the general concept and implementation of MSM, our construction and validation of MSM, and our transition paths and timescales among the PAC_1 receptor ECD open and closed states.

We used the program MSMBuilder 3.2.0 [40, 41] to build a reversible MSM (*see* **Note 2** for examples of execution). An MSM contains a set of state definitions and a transition matrix characterizing the kinetics on this state space. MSM can model long-time scale kinetics from much shorter trajectories by efficiently sampling transitions between these metastable states [42–45]. For MD simulations, the MSM approach first transforms a collection of MD trajectories into a discrete set of $S = \{1, \ldots, m\}$ microstates in the conformation space. A $m \times m$ transition matrix $\mathbf{T}(\tau)$ is computed, where each element T_{ij} measures the probability of system going from one microstate (i) to another (j) within an observation time interval τ (lag time), by $T_{ij} = c_{ij}/\sum_{k=1}^{m} c_{ik}$. Here, c_{ij} counts the number of times the system traverses from i to j at time τ. At

timescales slightly longer than the microstate lag time, *i.e.* $2\tau, \ldots, n\tau$, the transition counts become less between microstates and fewer microstates are kinetically connected in a transition matrix. For a Markov process, the vector of probabilities of the system to be in any of its microstate at time $(n\tau)$ must meet the *Chapman–Kolmogorov equation* [42, 46].

$$\mathbf{p}(n\tau) = \mathbf{p}(0)\mathbf{T}(n\tau) \approx \mathbf{p}(0)[\mathbf{T}(\tau)]^n \tag{1}$$

1. *Dataset preparation.* First, we prepared all trajectories using the C_α coordinates of residues 30-419 in the PAC_1 receptor, as the C_α atoms are often used to represent the overall protein structure. The first five and the last five amino acid residues were excluded, because they are highly dynamic in the N- and C-termini. The time evolutions of the receptor structures show that each PAC_1 receptor model reached a relatively stable structure after 200–500 ns; continuous structure relaxation improved the stability for another 1.5–2.5 μs (Fig. 2b). The last 1.5–2.5 μs of each microsecond simulation with stable final conformations were not included in building an MSM (*see* **Note 3**). Hence, the collected six shorter MD simulations of 20–50 ns each, and the first 200–550 ns of the microsecond long simulations, totaling 6324 configurations were used to build the MSM.
2. *Clustering.* Next, the MD trajectories were transformed into a dataset of microstates based on structural similarities. We used the *k*-centers algorithm [40] to group the dataset into 55 clusters by RMSD metric with a mean distance of ~0.34 Å and a

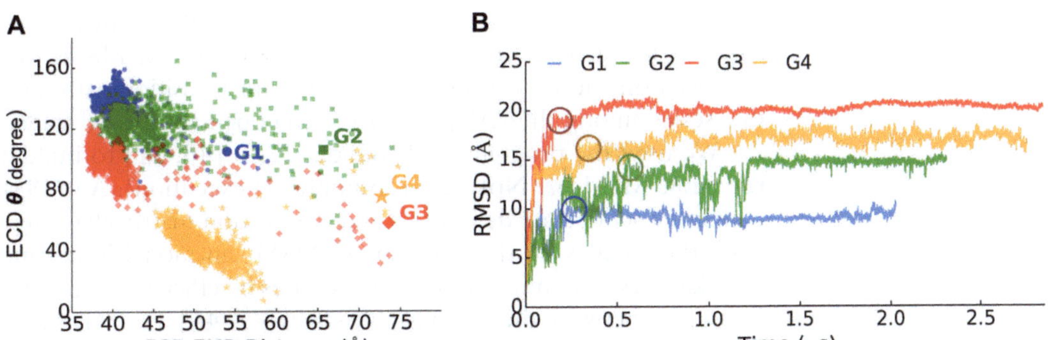

Fig. 2 (**a**) Plot of the ECD tilt angle (θ) against the ECD–7TM distance. Starting points are labeled with larger markers. (**b**) Time evolution of overall RMSD of the PAC_1 receptor. RMSDs were computed by backbone alignments on initial structures with standard deviations of 0.34–0.56 Å in the last ~1.5 μs. Each model of the PAC_1 receptor reached a relatively stable state after 200–500 ns, which had been continuously relaxed to demonstrate model stability for another 1.5–2.5 μs. The conformational states between which we calculated the shortest pathways are circled. Reprinted from Liao C, Zhao X, Brewer M, May V, Li J (2017) Sci Rep 7 (1):5427 with permission from Scientific Reports

maximum distance of ~0.56 Å, which were within the range of the RMSD standard deviations of the last ~1.5 μs in Fig. 2b. The small cluster number, different from the hundreds to thousands of intrinsic conformations of previous protein folding/unfolding studies [47, 48], is the result of limited conformational changes like the ECD rotation around the melting linker.

3. *Lumping and validation.* A series of transition matrices in the evolution of the observation interval (lag time) at τ, 2τ, …, $n\tau$ were constructed using maximum likelihood estimation. As the lag time is increased, fewer microstates are kinetically relevant (kinetically reach each other on timescales faster than the lag time) [49]. Thus, those kinetically relevant microstates can be lumped into macrostates (larger and coarser grained) using the *Perron-Cluster Cluster Analysis* (PCCA) method [50]. Validation was carried out to examine if the model at lag time = τ appears Markovian with increasing lag time = 2τ, …, $n\tau$ using both implied timescales and the *Chapman–Kolmogorov* test [46]. If the macrostate partitions were less robust along implied timescales or if the Markov model errors between the true probability density at time $n\tau$ and the probability density predicted by the Markov model at the same time were large, then a refinement of the partition [46] or an improvement of the initial dataset [49] would be necessary.

4. *Implied timescales.* The implied timescales as a function of the lag time and the eigenvalues of the transition matrix are shown in Fig. 3. Four macrostates were selected, given the number of the major gaps of the implied timescales as well as the number of eigenvalues of the transition matrix that were close to 1 [50]. For a transition matrix of microstates, the partition of

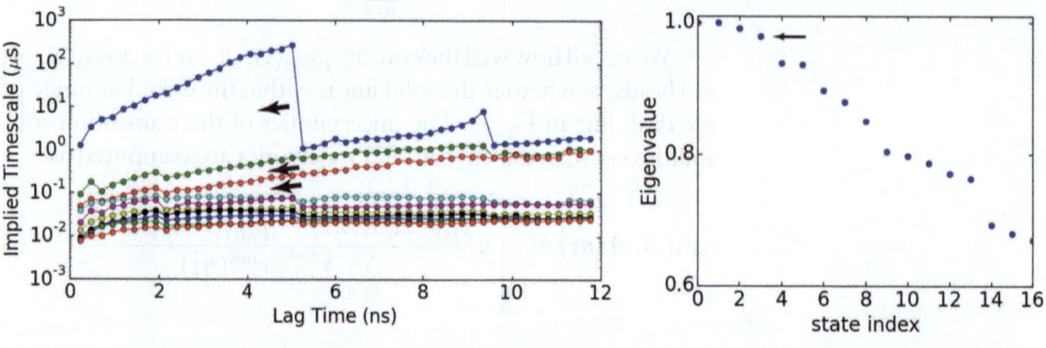

Fig. 3 *Left panel:* Implied timescales as a function of the lag time. There are three major gaps lasting from 2 to ~9 ns, implying four macrostate partitions (count as one more than the number of implied timescales above the major gap) [40]. *Right panel:* eigenvalues of the transition matrix at lag time of 4.8 ns. Only the first seventeen data are shown. There were four points close to 1. Reprinted from Liao C, Zhao X, Brewer M, May V, Li J (2017) Scientific Reports 7 (1):5427 with permission

four macrostates was calculated from the eigenfunction structure using PCCA [50]. Consistently, the conformations of G1, G2, G3, and G4 were lumped into the four macrostates, labeled as A, B, C, and D, respectively.

5. *Chapman–Kolmogorov test.* In general, given a set of states A that contains either an individual microstate or set of microstates, we compared the true probability density of $\mathbf{T}(n\tau)$ based on the transition counts (known as observed trajectory) and the probability density predicted by $[\mathbf{T}(\tau)]^n$ [46, 48]. The initial stationary distribution at time τ restricted to a set A is given by

$$w_i^A = \begin{cases} \dfrac{\pi_i}{\sum_{j\in A}\pi_j} & i\in A \\ 0 & i\notin A \end{cases}, \tag{2}$$

where π is the stationary probability of the $m \times m$ transition matrix $\mathbf{T}(\tau)$. The trajectory-based time-dependence of the probability after time $n\tau$ with starting distribution $\mathbf{w}^A$ is given by

$$p_{\mathrm{MD}}(A, A; n\tau) = \sum_{i\in A} w_i^A\, p_{\mathrm{MD}}(i, A; n\tau), \tag{3}$$

where $p_{\mathrm{MD}}(i, A; n\tau)$ is the trajectory-based estimate of the stochastic transition function given by

$$p_{\mathrm{MD}}(i, A; n\tau) = \frac{\sum_{j\in A} c_{ij}^{\mathrm{obs}}(n\tau)}{\sum_{j=1}^{m} c_{ij}^{\mathrm{obs}}(n\tau)}, \tag{4}$$

where $c_{ij}^{\mathrm{obs}}(n\tau)$ is the number of transition counts between states i and j at time $n\tau$.

Likewise, the probability to be at A *by* the Markov model is given by

$$p_{\mathrm{MSM}}(A, A; n\tau) = \sum_{i\in A} \left[\left(\mathbf{w}^A\right)^T \mathbf{T}^n(\tau)\right]_i, \tag{5}$$

We tested how well the equality $p_{\mathrm{MD}}(A, A; n\tau) = p_{\mathrm{MSM}}(A, A; n\tau)$ holds, as whether the solid line is within the error bar range of the dash line in Fig. 4. The uncertainties of the transition probabilities estimated from the MD trajectories are computed as:

$$\epsilon_{\mathrm{MD}}(A, A; n\tau) = \sqrt{n\frac{p_{\mathrm{MD}}(A, A; n\tau) - [p_{\mathrm{MD}}(A, A; n\tau)]^2}{\sum_{i\in A}\sum_{j=1}^{m} c_{ij}^{\mathrm{obs}}(n\tau)}}. \tag{6}$$

There are around 27 microstates that constitute the shortest and second shortest transition pathways between the closed and open states; they are identified as four subsets based on the macrostate division. The *Chapman–Kolmogorov* test of the four

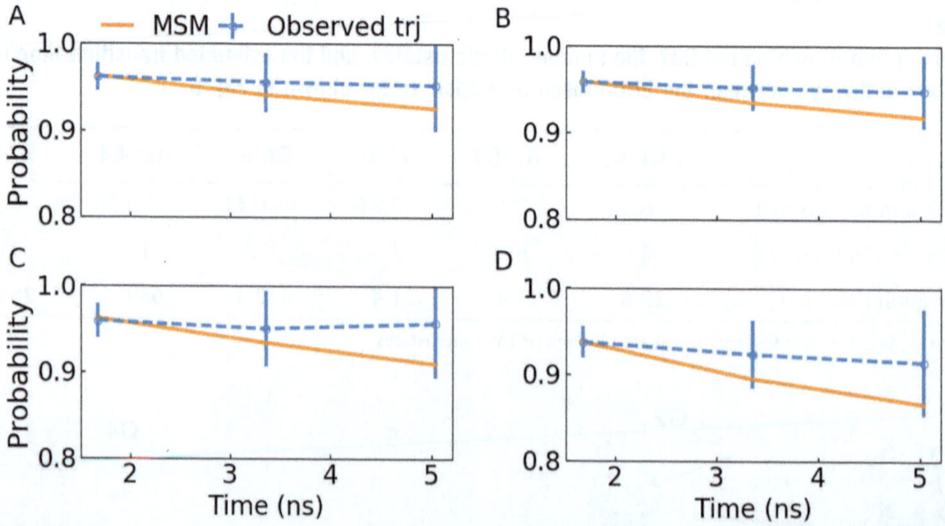

Fig. 4 Microstates, which constitute the shortest and second shortest transition pathways between the closed and open states, were divided into four subsets (**a**, **b**, **c**, and **d**) according to the macrostate division and examined by the *Chapman–Kolmogorov* test. The transition probabilities from MSM agreed well with the probabilities in the observed trajectories within statistical uncertainty. Reprinted from Liao C, Zhao X, Brewer M, May V, Li J (2017) Scientific Reports 7 (1):5427 with permission

subsets is shown in Fig. 4, all of which ensure the test within statistical uncertainty. The transition probabilities from MSM agree well with the probabilities in observed trajectory within statistical uncertainty at a lag time of 1.68 ns. Thus, we built a four-macrostate MSM with 94% data in use with the lag time of 1.68 ns.

We used transition path theory (TPT) [48, 51–53] to calculate the minimum transition net flux of the shortest pathway connecting the ECD closed states or from the ECD open states to the closed states in the transition matrix. For state space $S = \{1, \ldots, m\}$, we define the source set A, the target set B, and the intermediate set I. The rate of transitions observed from A → B per τ (time unit) is given by:

$$k_{\mathrm{AB}} = F/\left(\tau \sum_{i=1}^{m} \pi_i \left(1 - q_i^+\right)\right) \tag{7}$$

where F gives the total transition flux by $F = \sum_{i \in A} \sum_{j \notin A} \pi_i T_{ij} q_j^+$, π_i is the stationary probability at state i, and q_i^+ is a committor probability, $q_i^+ = 0$ for $i \in A$, $q_i^+ = 1$ for $i \in B$, $q_i^+ = \sum_{j \in S} T_{ij} q_j^+$ for $i \notin \{A, B\}$ [48, 51, 54, 55].

With the lag time divided by the minimum transition net flux, we obtained the time to travel from one set of states to the other. The minimum transition net flux, number of microstates, and the estimated transition time in the shortest pathway between the ECD closed and open states are summarized in

Table 1
The minimum transition net flux, the number of microstates, and the estimated transition time of the shortest pathways between the ECD closed and open states shown in Fig. 5

	G1–G2	G1–G3	G2–G3	G1–G4	G2–G4	G3–G4
Minimum net flux (10^{-5})	6.10	7.17	7.84	0.41	0.26	0.59
Number of microstates	3	5	3	13	12	11
Transition time (μs)	27.6	23.4	21.4	412.3	650	284.5

Note: G1, G2, G3, and G4 are represented by one or two microstates

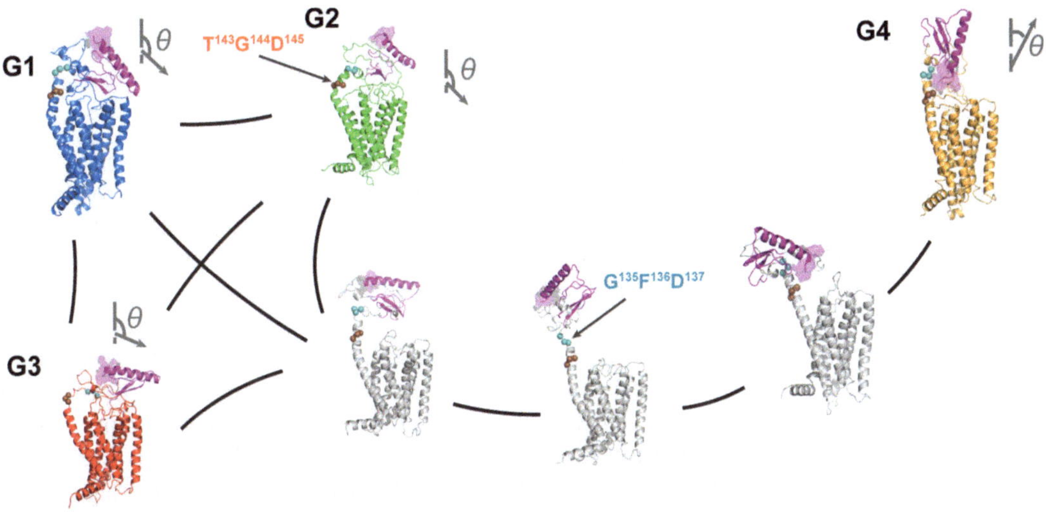

Fig. 5 Illustration of the PAC_1 receptor's conformational transition between the ECD open (G4) and closed (G1–G3) states with representative states from MSM and the transition-path theory. Vectors (showing the N-to-C direction of helix 1 in ECD) show the ECD orientations; and the ECD N-terminus is highlighted in a purple surface representation. Reprinted from Liao C, Zhao X, Brewer M, May V, Li J (2017) Scientific Reports 7 (1):5427 with permission

Table 1. The pathways connecting the closed states (G1, G2, and G3) are relatively short, but from the closed states to the open states (G4) it is rather remote, suggesting a clear partition in the conformational states. Regarding the minimum transition flux, the reversible transitions within the closed states are about 10–30 times faster than the transition from open states to a closed one. Key conformations in the shortest pathways among closed and open states are summarized in Fig. 5.

3 Conclusions

In summary, conformational ensembles with different roles in ligand binding and receptor activation can be discovered using long MD simulations. Using the dynamical network analysis,

shared features can be revealed in conformational ensembles of variant ECD orientations, comparing with a ligand-bound state. With the construction of Markov State Model, the transition network kinetically connecting various GPCR ensembles can be identified. Overall, long MD simulations combined with these structural and kinetic analyses are useful tools to assess conformational transitions for the Class B GPCRs, based on which mechanistic insight can be gained to guide future designs of therapeutics to target these receptors.

4 Notes

1. In the MD simulations run by the program NAMD [56], we used the technology of Langevin dynamics [57] with a low damping coefficient of 1 ps^{-1} for the temperature control and the Nose-Hoover Langevin piston pressure control [58, 59] with a piston period of 0.05 ps and a piston decay time of 0.025 ps (default settings). In the microsecond-long MD simulations performed on the Anton supercomputer, Berendesn thermostat/barostat were applied for the NPT ensemble.
2. Examples of Execution (Modules may change according to different versions of MSMBuilder):

```
### building microstates
msmb KCenters --inp 'trajectory_path/*.dcd' --transformed kcenters_rmsd.h5 --metric
rmsd --top trajectory_path/ca.pdb --n_clusters 55 --random_state 1746

### building a transition matrix (in python interface)
from msmbuilder.dataset import dataset
from msmbuilder.cluster import KCenters
from msmbuilder.msm import MarkovStateModel

ds = dataset('kcenters_rmsd.h5')
model = MarkovStateModel(lag_time=1, verbose=True).fit(ds)
for i in range(0, len(model.transform(ds))):
        print i, len(model.transform(ds)[i])
sample = model.draw_samples(ds, 5)

###-scan implied_timescales---
tscale=[]
n=[]
for i in range(1, 50):
        model = MarkovStateModel(lag_time=i, n_timescales=10, verbose=True).fit(ds)
        n.append(model.n_states_)
        tscale.append(model.timescales_)
```

```
### lumping microstates to macrostates
model.eigtransform(ds, right=True, mode='clip')
model.eigenvalues_

### choose number of macrostates base on model.eigenvalues_ that are closed to 1.
from msmbuilder.lumping import PCCAPlus
pcca = PCCAPlus.from_msm(model, n_macrostates=4)
macro= pcca.microstate_mapping_

###-stationary population—
model = MarkovStateModel(lag_time=1, verbose=True).fit(ds)
mydict = {i:model.populations_[i] for i in range(0, model.n_states_)}
top_pi=sorted(mydict, key=mydict.__getitem__, reverse=True)

### key attributes for Chapman-Kolmogorov test
model.mapping_
model.populations_
model.countsmat_
model.transmat_

### find the first five shortest paths
from msmbuilder.tpt import net_fluxes
from msmbuilder.tpt import paths

netflux = net_fluxes(int, lst, model)
topN = paths(int, lst, net, num_paths=5)
commitor = committors(int, lst, model)
```

3. The last 1.5–2.5 μs of each microsecond simulation with stable final conformations were not included in building an MSM. A few trial tests showed that large population of final states caused the MSM to trim off the initial regions of less population and keep only some final states of large population to be most kinetically relevant.

Acknowledgements

Computational resources were provided by Anton (PSC, NIH P41GM103712-S1), Stampede (XSEDE, NSF ACI-1053575), and Vermont Advanced Computing Core (VACC). J.L. and V.M. thank the UVM REACH grant and the National Institutes of Health (NIH) under Award Number R01GM129431 for the funding support.

References

1. Katritch V, Cherezov V, Stevens RC (2013) Structure-function of the G protein–coupled receptor superfamily. Annu Rev Pharmacol Toxicol 53:531–556. https://doi.org/10.1146/annurev-pharmtox-032112-135923
2. Sakmar TP (2017) Introduction: G-protein coupled receptors. Chem Rev 117:1–3. https://doi.org/10.1021/acs.chemrev.6b00686
3. Harmar AJ, Fahrenkrug J, Gozes I, Laburthe M, May V, Pisegna JR, Vaudry D, Vaudry H, Waschek JA, Said SI (2012) Pharmacology and functions of receptors for vasoactive intestinal peptide and pituitary adenylate cyclase-activating polypeptide: IUPHAR review 1. Br J Pharmacol 166:4–17
4. Culhane KJ, Liu Y, Cai Y, Yan EC (2015) Transmembrane signal transduction by peptide hormones via family B G protein-coupled receptors. Front Pharmacol 6:264
5. Bortolato A, Dore AS, Hollenstein K, Tehan BG, Mason JS, Marshall FH (2014) Structure of Class B GPCRs: new horizons for drug discovery. Br J Pharmacol 171:3132–3145. https://doi.org/10.1111/bph.12689
6. Zhang H, Qiao A, Yang D, Yang L, Dai A, de Graaf C, Reedtz-Runge S, Dharmarajan V, Zhang H, Han GW, Grant TD, Sierra RG, Weierstall U, Nelson G, Liu W, Wu Y, Ma L, Cai X, Lin G, Wu X, Geng Z, Dong Y, Song G, Griffin PR, Lau J, Cherezov V, Yang H, Hanson MA, Stevens RC, Zhao Q, Jiang H, Wang M-W, Wu B (2017) Structure of the full-length glucagon class B G-protein-coupled receptor. Nature 546:259–264. https://doi.org/10.1038/nature22363
7. Jazayeri A, Rappas M, Brown AJH, Kean J, Errey JC, Robertson NJ, Fiez-Vandal C, Andrews SP, Congreve M, Bortolato A, Mason JS, Baig AH, Teobald I, Doré AS, Weir M, Cooke RM, Marshall FH (2017) Crystal structure of the GLP-1 receptor bound to a peptide agonist. Nature 546:254–258. https://doi.org/10.1038/nature22800. http://www.nature.com/nature/journal/v546/n7657/abs/nature22800.html#supplementary-information
8. Zhang H, Qiao A, Yang L, Van Eps N, Frederiksen KS, Yang D, Dai A, Cai X, Zhang H, Yi C, Cao C, He L, Yang H, Lau J, Ernst OP, Hanson MA, Stevens RC, Wang MW, Reedtz-Runge S, Jiang H, Zhao Q, Wu B (2018) Structure of the glucagon receptor in complex with a glucagon analogue. Nature 553:106–110. https://doi.org/10.1038/nature25153
9. Hoare SRJ (2005) Mechanisms of peptide and nonpeptide ligand binding to class B G-protein coupled receptors. Drug Discov Today 10:417–427. https://doi.org/10.1016/S1359-6446(05)03370-2
10. Lappano R, Maggiolini M (2011) G protein-coupled receptors: novel targets for drug discovery in cancer. Nat Rev Drug Discov 10:47–60
11. Millar RP, Newton CL (2010) The year in G protein-coupled receptor research. Mol Endocrinol 24:261–274. https://doi.org/10.1210/me.2009-0473
12. Lebon G, Warne T, Edwards PC, Bennett K, Langmead CJ, Leslie AGW, Tate CG (2011) Agonist-bound adenosine A(2A) receptor structures reveal common features of GPCR activation. Nature 474:521–525. https://doi.org/10.1038/nature10136
13. Li JN, Jonsson AL, Beuming T, Shelley JC, Voth GA (2013) Ligand-dependent activation and deactivation of the human adenosine A (2A) receptor. J Am Chem Soc 135:8749–8759. https://doi.org/10.1021/Ja404391q
14. Dror RO, Arlow DH, Maragakis P, Mildorf TJ, Pan AC, Xu HF, Borhani DW, Shaw DE (2011) Activation mechanism of the beta(2)-adrenergic receptor. Proc Natl Acad Sci U S A 108:18684–18689. https://doi.org/10.1073/Pnas.1110499108
15. Yuan SG, Hu ZQ, Filipek S, Vogel H (2015) W246(6.48) opens a gate for a continuous intrinsic water pathway during activation of the adenosine A(2A) receptor. Angew Chem Int Ed 54:556–559. https://doi.org/10.1002/Anie.201409679
16. Liao C, Zhao X, Liu J, Schneebeli ST, Shelley JC, Li J (2017) Capturing the multiscale dynamics of membrane protein complexes with all-atom, mixed-resolution, and coarse-grained models. Phys Chem Chem Phys 19:9181–9188. https://doi.org/10.1039/C7CP00200A
17. Liao C, Zhao X, Brewer M, May V, Li J (2017) Conformational transitions of the pituitary adenylate cyclase-activating polypeptide receptor, a human class B GPCR. Sci Rep 7:5427. https://doi.org/10.1038/s41598-017-05815-x
18. Yang L, Yang D, de Graaf C, Moeller A, West GM, Dharmarajan V, Wang C, Siu FY, Song G, Reedtz-Runge S, Pascal BD, Wu B, Potter CS,

Zhou H, Griffin PR, Carragher B, Yang H, Wang MW, Stevens RC, Jiang H (2015) Conformational states of the full-length glucagon receptor. Nat Commun 6:7859. https://doi.org/10.1038/ncomms8859

19. Kean J, Bortolato A, Hollenstein K, Marshall FH, Jazayeri A (2015) Conformational thermostabilisation of corticotropin releasing factor receptor 1. Sci Rep 5:11954. https://doi.org/10.1038/srep11954. https://www.nature.com/articles/srep11954#supplementary-information
20. Song G, Yang D, Wang Y, de Graaf C, Zhou Q, Jiang S, Liu K, Cai X, Dai A, Lin G, Liu D, Wu F, Wu Y, Zhao S, Ye L, Han GW, Lau J, Wu B, Hanson MA, Liu Z-J, Wang M-W, Stevens RC (2017) Human GLP-1 receptor transmembrane domain structure in complex with allosteric modulators. Nature 546:312. https://doi.org/10.1038/nature22378
21. Singh R, Ahalawat N, Murarka RK (2015) Activation of corticotropin-releasing factor 1 receptor: insights from molecular dynamics simulations. J Phys Chem B 119:2806–2817. https://doi.org/10.1021/Jp509814n
22. Woolley MJ, Reynolds CA, Simms J, Walker CS, Mobarec JC, Garelja ML, Conner AC, Poyner DR, Hay DL (2017) Receptor activity-modifying protein dependent and independent activation mechanisms in the coupling of calcitonin gene-related peptide and adrenomedullin receptors to Gs. Biochem Pharmacol 142:96–110. https://doi.org/10.1016/j.bcp.2017.07.005
23. Shaw DE, Dror RO, Salmon JK, Grossman JP, Mackenzie KM, Bank JA, Young C, Deneroff MM, Batson B, Bowers KJ, Chow E, Eastwood MP, Ierardi DJ, Klepeis JL, Kuskin JS, Larson RH, Lindorff-Larsen K, Maragakis P, Moraes MA, Piana S, Shan Y and Towles B (2009) Millisecond-scale molecular dynamics simulations on Anton. Proceedings of the conference on high performance computing networking, storage and analysis, ACM, Portland, Oregon, pp. 1–11
24. Voelz VA, Bowman GR, Beauchamp K, Pande VS (2010) Molecular simulation of ab initio protein folding for a millisecond folder NTL9 (1-39). J Am Chem Soc 132:1526. https://doi.org/10.1021/ja9090353
25. Shelley MY, Selvan ME, Zhao J, Babin V, Liao C, Li J, Shelley JC (2017) A new mixed all-atom/coarse-grained model: application to melittin aggregation in aqueous solution. J Chem Theory Comput 13:3881–3897. https://doi.org/10.1021/acs.jctc.7b00071
26. Liao C, Zhang Z, Kale J, Andrews DW, Lin J, Li J (2016) Conformational heterogeneity of bax helix 9 dimer for apoptotic pore formation. Sci Rep 6:29502. https://doi.org/10.1038/srep29502. http://www.nature.com/articles/srep29502#supplementary-information
27. Liao C, Selvan ME, Zhao J, Slimovitch JL, Schneebeli ST, Shelley M, Shelley JC, Li J (2015) Melittin aggregation in aqueous solutions: insight from molecular dynamics simulations. J Phys Chem B 119:10390–10398. https://doi.org/10.1021/acs.jpcb.5b03254
28. Best RB, Zhu X, Shim J, Lopes PEM, Mittal J, Feig M, MacKerell AD (2012) Optimization of the additive CHARMM all-atom protein force field targeting improved sampling of the backbone ϕ, ψ and side-chain $\chi1$ and $\chi2$ dihedral angles. J Chem Theory Comput 8:3257–3273. https://doi.org/10.1021/ct300400x
29. Baker D, Agard DA (1994) Kinetics versus thermodynamics in protein folding. Biochemistry 33:7505–7509. https://doi.org/10.1021/bi00190a002
30. Earl DJ, Deem MW (2005) Parallel tempering: theory, applications, and new perspectives. Phys Chem Chem Phys 7:3910–3916
31. Zhang C, Ma JP (2010) Enhanced sampling and applications in protein folding in explicit solvent. J Chem Phys 132. https://doi.org/10.1063/1.3435332.Artn 244101
32. Martin HSC, Jha S, Coveney PV (2014) Comparative analysis of nucleotide translocation through protein nanopores using steered molecular dynamics and an adaptive biasing force. J Comput Chem 35:692–702. https://doi.org/10.1002/jcc.23525
33. Hamelberg D, Mongan J, McCammon JA (2004) Accelerated molecular dynamics: a promising and efficient simulation method for biomolecules. J Chem Phys 120:11919–11929. https://doi.org/10.1063/1.1755656
34. Pyrkosz AB, Eargle J, Sethi A, Luthey-Schulten Z (2010) Exit strategies for charged tRNA from GluRS. J Mol Biol 397:1350–1371. https://doi.org/10.1016/J.Jmb.2010.02.003
35. Alexander RW, Eargle J, Luthey-Schulten Z (2010) Experimental and computational determination of tRNA dynamics. FEBS Lett 584:376–386. https://doi.org/10.1016/J.Febslet.2009.11.061
36. Sethi A, Eargle J, Black AA, Luthey-Schulten Z (2009) Dynamical networks in tRNA: protein complexes. Proc Natl Acad Sci U S A 106:6620–6625. https://doi.org/10.1073/Pnas.0810961106
37. Humphrey W, Dalke A, Schulten K (1996) VMD: visual molecular dynamics. J Mol

Graph Model 14:33–38. https://doi.org/10.1016/0263-7855(96)00018-5

38. Glykos NM (2006) Software news and updates. Carma: a molecular dynamics analysis program. J Comput Chem 27:1765–1768. https://doi.org/10.1002/jcc.20482
39. Girvan M, Newman MEJ (2002) Community structure in social and biological networks. Proc Natl Acad Sci 99:7821
40. Bowman GR, Huang XH, Pande VS (2009) Using generalized ensemble simulations and Markov state models to identify conformational states. Methods 49:197–201. https://doi.org/10.1016/j.ymeth.2009.04.013
41. Bowman GR, Beauchamp KA, Boxer G, Pande VS (2009) Progress and challenges in the automated construction of Markov state models for full protein systems. J Chem Phys 131:124101. https://doi.org/10.1063/1.3216567
42. Chodera JD, Singhal N, Pande VS, Dill KA, Swope WC (2007) Automatic discovery of metastable states for the construction of Markov models of macromolecular conformational dynamics. J Chem Phys 126:155101. https://doi.org/10.1063/1.2714538.Artn 155101
43. Chodera JD, Swope WC, Pitera JW, Dill KA (2006) Long-time protein folding dynamics from short-time molecular dynamics simulations. Multiscale Model Sim 5:1214–1226. https://doi.org/10.1137/06065146x
44. Noé F, Fischer S (2008) Transition networks for modeling the kinetics of conformational change in macromolecules. Curr Opin Struct Biol 18:154–162. https://doi.org/10.1016/j.sbi.2008.01.008
45. Singhal N, Snow CD, Pande VS (2004) Using path sampling to build better Markovian state models: Predicting the folding rate and mechanism of a tryptophan zipper beta hairpin. J Chem Phys 121:415–425. https://doi.org/10.1063/1.1738647
46. Prinz JH, Wu H, Sarich M, Keller B, Senne M, Held M, Chodera JD, Schutte C, Noe F (2011) Markov models of molecular kinetics: generation and validation. J Chem Phys 134:174105. https://doi.org/10.1063/1.3565032.Artn 174105
47. Lane TJ, Bowman GR, Beauchamp K, Voelz VA, Pande VS (2011) Markov state model reveals folding and functional dynamics in ultra-long MD trajectories. J Am Chem Soc 133:18413–18419. https://doi.org/10.1021/ja207470h
48. Noe F, Schutte C, Vanden-Eijnden E, Reich L, Weikl TR (2009) Constructing the equilibrium ensemble of folding pathways from short off-equilibrium simulations. Proc Natl Acad Sci U S A 106:19011–19016. https://doi.org/10.1073/pnas.0905466106
49. Pande VS, Beauchamp K, Bowman GR (2010) Everything you wanted to know about Markov State Models but were afraid to ask. Methods 52:99–105. https://doi.org/10.1016/j.ymeth.2010.06.002
50. Noe F, Horenko I, Schutte C, Smith JC (2007) Hierarchical analysis of conformational dynamics in biomolecules: transition networks of metastable states. J Chem Phys 126:155102. https://doi.org/10.1063/1.2714539
51. Weinan E, Vanden-Eijnden E (2006) Towards a theory of transition paths. J Stat Phys 123:503–523. https://doi.org/10.1007/s10955-005-9003-9
52. Berezhkovskii A, Hummer G, Szabo A (2009) Reactive flux and folding pathways in network models of coarse-grained protein dynamics. J Chem Phys 130:205102. https://doi.org/10.1063/1.3139063.Artn 205102
53. Metzner P, Schutte C, Vanden-Eijnden E (2009) Transition path theory for Markov jump processes. Multiscale Mod Sim 7:1192–1219. https://doi.org/10.1137/070699500
54. Du R, Pande VS, Grosberg AY, Tanaka T, Shakhnovich ES (1998) On the transition coordinate for protein folding. J Chem Phys 108:334–350. https://doi.org/10.1063/1.475393
55. Bolhuis PG, Chandler D, Dellago C, Geissler PL (2002) Transition path sampling: throwing ropes over rough mountain passes, in the dark. Annu Rev Phys Chem 53:291–318. https://doi.org/10.1146/annurev.physchem.53.082301.113146
56. Phillips JC, Braun R, Wang W, Gumbart J, Tajkhorshid E, Villa E, Chipot C, Skeel RD, Kale L, Schulten K (2005) Scalable molecular dynamics with NAMD. J Comput Chem 26:1781–1802. https://doi.org/10.1002/jcc.20289
57. Toda M, Kubo R, Saitō N, Hashitsume N (1991) Statistical physics II: nonequilibrium statistical mechanics. Springer Science & Business Media, Berlin
58. Martyna GJ, DJ T, Klein ML (1994) Constant-pressure molecular-dynamics algorithms. J Chem Phys 101:4177–4189. https://doi.org/10.1063/1.467468
59. Feller SE, Zhang YH, Pastor RW, Brooks BR (1995) Constant-pressure molecular-dynamics simulation - the Langevin piston method. J Chem Phys 103:4613–4621. https://doi.org/10.1063/1.470648

Chapter 2

Molecular Dynamic Simulations to Probe Water Permeation Pathways of GPCRs

Katsufumi Tomobe, Eiji Yamamoto, and Kenji Yasuoka

Abstract

Rhodopsin is a light-driven G protein-coupled receptor mediating signal transduction in eyes. The molecular dynamics (MD) simulations are powerful computational tools to investigate molecular behavior of proteins and internal water molecules which are related to the function of proteins; however, the MD simulations of the rhodopsin require several technical setups for accurate calculations. This chapter discusses practical methods for setting up the MD simulations of the rhodopsin [preparation of initial systems, condition files for MD simulation package GROMACS, and data analysis]. The data analysis includes the root mean square deviation (RMSD) and mapping of accessibility of water molecules.

Key words Rhodopsin, G Protein-coupled receptor, Water pathway, Biological membrane, Molecular dynamics simulations

1 Introduction

G Protein-coupled receptors (GPCRs) are transmembrane proteins that mediate signal transduction via G proteins [1]. A light-driven GPCR protein, called rhodopsin, is one of the most important GPCR proteins, and has been used in many previous studies as a model protein to understand all GPCR proteins [2–4]. Rhodopsin is composed of transmembrane helixes parts and a light-sensitive chromophore (11-cis retinal). By absorption of light, the rhodopsin decays into the active state of the rhodopsin (called Opsin) and 11-trans retinal through several intermediate states, such as the Meta II state [5].

Molecular dynamics (MD) simulations are powerful computational methods to investigate molecular mechanism of proteins. This method is useful when researchers focus on dynamics of the proteins: conformational changes, fluctuations, and stability of the proteins [6–8]. Moreover, all-atom MD simulations have a great advantage that dynamics of water molecules can be followed. There have been several studies investigating rhodopsin using MD

Mario Tiberi (ed.), *G Protein-Coupled Receptor Signaling: Methods and Protocols*, Methods in Molecular Biology, vol. 1947, https://doi.org/10.1007/978-1-4939-9121-1_2, © Springer Science+Business Media, LLC, part of Springer Nature 2019

simulations [9]. Water molecules play important roles around biological materials, and interactions between water molecules and rhodopsin are also tightly related to its functions [9]. Using the all-atom MD simulations of the rhodopsin, we have examined the putative solvent pore on the cytoplasmic side of the rhodopsin, and the solvent pore was successfully identified in the Meta II state and Opsin [10]. Furthermore, the all-atom MD simulations provided trajectories of water molecules passing through the solvent pore, which led to finding the characteristic narrow regions in the solvent pore.

This chapter provides the practical methods for setting up the MD simulations of three states of rhodopsin with a 1-palmitoyl-2-oleoyl-sn-glycero-3-phosphocholine (POPC) bilayer and water molecules: dark-adapted, Meta II, and Opsin. These MD simulations were performed by using GROMACS 5.1.2 software [11] with the AMBERff99SB-ildn forcefield [12] for proteins, the Slipid forcefield for the membrane [14], and TIP3P water model [13]. Moreover, these procedures are applicable for MD simulations of other GPCR proteins.

2 Materials

1. GROMACS version 5.1.2 (*see* **Note 1**).
2. CHARMM-GUI via a web browser.
3. AMBERff99SB-ildn force field for rhodopsin.
4. Slipid for POPC membrane.
5. TIP3P water model for water molecules.

3 Methods

3.1 Preparation of the Lipid Bilayer and Water Slabs

Here, we focus on the method to build the system of the lipid bilayer by using the VMD software [15].

1. Build the initial coordinate of a lipid bilayer using the membrane builder in the visual molecular dynamics (VMD).
2. Select 'POPC' as a lipid and set *X* and *Y* length of the system.
3. Since the system obtained by the VMD is a mix of POPC lipid bilayer and water molecules, separate the structure data into POPC lipid bilayer (lipid.pdb) and water molecule data.
4. After preparation of two water slabs (water1.pdb and water2.pdb) by the VMD, combine lipid.pdb, water1.pdb, and water2.pdb.

3.2 Preparation of the Protein in the Membrane

The first step for preparation of the protein is to download the appropriate PDB files from the Research Collaboratory for Structural Bioinformatics (RCSB) web site (https://www.rcsb.org/). Select PDB ID numbers for the structures of the dark-adapted (1U19), Meta II (3PXO), and Opsin (3CAP) states of rhodopsin, respectively (protein.pdb). Then, prepare a system and force field by Tcl script using the VMD. The Tcl file provides the following information.

1. Load psfgen module and topology.
2. Load structures of the rhodopsin, lipid bilayer, and water molecules:

```
segment U1 {pdb ./protein.pdb}
segment U2 {pdb ./lipid.pdb}
segment U3 {pdb ./water1.pdb}
segment U4 {pdb ./water2.pdb}
coordpdb ./protein.pdb U1
coordpdb ./lipid.pdb U2
coordpdb ./water1.pdb U3
coordpdb ./water2.pdb U4
```

3. Select and delete lipids overlapping the protein:

```
set sellip [atomselect top "resname POPC"]
set lseglist [lsort -unique [$sellip get segid]]
 foreach lseg $lseglist {
 set selover [atomselect top "segid $lseg and within 0.8 of
 protein"]
 set resover [lsort -unique [$selover get resid]]
 foreach res $resover {
 delatom $lseg $res
 }
 }
```

4. Select and delete water molecules overlapping the protein:

```
set selwat [atomselect top "resname TIP3"]
set lseglist [lsort -unique [$selwat get segid]]
 foreach lseg $lseglist {
 set selover [atomselect top "segid $lseg and within 3.8 of
 protein"]
 set resover [lsort -unique [$selover get resid]]
 foreach res $resover {
 delatom $lseg $res
 }
 }
```

5. Write the structure and force field on files:

```
writepsf protein_and_membrane.psf
writepdb protein_and_membrane.pdb
```

3.3 MD Simulations

The MD simulations are composed of two steps: equilibration and production run (*see* **Note 2**).

1. Set the GROMACS condition file for the energy minimization, the algorithm for the energy minimization, energy step size, maximum number of steps, and energy groups are set:

```
integrator = steep
emstep = 0.001
nsteps = 5000
energygrps = system
```

2. Perform a short *NVT* and *NPT* simulations to equilibrate structures of the protein before the production run as the first step (*see* **Note 3**). During the *NVT* and *NPT* simulations, the position restraint is used to prevent large-scale reformation and disruption of the protein. In the GROMACS condition file for the *NVT* and *NPT* simulations, the position restraint is set by.

```
define = -DPOSRES
```

3. In the second step, perform a long *NPT* simulation as the production run. The GROMACS condition file for the equilibration is similar to that for the production run; however, the position restraint is removed and simulation time is extended. The GROMACS condition file contains the following parameters:

```
pcoupl = Parrinello-Rahman
pcoupltype = semiisotropic
```

3.4 Data Analysis: Root Mean Square Deviations (RMSDs)

RMSDs are an indicator of conformational stability of the protein (*see* **Note 4**).

1. Center the system in the box using the 'gmx trjconv' command.

```
gmx trjconv -s {TRAJECTORY}.tpr -f {REFERENCE}.xtc -n
rmsd.ndx -o temp.xtc -pbc mol -center -boxcenter tric
```

These options mean setting the center of the system on half of the sum of box vectors.

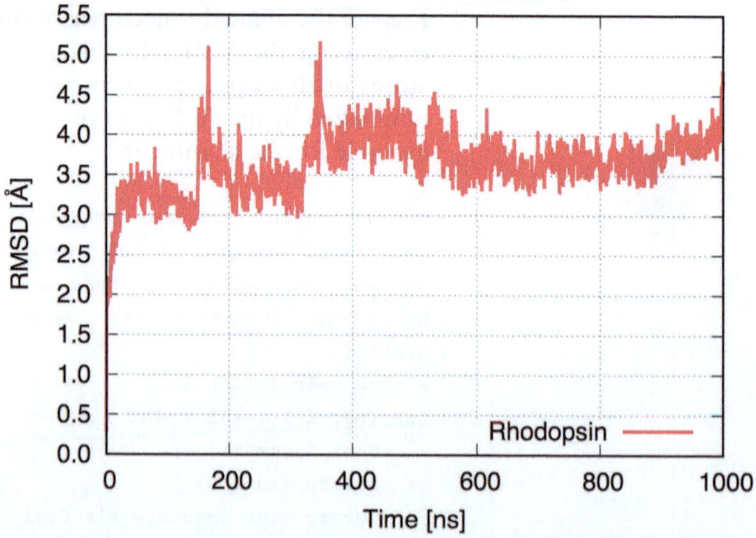

Fig. 1 Evolution of the root mean square deviation (RMSD) from the initial structure during the production run. The RMSD was calculated only using the helical parts of rhodopsin

2. Then, fit the trajectories to the reference:

```
gmx trjconv -s {TRAJECTORY}.tpr -f temp.xtc -o temp2.xtc
-fit rotxy+transxy
```

The option "`rotxy+transxy`" is to remove the rotation in the x-y plane.

3. Finally, calculate the RMSD as follows:

```
gmx rms -s {REFERENCE}.xtc -n rmsd.ndx -f temp2.xtc -xvg
none -o rmsd.xvg
```

For example, Fig. 1 shows the RMSD of a MD simulation of the rhodopsin. The RMSD has rapidly increased due to the relaxation from the initial structure; then, the RMSD converged on 4.0 Å.

3.5 Data Analysis: Accessibility of Water Molecules in the Rhodopsin

Rhodopsin has many internal water molecules, and there is a flux of water molecules between the bulk and its inside. Mapping the accessibility of water molecules shows regions where water molecules have accessed in rhodopsin during the MD simulation and reveals positions of the internal hydration sites. Moreover, this analysis can also identify the solvent pore between the bulk and internal hydration sites.

1. Prepare the aligned trajectories of rhodopsin and water molecules using the VMD. In this alignment, the RMSD of Cα atoms of rhodopsin between the reference and each frame is minimized by the "fit" function. The following Tcl code is one example for the alignment:

```
set nf [molinfo top get numframes]
set frame0 [atomselect top "name CA" frame 0]
set sel [atomselect top "name CA"]
set every [atomselect top "water or protein or (resname RTAL)"]
# rmsd calculation loop
for {set i 1 } {$i < $nf } { incr i } {
    $sel frame $i
    $every frame $i
    $every move [measure fit $sel $frame0]
}
```

2. After the alignment is complete, discretize the space in the rhodopsin like a mesh with less than 0.1 Å to reproduce a fine figure of the mapping. If at least one water molecule has accessed a box of the mesh, label the box. Finally, identify all regions where water molecules have accessed during the MD simulation.
3. Set drawing method as "QuickSurf" to obtain mapping of accessibility of water molecules in rhodopsin (Fig. 2). As shown in Fig. 2, the labeled boxes of the mesh are drawn as blue surfaces (*see* **Note 5**).

3.6 Data Analysis: Water Displacements in z-Coordinate Within the Pore

The accessibility of water molecules reveals important hydration sites and pores of water molecules in rhodopsin. Here, we analyze the trajectories of water molecules in the solvent pore (*see* **Note 6**).

1. In the first step, find water molecules which have passed through the solvent pore during the MD simulation. If interested in water molecules which have completely passed though the solvent pore, choose all water molecules interacting with atoms of the beginning and end parts of the solvent pore.
2. After having gathered all the water molecules, check the appropriateness of trajectories using the VMD visualization.
3. Using the aligned trajectories, extract only *z*-coordinates of the water molecules. This analysis provides more details about the permeation process of water molecules in the solvent pore. For instance, Fig. 3 shows water displacements in *z*-coordinate within the solvent pore. This water molecule has passed through the solvent pore from the bulk to the retinal binding pocket. In this case, the trajectory of a water molecule reveals

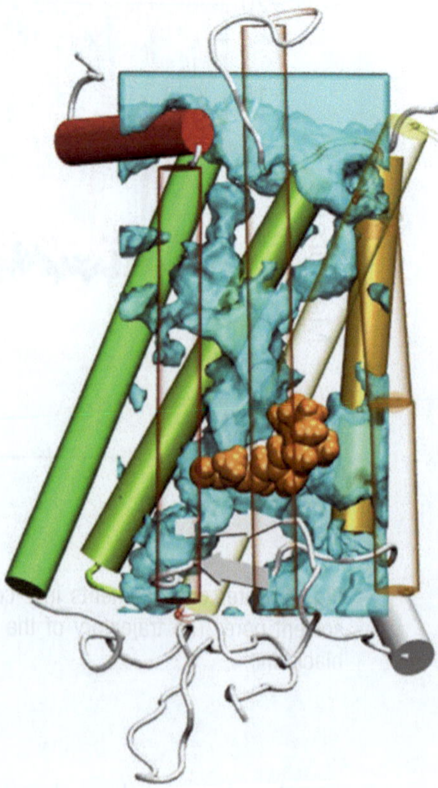

Fig. 2 Accessibility of water molecules. Cross sectional diagrams of water accessibility with the cross section taken at the middle of the rhodopsin. Blue surfaces represent places where water molecules have reached during the MD simulation. Red, yellow, and green cylinders represent the N terminus, the middle, and the C terminus of the protein molecules. The retinal is depicted as orange balls. Figure generated in a.vdw file format

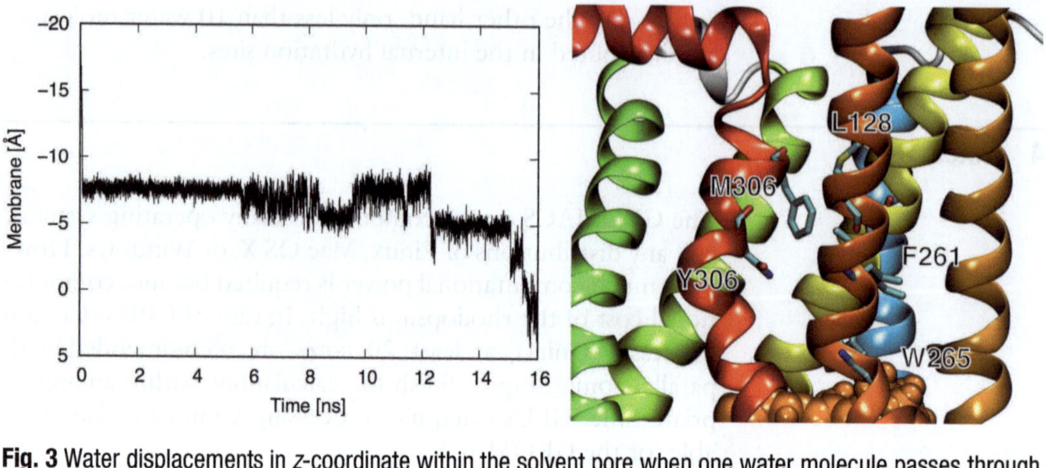

Fig. 3 Water displacements in *z*-coordinate within the solvent pore when one water molecule passes through the solvent pore. Left figure shows the trajectory of the water molecule which passed through the solvent pore. Right figure shows the structure of rhodopsin, and residues of two characteristic narrow gates are shown as cyan

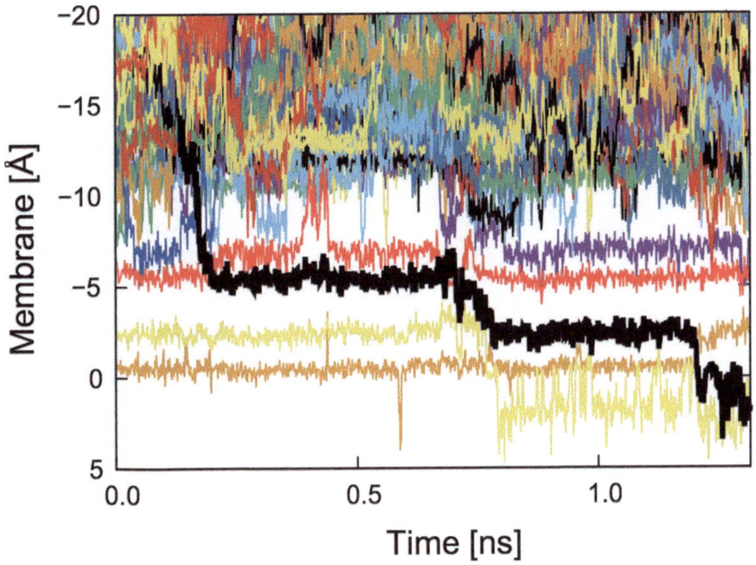

Fig. 4 Water displacements in *z*-coordinate of all water molecules within the solvent pore. The trajectory of the target water molecule is shown by a thick black line

that there are three characteristic positions distinguished by the two narrow regions. In these characteristic positions, *z*-coordinate of water molecules is stable (stable hydration sites). Thus, this analysis also provides stability of internal hydration sites. Figure 4 shows water displacements in *z*-coordinate of all water molecules within the solvent pore. This analysis clearly reveals the flow of water molecules in the solvent pore. The entrance of the solvent pore (less than -10 Å) is filled with bulk water, and most of the water molecules are blocked by the first narrow region. On the other hand, only less than 10 water molecules are distributed in the internal hydration sites.

4 Notes

1. The GROMACS can be compiled on many operating systems, e.g. any distributions of Linux, Mac OS X, or Windows. However, much computational power is required because computational cost of the rhodopsin is high. In case of CPUs (Central Processor Units), at least 20 cores are recommended with parallel computing to finish the calculations within an appropriate time. GPUs (Graphics Processing Units) are also available for the GROMACS.

2. In the first step, energy minimization is performed to remove an artificial force in the system. When a molecular system is artificially built, the initial system is not stable, and MD simulations may crash without the energy minimization due to divergence of the force.
3. *NVT* and *NPT* represent isothemal and isothermal–isobaric simulations, respectively. In *NVT* simulations, the number of atoms *N*, the volume *V*, and the temperature *T* are constant. In *NPT* simulations, the number of atoms *N*, the temperature *T*, and the pressure *P* are constant.
4. In the case of GPCRs, we are only interested in the helical parts of the rhodopsin, and we also remove the rotational and translational motions. Thus, the RMSDs are calculated only using backbone atoms of helical parts with protein fitting.
5. There are various visualization methods for the accessibility of water molecules. Herein, one of the easiest methods with the VMD is introduced. Labeled boxes of the mesh are listed in a new pdb file (list.pdb), which means the boxes of the mesh are regarded as imaginary particles. Using the VMD, these imaginary particles are shown with the rhodopsin.
6. The trajectories of water molecules in the solvent pore can reveal the accessibility of water molecules. These trajectories can clearly show structure and dynamics of the solvent pore.

Acknowledgments

This work is supported in part by MEXT (Ministry of Education, Culture, Sports, Science and Technology) Grant-in-Aid for the Program for Leading Graduate Schools, Keio University Doctorate Student Grant-in-Aid Program, and a Bilateral Program JSPS. E.Y. was supported by MEXT Grant-in-Aid for the "Building of Consortia for the Development of Human Resources in Science and Technology".

References

1. Palczewski K (2006) G protein-coupled receptor rhodopsin. Annu Rev Biochem 75:743–767
2. Kukura P, McCamant DW, Yoon S, Wandschneider DB, Mathies RA (2005) Structural observation of the primary isomerization in vision with femtosecond-stimulated Raman. Science 310:1006–1009
3. Frutos LM, Andruniów T, Santoro F, Ferré N, Olivucci M (2007) Tracking the excited-state time evolution of the visual pigment with multiconfigurational quantum chemistry. Proc Natl Acad Sci U S A 104:7764–7769
4. Röhrig UF, Guidoni L, Laio A, Frank I, Rothlisberger U (2004) A molecular spring for vision. J Am Chem Soc 126:15328–15329
5. Smith SO (2010) Structure and activation of the visual pigment rhodopsin. Annu Rev Biophys 39:309–328
6. Yamamoto E, Akimoto T, Hirano Y, Yasui M, Yasuoka K (2014) 1/f fluctuations of amino

acids regulate water transportation in aquaporin 1. Phys Rev E 89:022718

7. Tomobe K, Yamamoto E, Akimoto T, Yasui M, Yasuoka K (2016) Instability of buried hydration sites increases protein subdomains fluctuations in the human prion protein by the pathogenic mutation T188R. AIP Adv 6:055324
8. Lindorff-Larsen K, Piana S, Dror RO, Shaw DE (2011) How fast-folding proteins fold. Science 334:517–520
9. Sun X, Ågren H, Tu Y (2014) Functional water molecules in rhodopsin activation. J Phys Chem B 118:10863–10873
10. Tomobe K, Yamamoto E, Kholmurodov K, Yasuoka K (2017) Water permeation through the internal water pathway in activated GPCR rhodopsin. PLoS One 12:e0176876
11. Abraham MJ, Murtola T, Schulz R, Páll S, Smith JC, Hess B et al (2015) GROMACS: high performance molecular simulations through multi-level parallelism from laptops to supercomputers. SoftwareX 1:19–25
12. Lindorff-Larsen K, Piana S, Palmo K, Maragakis P, Klepeis JL, Dror RO et al (2010) Improved side-chain torsion potentials for the Amber ff99SB protein force field. Proteins 78:1950–1958
13. Jorgensen WL, Chandrasekhar J, Madura JD, Impey RW, Klein ML (1983) Comparison of simple potential functions for simulating liquid water. J Chem Phys 79:926–935
14. Jämbeck JP, Lyubartsev AP (2012) An extension and further validation of an all-atomistic force field for biological membranes. J Chem Theory Comput 8:2938–2948
15. Humphrey W, Dalke A, Schulten K (1996) VMD: visual molecular dynamics. J Mol Graph 14:33–38

Chapter 3

Expression and Purification of a Functional *E. coli* $^{13}CH_3$-Methionine-Labeled Thermostable Neurotensin Receptor 1 Variant for Solution NMR Studies

Fabian Bumbak, Ross A. D. Bathgate, Daniel J. Scott, and Paul R. Gooley

Abstract

Escherichia coli (*E. coli*) is the most widely used expression host for recombinant proteins due to high expression yields and straightforward molecular cloning. Directed evolution of G protein-coupled receptors (GPCRs) has made several of these difficult to express membrane proteins amenable to prokaryotic expression. Here, we describe a protocol for near complete $^{13}CH_3$-methionine labeling of a thermostable neurotensin receptor 1 (enNTS$_1$) variant in *E. coli* for solution NMR-based dynamics studies. Our expression strategy utilizes methionine biosynthesis pathway inhibition forcing *E. coli* to incorporate exogenous methionine with 96% efficiency at expression levels of 2.6 mg enNTS$_1$ per liter of expression culture containing 50 mg of $^{13}CH_3$-methionine. We also provide a 3-step purification protocol that produces final yields of 0.6 mg of functional Apo-state enNTS$_1$.

Key words Membrane protein, GPCR, Neurotensin receptor 1, Thermostable, *E. coli*, $^{13}CH_3$-methionine, Apo-state

1 Introduction

G Protein-Coupled Receptors (GPCRs) are targets for approximately 34% of all drugs approved by the US Food and Drug Administration (FDA) [1]. Conversely, over 73% of the human non-olfactory GPCRs remain undrugged, often despite clear linkage to disease. Important reasons for this are that high-throughput screening has failed to deliver optimal drug leads and a lack of structural knowledge about how ligands engage GPCRs has hindered compound optimization. Due to significant technological advances in GPCR crystallography [2], the past decade has seen a steep increase in available GPCR crystal structures of currently (August 2018) 364 unique receptor–ligand complexes involving 52 unique GPCRs (GPCRdb). Obtaining suitable amounts of recombinant and sufficiently pure GPCR remains one of the major bottlenecks for structural studies of GPCRs [3].

Mario Tiberi (ed.), *G Protein-Coupled Receptor Signaling: Methods and Protocols*, Methods in Molecular Biology, vol. 1947, https://doi.org/10.1007/978-1-4939-9121-1_3,

The introduction of isotope labels, which is necessary for nuclear magnetic resonance (NMR) studies requiring one or more NMR-active nuclei in addition to naturally abundant ^{1}H, generally puts an additional burden on the sample bottleneck [4]. Solution NMR techniques hold the promise of probing the conformational dynamics of GPCRs in solution, which is crucial for understanding activation and inactivation mechanisms of GPCRs upon ligand and effector protein interaction. When reconstituted in membrane mimetic environments (e.g., in detergent micelles, nanodiscs, or liposomes), integral membrane proteins (IMPs) such as GPCRs show macromolecular behavior whereby the molecular tumbling rate decreases with increasing size causing faster relaxation of the transverse magnetization (T_2) [5]. Short T_2 relaxation times lead to line broadening, confounding resonance assignments of NMR signals of completely ^{13}C- or ^{15}N-labeled macromolecules, thus forcing investigators to reduce the number of probes/signals via selective isotope labeling [4]. As ^{19}F is not naturally present in biomacromolecules, incorporation of ^{19}F into GPCRs has been applied to probe conformational dynamics with ^{19}F-NMR. These studies have utilized ^{19}F-TET (2,2,2-trifluoroethanethiol) and ^{19}F-BTFA (3-bromo-1,1,1-trifluoroacetone) labels chemically conjugated to either naturally occurring or engineered cysteine residues. Fluorine is particularly sensitive to changes in its chemical environment [6] and has been used to study light-induced conformational changes of endogenous as well as engineered ^{19}F-TET-labeled cysteine residues in bovine rhodopsin [7] which was further extended by measuring ^{19}F-^{19}F Nuclear Overhauser Effects (NOEs) between fluorine labels in proximity [8]. Recently, ^{19}F-NMR studies were instrumental in deconvoluting multi-state transitions of two endogenous cysteines at the intracellular ends of TM6 and TM7 in the beta-2 adrenergic receptor (β_2AR) [9–11] and an engineered cysteine at the intracellular end of TM6 in the adenosine 2A receptor (A_2AR) [12]. Another approach to reducing the spectral complexity is the introduction of NMR-active methyl groups in proteins. The three-fold symmetry of CH_3 groups and fast rotation about the S^{ε}-C^{δ} bond yield favorable relaxation properties that are exploited in ^{1}H–^{13}C heteronuclear multiple quantum coherence (HMQC) experiments for well-resolved NMR spectra of large protein complexes [13, 14]. Kofuku et al. [15] and Nygaard et al. [16] introduced $^{13}CH_3$-NMR to the GPCR field, expressing β_2AR in insect cells using methionine-deficient media supplemented with $^{13}CH_3$-methionine. $^{13}CH_3$-methionine labeling enabled the two groups to target site-specific probes in the interior of β_2AR that would not be accessible to labeling via chemical ligation [17]. Methionine is ideally suited for this purpose since it is of low natural abundance in proteins [18] and is often found in regions involved in ligand recognition [19]. The three equivalent protons and the internal diffusion of the methionine methyl groups

result in narrow linewidths, even for large macromolecules [20, 21]. Application of $^{13}CH_3$-methionine labeling in a largely deuterated background was extended to the μ-opioid receptor (μOR) yielding excellent sensitivity employing the ^{1}H-^{13}C HMQC experiment in 2,2-didecylpropane-1,3-bis-β-D-maltopyranoside (LMNG) micelles [22]. More recently $^{13}CH_3$-methionine labeling of a thermostabilized β1 adrenergic receptor (β_1AR) variant containing six endogenous and two engineered methionines facilitated insight into the conformational equilibria of the receptor ranging from apo-state over several ligand-bound states to G protein mimicking nanobody-bound states [23]. Other site-specific labels that have been used to study GPCR dynamics are ^{15}N-valine [24], $^{13}CH_3$-δ1-isoleucine [25], and reductively methylated (^{13}C-dimethyl) lysines [26, 27].

1.1 Bacterial Expression and Purification of Neurotensin Receptor 1

Here, we describe the overexpression of a $^{13}CH_3$-methionine-labeled thermostabilized neurotensin receptor 1 (NTS_1) variant [28] using a methionine biosynthesis pathway inhibition protocol [29, 30] targeted to the *E. coli* inner membrane [31]. *E. coli* is genetically facile, outperforms other expression systems in terms of cost (it grows in cheap media and is easy to handle), homogeneity of recombinant proteins (no post-translational modifications), and generation time (20 min doubling time under optimal growth conditions), while the expression of recombinant proteins starts immediately upon induction [32]. The choice of *E. coli* as an expression host for GPCRs is controversial as *E. coli* lacks the post-translational modification machinery needed for the proper processing of mammalian IMPs and its membrane composition is significantly different from mammalian membranes. However, several folded GPCRs have been expressed in *E. coli* [33–43] but NTS_1 remains the only GPCR that has led to crystal structures from *E. coli*-expressed protein [44]. The NTS_1 variant used in this protocol (termed enNTS_1 [28]) was derived from a mutant generated using a directed evolution method termed cellular high-throughput encapsulation, solubilization, and screening (CHESS) [45]. The most stable rNTS_1 (rat neurotensin receptor) variant generated using CHESS, NTS_1-H4 [46], led to the first crystal structure of a CHESS-stabilized protein (PDB 4BWB) [44] while the variant NTS_1-B5 [45] was selected as a template to generate enNTS_1. In contrast to NTS_1-H4, we showed that enNTS_1 retains the ability to signal and hence retains wild-type-like conformational dynamics [28].

Our expression protocol builds on existing methods for the recombinant expression of NTS_1 in *E. coli*, where it was established that N- and C-terminal fusion proteins, low concentrations (≤0.5 mM) of the inducer Isopropyl β-D-1-thiogalactopyranoside (IPTG), and low expression temperatures (≤22 °C) significantly enhance functional expression levels [33, 47]. Rather than

supplementing minimal media expression cultures with $^{13}CH_3$-methionine [48] we chose to adapt a methionine pathway inhibition protocol previously described for the incorporation of selenomethionine in *E. coli*-expressed proteins [30]. The methionine pathway inhibition protocol utilizes feedback inhibition and repression of aspartokinases in *E. coli* [49], which has been reported to result in high expression yields and incorporation efficiency of exogenous methionine without the need to use a methionine auxotrophic *E. coli* strain [50]. Mass spectrometry revealed that $^{13}CH_3$-methionine incorporation was 96% in enNTS$_1$ [28] expressed using the protocol described herein.

Furthermore, we present a generic 3-step purification protocol for a 212-fold enrichment of apo-state enNTS$_1$ yielding 0.6 mg per liter of binding competent receptor at a purity of 94% [28]. Our purification protocol uses immobilized metal-affinity chromatography (IMAC) to capture solubilized enNTS$_1$ from the cell lysate followed by human rhino virus (HRV) 3C protease cleavage of the N- and C-terminal fusion proteins and a second IMAC (reverse IMAC) to remove the fusion proteins which both contain His$_{10}$-tag sequences. Size exclusion chromatography (SEC) is used for final polishing to ensure homogeneity and monodispersity of enNTS$_1$. A similar purification strategy comprising two consecutive IMAC steps has led to a number of crystal structures of ligand-bound GPCRs such as CXCR4 chemokine [51], kappa-opioid (κ-OR) [52], glucagon (GCGR) [53], P2Y$_{12}$ [54, 55], smoothened (SMO) [56, 57], and 5-HT$_{1B}$ serotonin [58, 59] receptors. An alternative protocol for the purification of ligand-bound NTS$_1$ using ligand affinity chromatography has been published previously [60], where the NTS$_1$ agonist NT8-13 was linked to the column matrix via a HRV 3C protease cleavage site facilitating elution of agonist bound NTS$_1$ via proteolytic cleavage (and concomitantly cleaving N- and C-terminal fusion proteins). Alternatively, elution was achieved via the high-affinity NTS$_1$ antagonist SR142948 whereby the affinity of column-bound NT8-13 was decreased via mutations (P10A and I12A).

1.2 Introduction of the enNTS$_1$ Expression Construct

The enNTS$_1$ variant used in this protocol is based on the NTS$_1$-B5 variant described previously [45] containing a number of modifications (Table 1). Exposed cysteines were neutralized to reduce the possibility of receptor aggregation by means of intermolecular disulfide-bridge formation. Therefore, the cysteines in positions C278 (sequential rNTS$_1$ numbering), C386, and C388 were replaced with serines using site-directed mutagenesis. Cysteine C417 was removed by C-terminal truncation of the receptor after T416, thus removing the last eight residues of the receptor. L167$^{3.50}$ and V208$^{4.64}$ (superscripts represent Ballesteros–Weinstein numbering [61]) were back mutated to arginine and methionine respectively as occurring in wild-type rNTS$_1$. Residue

Table 1
Amino acid differences between wild-type rNTS$_1$ and CHESS-generated thermostable mutants. Only amino acid positions (sequential numbering) modified relative to the wild-type rat neurotensin receptor 1 (rNTS$_1$) are shown. NTS$_1$-H4 refers to the most stable CHESS-derived mutant crystallized previously (PDB 4BWB[241]), NTS$_1$-B5 refers to the CHESS-derived mutant used as a starting point to generate enNTS$_1$. The amino acids changed to generate enNTS$_1$ from NTS$_1$-B5 are highlighted in bold characters

Position	83	86	101	103	105	113	119	121	124	125	141	150	161	167	201	202	208
rNTS$_1$-wt	S	A	T	H	H	D	L	M	E	L	R	D	A	R	A	I	M
NTS$_1$-H4	G	L	R	D	Y	D	F	L	D	L	K	E	V	L	A	I	M
NTS$_1$-B5	G	L	R	D	Y	S	F	L	D	V	K	E	V	L	S	L	V
enNTS$_1$	G	L	R	D	Y	S	F	L	D	V	K	E	V	**R**	S	L	**M**

Position	213	234	235	240	253	260	262	263	275	278	305	332	342	354	358	362	386	388
rNTS$_1$-wt	R	V	K	V	I	I	N	K	G	C	H	C	F	T	F	S	C	C
NTS$_1$-H4	L	L	R	L	A	A	R	R	–	–	R	V	A	S	V	A	C	C
NTS$_1$-B5	L	L	R	L	A	I	R	R	G	C	R	V	A	S	V	A	C	C
enNTS$_1$	L	L	R	L	A	I	R	R	**A**	**S**	R	V	A	S	V	A	**S**	**S**

R167$^{3.50}$ is part of the highly conserved D/ERY motif which is directly involved in governing class A GPCR conformation and activation [62] and its restoration in the NTS$_1$ variant TM86V-Δ IC3A resulted in increased basal and agonist-dependent G protein signaling as well as β-arrestin 2-mediated desensitization [44]. The reinstatement of M208$^{4.64}$ was based on its involvement in ligand binding, which is evident from mutagenesis studies [63, 64] and crystal structures [44, 65]. The 374 amino acid enNTS$_1$ sequence (T42 to T416) was sub-cloned into the pQE-30-derived vector, pDS170 (Fig. 1).

The open reading frame of pDS170 encodes an N-terminal maltose-binding protein signal sequence (MBPss), followed by a His$_{10}$-tag, a maltose binding protein (MBP), a NNNNNNNNNNG linker and a HRV 3C protease site (LEVLFQGP), which is linked via a BamHI restriction site (resulting in additional residues GS) to residue T42 of the receptor. C-terminally, T416 of the receptor is linked via a NheI restriction site (resulting in additional residues AS) to an Avi-tag for in vivo biotinylation followed by a HRV 3C protease site, a GGSGGS linker and a monomeric ultra-stable green fluorescent protein (muGFP) [66], which is followed by a second His$_{10}$-tag. Cleaved enNTS$_1$ retains the C-terminal Avi-tag which can be used for various assays such as ligand binding assays in combination with streptavidin-coated magnetic beads as utilized in a recent

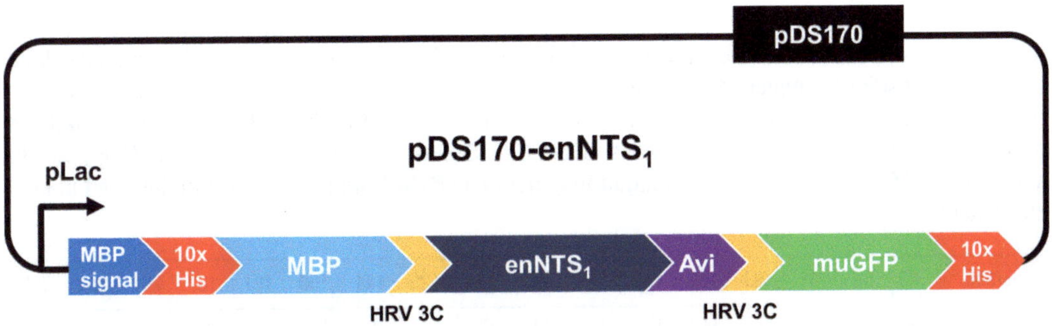

Fig. 1 Expression vector used for recombinant enNTS$_1$ expression. The expression vector pDS170 was constructed as described in Subheading 1.2 driving expression of the recombinant protein under the control of the lac operon. N-terminal maltose binding protein (MBP) including the MBP signal sequence is used for targeting the receptor to the plasma membrane of *E. coli* [77] and C-terminal monomeric ultra-stable green fluorescent protein (muGFP) [66] for improved expression and fluorescence-based quantification of expression. MBP and muGFP are both flanked by a His$_{10}$-tag sequence at their N- and C-terminus respectively. The multiple cloning site harboring enNTS$_1$ followed by an Avi-tag for in vivo biotinylation in *E. coli* is flanked by two human rhino virus (HRV) 3C protease cleavage sites

publication [28] or surface plasmon resonance (SPR) using streptavidin-coated sensor chips as shown for NTS$_1$-H4 [67, 68]. All molecular cloning work was carried out in *E. coli* DH5α cells.

2 Materials

2.1 NTS$_1$ Overexpression in E. coli

1. pDS170-enNTS$_1$ [28].
2. Chemically competent *E. coli* C43(DE3) cells.
3. Luria-Bertani (LB) broth: dissolve 10 g tryptone, 5 g yeast extract, 10 g NaCl in 900 mL deionized water. Adjust the volume to 1 L with deionized water. Sterilize by autoclaving. Keep at room temperature (RT).
4. LB-Agar plate containing 100 μg/mL ampicillin and 1% w/v glucose.
5. 100 mg/mL ampicillin stocks in deionized water. Sterilize by filtering and store at −20 °C in aliquots of 1 mL.
6. 20% (w/v) glucose in deionized water. Sterilize by autoclaving or filtering.
7. Sterile 15 mL round-bottom tubes.
8. Sterile 200 μL and 1 mL pipette tips and corresponding pipettes.
9. Centrifuge capable of spinning 15 mL and 50 mL tubes.
10. Minimal salts medium per liter: dissolve 4 g KH_2PO_4, 4 g K_2HPO_4, 3.5 g $Na_2HPO_4{\cdot}2H_2O$, 1 g NaCl, 2 g NH_4Cl in

750 mL deionized water. Adjust to pH 7.4 with 5 N NaOH (or NaOH pellets; approx. 7 pellets per liter). Adjust the volume to 1 L with deionized water. Distribute in shaker flasks (250 mL in 1 L flask for second pre-culture; 500 mL in 2 L shaker flasks for expression cultures). Sterilize by autoclaving. Keep at RT. The protocols for cell lysis, membrane solubilization and purification outlined in Subheadings 3.2 and 3.3 are designed to use cells from 3 L (6× 500 mL) of expression culture.

11. 1 M $MgSO_4$ in deionized water. Sterilize by filtering and keep at RT.
12. 100× Trace metal stock solution [69]: dissolve 0.6 g $FeSO_4{\cdot}7H_2O$, 0.6 g $CaCl_2{\cdot}2H_2O$, 0.12 g $MnCl_2{\cdot}4H_2O$, 0.08 g $CoCl_2{\cdot}6H_2O$, 0.07 g $ZnSO_4{\cdot}7H_2O$, 0.038 g $CuSO_4{\cdot}5H_2O$, 0.025 g $Na_2MoO_4{\cdot}4H_2O$, 0.635 g Na_2-EDTA·$2H_2O$, 2 mg H_3BO_3 in 100 mL deionized water. Sterilize by filtering and keep at RT.
13. 50 mg/mL Thiamine-HCl in deionized water. Sterilize by filtering and store at −20 °C as aliquots of 1 mL.
14. Amino acid mix (AA-mix) for methionine biosynthesis pathway inhibition (per L of expression medium): dissolve 50 mg of each $^{13}CH_3$-methionine, leucine, isoleucine, valine and 100 mg of each lysine hydrochloride, threonine, phenylalanine in 10 mL sterile filtered deionized water. Use 3× amounts of amino acids and water for 3 L of culture. Prepare the night before and incubate overnight on a turning wheel at 4 °C.
15. 1 M Isopropyl β-D-1-thiogalactopyranoside (IPTG) in deionized water. Sterilize by filtering and store at −20 °C in aliquots of 1 mL.
16. Centrifuge for separating large volumes (0.5 L centrifuge bottles).
17. Pellet wash buffer (50 mL): 25 mM HEPES, 100 mM NaCl. Adjust pH to 8. Sterilize by autoclaving or filtering and keep at 4 °C.
18. Spatula and/or 10 mL pipette.
19. Centrifuge capable of holding 50 mL Falcon® tubes.
20. Liquid nitrogen and −80 °C freezer.

2.2 Cell Lysis and Protein Purification

1. Deionized water (preferably MilliQ®).
2. 100 mM HEPES (MW 238.3 g/mol; use 23.83 g/L). Adjust pH to 8. Sterilize by autoclaving or filtering and keep at 4 °C.
3. 5 M NaCl (MW 58.44 g/mol; use 29.22 g/100 mL). Keep at RT.
4. Glycerol.

5. EDTA-free protease inhibitor tablets (Roche).
6. 100 mM phenylmethane sulfonyl fluoride (PMSF) in anhydrous ethanol or isopropanol. Store at −20 °C in aliquots of 1 mL.
7. Lysozyme.
8. DNAse.
9. Lysis buffer (50 mL): 4 mL of 5 M NaCl (400 mM), 10 mL glycerol (20% (v/v)), 500 μL of 1 M $MgCl_2$ (10 mM), top-up to 50 mL with 100 mM HEPES (approx. 50 mM HEPES after addition of cell pellets and detergents). Immediately before solubilization add 100 mg lysozyme, 1 scoop of DNAse (~10 mg), 1 EDTA-free protease inhibitor tablet, 100 μL of 100 mM PMSF (0.2 mM).
10. Sonicator with 6 mm microtip.
11. Detergent solution for solubilization: 1.6 g n-decyl β-D-maltoside (DM), 0.6 g 3-[(3-Cholamidopropyl)dimethylammonio]-1-propanesulfonate (CHAPS), 0.12 g Cholesterol hemisuccinate tris salt (CHS) in 30 mL deionized water. Leave rocking or on turning wheel overnight at 4 °C. Final concentrations after addition to sonicated cells will be 1.6% (DM), 0.6% (CHAPS), and 0.12% (CHS).
12. Centrifuge capable of spinning 20–50 mL volumes at 10,000 × *g* and 4 °C.
13. 50 mL syringe and 0.45 μm syringe filter.
14. 5 M imidazole (MW 68.08 g/mol; use 34.04 g/100 mL). Keep at RT.
15. 100 mL gravity-flow chromatography glass column (BioRad).
16. TALON® resin (Clontech).
17. IMAC equilibration buffer (50 mL): 12.5 mL of 100 mM HEPES (25 mM), 5 mL glycerol (10% (v/v)), 3 mL of 5 M NaCl (300 mM), 75 mg DM (0.15%), top-up to 50 mL with deionized water. Keep at 4 °C.
18. 1 M $MgCl_2$ ($MgCl_2 \cdot 6H_2O$ MW 203.31 g/mol; use 20.34 g/100 mL). Keep at RT.
19. IMAC wash buffer 1 (50 mL): 12.5 mL of 100 mM HEPES (25 mM), 5 mL glycerol (10% (v/v)), 5 mL of 5 M NaCl (500 mM), 75 mg DM (0.15%), 100 μL of 5 M imidazole (10 mM), 100 μL of 100 mM PMSF (0.2 mM), 203 mg (MW 507.18 g/mol) adenosine triphosphate (ATP; 8 mM), 500 μL of 1 M $MgCl_2$ (10 mM), top-up to 50 mL with deionized water. Keep at 4 °C.
20. Base buffer for remaining purification (300 mL): 75 mL of 100 mM HEPES (25 mM), 30 mL glycerol (10% (v/v)),

12 mL of 5 M NaCl (300 mM), 0.15 g n-dodecyl β-D-maltoside (DDM; 0.05%), top-up to 300 mL with deionized water. Keep at 4 °C.

21. IMAC wash buffer 2 (50 mL): 0.5 mL of 5 M NaCl (350 mM final concentration after addition of base buffer), 100 μL of 5 M imidazole (10 mM), 100 μL of 100 mM PMSF (0.2 mM), top-up to 50 mL with base buffer. Keep at 4 °C.
22. IMAC elution buffer (50 mL): 3.5 mL of 5 M imidazole (350 mM), 100 μL of 100 mM PMSF (0.2 mM), top-up to 50 mL with base buffer. Keep at 4 °C.
23. 100 kDa cut-off Amicon® centrifuge concentrator (Millipore).
24. PD10® de-salting column (GE Healthcare).
25. 1 M Na_2SO_4 (MW 142.04 g/mol; use 14.20 g/100 mL). Keep at RT.
26. 0.5 M tris(2-carboxyethyl)phosphine (TCEP) in deionized water pH-corrected to 7.5. Store at −20 °C in aliquots of 1 mL.
27. Human rhinovirus (HRV) 3C protease (His_{10}- or GST-tagged).
28. If GST-tagged HRV 3C protease is used: Glutathione Sepharose® 4B (GE Healthcare) and 15 mL gravity flow chromatography column.
29. 50 mL gravity-flow chromatography column.
30. IMAC wash buffer 3 (50 mL): 50 μL of 5 M imidazole (5 mM), top-up to 50 mL with base buffer. Keep at 4 °C.
31. Size Exclusion Chromatography buffer (>250 mL): prepare two solutions (a) 50 mM K_2HPO_4 (2.18 g), 100 mM NaCl (1.46 g), 0.02% DDM (50 mg), top-up to 250 mL with deionized water; (b) 50 mM KH_2PO_4 (0.68 g), 100 mM NaCl (0.59 g), 0.02% DDM (20 mg), top-up to 100 mL with deionized water. Add solution (b) to solution (a) until pH 7.4. Filter, degas and keep at 4 °C.
32. 30 kDa cut-off Amicon® centrifuge concentrators (Millipore).
33. NMR buffer (>15 mL): prepare two solutions (a) 50 mM K_2HPO_4 (109 mg), 100 mM NaCl (73 mg), top-up to 12.5 mL with D_2O; (b) 50 mM KH_2PO_4 (34 mg), 100 mM NaCl (29 mg), top-up to 5 mL with deionized water. Add solution (b) to solution (a) until pH 7.4. Filter with 0.22 μm syringe filter and keep at 4 °C.
34. Superdex 200 10/300 Increase size exclusion column (GE Healthcare).
35. Chromatography system, for example an Äkta pure system (GE Healthcare), equipped with a 0.5 mL sample loading loop.
36. Shigemi 5 mm symmetrical NMR microtubes (Sigma-Aldrich).

2.3 Amido Black 10B Protein Assay

1. 1 mg/mL BSA standard solution. Purchased in glass vials. Keep at 4 °C once vial is opened.
2. 1% sodium dodecyl sulfate (SDS) in 1 M Tris–HCl, pH 7.5 and keep at RT.
3. Solutions of 60% and 6% (w/v) trichloroacetic acid (TCA) in deionized water. Make-up 100% TCA by using e.g., 100 g TCA and adding deionized water to 100 mL then dilute to 60% and 6% TCA respectively and keep at RT.
4. Millipore HAWP 02400 (0.45 μm, 24 mm) type membrane filter discs.
5. Flat tweezers.
6. Millipore 1225 vacuum filtration sampling manifold.
7. 0.1% Amido Black10B staining solution (200 mL): dissolve 0.2 g Amido Black 10B in 200 mL methanol/glacial acetic acid/distilled water (45/10/45 vol. %). Make sure the dye is completely dissolved and keep at RT prior to use. The Amido Black 10B solution can be reused three to four times. Store in glass bottle at 4 °C.
8. 3× 200 mL of destaining solution: make up 600 mL of methanol/glacial acetic acid/distilled water (90/2/8 vol. %). Split in 3× 200 mL and keep at RT prior to use. The destaining solutions can be reused three to four times. Store in glass bottles at 4 °C.
9. Eluent solution: 25 mM NaOH, 0.05 mM EDTA in 50% aqueous ethanol.

3 Methods

3.1 $^{13}CH_3$-Methionine-Labeled $enNTS_1$ Overexpression in E. coli

1. First day: thaw 100 μL chemically competent *E. coli* C43(DE3) cells on ice and add 1 μL of pDS170-enNTS$_1$ (~100 ng/μL), gently flick tube a couple of times (do not vortex) and incubate on ice for 30 min. Heat shock at 42 °C for exactly 2 min and put the tube back on ice for another 2 min. Plate the whole transformation mix on a LB-Agar plate containing 100 μg/mL ampicillin and 1% w/v glucose. Incubate overnight at 37 °C.
2. Second day, early in the morning: inoculate 5 mL LB containing 100 μg/mL ampicillin and 1% w/v glucose with one colony from the transformation plate. Incubate in a shaking incubator at 37 °C and 225 rpm.
3. Later in the day: Complement the 250 mL minimal salts medium with 2 mM $MgSO_4$ (0.5 mL of 1 M $MgSO_4$ stock solution), 1% (v/v) 100× trace metal solution (2.5 mL), 2% (v/v) glucose (5 mL of 20% (v/v) glucose stock solution), 50 mg/L thiamine (250 μL of 50 mg/mL thiamine stock

solution), and 100 mg/mL ampicillin (250 μL of 100 mg/mL ampicillin stock solution).

4. Centrifuge the LB pre-culture at 1700 rcf and resuspend the pellet in 5 mL of the complemented minimal salts medium from the 250 mL in the 1 L flask.
5. Inoculate the complemented 250 mL minimal salts medium pre-culture with the resuspended cells and incubate overnight in a shaking incubator at 37 °C and 225 rpm.
6. Third day, early in the morning: Complement each of the 2 L flasks containing 500 mL minimal salts medium with 2 mM $MgSO_4$ (1 mL of 1 M $MgSO_4$ stock solution), 1% (v/v) 100× trace metal solution (5 mL), 2% (v/v) glucose (10 mL of 20% (v/v) glucose stock solution), 50 mg/L thiamine (500 μL of 50 mg/mL thiamine stock solution), and 100 mg/mL ampicillin (500 μL of 100 mg/mL ampicillin stock solution).
7. Measure the OD_{600} of the minimal salts medium pre-culture and calculate how much of pre-culture is needed to inoculate 500 mL of expression culture with a starting OD_{600} of 0.05. Multiply this volume by the number of flasks you will inoculate (e.g., multiply by six if you are inoculating 3 L of expression culture) (*see* **Note 1**).
8. Use 50 mL Falcon tubes to centrifuge the calculated volume of minimal salts medium pre-culture at 1700 rcf, discard the supernatant and resuspend the cells in 5 mL of freshly complemented minimal medium (from **step 6**) per 0.5 L of expression culture (i.e., 30 mL for 6× 0.5 L of expression culture).
9. Inoculate the 0.5 L minimal salts medium expression cultures with 5 mL of resuspended minimal medium pre-culture (this yields a theoretical starting OD_{600} of 0.05).
10. Incubate in a shaking incubator at 37 °C and 225 rpm.
11. Regularly measure OD_{600} until reaching 0.35–0.4 (this takes around 5–6 h).
12. Reduce temperature of the shaking incubator to 16 °C (*see* **Note 2**).
13. Cool the flasks with the expression cultures for 2 min on ice.
14. Add 5 mL of AA-mix per 0.5 L of culture.
15. Incubate for 15 min at 225 rpm in the shaker while it is cooling down (*see* **Note 3**).
16. Induce with 250 μM IPTG (125 μL of 1 M IPTG per 0.5 L of expression culture).
17. Incubate in the shaking incubator at 16 °C and 225 rpm for 12–16 h (*see* **Note 4**).

18. Fourth day: harvest cells by centrifugation for 10 min at 2600 rcf and 4 °C.
19. Remove as much of the medium as possible and scoop cells into 2× 50 mL Falcon tubes.
20. Wash cells with 2× 25 mL of pellet wash buffer (Subheading 2.1, **item 17**).
21. Centrifuge resuspended cells for 15 min at 3000 rcf and 4 °C, discard supernatant and continue with cell lysis and membrane solubilization (Subheading 3.2) or snap freeze pellets with liquid nitrogen and store at −80 °C. Frozen pellets can be kept for at least 4 months at −80 °C.

3.2 Cell Lysis and Membrane Solubilization

1. Continue with the washed cells or thaw pellet(s) on ice.
2. Complement the lysis buffer with lysozyme, DNAse, and protease inhibitors as outlined in Subheading 2.2, **item 9**.
3. Resuspend cells in each Falcon tube with 25 mL complemented lysis buffer (total volume ~30–35 mL each) (*see* **Note 5**).
4. Leave gently rocking for 30 min at 4 °C.
5. Sonicate first tube on ice (*see* **Note 6**). Repeat with second tube.
6. Add half of the detergent solution (Subheading 2.2, **item 11**) to each Falcon tube (total volume ~50 mL each) and top-up with 50 μL of 100 mM PMSF per Falcon tube.
7. Gently rock the solubilization mix for 2 h at 4 °C.
8. Centrifuge for 30 min at 12,000 rcf and 4 °C (*see* **Note 7**).
9. Filter supernatant with a 0.45 μm filter, add 5 M Imidazole to 10 mM (final) and add another 50 μL of 100 mM PMSF to each Falcon tube.

3.3 Purification of enNTS$_1$

3.3.1 First Pass IMAC (Capture)

1. While the solubilization mix is incubating (Subheading 3.2, **step 7**), prepare a column of 3 mL of TALON® resin by applying 6 mL of TALON® slurry (supplied as 50% (w/v) slurry in non-buffered 20% EtOH) to a 100 mL BioRad gravity-flow glass column equipped with a tap. Let the ethanol drain off and wash the resin 3× with at least 5 bead volumes (3× 15 mL) of deionized water. Repeat wash with 3× 5 bead volumes of IMAC equilibration buffer.
2. Transfer solubilization mix to the column with the equilibrated TALON® resin, close the column with the supplied lid and gently rock for 1.5 h at 4 °C (*see* **Note 8**).
3. Collect flow-through for analysis. Throughout the purification samples are retained for SDS-PAGE analysis (Fig. 2) (*see* **Note 9**).
4. Wash resin with 2× 25 mL of IMAC wash buffer 1. Incubate the 1st wash gently rocking for 15 min at 4 °C. Perform the

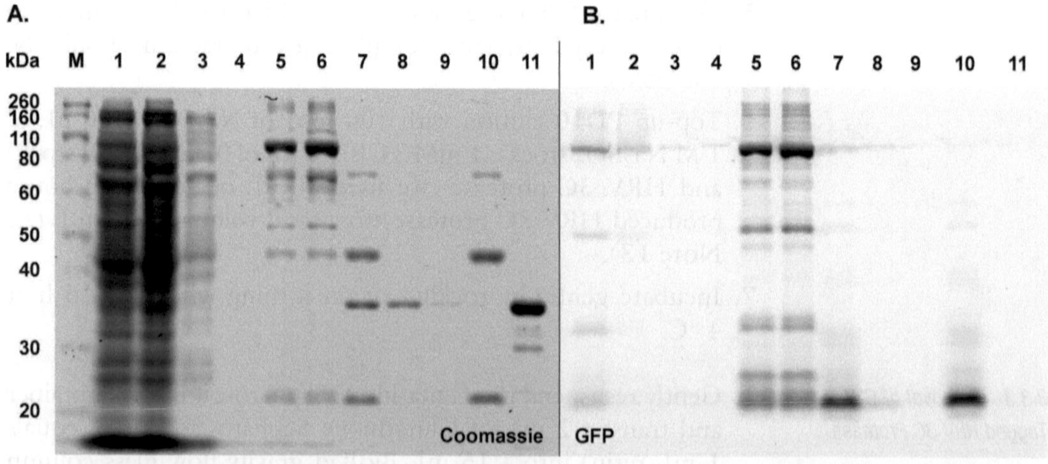

Fig. 2 SDS-Page analysis of a typical purification of $^{13}CH_3$-methionine-labeled enNTS$_1$. (**a**) 10% SDS-PAGE gel without prior heat denaturation stained with Coomassie brilliant blue staining. The image was taken using a high sensitivity infrared imager (Odyssey, Li-Cor). (**b**) GFP-fluorescence reading (Typhoon, GE Healthcare) the same SDS-PAGE gel prior to Coomassie blue staining. Lane M, Novex Sharp unstained protein standard (Thermo Fisher Scientific). Lane 1, DM/CHS/CHAPS solubilized *E. coli* cells prior to incubation with 1st pass IMAC resin (52.5 μg protein). Lane 2, flow-through of 1st pass IMAC (49.1 μg). Lane 3, wash of IMAC resin using DM- and ATP-containing buffer (9.3 μg). Lane 4, second wash of IMAC resin using DDM-containing buffer (0.25 μg). Lane 5, elution from IMAC resin (3.7 μg). Lane 6, elution from PD10 column (4.6 μg). Lane 7, cleaved enNTS$_1$/fusion protein mixture prior to reverse IMAC separation (2.1 μg). Lane 8, flow-through of 2nd pass IMAC containing enNTS$_1$ (0.56 μg). Lane 9, 2nd pass IMAC wash (0.34 μg). Lane 10, 2nd pass IMAC elution of His-tagged fusion proteins (2.3 μg). Lane 11, pooled SEC fractions containing enNTS$_1$ (2.5 μg)

2nd round without incubation but gently swirl up resin to ensure complete removal of contaminants. Keep the flow-through for analysis (*see* **Note 10**).

5. Wash resin with 2× 25 mL of IMAC wash buffer 2. Gently swirl up resin to ensure proper wash. Keep the flow-through for analysis (*see* **Note 11**).
6. Elute the His$_{10}$-tagged fusion protein by adding 3× 5 mL (5 bed volumes or 15 mL) of IMAC elution buffer. Each time gently swirl up resin to ensure complete elution.

3.3.2 Imidazole Removal and 3C Protease Cleavage

1. To remove imidazole, concentrate the elution to ~1 mL using a 100 kDa cut-off Amicon® concentrator (*see* **Note 12**).
2. While concentrating, drain the storage solution from a PD10® de-salting column, wash the column with 5× 5 mL of deionized water and subsequently equilibrate with 5× 5 mL of base buffer (Subheading 2.2, **item 20**).
3. Load the concentrated sample onto the PD10® de-salting column, let the buffer drip off and wait until the sample has completely entered the column.
4. Top up to 2.5 mL with base buffer (i.e., add 1.5 mL of base buffer if 1 mL of concentrated sample was added).

5. To elute the full-length receptor, add 3.5 mL of base buffer and collect everything (the complete green fraction should be collected).
6. Top-up PD10 elution with 100 mM of Na_2SO_4 (400 μL of 1 M Na_2SO_4 stock), 1 mM TCEP (8 μL of 0.5 M TCEP stock) and HRV 3C protease (we used 30 μL of 96 μM in-house produced HRV 3C protease) to a total volume of ~4 mL (*see* **Note 13**).
7. Incubate gently by rocking or on turning wheel for 16 h at 4 °C.

3.3.3 Removal of GST-Tagged HRV 3C Protease

1. Gently resuspend the Glutathione Sepharose® in the container and transfer 2 mL of Glutathione Sepharose® slurry (equals 1 mL resin) into a 15 mL BioRad gravity-flow glass column equipped with a tap. Let the storage solution drain off and wash the Glutathione Sepharose® with 2× 10 mL deionized water by gently resuspending the resin. Repeat with 2× 10 mL base buffer to equilibrate the resin (*see* **Note 14**).
2. Add the overnight cleavage mixture to the equilibrated Glutathione Sepharose®.
3. Incubate gently by rocking or on turning wheel for 1 h at 4 °C.
4. Drain and collect the supernatant.
5. Wash the Glutathione Sepharose® with 3× 5 mL base buffer by gently resuspending the resin (collect flow-through).

3.3.4 Second Pass IMAC (Fusion Protein Removal)

1. While the overnight cleavage mixture is incubating with the Glutathione Sepharose®, add 6 mL of TALON® slurry (supplied as 50% (w/v) slurry in non-buffered 20% EtOH), which equals 3 mL of TALON® resin, to a 50 mL BioRad gravity-flow glass column equipped with a tap. Let the ethanol drain off and wash the resin 3× with at least 5 bed volumes (3× 15 mL) of deionized water. Repeat wash with 3× 5 bed volumes of base buffer (Subheading 2.2, **item 20**).
2. Adjust the solution from **step 7**, Subheading 3.3.2 with Imidazole to 5 mM (~4 μL of 5 M Imidazole stock); or combine the eluted supernatant and Glutathione Sepharose® wash (~19 mL) from **steps 4** and **5** of Subheading 3.3.3 also adjusting with Imidazole to 5 mM (~19 μL of 5 M Imidazole stock). Add to the column with the equilibrated TALON® resin, close the column with the supplied lid and gently rock for 45 min at 4 °C (*see* **Note 15**).
3. Collect flow-through (contains cleaved receptor).
4. Wash resin with 3× 10 mL of IMAC wash buffer 3 by gently swirling up resin to ensure complete removal of receptor.

5. Elute the His_{10}-tagged fusion protein by adding 3× 5 mL (5 bed volumes or 15 mL) of elution buffer. Keep elution and TALON® resin for analysis.

3.3.5 Size Exclusion Chromatography (SEC)

1. Combine flow-through and wash from the 2nd pass IMAC (~34 mL) and concentrate to less than 0.45 mL using 30 kDa cut-off Amicon® concentrators (*see* **Note 12**).
2. While concentrating the 2nd pass IMAC flow-through and wash, or the day before, set up a Superdex 200 10/300 Increase column and perform a "pump wash basic" with sterile filtered and degassed MilliQ. Make sure to set the maximum pressure to 3 MPa. Wash column with 50 mL (2× bed volumes) of sterile filtered and degassed MilliQ. Equilibrate the column with 50 mL (2× bed volumes) sterile filtered and degassed SEC buffer (*see* **Note 16**).
3. Set-up the pre-loaded Superdex 200 10/300 Increase isocratic elution method at a constant flow of 0.75 mL/min and choose to collect 0.5 mL fractions.
4. Prior to loading into the sample loop, centrifuge the sample for 2 min at 10,000× rcf at 4 °C to pellet aggregated protein.
5. Load supernatant into a 0.5 mL sample loading loop (avoid loading bubbles and be careful not to suck up any pelleted material) and perform SEC.
6. Collect and pool the fractions corresponding to the peak eluting at 12.75 mL (*see* **Note 17**).
7. Determine the protein concentration using an appropriate assay (*see* **Note 18**).
8. Based on the molecular weight of the receptor (45 kDa for cleaved $enNTS_1$ including a C-terminal Avi-tag) the molarity of the sample can be determined (*see* **Note 19**).
9. Concentrate the sample to 0.5 mL using a 30 kDa Amicon® concentrator (*see* **Note 12**).
10. Top-up sample with 2 mL of NMR buffer (total volume of 2.5 mL) and concentrate again to 0.5 mL with regular resuspension (*see* **Note 20**).
11. Repeat two more times to obtain a final sample in 99% NMR buffer. $^{13}CH_3$-methionine incorporation efficiency can be accurately determined via LC-MS/MS analysis of tryptic $enNTS_1$ peptides [28] (Fig. 3) (*see* **Note 21**).
12. Split the sample into the desired number of aliquots, complement with ligands and/or signaling partner and top-up to 290 μL with NMR buffer (*see* **Note 22**).

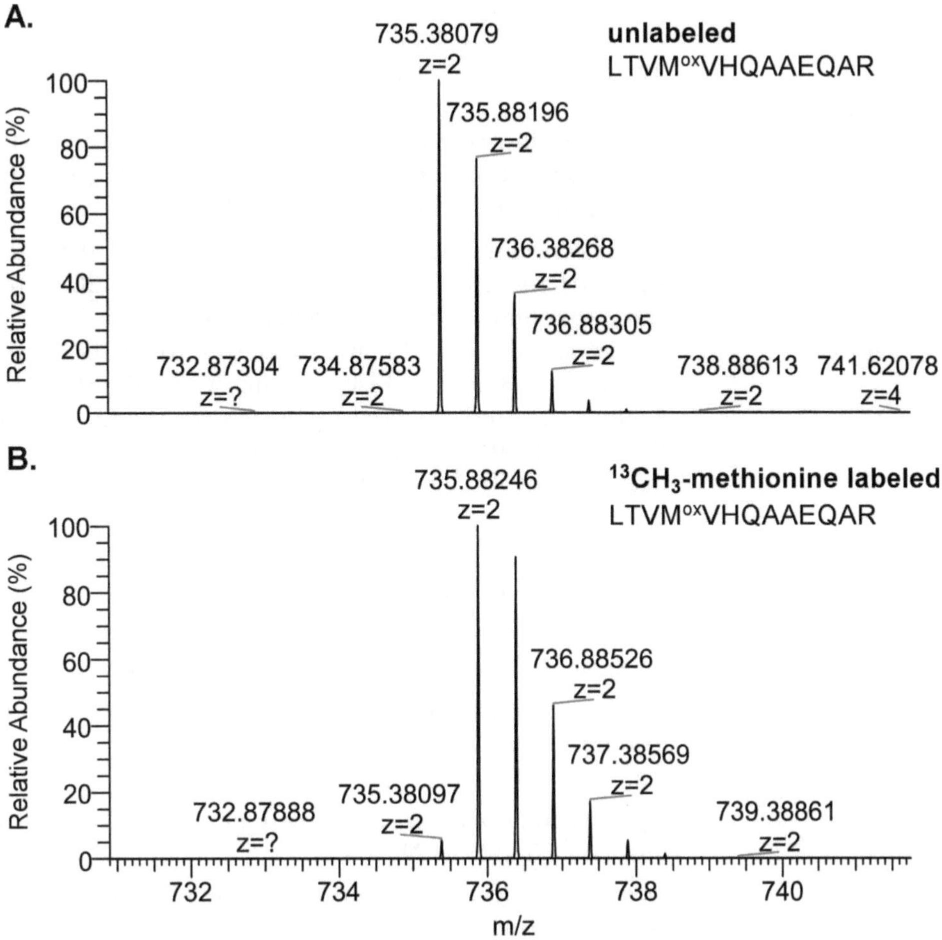

Fig. 3 Mass spectra for a doubly charged unlabeled and $^{13}CH_3$-methionine-labeled tryptic peptide. (**a**) Isotope series for unlabeled LTVM^{OX}VHQAAEQAR with isotope 0 at 735.3808 m/z and isotope 1 at 735.8820 m/z. (**b**) Isotope series for $^{13}CH_3$-methionine-labeled LTVM^{OX}VHQAAEQAR with isotope 0 at 735.8825 m/z. The isotope (−1) at 735.8810 m/z is attributed to the starting isotope (0) of a series of unlabeled isotopes overlapping with the $^{13}CH_3$-methionine-labeled isotope series

13. Transfer the final samples to 5 mm Shigemi NMR microtubes and cap or parafilm the tubes (*see* **Note 23**). A typical NMR spectrum is shown in Fig. 4.

3.4 Amido Black 10B Protein Assay

1. Dilute protein samples (containing 0.5–30 μg protein) to 0.27 mL with deionized water (i.e., use 1 μL of IMAC load and flow-through, and 10 μL of other samples and add deionized water to 0.27 mL in 1.5 mL Eppendorf tubes). We suggest preparing triplicates for each sample.
2. Make up standard solutions of 0, 0.5, 2, 5, 15, and 30 μg BSA in 0.27 mL deionized water (i.e., use 0, 0.5, 2, 5, 15, and 30 μL of 1 mg/mL BSA and add deionized water to 0.27 mL in

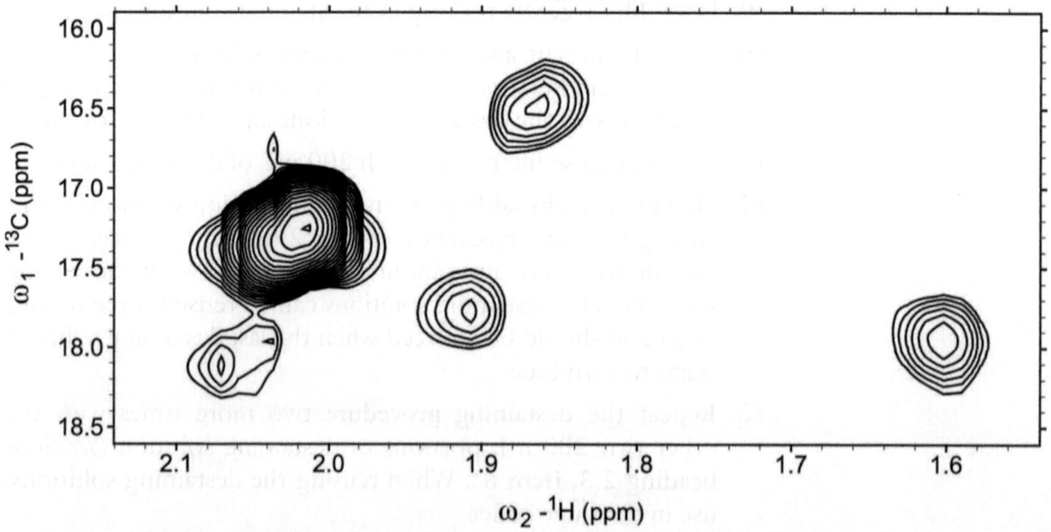

Fig. 4 ^{1}H-^{13}C SOFAST-HMQC spectrum of Apo-state $^{13}CH_3$-methionine-labeled enNTS$_1$. Spectrum showing the region containing $^{13}CH_3$-methionine chemical shifts only. The sample contained 28 μM enNTS$_1$ in 50 mM potassium phosphate, 100 mM NaCl, 0.02% DDM and 99% D_2O, pH 7.4. The spectrum was collected on a 800-MHz Bruker Avance II spectrometer equipped with a triple resonance cryoprobe with 25% non-uniform sampling (NUS) at 298 K with a ^{1}H spectral width of 12 ppm (1024 data points in t2) and a ^{13}C spectral width of 25 ppm (128 data points in t1), a relaxation delay of 400 ms and 2048 scans per t1 data point resulting in an acquisition time of 8.5 h. *Reprinted (adapted) from Biochimica et Biophysica Acta,* 1860(6), Bumbak, F., Keen, A.C., Gunn, N.J., Gooley, P.R., Bathgate, R.A.D., Scott, D.J., Optimization and $^{13}CH_3$ methionine labeling of a signaling competent neurotensin receptor 1 variant for NMR studies, 1372–1383, *Copyright (2018), with permission from Elsevier*

1.5 mL Eppendorf tubes). We suggest preparing triplicates for each standard.

3. Add 0.03 mL of 1% (w/v) SDS, 1 M Tris–HCl, pH 7.5 (final 0.1% SDS and 100 mM Tris–HCl) to each protein sample and standard solution and vortex thoroughly.
4. Add 0.1 mL of 60% (w/v) TCA (final 15% TCA) to each protein sample and standard solution and vortex thoroughly.
5. Incubate for at least 3 min at RT.
6. Number disc filters near the edges with a pencil and place in filtration manifold. Under vacuum, use a pipette with 1 mL tip to transfer each sample onto the filter labeled correspondingly (*see* **Note 24**).
7. After filtration, wash each filter with 2 mL of 6% (w/v) TCA under vacuum. Remove filters from the filtration manifold and collect (*see* **Note 25**).
8. Once all samples are filtered, place filters one by one in a container with 200 mL of Amido Black 10B staining solution (*see* **Note 26**).

9. Leave filters gently rocking in the staining solution for 3 min.
10. Carefully decant and save the staining solution. The staining solution can be reused several times and was usually changed together with the destaining solutions after three to four uses.
11. Carefully rinse filters once with 200 mL of deionized water.
12. Destain filters by adding 200 mL of destaining solution [methanol/glacial acetic acid/water (90/2/8 vol. %)], leave gently rocking for 1 min and carefully decant and save the destaining solution. The destaining solutions can be reused three to four times and should be replaced when the last destaining solution starts to turn blue.
13. Repeat the destaining procedure two more times with the other two 200 mL portions of destaining solution (*see* Subheading 2.3, **item 8**). When reusing the destaining solutions, use in the same order.
14. Wash filters once with 200 mL of deionized water by gently rocking for 2 min.
15. Remove the filters from the beaker and place on Kimwipes to remove excess water.
16. Excise the blue spot from each filter (or the equivalent sized area on the blank), cut into several smaller pieces and place them into a test tube containing 0.7 mL of eluent solution.
17. Incubate tubes for at least 20 min with occasional vortexing to elute the dye from the filters.
18. Read the absorbance of the eluates at 630 nm in 0.5 mL cuvettes against air (*see* **Note 27**).
19. Use the BSA standard sample values to generate a standard curve and determine the sample protein concentrations against this curve.

4 Notes

1. For the minimal salts medium pre-culture we generally observe OD_{600} values around 2.5 after 13 h of which we use 10 mL per 500 mL (or 60 mL per 3 L) of expression culture. If the OD_{600} value is different from 2.5, adjust the volume of minimal salts medium pre-culture used in Subheading 3.1, **step 7** accordingly.
2. Certain shakers take longer to cool down and it may be beneficial to have a second shaker pre-cooled to 16 °C.
3. This step adapts the culture to the presence of the added amino acids and ensures that methionine biosynthesis is inhibited prior to induction.

4. We found that an expression window between 12 h and 16 h post-induction results in the highest yield of recombinant $enNTS_1$. Longer incubation times can result in reduced yields due to protein degradation by proteases released from dead cells. Expect OD_{600} values between 2 and 4.
5. We generally apply vortexing with frequent periods of cooling on ice to resuspend *E. coli* cell pellets containing MBP-$enNTS_1$-muGFP fusion protein. For cell pellets containing α1 adrenergic receptor (α_1AR) fusion protein [70] however, it is crucial to avoid vortexing as it causes receptor truncation. We suggest assessing the suitability of vortexing should this protocol be applied to constructs different form MBP-$enNTS_1$-muGFP.
6. We use a Q Sonica Q500 sonicator with a 6.4 mm microtip set to 30% power and 12 cycles with 10 s on and 20 s off periods. Place the Falcon tube in a beaker with wet ice such as all the solution is immersed in ice. Make sure the microtip is always properly immersed and readjust depth during off periods to ensure even sonication. Avoid touching the walls of the Falcon tube and readjust depth if excessive foaming occurs.
7. Expect a very small and translucent pellet as cells should be completely solubilized at this point.
8. We found that an incubation time longer than 1.5 h does not increase the yield of recombinant protein. We advise to use more TALON® resin if a significant proportion of the receptor is detected in the flow-through.
9. The use of muGFP as a fusion protein provides the opportunity to add a second layer of SDS-PAGE gel analysis in addition to standard staining protocols (Fig. 2). To take advantage of GFP fluorescence, samples are prepared using standard SDS-PAGE loading buffers but without heat denaturation prior to loading the gel. muGFP retains its fluorescence in SDS at high concentrations (up to 8%) and at temperatures up to 63 °C [66]. After electrophoresis, the SDS-PAGE gel is analyzed using a laser scanner (we use a Typhoon FLA 9000 from GE Healthcare) equipped with a suitable filter (excitation at 490 nm/emission at 508 nm; Fig. 2b) prior to Coomassie brilliant blue staining (Fig. 2a) or other suitable staining method.
10. Incubation with IMAC wash buffer 1 was found to effectively remove the chaperone DnaK. DnaK preferentially binds to particular peptide sequences and is commonly co-purified with recombinant proteins [71]. ATP binding reduces the affinity of DnaK for its substrate which allows efficient removal of the contaminant.

11. The contribution to contaminant removal of this step is insignificant as its main purpose is to exchange the detergent from DM to DDM.
12. We suggest starting with centrifugation intervals of 5 min (3400 rcf at 4 °C) and resuspension of the solution between runs to avoid precipitation. Run times can then be increased or decreased as required.
13. It has been shown previously that HRV 3C protease has increased activity in the presence of sulfate [72]. We regularly add 100 mM Na_2SO_4 to stimulate HRV 3C protease activity. Try increasing the Na_2SO_4 concentration if HRV 3C protease activity is not satisfactory.
14. Skip this step when using His-tagged HRV 3C protease. Using the GST-tagged HRV 3C protease in combination with the Glutathione Sepharose® affinity purification step was found to have no negative effect on protein yield when compared to using His-tagged HRV 3C protease without an additional purification step. However, the use of His-tagged HRV 3C protease helps reduce the total purification time by ~2 h.
15. We found that capture of cleaved fusion proteins is efficient at an incubation time of 45 min. A longer incubation time tends to promote the cleaved receptor to stick to the TALON® resin and hence to decrease the yield of recombinant protein. We advise to use more TALON® resin if significant proportions of fusion proteins are detected in the flow-through.
16. At this step we reduce the DDM concentration to 400 μM DDM (0.02% w/v) to reduce the detergent background. 400 μM corresponds to more than double the CMC of DDM [73] and was found to be sufficient for the stability of $enNTS_1$.
17. The pre-loaded method for the Superdex 200 10/300 Increase column starts fraction collection at 6.71 mL and we generally collect fractions 13–15.
18. For rapid measurements we suggest using the DirectDetect® system from Millipore. For accurate measurements we suggest using the Amido Black 10B method introduced by Schaffner and Weissmann [74] considering some modifications suggested by Kaplan and Pedersen [75] (*see* Subheading 3.4). If the protein concentration is determined for individual purification steps, we strongly suggest using the Amido Black 10B method. The Amido Black 10B is time consuming and can be swapped for a more rapid assay, such as DirectDetect®, once the receptor is exchanged to phosphate buffer (SEC buffer) containing 400 μM DDM and no glycerol. In SEC buffer, values determined using DirectDetect® are generally about

10–15% higher compared to values determined using the Amido Black 10B method. The Amido Black 10B assay is a colorimetric assay that employs a protein precipitation step prior to staining. Protein precipitation allows to efficiently remove chemicals that may interfere with protein assays [76]. The Amido Black 10B assay is insensitive to HEPES, detergents, and glycerol—components we use in most of the described purification steps—that are found to give rise to inconsistent results using other protein estimation methods. Furthermore, the use of the Amido Black 10B assay allows direct comparison of our results with those of Reinhard Grisshammer and co-workers who have been using this assay as their standard protein concentration measurement method since their first report of NTS_1 expression in 1993 [77].

19. A suitable online calculator can be found under http://www.molbiol.ru/eng/scripts/01_04.html.
20. Our NMR buffer does not contain DDM as the detergent tends to concentrate even when using high molecular weight concentrators. A control experiment using 400 μM DDM showed similar DDM signal intensities compared to experiments using samples prepared as described herein.
21. To identify tryptic fragments, we searched mass spectra of unlabeled $enNTS_1$ (expressed in 2YT) against the sequence of $enNTS_1$. The same is done for $^{13}CH_3$-methionine-labeled $enNTS_1$ assuming a mass difference of 1.0034 Da between ^{12}C and ^{13}C per methionine residue. Figure 3 shows the MS/MS isotope series for the unlabeled and $^{13}CH_3$-methionine-labeled versions of the peptide LTVM^{OX}VHQAAEQAR (M^{OX} represents oxidized methionine). The unlabeled isotope series starts with a monoisotopic peak at 735.3808 m/z (isotope 0) followed by isotope 1 at 735.8820 m/z. The $^{13}CH_3$-methionine-labeled isotope series of the same peptide is shifted by the mass difference between ^{12}C and ^{13}C divided by the charge state (which was 2 in the given example) with isotope 0 at 735.8825 m/z. The isotope series of the $^{13}CH_3$-methionine-labeled peptide reveals a further peak (isotope -1) at 735.8810 m/z which we attribute to isotope 0 in the isotope series of the unlabeled peptide. Hence, isotope -1 in the isotope series of the $^{13}CH_3$-methionine-labeled peptide represents isotope 0 of unlabeled peptide species in the same sample. Based on the isotope 0 to isotope 1 ratio of the unlabeled peptide (in Fig. 3a), we can estimate the intensity contribution of unlabeled isotope 1 to the $^{13}CH_3$-methionine-labeled peptide isotope 0 (in Fig. 3b) as they must overlap. The intensity ratio between the corrected isotope 0 and isotope -1 in the spectrum of the $^{13}CH_3$-methionine-labeled peptide represents the ratio between $^{13}CH_3$-methionine-labeled and

unlabeled LTVM^{OX}VHQAAEQAR fragments and hence the $^{13}CH_3$-methionine incorporation efficiency.

22. Depending on the protein concentration determined in **steps** 7 and **8** (Subheading 3.3.5), up to four NMR samples may be prepared. We recommend enNTS$_1$ concentrations of at least 20 μM as determined using the DirectDetect® method.
23. Do not use the plunger as this will introduce bubbles.
24. Use flat tweezers to handle the filters. Pipette slowly to keep the filtration spot as small as possible. At this point, we recommend rinsing the Eppendorf tube with 0.5 mL of 6% (w/v) TCA and adding this wash on top of the filter spot to ensure complete transfer of precipitated protein.
25. The filtration manifold holds 12 filters simultaneously and you will be required to run several filtration/wash cycles as only the standard solutions add up to 18 samples when in triplicates. Therefore, make sure that all filters are labeled in advance and ample 6% TCA solution is available. Do not touch the top of the filter during the wash step and when removing from the filtration manifold.
26. Used 1 mL tip boxes work well as containers for the staining and destaining steps.
27. For the reagent blanks (no BSA or sample) absorbances between 0.05 and 0.08 can be expected.

Acknowledgments

This work was supported by NHMRC project grants 1081801 (D.J.S.) and 1081844 (R.A.D.B., P.R.G., D.J.S.); ARC equipment grant LE120100022. D.J.S. is an NHMRC Boosting Dementia Research Leadership Fellow. R.A.D.B. is an NHMRC Senior Research Fellow. Studies at The Florey Institute of Neuroscience and Mental Health were supported by the Victorian Government's Operational Infrastructure Support Program.

References

1. Hauser AS, Attwood MM, Rask-Andersen M et al (2017) Trends in GPCR drug discovery: new agents, targets and indications. Nat Rev Drug Discov 16:829–842
2. Grisshammer R (2017) New approaches towards the understanding of integral membrane proteins – A structural perspective on G protein-coupled receptors. Protein Sci 26:1493–1504
3. Maeda S, Schertler GF (2013) Production of GPCR and GPCR complexes for structure determination. Curr Opin Struct Biol 23:381–392
4. Verardi R, Traaseth NJ, Masterson LR et al (2012) Isotope labeling for solution and solid-state NMR spectroscopy of membrane proteins. Adv Exp Med Biol 992:35–62
5. Foster MP, Mcelroy CA, Amero CD (2007) Solution NMR of large molecules and assemblies. Biochemistry 46:331–340

6. Didenko T, Liu JJ, Horst R et al (2013) Fluorine-19 NMR of integral membrane proteins illustrated with studies of GPCRs. Curr Opin Struct Biol 23:740–747
7. Klein-Seetharaman J, Getmanova EV, Loewen MC et al (1999) NMR spectroscopy in studies of light-induced structural changes in mammalian rhodopsin: applicability of solution ^{19}F NMR. Proc Natl Acad Sci U S A 96:13744–13749
8. Loewen MC, Klein-Seetharaman J, Getmanova EV et al (2001) Solution ^{19}F nuclear Overhauser effects in structural studies of the cytoplasmic domain of mammalian rhodopsin. Proc Natl Acad Sci U S A 98:4888–4892
9. Kim TH, Chung KY, Manglik A et al (2013) The role of ligands on the equilibria between functional states of a G protein-coupled receptor. J Am Chem Soc 135:9465–9474
10. Liu JJ, Horst R, Katritch V et al (2012) Biased signaling pathways in beta2-adrenergic receptor characterized by ^{19}F-NMR. Science 335:1106–1110
11. Manglik A, Kim TH, Masureel M et al (2015) Structural insights into the dynamic process of β2-adrenergic receptor signaling. Cell 161:1101–1111
12. Ye L, Van Eps N, Zimmer M et al (2016) Activation of the A2A adenosine G-protein-coupled receptor by conformational selection. Nature 533:265–268
13. Tugarinov V, Ollerenshaw JE, Kay LE (2005) Probing side-chain dynamics in high molecular weight proteins by deuterium NMR spin relaxation: an application to an 82-kDa enzyme. J Am Chem Soc 127:8214–8225
14. Ruschak AM, Kay LE (2010) Methyl groups as probes of supra-molecular structure, dynamics and function. J Biomol NMR 46:75–87
15. Kofuku Y, Ueda T, Okude J et al (2012) Efficacy of the b_2-adrenergic receptor is determined by conformational equilibrium in the transmembrane region. Nat Commun 3:1045
16. Nygaard R, Zou Y, Dror RO et al (2013) The dynamic process of beta(2)-adrenergic receptor activation. Cell 152:532–542
17. Prosser RS, Kim T (2015) Nuts and bolts of CF_3 and CH_3 NMR toward the understanding of conformational exchange of GPCRs. Methods Mol Biol 1335:39–51
18. Duewel H, Daub E, Robinson V et al (1997) Incorporation of trifluoromethionine into a phage lysozyme: Implications and a new marker for use in protein ^{19}F NMR. Biochemistry 36:3404–3416
19. Gellman SH (1991) On the role of methionine residues in the sequence-independent recognition of nonpolar protein surfaces. Biochemistry 30:6633–6636
20. Dellavecchia MJ, Merritt WK, Peng Y et al (2007) NMR analysis of [methyl-13C] methionine UvrB from Bacillus caldotenax reveals UvrB-domain 4 heterodimer formation in solution. J Mol Biol 373:282–295
21. Bose-Basu B, Derose EF, Kirby TW et al (2004) Dynamic characterization of a DNA repair enzyme: NMR studies of [methyl-13C] methionine-labeled DNA polymerase beta. Biochemistry 43:8911–8922
22. Okude J, Ueda T, Kofuku Y et al (2015) Identification of a conformational equilibrium that determines the efficacy and functional selectivity of the μ-opioid receptor. Angew Chem Int Ed Engl 54:15771–15776
23. Solt AS, Bostock MJ, Shrestha B et al (2017) Insight into partial agonism by observing multiple equilibria for ligand-bound and Gs-mimetic nanobody-bound β1-adrenergic receptor. Nat Commun 8:1795
24. Isogai S, Deupi X, Opitz C et al (2016) Backbone NMR reveals allosteric signal transduction networks in the β1-adrenergic receptor. Nature 530:237–241
25. Clark LD, Dikiy I, Chapman K et al (2017) Ligand modulation of sidechain dynamics in a wild-type human GPCR. elife 6:e28505
26. Bokoch MP, Zou Y, Rasmussen SG et al (2010) Ligand-specific regulation of the extracellular surface of a G-protein-coupled receptor. Nature 463:108–112
27. Sounier R, Mas C, Steyaert J et al (2015) Propagation of conformational changes during m-opioid receptor activation. Nature 524:375–378
28. Bumbak F, Keen AC, Gunn NJ et al (2018) Optimization and 13CH3 methionine labeling of a signaling competent neurotensin receptor 1 variant for NMR studies. Biochim Biophys Acta 1860:1372–1383
29. Doublie S, Kapp U, Aberg A et al (1996) Crystallization and preliminary X-ray analysis of the 9 kDa protein of the mouse signal recognition particle and the selenomethionyl-SRP9. FEBS Lett 384:219–221
30. Van Duyne GD, Standaert RF, Karplus PA et al (1993) Atomic structures of the human immunophilin FKBP-12 complexes with FK506 and rapamycin. J Mol Biol 229:105–124
31. Schlegel S, Hjelm A, Baumgarten T et al (2013) Bacterial-based membrane protein production. Biochim Biophys Acta 1843:1739–1749

32. Grisshammer R, Tate CG (1995) Overexpression of integral membrane proteins for structural studies. Q Rev Biophys 28:315–422
33. Tucker J, Grisshammer R (1996) Purification of a rat neurotensin receptor expressed in Escherichia coli. Biochem J 317:891–899
34. Weiss HM, Grisshammer R (2002) Purification and characterization of the human adenosine A (2a) receptor functionally expressed in *Escherichia coli*. Eur J Biochem 269:82–92
35. Marullo S, Delavier-Klutchko C, Guillet J-G et al (1989) Expression of human [beta]1 and [beta]2 adrenergic receptors in E. Coli as a new tool for ligand screening. Nat Biotechnol 7:923–927
36. Ge B, Wang M, Li J et al (2014) Maltose binding protein facilitates functional production of engineered human chemokine receptor 3 in *Escherichia coli*. Process Biochem 50:285–293
37. Ren H, Yu D, Ge B et al (2009) High-level production, solubilization and purification of synthetic human GPCR chemokine receptors CCR5, CCR3, CXCR4 and CX3CR1. PLoS One 4:e4509
38. Wiktor M, Morin S, Sass HJ et al (2013) Biophysical and structural investigation of bacterially expressed and engineered CCR5, a G protein-coupled receptor. J Biomol NMR 55:79–95
39. Bertin B, Freissmuth M, Breyer RM et al (1992) Functional expression of the human serotonin 5-Ht1a receptor in *Escherichia coli*. Ligand-binding properties and interaction with recombinant G-protein alpha-subunits. J Biol Chem 267:8200–8206
40. Furukawa H, Haga T (2000) Expression of functional M2 muscarinic acetylcholine receptor in *Escherichia coli*. J Biochem 127:151–161
41. Link AJ, Skretas G, Strauch EM et al (2008) Efficient production of membrane-integrated and detergent-soluble G protein-coupled receptors in *Escherichia coli*. Protein Sci 17:1857–1863
42. Locatelli-Hoops S, Sheen FC, Zoubak L et al (2013) Application of halotag technology to expression and purification of cannabinoid receptor CB. Protein Expr Purif 89:62–72
43. Ma Y, Kubicek J, Labahn J (2013) Expression and purification of functional human Mu opioid receptor from E.coli. PLoS One 8:e56500
44. Egloff P, Hillenbrand M, Klenk C et al (2014) Structure of signaling-competent neurotensin receptor 1 obtained by directed evolution in Escherichia coli. Proc Natl Acad Sci U S A 111:E655–E662
45. Scott DJ, Pluckthun A (2013) Direct molecular evolution of detergent-stable g protein-coupled receptors using polymer encapsulated cells. J Mol Biol 425:662–677
46. Scott DJ, Kummer L, Egloff P et al (2014) Improving the apo-state detergent stability of NTS1 with CHESS for pharmacological and structural studies. Biochim Biophys Acta 1838:2817–2824
47. White JF, Trinh LB, Shiloach J et al (2004) Automated large-scale purification of a G protein-coupled receptor for neurotensin. FEBS Lett 564:289–293
48. Gelis I, Bonvin AMJJ, Keramisanou D et al (2007) Structural basis for signal-sequence recognition by the translocase motor SecA as determined by NMR. Cell 131:756–769
49. Stadtman ER, Cohen GN, Lebras G (1961) Feedback inhibition and repression of aspartokinase activity in *Escherichia coli*. Ann N Y Acad Sci 94:952–959
50. Walden H (2010) Selenium incorporation using recombinant techniques. Acta Crystallogr D Biol Crystallogr 66:352–357
51. Wu BL, Chien EYT, Mol CD et al (2010) Structures of the CXCR4 chemokine GPCR with small-molecule and cyclic peptide antagonists. Science 330:1066–1071
52. Wu H, Wacker D, Mileni M et al (2012) Structure of the human k-opioid receptor in complex with JDTic. Nature 485:327–332
53. Zhang H, Qiao A, Yang D et al (2017) Structure of the full-length glucagon class B G-protein-coupled receptor. Nature 546:259–264
54. Zhang J, Zhang K, Gao ZG et al (2014) Agonist-bound structure of the human P2Y12 receptor. Nature 509:119–122
55. Zhang K, Zhang J, Gao Z-G et al (2014) Structure of the human P2Y12 receptor in complex with an antithrombotic drug. Nature 509:115–118
56. Wang C, Wu H, Katritch V et al (2013) Structure of the human smoothened receptor bound to an antitumour agent. Nature 497:338–343
57. Wang C, Wu H, Evron T et al (2014) Structural basis for Smoothened receptor modulation and chemoresistance to anticancer drugs. Nat Commun 5:4355
58. Wang C, Jiang Y, Ma J et al (2013) Structural basis for molecular recognition at serotonin receptors. Science 340:610–614
59. Wacker D, Wang S, Mccorvy JD et al (2017) Crystal structure of an LSD-bound human serotonin receptor. Cell 168:377–389.e312

60. Egloff P, Deluigi M, Heine P et al (2014) A cleavable ligand column for the rapidisolation of large quantities of homogeneous and functional neurotensin receptor 1 variants from *E. coli*. Protein Expr Purif 108:106–114
61. Ballesteros JA, Weinstein H (1995) [19] Integrated methods for the construction of three-dimensional models and computational probing of structure-function relations in G protein-coupled receptors. In: Stuart CS (ed) Methods in neurosciences. Academic Press, San Diego, pp 366–428
62. Rovati GE, Capra V, Neubig RR (2007) The highly conserved DRY motif of class A G protein-coupled receptors: beyond the ground state. Mol Pharmacol 71:959–964
63. Labbe-Jullie C, Barroso S, Nicolas-Eteve D et al (1998) Mutagenesis and modeling of the neurotensin receptor NTR1. Identification of residues that are critical for binding SR 48692, a nonpeptide neurotensin antagonist. J Biol Chem 273:16351–16357
64. Barroso S, Richard F, Nicolas-Etheve D et al (2000) Identification of residues involved in neurotensin binding and modeling of the agonist binding site in neurotensin receptor 1. J Biol Chem 275:328–336
65. White JF, Noinaj N, Shibata Y et al (2012) Structure of the agonist-bound neurotensin receptor. Nature 490:508–513
66. Scott DJ, Gunn NJ, Yong KJ et al (2018) A novel ultra-stable, monomeric green fluorescent protein for direct volumetric imaging of whole organs using clarity. Sci Rep 8:667
67. Huber S, Casagrande F, Hug MN et al (2017) SPR-based fragment screening with neurotensin receptor 1 generates novel small molecule ligands. PLoS One 12:e0175842
68. Ranganathan A, Heine P, Rudling A et al (2016) Ligand discovery for a peptide-binding GPCR by structure-based screening of fragment- and lead-like chemical libraries. ACS Chem Biol 12:735–745
69. Cai M, Huang Y, Sakaguchi K et al (1998) An efficient and cost-effective isotope labeling protocol for proteins expressed in Escherichia coli. J Biomol NMR 11:97–102
70. Yong KJ, Vaid TM, Shilling PJ et al (2018) Determinants of Ligand subtype-selectivity at alpha1A-adrenoceptor revealed using saturation transfer difference (STD) NMR. ACS Chem Biol 13:1090–1102
71. Rial DV, Ceccarelli EA (2002) Removal of DnaK contamination during fusion protein purifications. Protein Expr Purif 25:503–507
72. Wang QM, Johnson RB (2001) Activation of human rhinovirus-14 3C protease. Virology 280:80–86
73. Vanaken T, Foxall-Vanaken S, Castleman S et al (1986) Alkyl glycoside detergents: synthesis and applications to the study of membrane proteins. Methods Enzymol 125:27–35
74. Schaffner W, Weissmann C (1973) A rapid, sensitive, and specific method for the determination of protein in dilute solution. Anal Biochem 56:502–514
75. Kaplan RS, Pedersen PL (1985) Determination of microgram quantities of protein in the presence of milligram levels of lipid with amido black 10B. Anal Biochem 150:97–104
76. Sapan CV, Lundblad RL (2015) Review of methods for determination of total protein and peptide concentration in biological samples. Proteomics Clin Appl 9:268–276
77. Grisshammer R, Duckworth R, Henderson R (1993) Expression of a rat neurotensin receptor in Escherichia coli. Biochem J 295:571–576

Chapter 4

A Combined Cell-Free Protein Synthesis and Fluorescence-Based Approach to Investigate GPCR Binding Properties

Anne Zemella, Theresa Richter, Lena Thoring, and Stefan Kubick

Abstract

Fluorescent labeling of de novo synthesized proteins is in particular a valuable tool for functional and structural studies of membrane proteins. In this context, we present two methods for the site-specific fluorescent labeling of difficult-to-express membrane proteins in combination with cell-free protein synthesis. The cell-free protein synthesis system is based on Chinese Hamster Ovary Cells (CHO) since this system contains endogenous membrane structures derived from the endoplasmic reticulum. These so-called microsomes enable a direct integration of membrane proteins into a biological membrane. In this protocol the first part describes the fluorescent labeling by using a precharged tRNA, loaded with a fluorescent amino acid. The second part describes the preparation of a modified aminoacyl-tRNA-synthetase and a suppressor tRNA that are applied to the CHO cell-free system to enable the incorporation of a non-canonical amino acid. The reactive group of the non-canonical amino acid is further coupled to a fluorescent dye. Both methods utilize the amber stop codon suppression technology. The successful fluorescent labeling of the model G protein-coupled receptor adenosine A_{2A} (Adora2a) is analyzed by in-gel-fluorescence, a reporter protein assay, and confocal laser scanning microscopy (CLSM). Moreover, a ligand-dependent conformational change of the fluorescently labeled Adora2a was analyzed by bioluminescence resonance energy transfer (BRET).

Key words Cell-free protein synthesis, G protein-coupled receptor, Protein modification, Non-canonical amino acids, Amber suppression, Confocal laser scanning microscopy

1 Introduction

G protein-coupled receptors (GPCRs) are involved in vital processes including the regulation of signaling pathways and the subsequent triggering of essential physiological responses. Up to date more than 50% of pharmaceutical drugs are targeting GPCRs and GPCR interaction partners [1]. Therefore, the development of novel production processes of GPCRs for structural and functional characterization remains essential. In a natural environment, membrane proteins are only present in small concentrations [2]. The overexpression of membrane proteins is usually performed in heterologous systems involving a complex vector design. Nevertheless,

Mario Tiberi (ed.), *G Protein-Coupled Receptor Signaling: Methods and Protocols*, Methods in Molecular Biology, vol. 1947, https://doi.org/10.1007/978-1-4939-9121-1_4,

overexpression obstacles such as low solubility, cytotoxicity, and low stability can occur [3]. In addition, detergents used to solubilize membrane proteins often induce altered ligand binding properties [4]. Cell-free protein synthesis based on translationally active lysates represents an approved alternative for membrane protein synthesis. Eukaryotic cell lysates based on insect, CHO, and human cell lines are predestined for membrane protein synthesis due to the presence of endogenous membrane structures [5]. These microsomes enable an integration of membrane proteins into a biological membrane. In addition, posttranslational modifications such as disulfide bridging, phosphorylation, and partly glycosylation are possible [6]. Furthermore, the CHO cell-free system is characterized by a high productivity of complex proteins [7].

In the context of protein modification and fluorescent labeling by using genetic code expansion [8], cell-free systems are of special interest due to their open reaction mode [9]. Exogenously prepared components can be easily and directly added to the cell-free protein synthesis reaction to incorporate a defined non-canonical amino acid site-specifically at the position of a redefined codon within the gene sequence [10]. In a first step, addressing of a defined amber stop codon can be verified by the addition of a precharged tRNA to the cell-free protein synthesis reaction. The tRNAs in this report are aminoacylated to a lysine residue that is coupled to a Bodipy TMR dye (BP). The tRNA recognizes either the amber stop codon or the phenylalanine codon. The lysine residue with the coupled fluorescent dye is subsequently site-specifically (BP-CUA) or statistically (BP-GAA) incorporated. The lack of a re-aminoacylating mechanism of the tRNA leads to the consumption of large amounts of precharged tRNA. Therefore, we describe an alternative system that is inspired by the natural aminoacylation of tRNAs in vivo.

The addition of non-canonical amino acids in vivo that usually do not permeate the cell wall can be realized in cell-free systems due to the missing cell wall. Even high concentrations of non-canonical amino acids can be added to the reaction. Moreover, the suppressor tRNA and the modified aminoacyl-tRNA-synthetase can be applied to the cell-free reaction in defined concentrations enabling an identification of optimal concentrations for highest suppression efficiency. Here, the described experimental setup is based on an evolved aminoacyl tRNA-synthetase (aaRS) derived from *E.coli* [11, 12] and a modified suppressor tRNA recognizing the UAG amber stop codon [13]. The combination of the aaRS, tRNA, and a phenylalanine-derivate (p-propargyloxy-L-phenylalanine, pPa) with an alkyne group in para-position resulted in an efficient incorporation of the non-canonical amino acid into the model protein adenosine A_{2A} receptor (Adora2a). The human Adora2a is expressed in almost every tissue of the body. Primarily, the membrane protein

can be detected in the brain, cells of the immune system, leucocytes, and platelets. In peripheral tissues, the Adora2a is involved in inflammation reactions and affects the control of the cardiovascular system [14]. The interaction of the Adora2a with different neurotransmitters in the brain is of highest interest since the receptor displays an important role in the regulation of dopamine and glutamate release thereby making it a promising therapeutic target for the treatment of depression, insomnia, and Parkinson's disease [15]. Consequently, a better understanding of the detailed interactions between membrane proteins and their specific ligands is mandatory. Information on the structural and functional relationship of a defined GPCR might be gained by a visualization of the target membrane protein in a nature-like lipid environment [14]. A highly preferred method to determine GPCR interaction is based on the spatial proximity of two fluorescent moieties. Interaction partners can be determined by an intermolecular fluorescence resonance energy transfer (FRET) and specific conformation changes by an intramolecular FRET. For the Adora2a, intermolecular as well as intramolecular FRETs were established to analyze the receptors' behavior [16, 17]. In this way, the conformational rearrangement of the helices III and VI induced by agonist binding was elucidated [18]. A coupled fluorescent dye in the region of amino acid 215 will be excited by a C-terminally fused NanoLuciferase (Nluc) resulting in a detectable bioluminescence resonance energy transfer (BRET). A signal change of the BRET value after the addition of a ligand indicates a conformational change of the Adora2a. The main objective of this chapter is to describe a cell-free protein synthesis-based alternative methodology to the standard protein visualization methods typically used in vivo. Our cell-free procedure is based on the site-specific and statistical incorporation of a precharged tRNA and a site-specific incorporation of a non-canonical amino acid into the target protein. The cell-free synthesized protein is subsequently labeled by using a chemoselective reaction. Moreover, we adapted an intramolecular BRET for the Adora2a to show a ligand-dependent conformational change. In addition, we present the preparation of the evolved aminoacyl-tRNA-synthetase and the suppressor tRNA.

2 Materials

2.1 Preparation of Enhanced Orthogonal Aminoacyl-tRNA-Synthetase

1. Coding sequence for the modified tyrosyl-tRNA-synthetase (eAzFRS, including the mutations Thr37, Ser182, Ala183, and Arg265 [11, 12] and a C-terminal Strep-Tag) from *E.coli.*
2. *E.coli* expression system (RTS 500 *E.coli* HY Kit, biotechrabbit).
3. 100 mM Isopropyl β-D-1-thiogalactopyranoside (IPTG).

4. Gravity flow Strep-Tactin® superflow mini-column (0.2 ml).
5. Strep-Tactin® Purification Buffer Set: 10× Washing Buffer (1 M Tris–Cl, pH 8.0, 1.5 M NaCl, 10 mM EDTA), 10× Elution Buffer (1 M Tris–Cl, pH 8.0, 1.5 M NaCl, 10 mM EDTA, 25 mM Desthiobiotin) and 10× Regeneration Buffer (1 M Tris–Cl, 1.5 M NaCl, 10 mM EDTA, 10 mM HABA (hydroxyl-azophenyl-benzoic acid)).
6. Zeba™ Spin Desalting Columns (7 K MWCO, 0.5 ml).
7. Amicon® Ultra Centrifugal Filters (10 K device, 0.5 ml).
8. Synthetase storage buffer: 50 mM HEPES pH 7.6, 10 mM KOAc, 1 mM $MgCl_2$, 4 mM DTT.
9. Thermomixer with a microtiter plate adapter and a RTS 500 adapter.

2.2 Preparation of Suppressor tRNA

2.2.1 Generation of PCR Product

1. Vector containing the nucleotide sequence of tRNATyrCUA (SupF Gene).
2. tRNATyrCUA-specific forward primer (5′ CgA gCT CgC CCA CCg gAA TTC 3′) and 2′-OMe reverse primer (5′ Tgg Tgg Tgg ggg AAg gAT TCg 3′).
3. 0.2 ml PCR tubes.
4. PCR cycler.
5. Taq DNA polymerase.
6. Taq buffer.
7. dNTPs.
8. 25 mM $MgCl_2$.
9. Agarose gel electrophoresis chambers.
10. Agarose.
11. Rotiphorese 10× TBE buffer.
12. DNA stain.
13. DNA ladder.
14. PCR Purification Kit.

2.2.2 Generation of RNA Transcript

1. T7 RNA Polymerase (f.c. 1 U/μl, Agilent).
2. 5× NTP mix containing 18.75 mM ATP, 18.75 mM CTP, 18.75 mM UTP and 7.5 mM GTP.
3. 5× transcription buffer: 400 mM HEPES-KOH, 0.5 mM Spermidine, 50 mM DTE and 75 mM $MgCl_2$.
4. DNAseI (1 U per μg plasmid DNA).
5. 10× MOPS buffer: 200 mM MOPS, 50 mM NaOAc, 10 mM EDTA, pH 8.0.

6. MOPS sample buffer: 8% (v/v) formaldehyde, 12 ml formamide, 2.4 ml 10× MOPS buffer, 0.05% (v/v) bromophenol blue to a total volume of 24 ml.

2.2.3 RNA Isolation and Folding

1. TRIzol reagent.
2. High Performance Liquid Chromatography (HPLC) grade Chloroform.
3. HPLC grade Isopropyl.
4. 75% Ethanol.
5. Cooled centrifuge.
6. Nanodrop 2000c.

2.3 Cell-Free Protein Synthesis and Fluorescent Labeling of Modified Adora2a

1. Coding sequence for the modified adenosine A_{2A} receptor (Uniprot: P29274, amino acids 1-340, with and without amber stop codon at nucleotide sequence coding for P215) with a receptor C-terminally fused Nanoluciferase (Fig. 1). Flanking sequences containing 5′ regulatory sequences (T7 RNA polymerase promotor, Cricket paralysis virus (CRPV) IGR IRES sequence (Genbank accession no. AF218039, nucleotides 6025-6216)) and 3′ regulatory sequences (T7 terminator), cloned with the coding sequence into a plasmid (BioCat or Thermo Fisher Scientific).
2. 1.5 ml reaction tubes.
3. CHO lysate prepared as described [19, 20] (*see* **Note 1**).
4. 10× translation mix: 300 mM HEPES-KOH (pH 7.6), 2250 mM KOAc, 2.5 mM spermidine, 1 mM of each canonical amino acid (Merck) and 39 mM $Mg(OAc)_2$.
5. 5× energy: 100 mM creatine phosphate, 1.5 mM GTP, 1.5 mM CTP, 1.5 mM UTP 8.75 mM ATP and 0.5 mM $m^7G(ppp)G$ cap analogue.
6. 100 μM Polyguanylic acid (polyG, IBA).
7. T7 RNA polymerase (f.c. 1 U/μl).
8. ^{14}C-leucine.
9. 100 μM Bodipy-TMR-lysine-tRNACUA (BP-CUA, biotechrabbit).
10. 100 μM Bodipy-TMR-lysine-tRNAGAA (BP-GAA, biotechrabbit).
11. 100 μM eAzFRS.
12. 100 μM tRNATyrCUA.
13. 100 mM p-propargyloxy-L-phenylalanine (pPa, Iris Biotech).
14. 5 mM Copper(II) sulfate ($CuSO_4$).

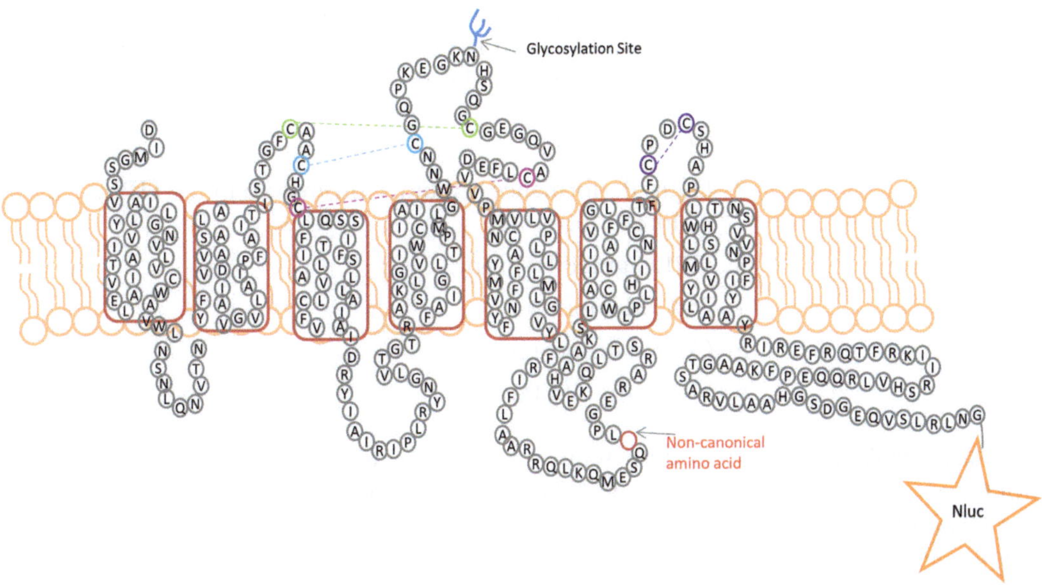

Fig. 1 Schematic illustration of the Adora2a receptor with a C-terminally fused Nluc. Arrows indicate the glycosylation site and the position of the non-canonical amino acid. Disulfide bridges are indicated by the colors green, blue, pink, and purple

15. 2.5 mM Tris(3-hydroxypropyltriazolylmethyl)amine (THPTA, Iris Biotech).
16. 80 mM Sodium ascorbate (NaAsc).
17. Phosphate-buffered saline (PBS).
18. 100 μM Sulfo-Cy5-azide (Lumiprobe).

2.4 Analysis of De Novo Synthesized Proteins

1. Trichloroacetic acid (TCA).
2. Water bath.
3. Glass fiber filters.
4. Acetone.
5. Scintillation vials.
6. Scintillation cocktail.
7. Scintillation counter.
8. SDS-PAGE Sample buffer: 1× LDS buffer containing 106 mM Tris–HCl, 141 mM Tris base, 2% LDS, 10% glycerol, 0.51 mM EDTA, 0.22 mM SERVA Blue G, 0.175 mM Phenol Red, pH 8.5 with 50 mM DTT.
9. SDS-PAGE gels.
10. Fluorescently labeled protein ladder for SDS-PAGE.
11. Fluorescence/phosphorimager.
12. Gel dryer.

13. Phosphorscreens.
14. Adenosine.
15. Nano-Glo® Luciferase Assay System.
16. 96-well microtiter plate.
17. Multimode Microplate Reader Mithras^2 LB 943.
18. μ-Ibidi-Slide (μ-Slide 18 well, flat, Ibidi).
19. Confocal laser scanning microscope.

3 Methods

3.1 Preparation of Enhanced Orthogonal Aminoacyl-tRNA-Synthetase

1. For prokaryotic cell-free synthesis, the eAzFRS gene should be cloned into a vector containing a T7 promotor, ribosomal binding site, and T7 terminator such as pIX3.0, pIVEX2.3d, and pIVEX2.4d vectors or alternatively containing a T5 promotor such as pQE2 vectors as used in this protocol. eAzFRS is synthesized in a cell-free system using an *E. coli* lysate in a dialysis mode. A typical 1.1 ml reaction is composed of 0.525 ml *E.coli* lysate, 0.225 ml reaction mix, 0.27 ml amino acids without methionine, 30 μl methionine, 11 μl IPTG for the induction of the protein expression pQE2 vector, 39 μl template containing 110 μg plasmid DNA.
2. The surrounding feeding mixture contains 7990 μl feeding mix, 110 μl IPTG, 2650 μl amino acids without methionine and 300 μl methionine (*see* **Note 2**).
3. Fill the reaction solution into the reaction compartment (marked through the red lid).
4. Fill the feeding mix into the feeding chamber (marked through the colorless lid).
5. Insert the prepared chamber into the RTS 500 adapter in a thermomixer. The reaction time is 20 h at 30 °C and a shaking speed of 900 rpm.
6. For the separation of aggregated proteins from soluble eAzFRS a centrifugation step at 16,000 × *g* for 10 min at 4 °C is recommended.
7. Equilibrate two Strep-Tactin columns with 400 μl of 10× washing buffer and add 500 μl of the supernatant of the cell-free reaction to each column.
8. After the supernatant has completely entered the column, wash each column 5× with 200 μl washing buffer (*see* **Note 3**).
9. Elute the protein 6× with 100 μl elution buffer and collect the fractions.
10. Elution fractions containing the target protein are pooled.

11. Regenerate the column with 3× 1 ml of 1× regeneration buffer and remove the regeneration buffer 2× with 800 μl of 1× washing buffer. Store the column in 2 ml washing buffer at 4 °C.
12. The combined elution fractions are applied to Zeba™ Spin Desalting Columns to exchange the elution buffer of the strep-tag purification to a synthetase storage buffer. Therefore, remove the storage solution of the Zeba™ Spin Desalting Column by centrifugation at 1500 × *g* for 1 min. Add 300 μl of the synthetase storage buffer to the resin bed and centrifuge at 1500 × *g* for 1 min. Repeat this **step 2**×. Place the column in a new collection tube and apply 100 μl of the pooled synthetase solution to each column. Centrifuge at 2000 × *g* for 2 min and collect the synthetase.
13. The concentration of the synthetase can be performed with Amicon® Ultra Centrifugal Filters. Add up to 500 μl of the synthetase solution to the concentrator and centrifuge at 14,000 × *g* for 10 min and 4 °C. Collect the concentrated sample and determine the concentration by NanoDrop measurement using the molecular mass (48.6 kDa) and the extinction coefficient (54.3) (*see* **Note 4**).
14. The synthetase can be stored at −80 °C after shock freezing in liquid nitrogen.

3.2 Preparation of Suppressor tRNA

3.2.1 Generation of PCR Product

1. For specific and homogenous 3′-ends of the suppressor tRNA, an additional PCR step before transcription reaction is included. Therefore, the reverse primer contains a 2′-OMe-group to prevent unspecific nucleotides at the 3′-end of the tRNA that can be added by the T7 polymerase during transcription reaction. Amplify the template by pipetting in a PCR tube final concentrations of 1× Taq Buffer, 0.2 mM dNTP mix, 0.5 μM forward primer, 0.5 μM reverse primer, 2.5 mM $MgCl_2$, 0.01 ng/μl plasmid and 0.04 U/μl Taq DNA polymerase. Fill the reaction with water to a final volume of 250 μl (*see* **Note 5**).

3.2.2 Generation of RNA Transcript

1. Thaw the components for in vitro transcription on ice and pipette the reaction at room temperature. Mix 1× transcription buffer, 1× NTP mix, 1 U/μl T7 RNA Polymerase and 8 ng/μl template DNA. Fill the reaction with water to the final volume of 500 μl. Incubate the reaction for 3–6 h at 37 °C and 500 rpm.
2. Centrifuge the RNA at 12,000 × *g* for 1 min und use the supernatant for the DNAseI treatment (*see* **Note 6**) Add 1 U DNAseI per 1 μg DNA. Incubate for 10 min at 37 °C and 500 rpm.

3.2.3 RNA Isolation and Folding

1. Handle the TRIzol and chloroform reagent with care and use a fume hood. Add a threefold volume of TRIzol to the transcription reaction and mix carefully. Incubate for 5 min at room temperature. Add 200 μl chloroform for 1 ml TRIzol and mix carefully for 15 s by inverting. Incubate for 3 min at room temperature. Centrifuge at 12,000 × *g* for 15 min at 4 °C. Isolate the aqueous phase (*see* **Note 7**). Add 500 μl isopropyl for 1 ml TRIzol and mix carefully. Incubate overnight at 4 °C.
2. Centrifuge at 15,000 × *g* at least for 1 h at 4 °C and discard the supernatant. Overlay the pellet with 1 ml 75% ethanol for 1 ml TRIzol and incubate for 30 min at −20 °C. Centrifuge at 7500 × *g* for 10 min and 4 °C. Discard the supernatant and air dry the pellet. Solve the pellet in water. Measure concentration using a NanoDrop and adjust the concentration to 100 μM.
3. Fold the tRNA by slowly decreasing the temperature from 80 to 25 °C in a PCR cycler. The tRNA can be stored at −80 °C after shock freezing in liquid nitrogen.

3.3 Cell-Free Protein Synthesis

1. Thaw all required components for cell-free protein synthesis on ice (*see* **Note 8**).
2. The cell-free protein synthesis is performed in a coupled mode where transcription and translation reaction take place in one vessel. A standard reaction is composed of 40% CHO lysate, 10 μM polyG, 1× translation mix, and 1× energy mix. Add plasmid at a concentration of 40 nM (*see* **Note 9**). Add 1 U/μl T7 RNA polymerase for the transcription reaction and ^{14}C-leucine (specific radioactivity of 66.67 dpm/pmol) for further analysis of de novo synthesized proteins. Fill the reaction with water to the final volume of 50 μl.

3.3.1 Fluorescent Labeling with Bodipy-TMR-Lysine

1. Thaw the Bodipy-TMR-lysine-tRNACUA (BP-CUA) and Bodipy-TMR-lysine-tRNAGAA (BP-GAA) on ice and keep it dark (*see* **Note 10**).
2. Pipette 2 μM of the precharged tRNA to the cell-free reaction and incubate the prepared cell-free protein synthesis reaction by 27 °C for 3 h and shaking at 600 rpm (*see* **Note 11**). Cover the thermomixer with a lid or aluminum foil to prevent the reaction of any light.
3. Take a 5 μl aliquot of the translation mixture for SDS-PAGE. Centrifuge the translation reaction for 15 min at 16,000 × *g* and 4 °C. Take a 5 μl of the supernatant and resuspend the pellet (microsomal fraction) in an equal volume PBS in comparison to the volume of the translation reaction. This step is required for the analysis of the localization of the synthesized

Table 1
Expectations of cell-free protein synthesis and fluorescent labeling of different Adora2a constructs with Bodipy-tRNA directed to phenylalanine codons (GAA) or directed to the amber stop codon (CUA)

Construct	Precharged tRNA	Expectation	
		Synthesis	Fluorescence
Adora2a	BP-GAA	Full-length protein	Highly fluorescently labeled protein with statistical incorporation of ncAA
Adora2a	BP-CUA	Full-length protein	The full-length protein is not labeled with the fluorescent dye and the ncAA is not incorporated.
Adora2a_amb	BP-CUA	Full-length protein and partial termination product	The full-length protein is fluorescently labeled. The fluorescence signal has a lower intensity in comparison to the statistically labeled Adora2a due to the site-specific incorporation of the ncAA at one defined position. Termination product is not labeled.
Adora2a_amb	BP-GAA	Termination product	The termination product is highly fluorescently labeled due to the statistical incorporation of the ncAA. No full-length product is visible.

protein. The expectations of the incorporation of the different Bodipy-tRNAs are described in Table 1 (*see* **Note 12**).

4. An example of the performance of the fluorescent labeling of Adora2a with or without incorporated non-canonical amino acid determined by in-gel-fluorescence and autoradiography, and the measurement of the reporter protein are shown in Fig. 2. In-gel-fluorescence and autoradiography (Fig. 2a, c) showed the expected results as described in Table 1. The reporter protein assay (Fig. 2b) is based on the activity of a receptor C-terminally fused Nanoluciferase (Nluc). Nluc activity can only be measured after the translation of the full-length fusion protein. The Adora2a synthesized by using the DNA construct without an amber stop codon (Adora2a + BP-GAA/ BP-CUA) resulted in a fusion protein with a detectable Nluc activity. The fusion protein translated from the amber stop codon DNA construct (Adora2a_amb + BP-CUA) showed a fivefold reduced Nluc activity. This result implicates a rather low incorporation efficiency of the precharged tRNA addressing the amber stop codon. The combination of Adora2a_amb with BP-GAA showed no luciferase activity since only the termination product was translated. Protein yields were calculated by scintillation counting and resulted in approximately 20 μg/ml in translation mixture. In the microsomal fraction and the supernatant, approximately 6 μg/ml and 14 μg/ml were detected, respectively. A similar Adora2a distribution was

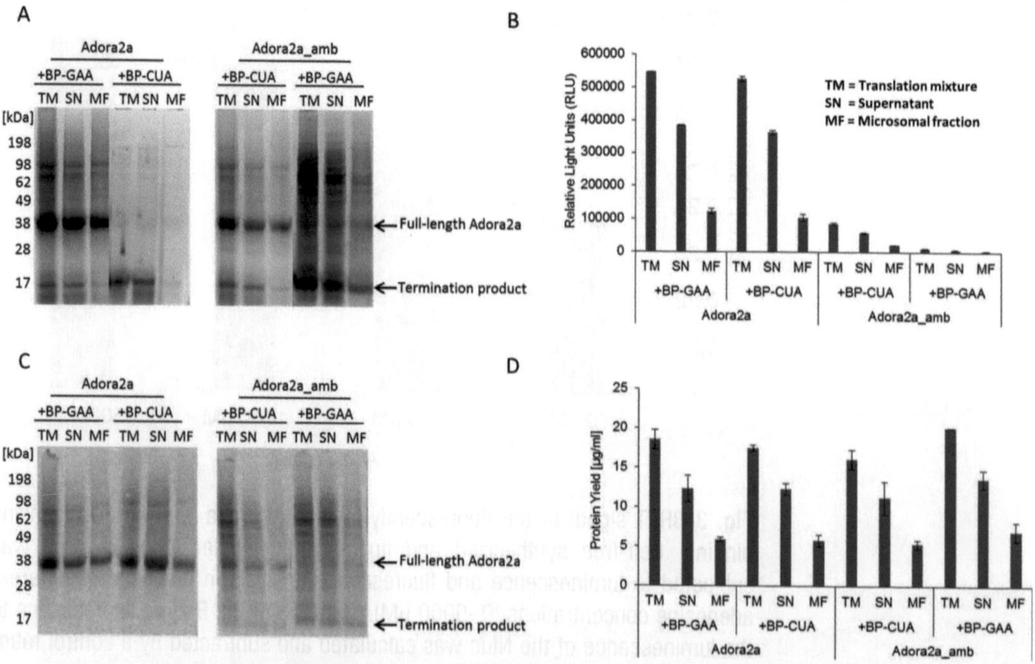

Fig. 2 Analysis of fluorescent labeling by using precharged tRNAs. (**a**) In-gel-analysis of Adora2a by statistical and site-specific incorporation of a fluorescently labeled amino acid. The translation mixture (TM) was fractionated into supernatant (SN) and microsomal fraction (MF). (**b**) Analysis of the translation of a reporter protein (Nluc) fused to the C-terminus of Adora2a by measuring the luminescence of the Nluc. (**c**) Autoradiographic analysis of de novo synthesized Adora2a by incorporation of ^{14}C-leucine during the cell-free synthesis reaction. (**d**) Determination of protein yield of de novo synthesized Adora2a by scintillation counting

obtained by in-gel-fluorescence, luciferase activity, and autoradiography. Previously reported distributions of membrane proteins are comparable to the here described distribution [7]. A comparable protein yield was calculated for the termination product.

3.3.2 Analysis of Ligand-Dependent Conformational Change Using a BRET Assay

1. Site-specifically label the Adora2a_amb using the precharged tRNA Bodipy-TMR-lysine-tRNACUA as described in Subheading 3.3.1, **steps 1–3**.
2. Resuspend the microsomal fraction of the Adora2a_amb in PBS. 5 µl aliquots of resuspended Adora2a_amb were mixed with 5 µl adenosine in PBS with final concentrations of 0 µM, 100 µM, 1000 µM, and 5000 µM adenosine (*see* **Note 13**).
3. 10 µl *samples* were applied for the luminescence and fluorescence measurement. In a first step, the luminescence of the Nluc was detected using an OD2 filter. In a second step, the fluorescence of the coupled Bodipy dye, excited by the Nluc emission was detected.
4. As a *control 5 µl of the resuspended microsomal fraction of* the full-length Adora2a protein without any fluorescent label is

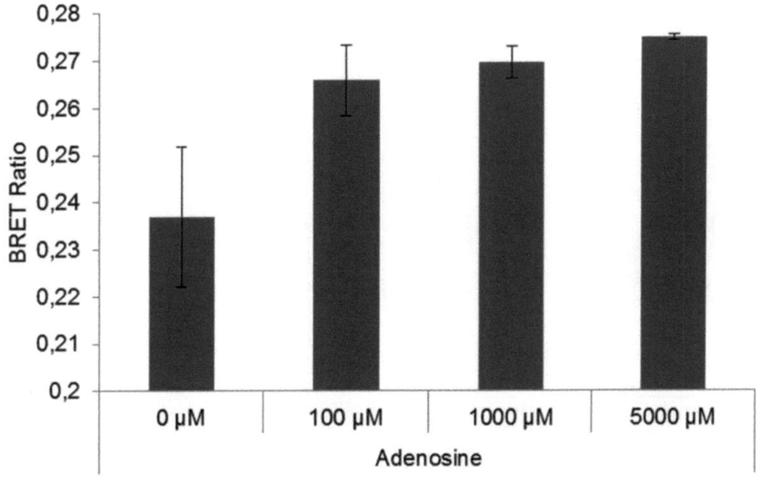

Fig. 3 BRET signal of the fluorescently labeled Adora2a_amb after adenosine binding. Cell-free synthesized and fluorescently labeled Adora2a_amb was subjected to luminescence and fluorescence analysis in presence of different adenosine concentrations (0–5000 μM). The ratio of the Bodipy fluorescence to the luminescence of the Nluc was calculated and subtracted by a control ratio

treated with the same concentrations of adenosine (**step 2**) to determine background fluorescence caused by the broad emission spectrum of the Nluc and possible interactions of adenosine with the Nluc.

5. The BRET ratio is calculated as follows:

$$\mathrm{BRET}_{\mathrm{ratio}} = \frac{\mathrm{Fluorescence\ (sample)}}{\mathrm{Luminescence\ (sample)}} - \frac{\mathrm{Fluorescence\ (control)}}{\mathrm{Luminescence\ (control)}}$$

6. Herein, the calculated BRET ratio (Fig. 3) showed a change in the relation of the fluorescence of the Bodipy in comparison to the Nluc luminescence after the addition of different adenosine concentrations. Only a minimal increase of the BRET ratio can be seen after addition of higher adenosine concentrations (above 100 μM) indicating that at lower concentrations all receptors are occupied with adenosine. The result indicates a conformational change of the helix III and the connected third intracellular loop that is expected for the Adora2a.

3.3.3 Site-Specific Incorporation of a Non-canonical Amino Acid with Subsequent Fluorescent Labeling and Microscopic Analysis

1. Additional components are required for the recharging of the suppressor tRNA and a subsequent incorporation of a non-canonical amino acid. Therefore, add the p-propargyloxy-L-phenylalanine, tRNATyrCUA, and eAzFRS in a specific order (Table 2) (*see* **Note 14**).

Table 2
Pipetting order of a standard cell-free reaction with orthogonal components for the incorporation of non-canonical amino acids

Order	Components	Final concentration
1	Water	
2	PolyG	10 µM
3	Translation mix	1×
4	pPa	2 mM
5	DNA-template	40 nM
6	Lysate	40%
7	tRNATyrCUA	3 µM
8	eAzFRS	3 µM
9	^{14}C-leucine	66.67 dpm/pmol
10	T7-RNA-Polymerase	1 U/µl
11	Energy mix	1×

2. Incubate the prepared cell-free reaction at 27 °C for 3 h by gentle shaking at 600 rpm. Keep the reaction in the dark (*see* **Note 15**).
3. The membrane protein is translocated and integrated into the microsomal membrane during the cell-free reaction. Therefore, separate the microsomal fraction by centrifugation at 16,000 × *g* for 15 min at 4 °C. Resuspend the pellet fraction in PBS. Use an equal volume of the cell-free reaction for resuspension.
4. For the labeling reaction prepare the labeling mix as follows: combine 200 µM $CuSO_4$ with 600 µM THPTA, 5 mM NaAsc, PBS and a final concentration of 3 µM Sulfo-Cy5-azid to a final volume of 5 µl. Add 5 µl of the resuspended microsomal fraction to the labeling mix. Incubate the labeling reaction at room temperature for 1 h. Keep the reaction dark (*see* **Note 16**).
5. Centrifuge the labeling reaction for 15 min at 16,000 × *g* and 4 °C. Discard the supernatant and resuspend the pellet in 10 µl PBS. This step removes excess fluorescent dye to decrease the background signal in the subsequent fluorescent analyses.
6. The analysis of the synthesis of the full-length protein by using the Adora2a construct with or without an amber stop codon was performed by autoradiography and a reporter protein assay (Fig. 4). The autoradiography shows for both synthesis reactions a similar band pattern (Fig. 4a). Interestingly, the

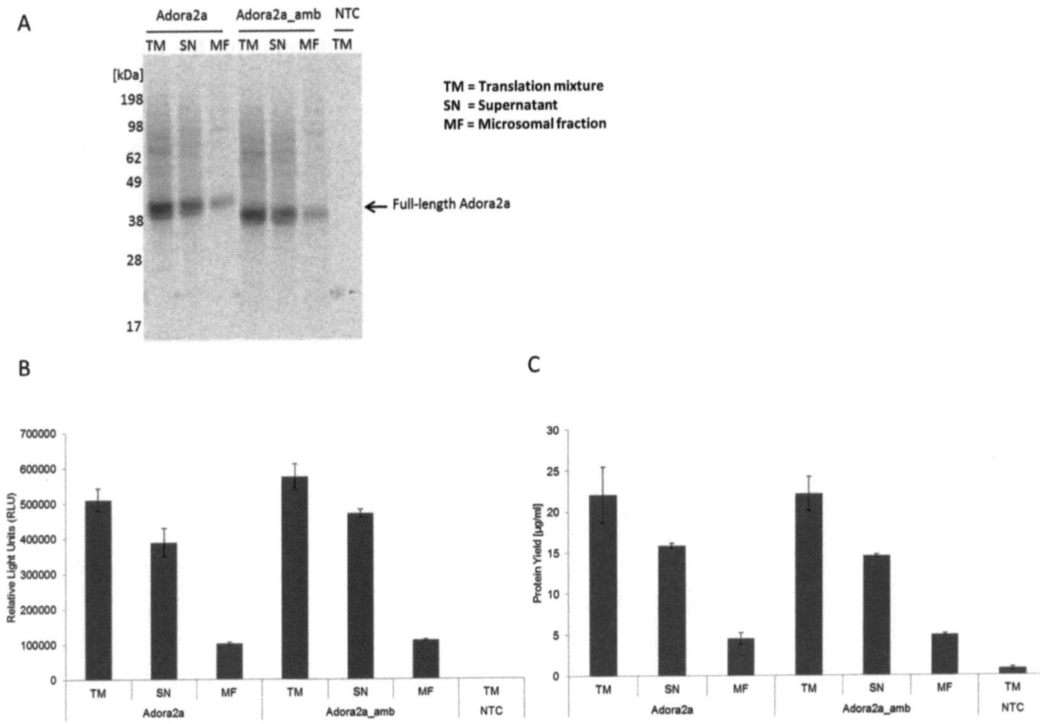

Fig. 4 Analysis of incorporation efficiency using an orthogonal system. (**a**) Autoradiography of synthesized Adora2a after incorporation of ^{14}C-leucine. Translation mixture (TM) was fractionated into supernatant (SN) and microsomal fraction (MF). (**b**) Schematic illustration of subsequent labeling reaction after reaching a high incorporation of the non-canonical amino acid. (**c**) Analysis of the translation of a reporter protein (Nluc) fused to the C-terminus of Adora2a by measuring the luminescence of the luciferase. NTC = no template control. (**d**) Protein yield of synthesized Adora2a, determined by scintillation counting

incorporation of pPa led to a comparable band signal as obtained for the full-length protein translated from the DNA construct without an amber stop codon. This result implicates a high incorporation efficiency of the non-canonical amino acid. In addition, no termination product is detected in the autoradiograph. The high incorporation efficiency is the basis for a further coupling reaction to a fluorescent dye. In addition, this result is supported by the Nluc assay (Fig. 4b). The measured luciferase activity of the suppression product reaches up to 80% of the luciferase activity of the full-length product. The protein yield and protein distribution (Fig. 4c) is comparable to the previously described results (Fig. 2, *see* Subheading 3.3.1, **step 4**).

7. Fluorescently labeled GPCRs should be detectable in microsomal structures (Fig. 5). It is recommended to visualize the labeled sample first by in-gel-fluorescence (Fig. 5, left panel).

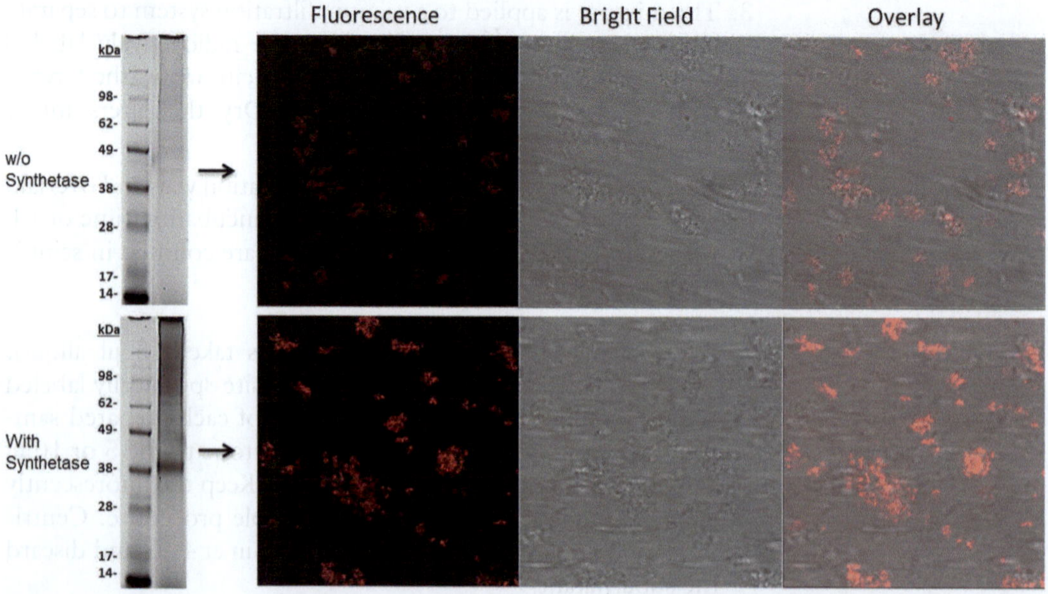

Fig. 5 Fluorescence analysis of labeled Adora2a. Left side: In-gel-fluorescence of synthesized and with Sulfo-Cy5-azide labeled Adora2a in absence or presence of the modified synthetase during the cell-free protein synthesis reaction. Right side: confocal laser scanning microscopy of the control sample (without addition of the modified synthetase during cell-free protein synthesis) and the labeled Adora2a (cell-free protein synthesis in presence of the synthetase)

The synthesis reaction in the presence of all orthogonal components led to a specific band at the expected molecular weight. The control reaction without addition of the modified synthetase resulted in no visible band. The success of the microscopic analysis highly correlates to the quality of the in-gel-fluorescence. High background fluorescence during in-gel-analysis often results in unspecific staining of the microsomes. The microscopic analysis clearly shows a difference in the fluorescence intensity of the labeled Adora2a and the unspecific staining of the microsomes (right panel).

3.4 Analysis of De Novo Synthesized Fluorescent Proteins

3.4.1 TCA Precipitation and Scintillation Counting

1. After the reaction is completed collect 2 × 3 μl of the translation mixture. Centrifuge the remaining mix at 16,000 × *g* for 15 min and 4 °C and collect 2 × 3 μl of the supernatant. Resuspend the microsomal fraction in an equal volume of PBS in comparison to the volume of the translation mixture. Collect 2 × 3 μl of the microsomal fraction.
2. Mix each aliquot with 3 ml TCA and incubate in a water bath at 80 °C for 15 min. Store the aliquots for 30 min on ice or overnight at 4 °C.

3. The mixture is applied to a vacuum filtration system to separate non-incorporated ^{14}C-leucine from the radioactively labeled protein. Filters with the collected protein are washed twice with TCA and twice with acetone. Dry the filters under the hood.
4. The filters are transferred into the scintillation vials and overlaid with 3 ml scintillation cocktail. After an incubation time of 1 h with gentle shaking, scintillation vessels are counted in scintillation counter.

3.4.2 In-Gel-Fluorescence and Autoradiography

1. For preparation of SDS-PAGE samples take a 5 μl aliquot (Subheading 3.3.1, **steps 1–3**) or 10 μl site-specifically labeled aliquot (Subheading 3.3.3, **steps 1–5**) of each prepared sample. Add 45 μl water and 150 μl cold acetone to the 5 or 10 μl aliquots and incubate for 15 min on ice. Keep the fluorescently labeled samples in dark during the whole procedure. Centrifuge the samples at 16,000 × *g* for 10 min at 4 °C and discard the supernatant.
2. Dry the pellets for 1 h at 45 °C in a thermo mixer with a shaking speed of 1000 rpm.
3. Resuspend the dried pellets in 20 μl SDS-PAGE sample buffer and load the samples on a prepared 10% SDS-PAGE gel. Use a ladder with fluorescently labeled bands. Run the gel.
4. Transfer the gel to the fluorescence imaging system and detect the labeled protein bands. For Bodipy-TMR-lysine use a 532 nm laser and a 580 nm emission filter. Sulfo-Cy5 can be detected with extinction at 633 nm and emission at 670 nm.
5. Afterwards dry the gel for 60 min at 70 °C using a unigeldryer. The dried gels are exposed on a phosphorscreen for minimal 3 days and read out using a multi-mode imager.

3.4.3 Confocal Laser Scanning Microscopy

1. For confocal laser scanning microscopy use 5 μl of the fluorescently labeled protein in the microsomal fraction and dilute the sample in 20 μl PBS. Add the mixture to a μ-Ibidi-Slide.
2. Fix the slide. Use a plan-apochromat objective with a 60× or 100× magnification. Microsomal structures usually have a diameter of 1–10 μm.
3. Adjust the beam path to the coupled fluorescent dye. Standard dyes such as Cy5 and FITC usually have a preset configuration. Cy5 is excited at 633 nm and the emission is detected with a long-pass filter above a wavelength of 670 nm.
4. Adjust the microscope settings (laser intensity, gain master, focus, pinhole) according to the individual sample (*see* **Note 17**).

4 Notes

1. Preparation of CHO lysate: The cultivation was carried out in a Biostat B-DCU II bioreactor (Sartorius Stedium Biotech GmbH) at 37 °C with a chemical defined and serum-free media. Cells were harvested at a density of 3.5–5 × 10^6 cells/ml by centrifugation at 200 × *g* for 5 min. The cell pellet was resuspended in 40 mM HEPES-KOH, pH 7.5, 100 mM NaOAc and 4 mM DTT. The cell suspension was passed through a 20-gauge needle to mechanically disrupt the cell membrane. Nuclei and cell debris were removed by a centrifugation step at 10,000 × *g* for 10 min. Raw lysate is applied to equilibrated Sephadex G-25 columns. Elution fractions with the highest measured RNA concentration were pooled and treated with micrococcal S7 nuclease to digest endogenous mRNA. The inactivation of the calcium-dependent nuclease was performed by adding 6.7 mM EGTA to complex the calcium ions. The lysate was further supplemented with creatine kinase (f.c. 100 μg/ml) to ensure the regeneration of ATP out of creatine phosphate. The prepared lysate was shock frozen in liquid nitrogen and stored at −80 °C.
2. The cell-free reaction in a dialysis mode is performed in a two-chamber device. The reaction chamber (1.1 ml, red lid) and the feeding chamber (11 ml, colorless lid) are separated by a semipermeable membrane with a molecular weight cut-off of 10 kDa. Whereas inhibitory byproducts such as accumulating phosphates are removed, amino acids and energy components are delivered to the reaction chamber.
3. It is recommended to collect samples of the translation mix, supernatant, purification steps including flow through, washing fractions and elution fractions as well as buffer exchange procedure and concentration. The aliquots can be diluted in SDS-PAGE sample buffer and loaded to the SDS-PAGE in order to monitor the purity of the aminoacyl-tRNA-synthetase during the preparation.
4. Concentrate the synthetase to a concentration of 5 g/l to ensure a minimal final concentration of 100 μM. If necessary repeat the concentration step.
5. The PCR product is purified with the QIAquick PCR Purification Kit and the concentration is determined by using a NanoDrop 2000c. For further analysis prepare a 1% (w/v) agarose gel and load 1 μl of the PCR product. The expected band size is 123 bps.

6. The prepared RNA can be analyzed by gel-electrophoresis run in 1× TBE buffer (Subheading 2.2.1, **item 11**). Therefore, prepare a 2% (w/v) agarose gel. Mix 2 μl of the RNA with 6 μl MOPS sample buffer (Subheading 2.2.2, **item 6**) and load the sample to the agarose gel. Use a RNA ladder. The expected band size is around 200 bps.
7. After centrifugation three phases are present: on top the aqueous phase with approximately 50% of the total volume, containing the RNA; a middle interphase that is nearly invisible and below the red phenol/chloroform phase. Try to isolate only the aqueous phase.
8. It is important that the components are dissolved completely to ensure the correct concentration. Vortex the components and store them in aliquots to avoid repeated thaw and freeze cycles.
9. The plasmid concentration can be varied and can be dependent on the chosen vector backbone. It is recommended to apply different plasmid concentrations in the range of 20–100 nM to the cell-free protein synthesis reaction.
10. The fluorescent dye is susceptible to light. An illumination will decrease the fluorescence intensity of the dye. Use colored tubes or wrap the tube with aluminum foil. In addition, keep in mind that the precharged tRNACUA will address the amber stop codon whereas the precharged tRNAGAA will address statistically phenylalanine codons.
11. The optimal temperature for cell-free protein synthesis reaction in a CHO lysate is 30 °C. However, a temperature series is recommended to determine the optimal conditions for an individual protein in terms of folding and activity. The Adora2a-Nluc construct showed the highest Nluc activity at 27 °C.
12. It is recommended to analyze the incorporation efficiency of the non-canonical amino acid. With a low efficiency the subsequent labeling reaction will yield as well in a low amount of labeled protein. Therefore, two different methods can be utilized.

    (a) Adding a reporter protein downstream of the amber stop codon. The reporter protein will only be translated if the amber stop codon is addressed by the tRNA and the non-canonical amino acid is transferred to the polypeptide chain. The intensity of the reporter protein signal is directly correlated to the amount of full-length protein and in conclusion to the incorporation efficiency. A comparison to a DNA construct without amber stop codon is possible.

(b) Determine the amount of full-length protein with autoradiography. If the termination product has an adequate amount of incorporated ^{14}C-leucine during the cell-free protein synthesis reaction and a detectable size, an additional band should be visible in the autoradiography. The ratio of full-length product and termination product can be calculated.

13. It is recommended to evaluate different agonist concentrations because at a defined concentration the binding sites of the cell-free synthesized Adora2a should be completely occupied resulting in a saturation effect in the BRET signal. The saturation effect supports the specific conformational rearrangement after ligand binding.
14. The non-canonical amino acid is solubilized in 0.5 M NaOH. The alkaline pH of NaOH will shift the pH of the lysate that might result in an inactivation of enzymes involved in protein synthesis. Therefore, dilute the ncAA in translation mix, water and components that are not affected by an alkaline pH.
15. Reactive groups of ncAA are often instable and susceptible to illumination. If the reactive group is inert a following coupling to a fluorescent dye is not possible. Therefore, keep the ncAA in the dark by using colored or wrapped tubes.
16. Copper has a toxic effect on proteins due to oxidative damage, but it is necessary as catalyst for the copper(I)-catalyzed alkyne-azide cycloaddition (CuAAC). It is recommended to use different copper concentrations for the labeling reaction. For the Adora2a_amb construct we have seen a highest labeling efficiency with 200 μM copper. The lowest copper concentration resulting in a detectable fluorescent band was 50 μM. Keep in mind that the concentration of THPTA has to be adjusted as well (threefold concentration of THPTA to $CuSO_4$). In addition, it is recommended to adjust the incubation time of the labeling reaction. In general, CuAAC is a fast and efficient click reaction. A decrease in incubation time might enhance proteins activity.
17. The same microscopic settings that are chosen for the fluorescently labeled sample should be applied to the negative control to exclude an unspecific labeling of microsomal structures. Fluorescent dyes are usually highly hydrophobic and tend to stick unspecific to lipid membranes. It is recommended to evaluate the unspecific binding of different dyes to figure out the most suitable dye for certain applications. For our images a pinhole of 1 Airy unit (118 μm), laser intensities of 22% (bright field) and 8% (fluorescence) resulted in the optimal recording of fluorescently labeled Adora2a.

Acknowledgments

This work is supported by the European Regional Development Fund (EFRE), the German Ministry of Education and Research (BMBF, No. 031B0078A), and the German Research Foundation (DFG Priority Programme 1623).

References

1. Sarramegn V, Muller I, Milon A et al (2006) Recombinant G protein-coupled receptors from expression to renaturation: a challenge towards structure. Cell Mol Life Sci 63 (10):1149–1164. https://doi.org/10.1007/s00018-005-5557-6
2. Serebryany E, Zhu GA, Yan EC (2012) Artificial membrane-like environments for in vitro studies of purified G-protein coupled receptors. Biochim Biophys Acta 1818 (2):225–233. https://doi.org/10.1016/j.bbamem.2011.07.047
3. Andréll J, Tate CG (2012) Overexpression of membrane proteins in mammalian cells for structural studies. Mol Membr Biol 30 (1-2):52–63. https://doi.org/10.3109/09687688.2012.703703
4. Jamshad M, Charlton J, Lin Y et al (2014) G-protein coupled receptor solubilization and purification for biophysical analysis and functional studies, in the total absence of detergent. Biosci Rep 35(2):e00188. https://doi.org/10.1042/BSR20140171
5. Brödel AK, Sonnabend A, Roberts LO et al (2013) IRES-mediated translation of membrane proteins and glycoproteins in eukaryotic cell-free systems. PLoS One 8(12):e82234. https://doi.org/10.1371/journal.pone.0082234
6. Zemella A, Thoring L, Hoffmeister C et al (2015) Cell-free protein synthesis: pros and cons of prokaryotic and eukaryotic systems. Chembiochem 16(17):2420–2431. https://doi.org/10.1002/cbic.201500340
7. Thoring L, Dondapati SK, Stech M et al (2017) High-yield production of "difficult-to-express" proteins in a continuous exchange cell-free system based on CHO cell lysates. Sci Rep 7:11710. https://doi.org/10.1038/s41598-017-12188-8
8. Daggett KA, Sakmar TP (2011) Site-specific in vitro and in vivo incorporation of molecular probes to study G-protein-coupled receptors. Mol Divers 15(3):392–398. https://doi.org/10.1016/j.cbpa.2011.03.010
9. Quast RB, Mrusek D, Hoffmeister C et al (2015) Cotranslational incorporation of non-standard amino acids using cell-free protein synthesis. FEBS Lett 589(15):1703–1712. https://doi.org/10.1016/j.febslet.2015.04.041
10. Quast RB, Ballion B, Stech M et al (2016) Cell-free synthesis of functional human epidermal growth factor receptor: investigation of ligand-independent dimerization in Sf21 microsomal membranes using non-canonical amino acids. Sci Rep 6:34048
11. Chin JW, Cropp TA, Anderson JC et al (2003) An expanded eukaryotic genetic code. Science 301(5635):964. https://doi.org/10.1126/science.1084772
12. Takimoto JK, Adams KL, Xiang Z et al (2009) Improving orthogonal tRNA-synthetase recognition for efficient unnatural amino acid incorporation and application in mammalian cells. Mol BioSyst 5(9):931–934. https://doi.org/10.1039/B904228H
13. Edwards H, Schimmel P (1990) A bacterial amber suppressor in Saccharomyces cerevisiae is selectively recognized by a bacterial aminoacyl-tRNA synthetase. Mol Cell Biol 10 (4):1633–1641
14. Aren W, Dierckx Rudi AJO, Xiaoyun Z et al (2017) Potential therapeutic applications of adenosine A2A receptor ligands and opportunities for A2A receptor imaging. Med Res Rev 38(1):5–56. https://doi.org/10.1002/med.21432
15. Chen J, Eltzschig HK, Fredholm BB (2013) Adenosine receptors as drug targets–what are the challenges? Nat Rev Drug Discov 12:265
16. Hoffmann C, Gaietta G, Bünemann M et al (2005) A FlAsH-based FRET approach to determine G protein–coupled receptor activation in living cells. Nat Methods 2:171
17. Meritxell C, Javier B, Daniel M et al (2003) Homodimerization of adenosine A2A receptors: qualitative and quantitative assessment by fluorescence and bioluminescence energy transfer. J Neurochem 88(3):726–734.

https://doi.org/10.1046/j.1471-4159.2003.02200.x

18. Bissantz C (2003) Conformational changes of G protein-coupled receptors during their activation by agonist binding. J Recept Signal Transduct Res 23(2-3):123–153. https://doi.org/10.1081/RRS-120025192
19. Brödel AK, Wüstenhagen DA, Kubick S (2015) Cell-free protein synthesis systems derived from cultured mammalian cells. In: Owens RJ (ed) Structural proteomics: high-throughput methods. Springer New York, New York, NY, pp 129–140
20. Thoring L, Wüstenhagen DA, Borowiak M et al (2016) Cell-free systems based on cho cell lysates: optimization strategies, synthesis of "difficult-to-express" proteins and future perspectives. PLoS One 11(9):e0163670. https://doi.org/10.1371/journal.pone.0163670

Part II

Tweaking Ligands for GPCR Studies

Chapter 5

Furan Cross-Linking Technology for Investigating GPCR–Ligand Interactions

Marleen Van Troys, Willem Vannecke, Christophe Ampe, and Annemieke Madder

Abstract

Interactions between G protein-coupled receptors and their ligands hold extensive potential for drug discovery. Studying these interactions poses technical problems due to their transient nature and the inherent difficulties when working with G protein-coupled receptors (GPCR) that are only functional in a membrane setting. Here, we describe the use of a furan-based chemical cross-linking methodology to achieve selective covalent coupling between a furan-modified peptide ligand and its native GPCR present on the surface of living cells under normal cell culture conditions. This methodology relies on the oxidation of the furan moiety, which can be achieved by either addition of an external oxidation signal or by the reactive oxygen species produced by the cell. The cross-linked ligand–GPCR complex is subsequently detected by Western blotting based on the biotin label that is incorporated in the peptide ligand.

Key words Furan, Receptor, GPCR, Chemical cross-linking, Peptide, Solid phase peptide synthesis, Western blot, Singlet oxygen, Reactive oxygen species

1 Introduction

The process of cell signaling is crucial for the viability of multicellular organisms. Extracellular ligands, acting as chemical messengers, bind specifically to receptor proteins expressed at the cell surface. This leads to their activation and initiates a variety of signal cascades in the cell. G protein-coupled receptors (GPCR) form the largest superfamily of cell surface receptors, further subdivided into the rhodopsin (701 members encoded in the human genome), adhesion (24 members), frizzled/taste (24 members), glutamate (15 members), and secretin family (15 members) [1–3]. GPCRs are seven-pass transmembrane proteins which bind or react to a variety of extracellular signals, in the form of light energy, peptides/proteins, lipids, or sugars. The fact that more than 120 GPCRs already form targets of approved drugs [4, 5] underscores their pharmaceutical potential. Nonetheless, around one-third of the GPCR

Mario Tiberi (ed.), *G Protein-Coupled Receptor Signaling: Methods and Protocols*, Methods in Molecular Biology, vol. 1947, https://doi.org/10.1007/978-1-4939-9121-1_5,

with non-olfactory ligands still are orphan GPCRs, meaning that the natural ligand is unknown. This makes them an untapped source for novel drug discovery. Conversely, a multitude of orphan peptides exists that are candidate ligands for receptors, including GPCRs.

Identifying new and characterizing known ligand–receptor interactions is essential in understanding GPCR biology in disease settings [6]. However, due to the dynamic and transient interaction between (peptide) ligands and GPCRs, the study of the ligand binding and receptor activation of GPCRs is not straightforward. One way to trap protein interactions is by using chemical cross-linking: the formation of a covalent bond between the interacting partners. Several strategies have been developed to accomplish effective and even site-selective cross-linking of proximate binding partners and this has previously been applied for GCPRs [7, 8]). Nevertheless, it has been pointed out that the toolkit for cross-linking is still limited [9]. In this context, our research group developed a new furan-oxidation-based cross-link methodology as a new way of site-specific and selective cross-linking [10–13]. The required furan moiety can be easily incorporated in peptides via standard chemical synthesis using a commercially available amino acid analogue. Upon oxidation, the inert aromatic furan moiety is converted into a reactive 4-oxobutenal which will rapidly react with nucleophilic side-chains, such as those from lysine or cysteine residues, provided these are in close proximity. We have recently shown that this furan-based cross-linking technology can be applied to covalently couple a furan-modified peptide ligand to its cognate native GPCR receptor (in casu GPR54), present at the cell surface of living cells [14] (Fig. 1a).

As schematically shown in Fig. 1b, we here describe the complete procedure starting from the synthesis of the furan-modified peptide ligand (1) and the in situ cross-linking of the furan peptide ligand to the GPCR receptor in the plasma membrane of cells in culture (2), to the detection of the formed cross-link using Western blotting (3).

2 Materials

2.1 Peptide Synthesis

1. N-Fmoc protected amino acids including Fmoc-ß-(2-furyl)-Ala-OH (further abbreviated as Fmoc-FurAla). The latter is used for incorporation of a furan moiety in the peptide. Store at −20 °C.
2. D-biotin or PEGylated biotin. Store at −20 °C.
3. 2-(1*H*-Benzotriazol-1-yl)-1,1,3,3-tetramethyluronium hexafluorophosphate (HBTU) as coupling reagent. Store at −20 °C.

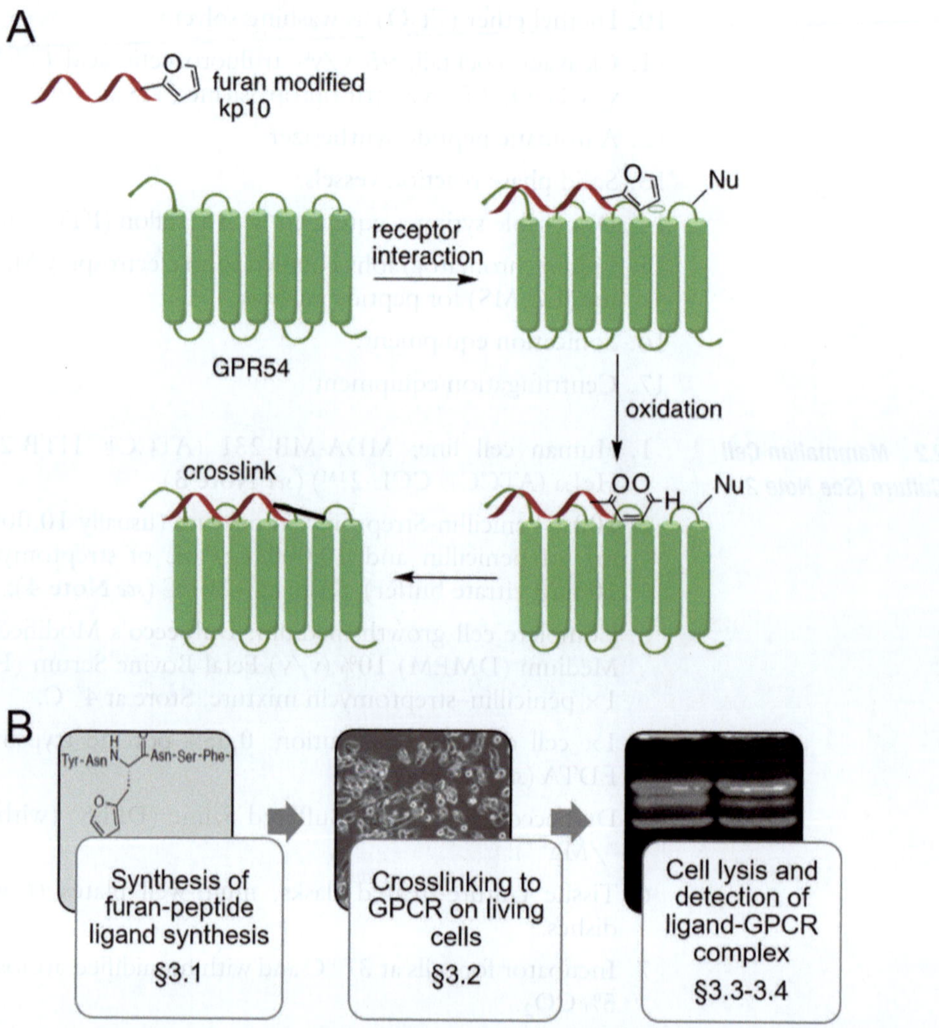

Fig. 1 Principle and overview of steps in cross-linking a furan-modified peptide ligand to a GPCR receptor. A: Principle: Upon formation of the complex between the peptide and the seven-pass transmembrane receptor (here GPR54) on the extracellular side of the living cells, the furan moiety is oxidized and reacts with a nucleophilic residue of GPR54 that is in close proximity, resulting in irreversible covalent bond formation. B. The three parts of the described protocol: ligand synthesis, ligand–receptor cross-link formation, detection

4. Rink-amide-ChemMatrix® resin for solid phase peptide synthesis (SPPS) (*see* **Note 1**).
5. A 0.2 M solution of N,N-Diisopropylethylamine (DIPEA) in N-methyl-pyrrolidone (NMP) as coupling base.
6. Freshly prepared 40 v/v % and 20 v/v% piperidine/DMF solutions as deprotection mixture.
7. Peptide-grade dimethylformamide (DMF) as coupling solvent.
8. Dichloromethane (DCM) as coupling/washing solvent.
9. Methanol (MeOH) as washing solvent.

10. Diethyl ether (Et_2O) as washing solvent.
11. Cleavage cocktail: 95 v/v% trifluoroacetic acid (TFA), 2.5% v/v H_2O, 2.5% v/v triisopropylsilane (TIS).
12. Automatic peptide synthesizer.
13. Solid phase reaction vessels.
14. Disposable syringes equipped with a Teflon (PTFE) frit.
15. Liquid chromatography coupled to an electrospray Mass detector (LC-MS) for peptide analysis.
16. Sonication equipment.
17. Centrifugation equipment.

2.2 Mammalian Cell Culture (See Note 2)

1. Human cell line: MDA-MB-231 (ATCC® HTB-26™) or HeLa (ATCC® CCL-2™) (*see* **Note 3**).
2. 100× Penicillin-Streptomycin mixture (usually 10,000 units/mL of penicillin and 10,000 μg/mL of streptomycin in a 10 mM citrate buffer). Store at −20 °C (*see* **Note 4**).
3. Complete cell growth medium: Dulbecco's Modified Eagle's Medium (DMEM) 10% (v/v) Fetal Bovine Serum (FBS) and 1× penicillin–streptomycin mixture. Store at 4 °C.
4. 1× cell dissociation solution: 0.05% porcine trypsin, 0.02% EDTA (*see* **Note 2**).
5. Dulbecco's Phosphate-Buffered Saline (DPBS) (without Ca^{2+}/Mg^{2+}).
6. Tissue culture-treated flasks, multi-well plates (6-wells) or dishes.
7. Incubator for cells at 37 °C and with humidified atmosphere of 5% CO_2.
8. Cell culture hood.
9. Basic bright field or phase contrast microscope for cell inspection.
10. Equipment for automated or manual cell counting (*see* **Note 2**).

2.3 Chemical Cross-Linking

1. Fresh 1 mM solution of N-bromosuccinimide (NBS, 99%) in sterile MilliQ water.
2. Fresh Rose Bengal stock solution in MilliQ water (max. solubility 100 mg/mL) (*see* **Note 5**).
3. Cold white light source.

2.4 Cell Collection, Cell Lysis, and Sample Preparation

1. Cell scrapers.
2. Cell lysis buffer: 7 M urea, 2 M thiourea, 1% (w/v) CHAPS (3-[(3-cholamidopropyl)dimethylamino]-1-propanesulfonate) in MilliQ water. Store in aliquots at −20 °C (*see* **Note 6**).

3. 1 M Dithiothreitol (DTT) in water.
4. Protease-inhibitor stock solution: 1 mg/mL leupeptin, 1 mg/mL aprotinin and 1 mg/mL antipain.
5. Ultrasonic liquid processer for sonication in small volumes (200 μL).
6. Dye reagent 10× concentrate for Bradford protein assay (store at 4 °C).
7. Bovine Serum Albumin (BSA) standard stock solution in MilliQ water (1.4 mg/mL, store in aliquots at −20 °C).
8. Disposable cuvettes.
9. Spectrophotometer.
10. 5× Laemmli sample buffer: 0.6 M Tris–HCl pH 6.8, 50% glycerol, 20% (w/v) sodium dodecyl sulfate (SDS), 0.005% (w/v) bromophenol blue, 250 mM DTT. Store in aliquots at −20 °C.

2.5 Denaturing Gel Electrophoresis

Proteins are analyzed using discontinuous SDS-PAGE (sodium dodecyl sulfate polyacrylamide gel electrophoresis). Pre-cast ready to use SDS PAGE gels are commercially available. Alternatively, hand-cast gels can be prepared using standard protocols (*see* **Note 7**).

1. 10% SDS-PAGE mini-gel (*see* **Note 8**).
2. 5× Tris glycine electrophoresis buffer: 15 g Tris–HCl, 72 g glycine, and 5 g sodium dodecyl sulfate (SDS) in 1 L distilled water. Store at room temperature.
3. Commercially available SDS-PAGE molecular weight marker. Store at −20 °C.
4. Electrophoresis cell for mini-gel (Mini-PROTEAN II electrophoresis cell or equivalent).

2.6 Western Blotting (See Note 9)

1. PVDF membrane (Hybond™ P PVDF Membrane or similar) (*see* **Note 9**).
2. Whatman 3MM filter paper.
3. Wet/Tank Blotting System and accessories.
4. 10× Western blot transfer buffer: 30 g Tris–HCl and 112 g glycine in 1 L of distilled water. Store at room temperature.
5. Odyssey® Blocking Buffer. Store at 4 °C (*see* **Note 10**).
6. 10× Phosphate buffered saline (PBS): 80 g NaCl, 28.6 g $Na_2HPO_4.12H_2O$ and 2.76 g $NaH_2PO_4.H_2O$ in 1 L deionized water with pH adjusted to 7.6. Store at room temperature.
7. 1× PBS-Tween buffer: 1 mL Tween20 (polyoxyethylenesorbitan monolaurate) and 100 mL 10× PBS in 1 L deionized water. Store at room temperature.

8. Primary antibody against GPR54: rabbit anti-GPR54 (Alomone Labs).
9. Secondary antibody: IRDye® 800CW Goat anti-Rabbit antibody (Licor Biosciences) (*see* **Note 11**).
10. IRDye 680 Streptavidin (Licor Biosciences) [15] (*see* **Note 11**).
11. Odyssey Infrared Imaging System (Licor Biosciences).

3 Methods

3.1 Design, Synthesis. and N-Terminal Biotin Labeling of Furan-Containing Peptide Ligand

1. Design the sequence of the peptide containing the furylalanine-containing peptide, i.e., establish the optimal position of the residue containing the furan moiety (*see* **Note 12**). Perform standard SPPS (start with **step 2**) which proceeds from the C-terminal toward the N-terminal amino acid. In the example biotin or PEGylated biotin was added to the amino-terminal amino acid in the final step of the synthesis (**step 6**). Synthesize also a non-furylalanine-containing peptide without biotin label (for specificity testing) and/or a biotin labeled peptide in which the original sequence is scrambled (negative control) (*see* **Note 13**).
2. Weigh the appropriate amount of Rink-amide-ChemMatrix® resin in a reaction vial for peptide synthesis and swell the resin in DMF (10 mL/g of resin) (*see* **Note 1**). In [14] peptide synthesis was in general carried out on a 50 mmole scale yielding on average between 2 and 5 mg of pure peptide.
3. Couple the first amino acid by dissolving it in the recommended solvents and adding it to the drained swollen resin. The resin is shaken for 2–3 h at room temperature. The specific procedure will depend on the type of linker used (e.g., Rink Amide, HMPB, 2-chlorotrityl chloride, etc.) (*see* **Note 14**).
4. Remove the Fmoc protecting group from the first amino acid by treating the resin with a 40 v/v % piperidine/DMF solution for 3 min, followed by a 12-min treatment with a 20% piperidine/DMF solution (10–12 mL/g of resin).
5. Couple the subsequent amino acids. If done manually, prepare a solution of 5 eq (with respect to the resin loading) of amino acid in DMF, add 5 eq of coupling reagent HBTU and 10 eq of DIPEA and add it to the resin. Let react for approximately 1 h at room temperature under continuous shaking. If an automated peptide synthesizer is used, follow the instructions of the manufacturer. Note that Fmoc-L-2-furylalanine can be coupled in the same manner as the other amino acids to the peptide. After each coupling and after each Fmoc deprotection, wash the resin extensively with DMF. To achieve a higher efficiency,

coupling can be repeated at each cycle prior to removal of the Fmoc group (*see* **step 4**).

6. N-terminal biotinylation (with or without PEG-linker, *see* **Note 15**) is carried out after Fmoc removal of the N-terminal (i.e., the last coupled) amino acid from the synthesized peptide. Dissolve 10 eq. D-biotin in DMF-DMSO solution (1:1). Slightly heat the solution to dissolve the biotin. Add 10 eq of HBTU and 20 eq of DIPEA. Add the solution to the resin and shake overnight at room temperature. Repeat the procedure (if coupling is not complete).
7. Wash the resin with DMF, MeOH, DCM, and Et_2O (using a quantity of solvent twice the volume of the amount of resin used). Dry the resin under a stream of nitrogen.
8. Add freshly prepared cleavage cocktail to the dry resin. Let react for 2 h at room temperature.
9. Collect the reaction solution in a centrifuge tube (15 mL) after filtering off the resin. Evaporate the majority of the TFA using pressurized air until the peptide starts precipitating. Add 5 mL of ice-cold Et_2O or MTBE (methyl-tert-butylether) and sonicate the mixture during 30 s. Centrifuge the precipitated peptide and remove the supernatant to which a fresh volume of Et_2O or MTBE is added. Repeat this process of sonication and centrifugation twice. After the last precipitation step, remove all Et_2O or MTBE under N_2 stream and store the peptide at −20 °C.
10. Check purity of the peptide using LC-MS. If necessary, perform High Pressure Liquid Chromatography (HPLC) purification (*see* **Note 16**).
11. Lyophilize the peptides for long-term storage (*see* **Note 17**).

3.2 In Situ Cross-Linking of the Furan-Containing Peptide Ligand–GPCR Complex

3.2.1 Preparatory Steps

1. Grow the mammalian cells endogenously expressing the GPCR in one or more 75 cm^2 sized culture flasks for at least two passages of subculturing using standard procedures (*see* **Note 2**).
2. Weigh a couple of mg of the synthesized furan peptide to prepare a working peptide stock in DMSO: the final recommended stock concentration is 0.5–1 mM in 100% DMSO (*see* **Note 18**). Prepare control peptides (*see* **Note 13**) in a similar manner. Determine the exact peptide concentration using absorption spectroscopy and store the peptide stocks at 20 °C (*see* **Note 18**).
3. Prepare a 6-well plate with cells of the mammalian cell line expressing GPR54 receptor and grow to full confluence. To obtain this, seed e.g., approximately 0.5–1 × 10^6 cells in each well in growth medium (Subheading 2.2, **item 2**) (*see* **Notes 2** and **19**) the day before the experiment and incubate overnight in a humidified incubator at 37 °C and 5% CO_2.

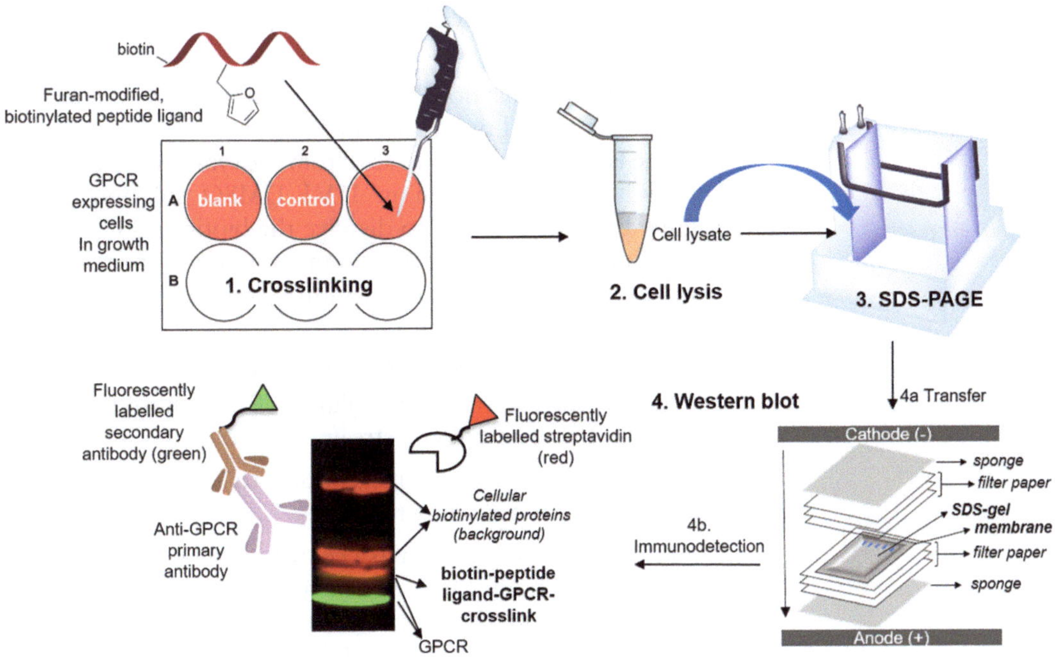

Fig. 2 Overview of cross-linking experiment. The furan-modified peptide additionally labeled with biotin is added to GPCR expressing cell line (Subheading 3.2). A blank reaction in which no peptide is added and a control reaction in which a biotinylated furan-containing scrambled (randomized) peptide is added are performed in parallel. After cross-link formation, the cells are lysed and samples are prepared (Subheading 3.3). Using SDS-PAGE the proteins are separated according to their molecular weight. Subsequently, they are transferred to a membrane (Subheadings 3.4, **steps 1** and **4**). To visualize the cross-linked complex formed between the GPCR receptor and biotinylated peptide, dual immunodetection is used (*see* Fig. 3)

3.2.2 Cross-Linking of Peptide to GPR54 Receptor (Fig. 2, Top Left)

1. Dilute the biotinylated furan-containing peptide ligand or the control peptides to the desired concentration in growth medium containing the normal serum level or free-serum conditions (*see* **Note 20** for effect of serum on crosslinking efficiency). The recommended starting concentration for furan peptide ligand is 0.5–10 μM (or higher depending on the tested interaction, *see* **Note 20**) and the final DMSO concentration is below toxic levels for the cells (*see* **Note 21**).
2. Remove the growth medium from the cells in the wells of the 6-well plate and carefully wash with 2 mL DPBS (take care not to disturb the cell layer and do not leave the cells without medium, work one well at a time). Remove the DPBS and replace by 2 mL of the growth medium containing the peptide of interest or a control peptide at the desired concentration (*see* **step 4** above). Include also a control well ('blank', negative control) in which no peptide is added to assist interpretation of the Western blot (*see* **Notes 13** and **22**) (Fig. 3).

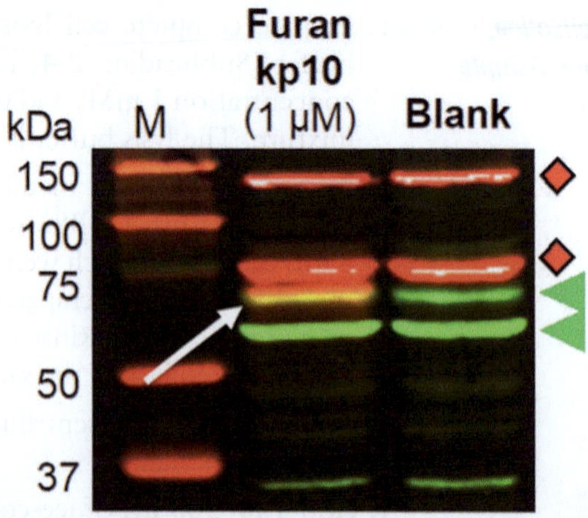

Fig. 3 Western blot analysis of cross-linking furan kp10 peptide to the GPR54 receptor on cells of the MDA-MB-231 cell line. Biotinylated proteins (red) and receptor (GPR54, green) are visualized. The first lane contains the molecular weight markers. The second lane shows the result of incubating the cells with 1 μM of the furan-modified peptide ligand. In the blank reaction (lane 3) no peptide was added to the cell culture. In this lane, the positions of the endogenous receptor signals (green arrowheads) are visible and the red background signals are from endogenously biotinylated carboxylases (75 and 130 kDa, red diamonds, [14]). The lower green band is a form of GPR54 which does not react with the GPCR ligand (compare with lane 2), mostly likely because it is not presented on the cell surface. The peptide interacts and covalently couples to the less abundant 72 kDa glycosylated and membrane-presented form of GPR54 (lane 2). The position of the cross-linked biotinylated peptide–receptor complex (orange-yellow band) overlaps with the 72 kDa receptor signal and is indicated by an arrow in the second lane

3. Incubate the plate at 37 °C and 5% CO_2 for 30 min (or more depending on the tested interaction). When relying on oxidation of the furan by endogenous ROS activity (*see* **Notes 20** and **23**), proceed to **step 1** of Subheading 3.3. When using NBS as oxidant, proceed to **step 4** of Subheading 3.2.2 or, for photosensitizer-induced oxidation of furan, to **step 5** of Subheading 3.2.2.
4. Add 1 eq of NBS (versus peptide) to the reaction mixture present in the well of the 6-well plate (1.5 mL medium + peptide). Incubate for an additional 30 min at 37 °C and 5% CO_2 (proceed to **step 1** of Subheading 3.3).
5. Add Rose Bengal (*see* Subheading 2.3, **item 2**) from a freshly prepared solution in MilliQ water to the reaction mixture in each well of the 6-well plate, to a final concentration of 10 μM. Irradiate the 6-well for 15 min with a cold white light source at room temperature (*see* **Note 5**).

3.3 Cell Collection, Cell Lysis, and Sample Preparation

1. Prepare complete cell lysis buffer by adding to the cell lysis buffer (Subheading 2.4, **item 2**) dithiothreitol (DTT) (final concentration 1 mM) and 0.1% (v/v) of the protease–inhibitor mixture. The lysis buffer is kept on ice (*see* **Note 6**).
2. Remove the growth medium from the 6-well plate and wash once with 2 mL DPBS. Remove the wash solution, again add 2 mL of DPBS to each well and gently detach the cells from the surface using a cell scraper. Collect the cells in a 15 mL tube at room temperature. Rinse each well with an additional 2 mL of DPBS and collect in the same tube.
3. Pellet the cells by centrifugation for 3 min at 400 × *g* and remove supernatant.
4. Add 100–200 μL of ice-cold cell lysis buffer and mix well by pipetting. To prevent protease activity (inherent to the procedure of lysing cells), keep samples at 0–4 °C from this point onward. After 20 min on ice, ensure complete lysis by sonicating the samples (e.g., 1 min using a sonicator with 2 mm microtip) while keeping the samples below 4 °C.
5. Spin samples at >10,000 × *g* for 10 min (4 °C). The supernatant is kept and termed 'cell lysate' below.
6. Determine the concentration of cell lysates using the Bradford assay (*see* **Note 24**). The total protein concentration in the lysate usually ranges between 2 and 8 mg/mL depending on the cell type and starting cell confluence.
7. The samples for SDS-PAGE are prepared by combining a volume of cell lysate containing 40–80 μg protein and 0.2% (v/v) of 5× Laemmli sample buffer (Subheading 2.4, **item 4**). Do not heat the samples (*see* **Note 6**). Freeze remaining cell lysate (−80 °C) and samples (−20 °C) or proceed to **step 1** of Subheading 3.4.

3.4 Detection of Cross-Linked Peptide Ligand–GPCR Complex

1. Mount the SDS-polyacrylamide gel (*see* **Note 7**) in the electrophoresis system and add 1× Tris glycine electrophoresis buffer. Load the samples prepared in **step 7** of Subheading 3.3 together with a molecular weight marker and start the separation of the protein mixtures by electrophoresis (Fig. 2 top right, *see* **Note 9**). Continue the electrophoresis until the bromophenol dye (from the Laemmli buffer) front reaches the bottom of the gel.
2. Prepare 1 L of 1× Western blot transfer buffer.
3. Wearing gloves, prepare, in a tray filled with transfer buffer, the Western blot sandwich for transfer of the SDS-PAGE separated proteins to a membrane. The blot sandwich (Fig. 2 bottom right) consists of a sponge, three sheets of Whatman 3 MM filter paper, the gel, the PVDF membrane (cut to the size of the gel),

three sheets of Whatman paper and a second sponge. All parts of this assembly (except the gel) are pre-wetted in Blot buffer before mounting. Ensure a tight contact between gel and membrane and remove air bubbles that get trapped in the assembly by wiping the surfaces at each step. Note that after SDS-PAGE, the detergent SDS is bound to the proteins in the gel, rendering them negatively charged. Place the cassette with the transfer sandwich vertically in the blot system ensuring the membrane is oriented to the anode (positive pole) (Fig. 2, bottom right), add a cooling unit and fill the tank with blot buffer. Blotting is performed during 1 h at 50 V or overnight at 25 V (*see* **Note 9**).

4. The sandwich is disassembled and the membrane is blocked (using 10 mL of a 1:1 (v/v) solution of blocking buffer in PBS) for 1 h at room temperature with constant shaking on a rocking platform (*see* **Note 25**).
5. Prepare a dilution (minimal volume 2–3 mL) of the primary anti GPR54 antibody 1:1000 (as recommended by the supplier) in a solution of 50% (v/v) 1× PBS-Tween, 50% (v/v) Odyssey blocking buffer (*see* **Note 26**).
6. Incubate the membrane, using constant rotation, with the primary antibody solution (*see* **step 5** in Subheading 3.4) during 1.5 h at room temperature or overnight at 4 °C.
7. The membrane is washed with 1× PBS-Tween buffer (3 × 5 min).
8. For signal development, the membrane is subsequently incubated with IRDye-conjugated secondary antibodies (diluted to 1:10000; to detect GPR54) and IRDye 680 Streptavidin (diluted 1:5000; to detect biotinylated proteins) in 50% (v/v) 1× PBS-Tween, 50% (v/v) Odyssey blocking buffer during 1 h at room temperature in the dark (e.g., use box or aluminum foil to shield off light) with constant shaking (*see* **Note 11**) (Fig. 2 bottom right).
9. The membrane is washed with 1× PBS-Tween buffer (3 × 5 min), 1× PBS (1 × 5 min), and MilliQ (1 × 5 min). Protect the membrane from light.
10. Scan the membrane (wet or air dried) at 685 nm ('700 channel') and 800 nm with an Odyssey Infrared Imaging System to record fluorescence from the IRDyes and to obtain a result such as shown in Fig. 3. The molecular marker proteins (lane 'M' in Fig. 3, red color) will be visual in the '700 channel'. Images are processed and quantified using the ImageStudio-Lite software freely downloadable from LI-COR Biosciences.

4 Notes

1. High purity of synthesized peptides can be achieved using 100% PEGylated resins, such as ChemMatrix® [16], that are now accepted to offer advantages over traditional PEG-based polystyrene resins for SPPS of longer peptides. ChemMatrix® is a patented, 100% PEG (polyethylene glycol)-based resin developed by PCAS BioMatrix. Information on the properties of ChemMatrix® including the different types available for the synthesis of peptide acids or amides, etc. can be found on the following websites: http://www.pcasbiomatrix.com/ChemMatrix®.html and http://www.biotage.com/product-group/ChemMatrix®-resins. These also include technical notes detailing starting protocols to use with each type of the ChemMatrix® Resin. The linker to which the first amino acid is coupled depends on the specific envisioned peptide sequence and the desired C-terminal functionality. HMPB-(Hydroxymethylbenzoic acid)- or Trityl OH-ChemMatrix® are generally used for C-terminal carboxylic acids, while Rink-Amide-ChemMatrix is applicable when C-terminal carboxamide peptides are desired. The standard loadings of commercial ChemMatrix® resins are 0.4–0.6 mmol/g of resin.
2. Mammalian cells are generally cultured in a Biosafety Laboratory Level 2. Standard procedures need to be followed during the culturing of mammalian cells. The ATCC® Animal Cell Culture Guide contains protocols for the initiation, maintenance, and cryopreservation of continuous cell lines and for prevention of contamination. Similar cell culture learning centers are available from vendors, e.g., Gibco Cell Culture Basics Handbook (Thermofisher) or Cell Culture Manual (Sigma). For adherent cells, subculturing entails dissociating the contacts between the cells and between cells and substrate, resuspending the cells in growth medium and returning them in part (1/10–1/5 depending on cell type, level of confluence before dissociation, etc.) in the growth medium present in the same or new culture flask. Cell dissociation buffers are commercially available. These contain a Ca^{2+} chelator and/or the protease trypsin (trypsin is not needed for all cell lines) and are used in DPBS on cells prewashed with DPBS. Trypsin is inactivated by adding growth medium. The number of cells/mL in a cell suspension (after dissociation) is counted in an automated cell counter or manually. We typically count manually using a Bürker-Türk chamber (BRAND) in which the number of cells in one large square of 1 mm^2 multiplied by 10^4 results in the cell number/mL in the counted cell suspension.

3. As detailed in [14], furan-based cross-linking was already performed on different adherent cell types: MDA-MB-231 (ATCC® HTB-26™), MCF-7 (ATCC® HTB-22™), NIH3T3 (ATCC® CRL-1658™), HeLa (ATCC® CCL-2™), and HEK293 (ATCC® CRL-1573™). Although we only applied the procedure to adherent cells, we see no reasons why it should not work on non-adherent (suspension) cells. Final furan-modified peptide concentration may, however, need optimizing when used with non-adherent cells. The choice of medium and serum concentration is dependent on the specific cell line used, e.g., Dulbecco's Modified Eagle's Medium (DMEM) or RPMI 1640 Medium. Cells lines and information on their optimal culturing condition are available from ATCC (https://www.lgcstandards-atcc.org). Pyruvate, which is a component of many conventional growth media, is a described Reactive Oxygen Species (ROS)-scavenger and should preferentially be omitted from the assay medium when furan-oxidation cross-link is performed without adding NBS or the photosensitizer Rose Bengal (*see* **Note 23**). Media without pyruvate are commercially available.
4. For use of antibiotics in mammalian cell culture, *see*: https://www.sigmaaldrich.com/technical-documents/articles/biology/antibiotics-in-cell-culture.html. Antibiotics are typically not added in the medium in which the actual cross-linking (Subheading 3.2.2, **step 2**) is done.
5. The xanthene dye Rose Bengal (4,5,6,7-tetrachloro-2′,4′,5′,7-′-tetraiodofluorescein) is an effective photosensitizer with an intense absorption band in the range 480–550 nm and a high quantum yield of 0.76. It produces singlet oxygen with high yields from oxygen [17, 18]. We recently demonstrated that singlet oxygen (generated by 10 μM of this compound) is suitable for oxidation of furan-containing peptides to allow site-specific labeling of the peptide with a set of fluorophores [19]. For irradiation a cold white light source, e.g., a Euromex 100 W cold light microscope lamp was used.
6. Transmembrane proteins such as GPCRs are not easily recovered upon cell lysis. The procedure here requires isolation of a modified version of GPR54, e.g., the covalently coupled peptide–receptor complex. We validated that the cell lysis buffer that contains both urea/thiourea and a zwitterion and that is traditionally used for total protein extraction for 2D-gel electrophoresis [20], is efficient for extraction of GPR54. Refrain from heating cell lysates containing urea/thiourea as the heat-induced decomposition of urea leads to carbamylation of proteins. The latter will distort the SDS-PAGE separation and/or subsequent identification. A protease inhibitor mixture is needed to prevent protein degradation in the cell lysate.

Table 1
Range of molecular weight of protein separated in SDS-PAGE of specific %

Single % gel	kDa
8%	40–250
10%	30–200
12%	20–150
16%	5–60
Gradient gel	**kDa**
10–20%	5–100
4–20%	5–300
8–16%	10–200
4–12%	30–300

Protease inhibitor cocktails are available from different vendors.

7. Precast SDS-PAGE gels are available from many vendors (*see* https://www.biocompare.com/Protein-Biochemistry/24770-SDS-PAGE-precast-gels/). They are used because they yield uniform results (compared to hand cast gels), save time, and prevent contact with un-polymerized acrylamide which is a strong neurotoxin. Precast SDS-PAGE gels for Western Blotting are available as single % gels or as gradient gels (*see* Table 1 for separated protein range in relation to %). Given the short shelf-life and high cost of precast gels, hand casting may still be preferred. We refer to the Bio-One Laemmli-SDS-PAGE protocol [21] for composition of the solutions and the steps to prepare a 10% separation gel and a 5% stacking gel. Similar protocols can be found at sites of vendors of electrophoresis products.
8. Use a mini-gel with appropriate separation range for the GPCR receptor of interest. For analysis of the ligand–GPR54 crosslink in [14] we used a 10% gel. Many GPCR are glycosylated and this modification has been reported to affect receptor folding, trafficking, and function. Since glycosylation obviously affects the mass and thus mobility of the receptor protein during SDS-PAGE, take this into account in choosing the SDS-PAGE % (Table 1). The GPR54 kisspeptin receptor studied in [14] is present in different subpopulations in mammalian cell lysate representing different states of glycosylation [22]. Depending on the cell type and the antibody, up to three bands (with apparent MW of 37 kDa, 54 kDa, and 72 kDa) are observed on Western blot (*see* Supplementary

Fig. 9 in [14]). The furan peptide ligand specifically crosslinked to the 72 kDa receptor subspecies which is a glycosylated and plasma-membrane presented form.

9. Western blotting is a standard (semi)-quantitative detection technique for proteins. The proteins of interest are transferred ('blotted') from the gel to a membrane by electrophoresis and are detected using primary antibodies that are in turn targeted by labeled secondary antibodies. Various detection methods based on differently labeled secondary antibodies are possible. We refer to https://www.proteinatlas.org/learn/method/western+blot for a historical and general overview of the method and the traditional detection methods. **Note 11** details the detection method using secondary antibodies labeled with near infrared fluorophores used in this protocol. Polyvinylidene fluoride (PVDF) or, alternatively, nitrocellulose membranes with standard pore size of 0.45 μm (for proteins >20,000 dalton) are available from many brands and vendors. Depending on the detection method (and especially for fluorescence-based detection), the membrane can be a source of background. We typically use PVDF membrane (Hybond™ P PVDF Membrane or similar) for the detection method described in this protocol. The Western blot sandwich is shown schematically in Fig. 2, bottom left and in https://www.proteinatlas.org/learn/method/western+blot. Useful video protocols for the entire Western Blot process are available on-line, e.g., https://www.youtube.com/watch?v=VgAuZ6dBOfs from Biorad. For a source of information and troubleshooting on the electrophoresis step within the Western blot procedure, *see* amongst other: http://www.bio-rad.com/en-be/applications-technologies/performing-protein-electrophoresis?ID=LUSPFBNEL.

10. The blocking step prevents unspecific binding of the antibodies to the membrane and is usually done using solutions (e.g. DPBS) containing BSA (5% w/v). BSA can be replaced by 5.0% (w/v) non-fat dry milk powder. Blocking solutions can be prepared or purchased ready-made. Verify beforehand that the blocking method is compatible with the detection method and the primary antibody used.

11. IRDye fluorescent dyes have absorption and emission wavelengths in the near infrared spectrum, between 680 and 800 nm. Details on IRDye secondary antibodies and the advantages of their use in regard to sensitivity, multiplexing, dynamic range, signal stability, and to perform Western blots in a quantitative manner are documented on https://www.licor.com/bio/. Biotin-labeled proteins can be detected on Western blot relying on the tight biotin–streptavidin interaction (IRDye labeled from Licor). The use of IRDye-based reagents for detection requires a compatible downstream detection

system (Subheading 2.6, **item 11** and Subheading 3.4, **step 9**). Alternative detection systems use enzyme-coupled secondary antibodies or enzyme-conjugated streptavidin which in the presence of a substrate generate chemiluminescence [23].

12. Oxidation of the furan moiety within the peptide generates an electrophilic moiety prone to react with a nucleophilic side chain residue of the targeted receptor. It has been previously shown [10] that proximity between the oxidized furan moiety and the nucleophile is a prerequisite for efficient reaction. Therefore, in the case of a known peptide–ligand/receptor interaction, the position of the furan moiety should be chosen so as to maximize proximity with respect to potential nucleophilic side chains (preferentially Lys but also potentially Cys, Ser, Thr, Tyr, Trp, His, or combinations thereof) in the binding site of the targeted receptor (based on, e.g., X-ray information or mutation data). In the case of an unknown interaction it might be advisable to synthesize a series of peptides incorporating the furylalanine moiety at different positions (FurAla scan) to maximize the chance of cross-link formation. In [14] it was shown that position of the furan-moiety strongly influences the potential for cross-linking.
13. Peptide ligand controls: *first*, a receptor peptide ligand containing the wild type (WT) sequence needs to be synthetized without the biotin label to use in a competition experiment to evaluate the specificity of the established ligand–receptor cross-linking. In [14], the WT peptide was added at 10 molar eq. versus the furan-containing peptide ligand (e.g., 10 μM) (Subheading 3.2.5). WT and furan-containing peptide ligand were added simultaneously to the cultured cells. Specificity is indicated when the biotin signal corresponding to the MW of the furan peptide–GPCR cross-link complex is absent or strongly decreased in the presence of excess of the unlabeled WT peptide (*see* Fig. 3a, b, e in [14]). *Second*, a synthesized peptide in which the ligand amino acid sequence is randomized but the furyl-alanine in the same position as in (at least one of) the furan-containing ligand is used as negative control in a parallel reaction. This 'scrambled' peptide should not give the signal corresponding to the expected MW for the ligand–GPCR complex on Western blot (*see* Fig. 3e, peptide 7 in [14]).
14. We here detail our procedures for coupling to Rink resin, HMPB resin, and 2-chlorotrityl chloride resin; alternative and additional protocols for other ChemMatrix resins can, e.g., be found at http://www.biotage.com/product-group/chemmatrix-resins. *Rink resin*: For coupling of the first amino acid to ChemMatrix Rink Amide resin, the resin is pre-swollen in DCM for 30 min and filtered off. A mixture of

5 eq of amino acid in DMF (0.5 M), 5 eq of HBTU in DMF (0.5 M), and 10 eq of DIPEA in NMP (0.2 M) is added to the resin. The reaction is shaken for 3 h. The resin is filtered off and washed with DMF, MeOH and DCM. The coupling is repeated using the same protocol. *HMPB resin*: For coupling of the first amino acid to ChemMatrix HMPB resin, the resin is pre-swollen in DCM for 30 min and filtered off. Ethyl (*2Z*)-2-cyan-2-hydroxyimino acetate (Oxyma, 5 eq), DIC (5 eq), and DMAP (0.3 eq) are added to a solution of amino acid (5 eq) in DCM:DMF (2:1). After 30 min of pre-activation, this mixture is added to the resin. The reaction is shaken for 3 h. The resin is filtered off and washed with DMF, MeOH, and DCM. The coupling is repeated using the same protocol. *2-chlorotrityl chloride resin*: For coupling of the first amino acid to 2-chlorotrityl chloride resin, the resin is swollen in dry DCM for 30 min under inert atmosphere. In a dry reaction flask, the amino acid (0.6 eq) is dissolved in dry DCM and DIPEA (2.4 eq) is added and the resulting solution kept under inert atmosphere. The amino acid solution is added to the drained resin and this is shaken for 2 h under inert atmosphere. After the reaction, the resin is washed 3 × 5 min with DCM/MeOH/DIPEA (17:2:1) and finally, with 3 × DCM, 3 × DMF, 3 × MeOH, 3 × DCM and 3 × Et_2O and dried in vacuo.

15. The biotin tag on the peptide ligand is used for detection of the cross-linked ligand–GPCR complex by Western blot (Subheading 3.4). The position of biotin in the peptide ligand is strongly dependent on the ligand of interest as it ought not affect the interaction with the receptor. In [14] on GPR54, the kisspeptin ligand kp-10 was labeled N-terminally. Addition of a $(PEG)_4$-linker to the biotin tag can further reduce steric hindrance by this label in the subsequent peptide–GPCR interaction and cross-linking [14]. The choice of biotin as tag on the peptides is based on the small size of this tag and on the available streptavidin-based detection tools (*see* **Note 11**). Biotinylation of the peptide ligand has as disadvantage that in the detection step endogenously biotinylated proteins, which are relatively abundant, are also detected on blot (*see* **Note 21** and Fig. 3). Alternative epitope tags such as FLAG, HA, V5, Myc, or His_6 can be used.
16. Purity of the peptides depends on the synthesized sequence. Typical procedures for purification in function of quality assessment of synthesized peptides include: *LC-MS analysis* can be performed on an Agilent 1100 Series HPLC instrument equipped with a Phenomenex Kinetex C18 column (150 mm × 4.6 mm, 5 μm) at 35 °C, using a flow rate of 1.5 mL/min. The column is eluted with a gradient, starting

with 100% H_2O containing 5 mM NH_4OAC up to 100% acetonitrile. A G1946C electrospray mass detector is directly coupled to the column. Alternatively, reversed phase *HPLC analysis* can be performed on an Agilent 1100 Series HPLC instrument, equipped with a Phenomenex Luna C18(2) column (250 mm × 4.6 mm, 5 μM) at 35 °C, using a flow rate of 1 mL/min. The column is eluted with a gradient, starting with 100% H_2O containing 0.1% TFA up to 100% acetonitrile. The collected fractions are subsequently analyzed on a Voyager-DE STR Biospectrometry Workstation (Applied Biosystems), with a-cyano-4-hydroxycinnamic acid as matrix. For *semi-preparative purification*, a Phenomenex Luna C18(2) at 35 °C, using a flow rate of **4.5** mL/min can be used on the same HPLC instrument. The column is eluted with a gradient, using following solvent systems: H_2O containing 0.1% TFA and CH_3CN. *Preparative purification* of the synthesized peptide is typically performed on an Agilent 218 solvent delivery system with a UV–VIS dual wavelength detector using a Phenomenex column (AXIA packed Luna C18(2), 250 × 21.2 mm, 5 μm particle size, 35 °C) with a flow rate of 17 mL/min. The column is eluted with a similar solvent system as for semi-preparative purification.

17. A Heto Drywinner freeze dryer was used here in combination with a Thermoelectron corporation Savat SPD111V speedvac concentrator.

18. When preparing the stock solutions in DMSO, ensure a sufficiently high peptide concentration taking into account that in the actual reaction with the cells the concentration of the peptide is ~0.5–10 μM and the DMSO concentration should be below toxic levels for the cells. This implies a stock solution 300- to 1000-fold higher than the final peptide concentration in the reaction (*see* **Notes 20** and **21**). Because of the presence of salts and counter ions, peptide concentration determination based on weighing lyophilized peptide appears less accurate than based on absorption. We estimate the molar extinction coefficient ε_M in M^{-1} cm^{-1} for each peptide at 280 nm (e.g., using the ExPASy ProtParam tool at https://web.expasy.org/protparam/) and measure, using 2 μL, the concentration in a NanoDrop 2000 Microvolume Spectrophotometer using the sample type setting 'Other protein (E & MW)' under 'Protein'. Since DMSO also absorbs at 280 nm, use it as blank measurement. Alternatively, to avoid possible effects of DMSO on the measurement, dilute the peptide to 10/90 DMSO/H_2O and measure versus this solvent as blank.

19. The cells should preferably be fully confluent at the start of the experiment with the pH of the culture at physiological level (pH is evaluated using the phenol red indicator present in the

growth medium, at physiological pH the medium colors orange-red). For HeLa cells, we seed 500,000–800,000 cells per well of a 6-well plate approximately 18 h before start of the experiment. For the smaller MDA-MB-231 or HEK293 cells 750,000–1 $\times$ 10^6 cells can be seeded per well. For cell counting: *see* **Note 2**.

20. Prevent local precipitation of the peptide upon diluting into the aqueous growth medium by mixing (vortex) while or immediately after adding the peptide. The peptide ligand is added in large excess to enhance the efficiency of the cross-link reaction. We use 0.5–10 μM peptide as starting concentration, but optimization by titration is recommended when exploring new interactions (a peptide concentration in the range of 1–100 μM is recommended). Both the affinity of the peptide ligand as well as the amount of presented receptor will determine the amount of cross-linking which is further influenced by the proximity between the oxidized furan moiety and the reacting nucleophiles at the peptide–receptor interface. When working in normal growth medium as described, the serum does decrease the efficiency of the crosslinking reaction (*see* Supplementary Fig. 12 in [14]. In studying new interactions, it may therefore be optimal to also test the efficiency of the crosslinking in serum-free conditions and longer incubation times. It was observed that highest cross-linking efficiency is observed at higher concentrations and longer incubation time of the peptide and under free-serum conditions.
21. It is important to ensure that the final concentration of DMSO in this solution does not exceed 0.1–0.5% v/v as higher amounts of DMSO generate adverse effects on cell viability and/or affect cell function. To control for solvent effects, the solvent (DMSO) concentration should be identical in all conditions that are compared, including the blank and peptide control conditions.
22. Figure 3 demonstrates the importance of a blank reaction, i.e., a reaction where the entire procedure is carried out without adding peptide. This blank sample is used to identify non-relevant signals on the Western blot including unspecific signals from the reaction of the labeled secondary antibody or, in case of probing biotin, signals of endogenous biotinylated proteins (indicated by red diamonds in Fig. 3, *see* also Supplementary Fig. 9 in [14]). In addition, the blank reaction shows the signals (green arrows in Fig. 3) corresponding to the GPR54 receptor in the absence of ligand interaction and cross-linking. Figure 3 demonstrates that the cross-linked peptide–ligand–GPR54 complex overlaps with one of the GPR54 signals (yellow signal).

23. The covalent coupling of the peptide ligand to the GPR54 receptor requires oxidation of the furan moiety in the peptide to a 4-oxo-enal moiety that will react with nucleophilic groups on the receptor. This oxidation can be induced by addition of an oxidant, such as N-bromosuccinimide (NBS) (Subheading 3.2.2, **step 4**) or singlet oxygen $^1O_2$ (generated in the presence of a photosensitizer, like Rose Bengal, by visible light, Subheading 3.2.2, **step 5**). We previously showed, however, that cells themselves can also provide the oxidant in an efficient manner (*see* Fig. 6 in [14]). Based on inhibitor studies, this 'endogenous oxidation' of the furan moiety in the peptide is dependent on reactive oxygen species (ROS) produced by the cells and on the activity of NADPH oxidase enzymes (NOX). When using this oxidation strategy for furan-based cross-linking, only the furan-containing peptide needs to be added to the cells. We observed endogenous oxidation in all cells listed in **Note 3** for the cross-linking of the kp10–GPR54 complex. We expect that the efficiency of this experimental approach varies for other cells or other ligand–GPCR interactions as the proximity of the ROS generating system can vary.

24. The total protein concentration in the cell lysate is determined using the Bradford assay [24]. This method relies on measuring (in a spectrophotometer or a NanoDrop Instrument) the absorbance shift at 595 nm that occurs when Coomassie Blue binds to proteins. A concentration range of 0.1–0.5 mg/mL bovine serum albumin protein is used as standard series (typically 5 standards). Dilute your sample to give a measurement within this range. Both BSA-solution and the Coomassie reagent are commercially available. We refer to http://www.protocol-online.org/.../Protein_Quantitation/Bradford_Protein_Assay for a collection of more detailed protocols. Ensure that the cell lysates are sufficiently diluted before measurement as the assay is sensitive to some buffer components including urea and DTT (*see* Table 1 in www.bio-rad.com/webroot/web/pdf/lsr/literature/4110065A.pdf). Alternative methods for total protein concentration include Bicinchoninic acid (BCA) assay, Pierce 600 protein assay, and Lowry assay. Exact protein quantitation in the lysates to ensure equal protein loading on SDS-PAGE which is important for (semi)quantitative comparison of signals on Western blot. In addition, when using biotin for detection, the background signal of endogenously biotinylated proteins (Fig. 3, *see* **Notes 13** and **21**) can be used to evaluate equal loading on the gel.

25. Manipulate the membrane only using tweezer and wear gloves to avoid increasing background.

26. Given the high cost of primary and secondary antibody, it is opportune to incubate the membrane in customized small

boxes or in small plastic poaches cut to size and sealed using a sealing machine. In this way the incubation volume can be limited to a couple of mL. Tween-20 is a non-ionic detergent that is added to this incubation step (and to subsequent washing steps) to reduce nonspecific binding of the antibodies.

References

1. Huang W, Manglik A, Venkatakrishnan AJ, Laeremans T, Feinberg EN, Sanborn AL, Kato HE, Livingston KE, Thorsen TS, Kling RC, Granier S, Gmeiner P, Husbands SM, Traynor JR, Weis WI, Steyaert J, Dror RO, Kobilka BK (2015) Structural insights into μ-opioid receptor activation. Nature 524:315–321
2. Stevens RC, Cherezov V, Katritch V, Abagyan R, Kuhn P, Rosen H, Wüthrich K (2013) The GPCR network: a large-scale collaboration to determine human GPCR structure and function. Nat Rev Drug Discov 12 (1):25–34. https://doi.org/10.1038/nrd3859
3. Fredriksson R, Lagerström MC, Lundin LG, Schiöth HB (2003) The G-protein-coupled receptors in the human genome form five main families. Phylogenetic analysis, paralogon groups, and fingerprints. Mol Pharmacol 63 (6):1256–1272. https://doi.org/10.1124/mol.63.6.1256
4. Shemesh R, Toporik A, Levine Z, Hecht I, Rotman G, Wool A, Dahary D, Gofer E, Kliger Y, Soffer MA, Rosenberg A, Eshel D, Cohen Y (2008) Discovery and validation of novel peptide agonists for G-protein-coupled receptors. J Biol Chem 283:34643–34649
5. Sriram K, Insel PA (2018) GPCRs as targets for approved drugs: How many targets and how many drugs? Mol Pharmacol. https://doi.org/10.1124/mol.117.111062
6. Hauser AS, Chavali S, Masuho I, Jahn LJ, Martemyanov KA, Gloriam DE, Babu MM (2018) Pharmacogenomics of GPCR drug targets. Cell 172(1–2):41–54.e19. https://doi.org/10.1016/j.cell.2017.11.033
7. Corgiat BA, Nordman JC, Kabbani N (2014) Chemical crosslinkers enhance detection of receptor interactomes. Front Pharmacol 4:171. https://doi.org/10.3389/fphar.2013.00171
8. Grunbeck A, Sakmar TP (2013) Probing G protein-coupled receptor-ligand interactions with targeted photoactivatable cross-linkers. Biochemistry 52(48):8625–8632. https://doi.org/10.1021/bi401300y
9. Scalabrin M, Dixit SM, Makshood MM, Krzemien CE, Fabris D (2018) Bifunctional cross-linking approaches for mass spectrometry-based investigation of nucleic acids and protein-nucleic acid assemblies. Methods 144:64–78. https://doi.org/10.1016/j.ymeth.2018.05.001
10. Carrette LLG, GysselsE D, Laet N, Madder A (2016) Furan oxidation based cross-linking: a new approach for the study and targeting of nucleic acid and protein interactions. Chem Commun 52:1539–1554. https://doi.org/10.1039/c5cc08766j
11. Op de Beeck M, Madder A (2012) Sequence specific furan based DNA crosslinking with visual light. J Am Chem Soc 134 (26):10737–10740
12. Stevens K, Madder A (2009) Furan-modified oligonucleotides for fast, high-yielding and site-selective DNA inter-strand cross-linking with non-modified complements. Nucleic Acids Res 37(5):1555–1565
13. Halila S, Velasco T, De Clercq P, Madder A (2005) Fine-tuning furan toxicity: fast and quantitative DNA interchain cross-link formation upon selective oxidation of a furan containing oligonucleotide. Chem Commun 7:936–938
14. Vannecke W, Ampe C, Van Troys M, Beltramo M, Madder A (2017) Cross-linking furan-modified kisspeptin-10 to the KISS receptor. ACS Chem Biol 12(8):2191–2200. https://doi.org/10.1021/acschembio.7b00396
15. https://www.licor.com/bio/products/reagents/irdye/streptavidin/. Accessed 19 May 2018
16. García-Martin F, Albericio F (2008) Solid supports for the synthesis of peptides–from the first resin used to the most sophisticated in the market. Chem Today 26:29–34
17. Redmond RW, Gamlin JN (1999) A compilation of singlet oxygen yields from biologically relevant molecules. Photochem Photobiol 70:391–475
18. Kochevar IE, Lambert CR, Lynch MC, Tedesco AC (1996) Comparison of photosensitized plasma membrane damage caused by singlet oxygen and free radicals. Biochim Biophys Acta 1280:223–230

19. Antonatou E, Hoogewijs K, Kalaitzakis D, Baudot A, Vassilikogiannakis G (2016) Madder a singlet oxygen-induced furan oxidation for site-specific and chemoselective peptide ligation. Chem Eur J 22(25):8457–8461. https://doi.org/10.1002/chem.201601113
20. Rabilloud T, Adessi C, Giraudel A, Lunardi J (1997) Improvement of the solubilization of proteins in two-dimensional electrophoresis with immobilized pH gradients. Electrophoresis 18(3–4):307–316
21. He F (2011) Laemmli-SDS-PAGE. Bio Protocol Bio101:e80. https://doi.org/10.21769/BioProtoc.80
22. Misu R, Oishi S, Setsuda S, Noguchi T, Kaneda M, Ohno H, Evans B, Navenot JM, Peiper SC, Fujii N (2013) Characterization of the receptor binding residues of kisspeptins by positional scanning using peptide photoaffinity probes. Bioorg Med Chem Lett 23:2628–2631
23. Alegria-Schaffer A (2014) Western blotting using chemiluminescent substrates. Methods Enzymol 541:251–259. https://doi.org/10.1016/B978-0-12-420119-4.00019-7
24. Bradford MM (1976) A rapid and sensitive method for the quantitation of microgram quantities of protein utilizing the principle of protein-dye binding. Anal Biochem 72:248–254

Chapter 6

Optical Regulation of Class C GPCRs by Photoswitchable Orthogonal Remotely Tethered Ligands

Amanda Acosta-Ruiz, Johannes Broichhagen, and Joshua Levitz

Abstract

G protein-coupled receptors (GPCRs) respond to a wide range of extracellular cues to initiate complex downstream signaling cascades that control myriad aspects of cell function. Despite a long-standing appreciation of their importance to both basic physiology and disease treatment, it remains a major challenge to understand the dynamic activation patterns of GPCRs and the mechanisms by which they modulate biological processes at the molecular, cellular, and tissue levels. Unfortunately, classical methods of pharmacology and genetic knockout are often unable to provide the requisite precision needed to probe such questions. This is an especially pressing challenge for the class C GPCR family which includes receptors for the major excitatory and inhibitory neurotransmitters, glutamate and GABA, which signal in a rapid, spatially-delimited manner and contain many different subtypes whose roles are difficult to disentangle. The desire to manipulate class C GPCRs with spatiotemporal precision, genetic targeting, and subtype specificity has led to the development of a variety of photopharmacological tools. Of particular promise are the photoswitchable orthogonal remotely tethered ligands ("PORTLs") which attach to self-labeling tags that are genetically encoded into full length, wild-type metabotropic glutamate receptors (mGluRs) and allow the receptor to be liganded and un-liganded in response to different wavelengths of illumination. While powerful for studying class C GPCRs, a number of detailed considerations must be made when working with these tools. The protocol included here should provide a basis for the development, characterization, optimization, and application of PORTLs for a wide range of GPCRs.

Key words Optogenetics, Photopharmacology, Tethered ligands, G protein-coupled receptors, Metabotropic glutamate receptors, PORTLs, Neuromodulation, SNAP tag

1 Introduction

G protein-coupled receptors (GPCRs) provide one of the primary molecular mechanisms for the intracellular response to extracellular signals in a wide range of physiological systems and serve as the largest family of drug targets in biology [1]. The major conserved structural feature of GPCRs is a seven helix transmembrane domain (TMD) which undergoes conformational changes following ligand binding that permits coupling with intracellular heterotrimeric G proteins to initiate diverse signaling cascades [2]. Of particular

Mario Tiberi (ed.), *G Protein-Coupled Receptor Signaling: Methods and Protocols*, Methods in Molecular Biology, vol. 1947, https://doi.org/10.1007/978-1-4939-9121-1_6,

importance in the nervous system is the class C GPCR family which contains receptors for neurotransmitters [3, 4], ions [5], and hormones [6, 7]. Class C GPCRs are characterized structurally by the presence of large, bi-lobed extracellular ligand binding domains (LBD) which mediate constitutive dimerization. The largest subfamily of class C GPCRs are the eight metabotropic glutamate receptors (mGluR1-8) which show distinct, yet overlapping localization and physiological roles [3, 4]. Despite great progress over the last two decades of research, mechanistic understanding of class C GPCRs at the biophysical, cellular, and neurophysiological levels remains incomplete. A deeper understanding of the activation and signaling mechanisms of class C GPCRs could provide key insight into the fundamental roles of G protein signaling in neurobiology, as well as lead to the development of improved therapeutics for many conditions.

Massive effort has been devoted to the development of an extensive repertoire of synthetic compounds for manipulating GPCRs in both a basic research and clinical context [1]. However, classical pharmacology has substantial shortcomings which limit the biological insight that can be gleaned from its use. For instance, pharmacological compounds often lack complete subtype specificity within their receptor subfamily. In the case of mGluRs, while many drugs can distinguish between groups there is a paucity of drugs that target a single subtype with strong specificity [3]. Furthermore, pharmacological compounds also lack the ability to be applied and removed with spatial and temporal control. This is an especially crucial point for class C GPCRs which respond to precise patterns of neurotransmitter release and whose native signaling may not be well-modeled by tonic application of drugs on slow time scales. In addition, pharmacological approaches lack genetic targeting, preventing the study of cell-type specific responses to receptor activation within tissue. Genetic knockouts have been used to study receptor function, but they also lack temporal control and cell-type targeting, and developmental compensation can occur and confound interpretation. Thus, to determine GPCR function it has become widely accepted that tools to facilitate spatiotemporally precise, genetically targeted activation with subtype specificity are needed [8].

As a way to improve spatiotemporal precision, caged compounds for neurotransmitter receptors have been developed. Caged compounds, including those for glutamate and GABA, are ligands masked with a photo-labile group which prevent the compound from functioning as an agonist until it is 'uncaged' with illumination by a specific wavelength of light over a given area [9, 10]. A related family of "reversibly caged" compounds has also been developed, including those for glutamate, which can be reversed with a second wavelength of light [10, 11]. While especially useful for studies of synaptic plasticity at individual dendritic

spines, a number of limitations exist for these approaches. Most importantly, the specificity of the caged compound is limited by the specificity of the compound that is being uncaged. Thus, for example, caged glutamate is able to potentially activate any ionotropic or metabotropic GluR in the vicinity of the illumination which requires a cocktail of antagonists if a single receptor is to be isolated. Even when more subtype-specific ligands are caged, as has been done for orthosteric and allosteric compounds targeting mGluRs [4], full subtype specificity remains elusive. In addition, caged compounds are soluble molecules whose binding and un-binding kinetics are subject to diffusion, limiting their temporal fidelity. Finally, while it is advantageous that such compounds work on native receptors, they are not genetically targeted in any way, limiting their utility for cell-type-specific probing.

A powerful approach to manipulating GPCRs with genetic targeting has been through designer receptors exclusively activated by designer drugs ("DREADDs") [12]. This approach takes advantage of a combination of synthetic, non-native drugs, and an engineered GPCR that can bind the synthetic drugs but are no longer sensitive to their native ligand. Upon activation, the DREADD activates G proteins to drive the relevant signaling pathways. For example, the m4 muscarinic acetylcholine receptor was converted into a DREADD by a combination of Y113C and A203G mutations and can be activated with the ligand clozapine N-oxide (CNO), but not by acetylcholine [13]. Genetically encoding and expressing the DREADDs allows cell-type specificity, but reliance on a soluble drug precludes efficient subcellular targeting or temporal control. Importantly, because this approach relies on mutated receptors, it is unclear how much receptor identity is maintained, challenging the use of DREADDs for understanding specific receptors rather than the general consequences of activation of a particular G protein family. Last but not least, careful control experiments have to be conducted with DREADDs to delineate the role of the introduced receptors versus background effects of the CNO ligand which is hydrolyzed to clozapine and can have effects on native receptors [14].

An alternative to DREADDs that allows both genetic targeting and optical control has been to use intrinsically light-sensitive GPCRs from the opsin family, including rhodopsin and melanopsin [8, 15]. Similar to DREADDs, exogenous expression of opsins, which are typically not expressed outside of the retina, allows general G protein pathways to be manipulated but does not facilitate the study of specific GPCRs. A number of studies have used chimera-based approaches to try to mimic the native signaling of specific GPCRs. These proteins maintain the light-sensitive transmembrane core of rhodopsin or melanopsin but are combined with the cytoplasmic loops and/or C-termini of a GPCR of interest [16–19]. However, it remains unclear if the signaling initiated by

these proteins is able to fully recapitulate the complex signaling properties of specific receptors and many photophysical properties of opsins, including desensitization and slow kinetics [20, 21], challenge the widespread use of these tools.

The development of Photoswitchable Tethered Ligands (PTLs) in the late-2000s provided a new way to activate various types of receptors that addresses the spatiotemporal limitations of classical pharmacology while maintaining the ability to manipulate full length receptors with subtype specificity [10, 22]. PTLs contain a chemical handle for tethering, a photoswitchable molecule, usually an azobenzene, and a ligand head group. As such, a maleimide (M) for attachment to a cysteine was first conjugated to an azobenzene (A) bearing a quaternary ammonium group (Q), resulting in MAQ which was used for the light-dependent blockage of a voltage-gated potassium channel termed "SPARK" [23]. Since the attachment mechanism uses a genetically encoded cysteine substitution on the receptor of interest, it can thus afford both cell-type and receptor subtype specificity. However, the apparent reactivity of maleimides toward other free thiols can be considered as a drawback. The first PTLs for glutamate receptors used a maleimide-azobenzene-glutamate (MAG) family of PTLs (*see* Table 2) to covalently bind a cysteine residue introduced close to the binding site of an ionotropic GluR of the kainate receptor family ("LiGluR") [24]. In 2013, this approach was successfully expanded to GPCRs with the development of light-activated mGluRs ("LimGluRs") [20]. Again, a maleimide-azobenzene-glutamate ("D-MAG$_n$") PTL was used, however, this version had different stereochemistry than the compounds used for LiGluR. Depending on the position of the cysteine substitution on the LBD and the length of the D-MAG variant, photoagonism was established for mGluR2 ("LimGluR2") and mGluR3 ("LimGluR3") and photoantagonism was established for mGluR2 ("LimGluR2-Block") and mGluR6 ("LimGluR6-Block").

While a powerful approach to manipulating mGluRs that has seen application in a number of biophysical and physiological contexts [10, 25–28], PTLs have shortcomings including, most importantly, their reliance on cysteine-maleimide chemistry which is inefficient in aqueous solutions, is prone to promiscuity and is incompatible with the reducing cytosolic environment for intracellular applications. In addition, identifying an optimal attachment position for cysteine substitution can be extremely challenging, providing a limiting step for the development of PTL-based control of other GPCRs. This issue has been tackled by in silico predictions [20, 29], but remains extremely difficult. To overcome these issues, photoswitchable orthogonal remotely tethered ligands (PORTLs) were recently developed with a focus on their application to mGluRs [30, 31]. Rather than attaching the photoswitchable ligand directly to the receptor of interest, PORTLs attach to the

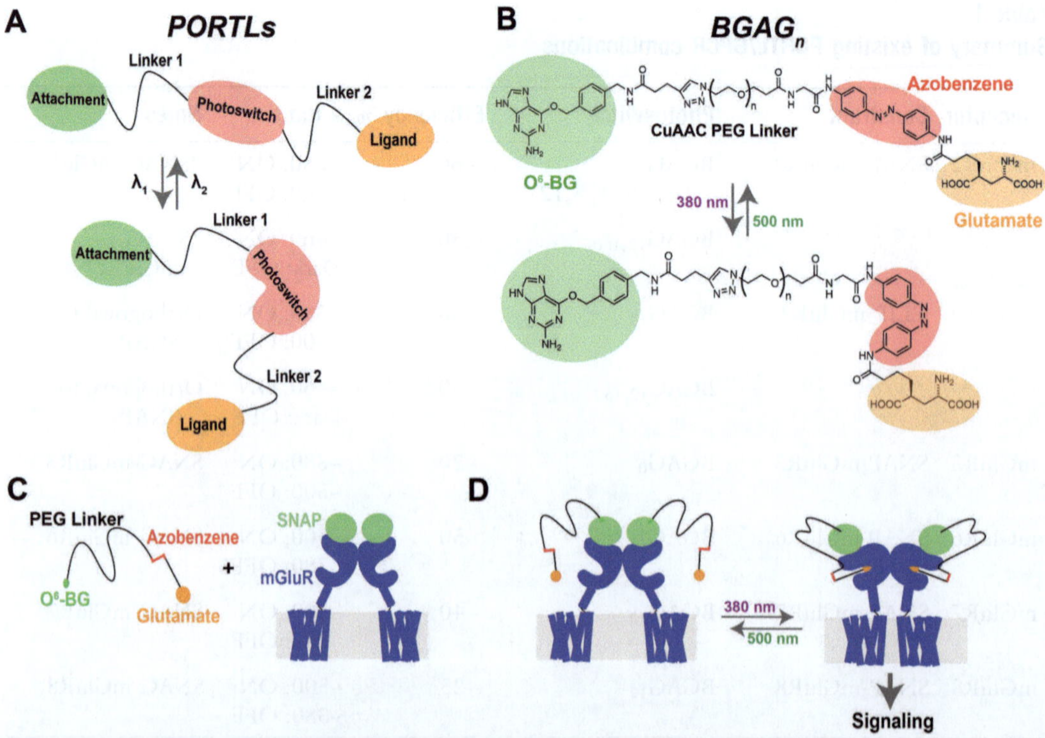

Fig. 1 Principle of PORTL-mediated Optical Control of Class C GPCRs. (**a**) Schematic showing general PORTL design. The relaxed *trans* state (top) of an azobenzene-based PORTL can isomerize to the *cis* state (bottom) panel by exposure to light of wavelength λ_1. Exposure to light of wavelength λ_2 or thermal relaxation drives isomerization back to the *trans* state. (**b**) Chemical structure of BGAG$_n$, a PORTL for SNAP-tagged mGluRs in the *trans* (top) and *cis* (bottom) configurations. (**c**) Schematic showing the general PORTL and SNAP-tagged mGluR design. In principle, SNAP tags may be replaced with CLIP-tags, Halo-tags, or other protein tags, and mGluRs may be replaced with any GPCR. (**d**) Schematic showing general principle of PORTL-mediated optical control of a class C GPCR. Light-induced (~380 nm) isomerization of BGAG$_n$ to the *cis* state allows the glutamate moiety to bind and activate the mGluR, which is reversed by either thermal relaxation or visible light (≥500 nm) illumination

receptor through a self-labeling protein tag, such as SNAP or CLIP, which is typically fused to the extracellular N-terminus of the receptor of interest located next to the LBD (Fig. 1). These tags offer several advantages over maleimide-thiol chemistry, including high selectivity and orthogonality to native reactions, high rate constants (k_2~10^6 M^{-1} s^{-1}), and the extreme stability of the tethering group [32]. Figure 1 outlines the general design of PORTLs and a depiction of the principle in the context of mGluRs. SNAP-targeting PORTLs attach to a receptor via the covalent binding between a genetically encoded SNAP tag and an O^6-benzylguanine (BG) moiety with high selectivity. Following the BG attachment moiety, a PORTL is made up of a polyethylene glycol (PEG) linker, which provides flexibility and appropriate length and is both highly water-soluble and chemically inert, an azobenzene group, which

Table 1
Summary of existing PORTL/GPCR combinations

Receptor	Construct	Photoswitch	Efficiency %	λ (nm)	Notes
mGluR2	SNAP-mGluR2	$BGAG_n$ ($n = 0, 4, 8, 12$)	~60	~380: ON ~500: OFF	"SNAG-mGluR2"
		$BGAG_{12,460}$	~50	~460: ON Dark: OFF	"$SNAG_{460}$-mGluR2"
	CLIP-mGluR2	$BCAG_{12}$	~60	~380: ON ~500: OFF	Orthogonal to SNAP
		$BCAG_{12,460}$	~50	~460: ON Dark: OFF	Orthogonal to SNAP
mGluR3	SNAP-mGluR3	$BGAG_0$	~20	~380: ON ~500: OFF	SNAG-mGluR3
mGluR6	SNAP-mGluR6	$BGAG_{12}$	~50	~500: ON ~380: OFF	SNAG-mGluR6
mGluR7	SNAP-mGluR7-N74K	$BGAG_{12}$	~40	~500: ON ~380: OFF	SNAG-mGluR7
mGluR8	SNAP-mGluR8	$BGAG_{12}$	~25	~500: ON ~380: OFF	SNAG-mGluR8

mediates photoisomerization, and a short-linker attached to a ligand, such as glutamate. The configuration of the azobenzene determines whether the ligand at the end of the photoswitchable compound is able or unable to serve as an agonist (Fig. 1a). The family of compounds developed for SNAP-mGluRs are called "BGAGs" for the *b*enzyl*g*uanine-*a*zobenzene-*g*lutamate components (Fig. 1b, c). The most robust photoswitching observed with the currently available tools has been with BGAG variants in combination with SNAP-mGluR2, producing the "SNAG-mGluR2" tool which provides rapid, reversible, and high-efficiency optical control of mGluR2 (Fig. 1d). PORTL-based optical control has also been established for other mGluR subtypes with a range of BGAG variants (Table 1).

The modularity of the synthesis of PORTLs has allowed for developing and testing versions that vary in linker length, protein tag, and azobenzene modifications. The linker lengths that have been developed for BGAG range from "short-BGAG" (i.e. direct urea linkage between BG and a glycine-spaced azobenzene) to 0, 4, 8, 12, or 28 PEG repeat units (Table 2). SNAP-mGluR2 shows a

Table 2
Chemical structures of mGluR-targeting PTLs, PORTLs and related compounds

PTLs	
L-MAG$_n$	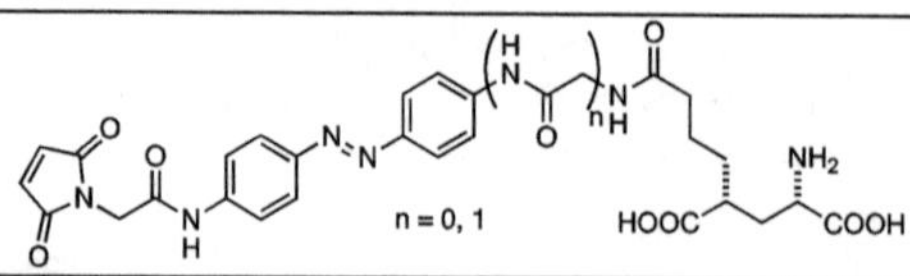
D-MAG$_n$	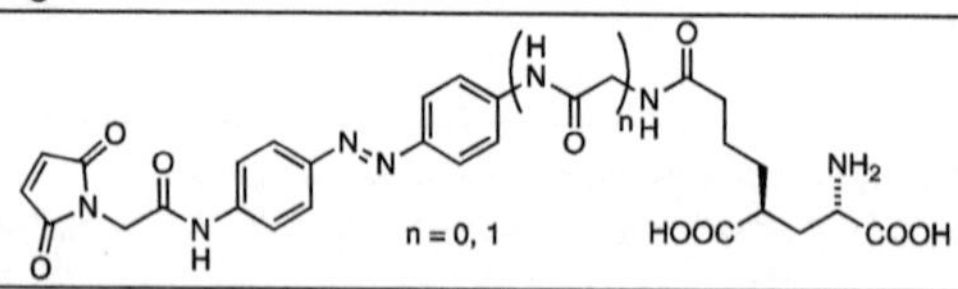
PORTLs	
AG$_{12}$ (untethered control compound)	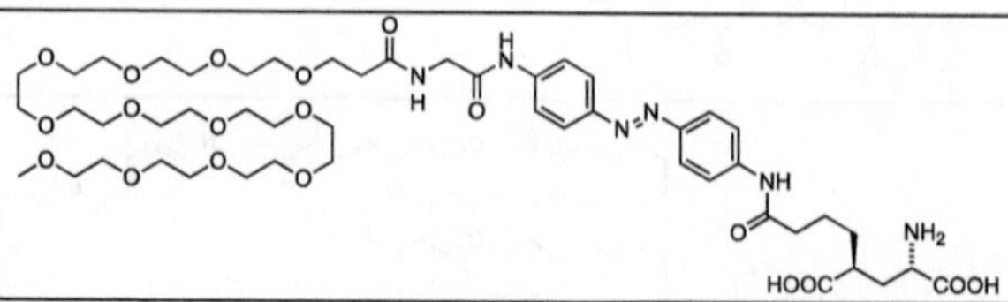
BGAG$_{short}$	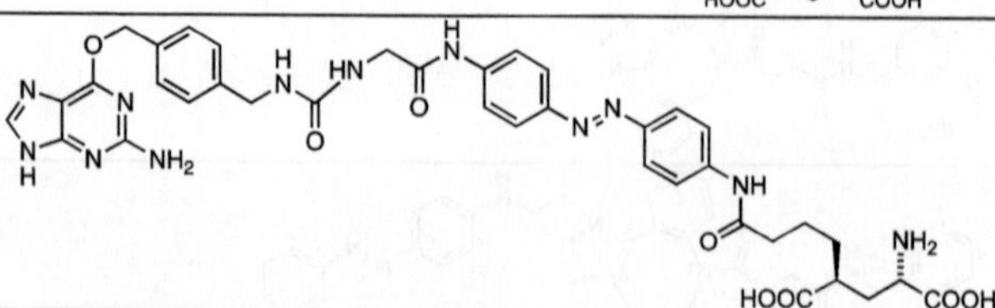
BGAG$_{short,460}$	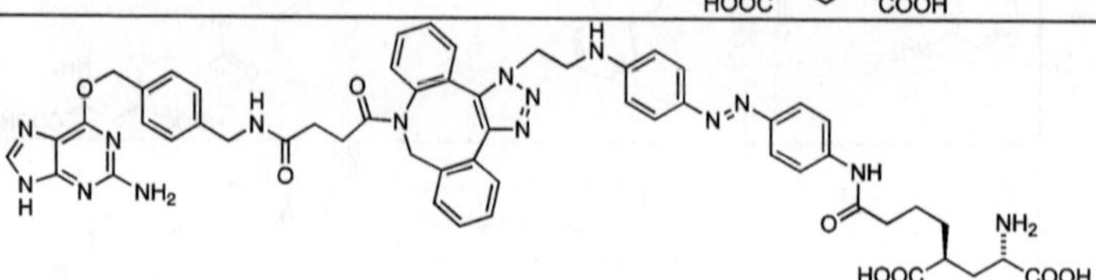
BGAG$_0$	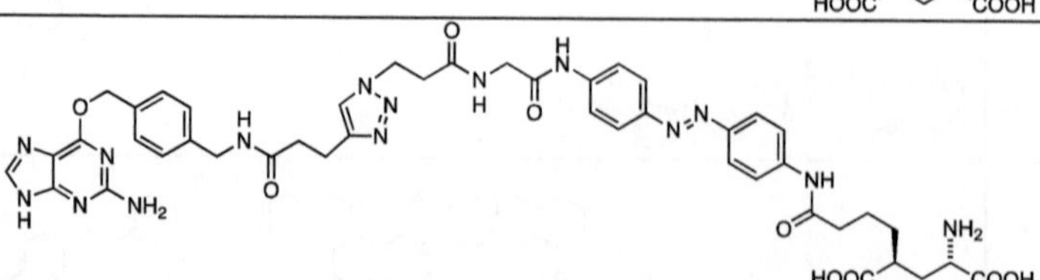
BGAG$_4$	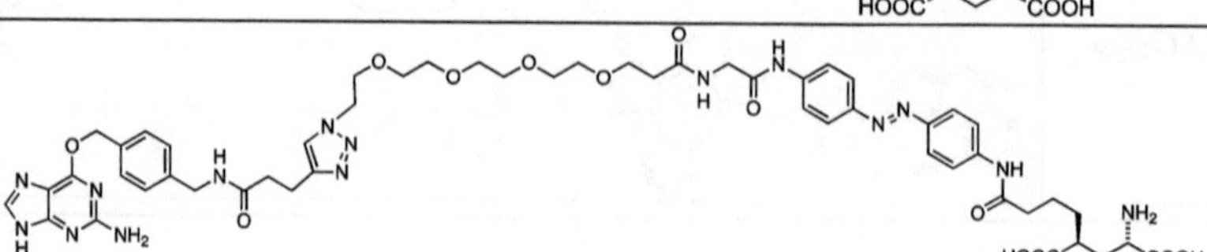
BGAG$_8$	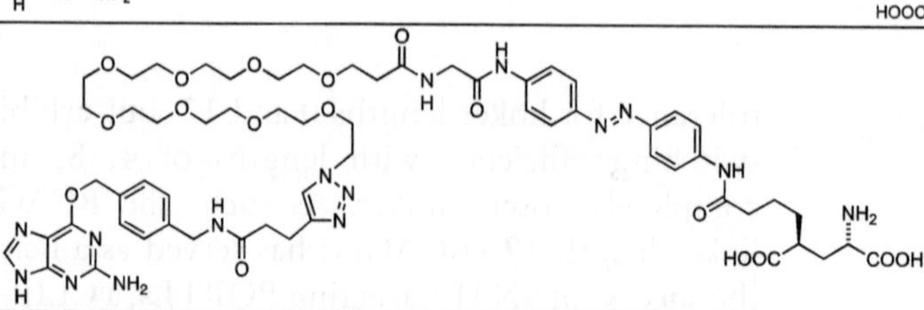

(continued)

Table 2
(continued)

BGAG$_{12}$	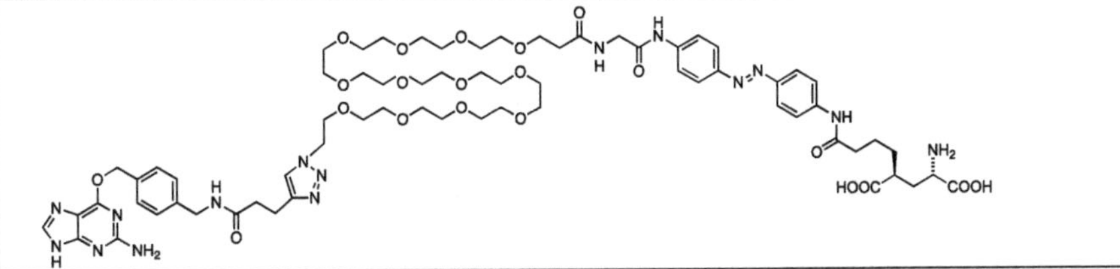
BGAG$_{28}$	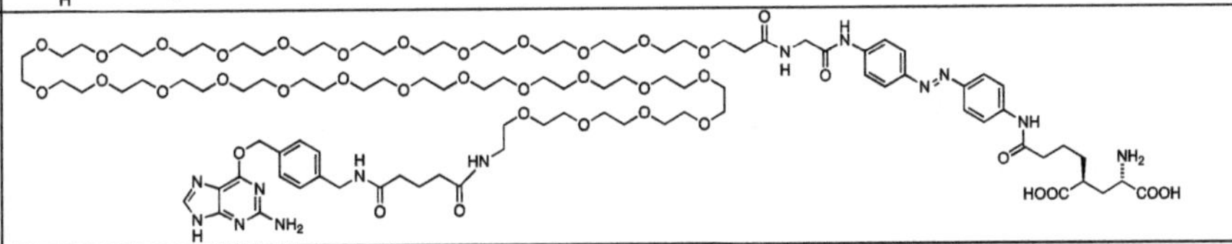
BGAG$_{12,460}$	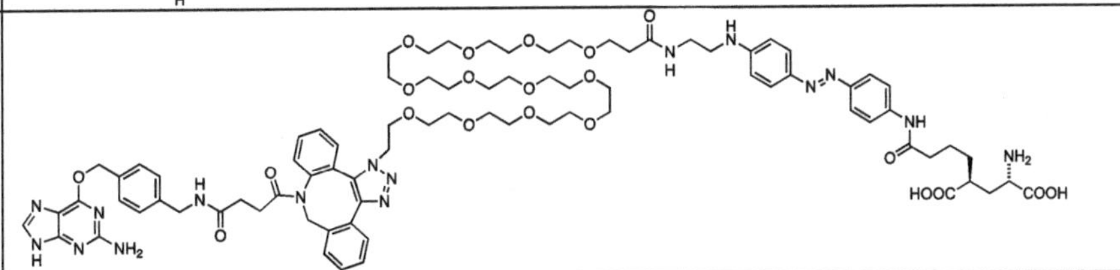
BCAG$_{short,460}$	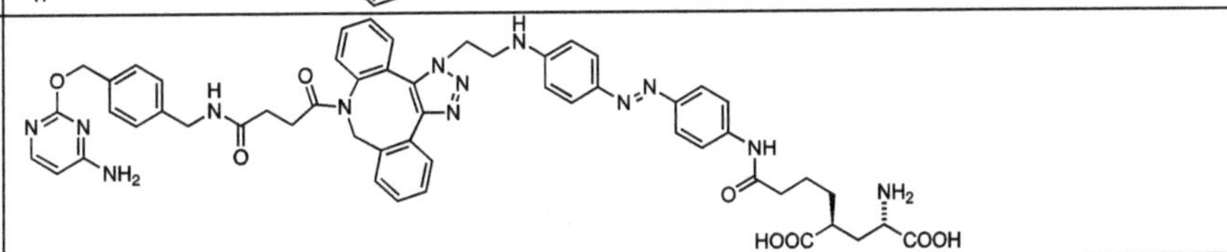
BCAG$_{12}$	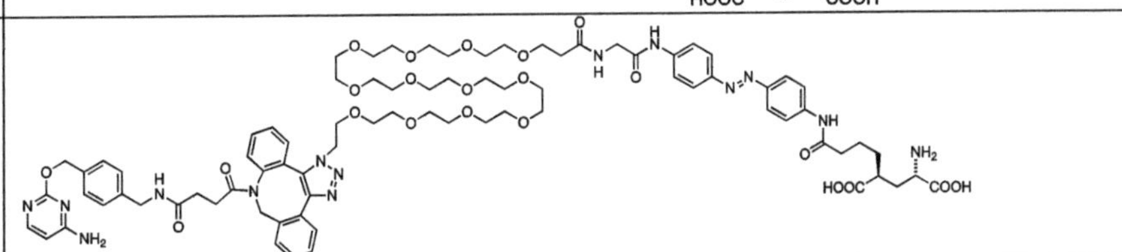
BCAG$_{12,460}$	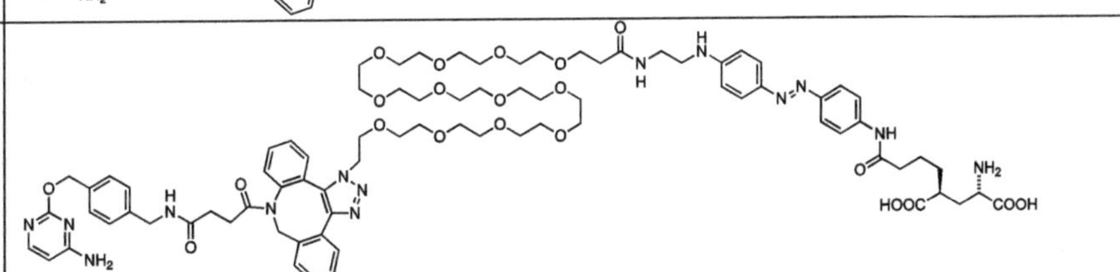

tolerance for linker lengths 0 and 12, but exhibits reduced photoswitching efficiency with lengths of 4, 8, and 28, as well as completely absent linker. As such, the BGAG compound with linker length 12 (BGAG$_{12}$) has served as a benchmark. Based on the success of SNAP-targeting PORTLs, a CLIP-binding version of BGAG has also been developed, termed BCAG, which uses an O^2-benzylcytosine (BC) moiety to attach the photoswitchable compound to the receptor [31]. These compounds show similar

photoswitching efficiency to their BG counterparts. Finally, spectral tuning can also be achieved with PORTLs, as has also been done with PTLs. Modified azobenzenes that permit photoagonism under 460 nm light and spontaneously relax in the dark back to their inactive form have been developed for both BGAG ("$BGAG_{n,460}$") and BCAG ("$BCAG_{n,460}$"). Apart from the advantage that laboratories can now utilize these tools with simple illumination of a single wavelength, these compounds bear other advantages such as the ability to use visible light which is less toxic and penetrates tissue deeper. The development of both orthogonal PORTL labeling, with SNAP or CLIP-targeting variants, and orthogonal wavelengths, with both standard azobenzene and red-shifted azobenzene cores, permits experiments targeting multiple receptor subtypes or variants within the same preparation [31].

In general, the PORTL technique should be adaptable beyond mGluRs to other GPCRs and to other forms of pharmacology other than agonism (i.e. antagonism or allosteric modulation). Soluble photoswitchable ligands have been developed for a number of class A and B GPCRs [33–39] and recently SNAP-tethered, non-photoswitchable agonists have been developed for both the GLP-1R (class B) and the GHS-R1a (class A) [40]. In addition, "*drugs acutely restricted by tethering*" (DART) describes an approach where a non-photoswitchable antagonist is coupled to a protein tag fused to a transmembrane domain for the manipulation of a subset of neurons [41]. These studies point to the feasibility of PORTL-based designs on a wide range of GPCRs. Finally, a major long-term goal of the PORTL approach is to adapt the technique for optical control of native receptors without the need for heterologous receptor (over)expression. As such, genome editing techniques like CRISPR/Cas9 offer the possibility of fusing labeling tags to endogenous proteins [42]. Most recently we have shown in a proof-of-principle study that BGAG-conjugated SNAP-tagged anti-GFP nanobodies can target GFP-tagged mGluR2 with low nanomolar binding constants and allow them to be optically controlled [43]. This exciting result indicates that the PORTL attachment domain can be further outsourced beyond the target receptor itself, opening up many exciting design possibilities that can allow targeting of native receptors.

While PORTL-mediated control of GPCRs can be applied in various experimental conditions, the protocol below uses an ion channel reporter (i.e. GIRK) in conjunction with patch clamp electrophysiology as a readout for $G_{i/o}$ activation of a GPCR. This approach provides a robust, repeatable, receptor-proximal readout with high temporal resolution that is orthogonal to light. These properties make such experiments ideally suited for PORTL development and refinement [30, 31], biophysical studies of receptor mechanisms [26], and is easily adapted to primary cells and

tissues for studies of receptor physiology [44]. However, PORTL-mediated control of GPCRs can be applied in various experimental conditions, including imaging-based readouts and experiments in native systems, such as cultured neurons or brain slices, where endogenous reporters (i.e. ion channels, kinases, gene expression) or physiological responses (i.e. neuronal firing, synaptic strength) further downstream of receptor activation may be assayed. Wherever possible, we will point out modifications that allow other GPCRs, PORTLs, downstream reporters, or cell types to be used to allow the full breadth of potential applications of these tools to be explored.

The focus of this protocol is on N-terminally SNAP-tagged mGluR2 and labeling with $BGAG_n$ variants, the current standard for high-efficiency PORTL-mediated optical control of a GPCR. Notably, PORTL-based optical control of mGluR2 has been comprehensively characterized in vivo in the retina in a mouse model of blindness using ex vivo electrophysiology in whole-mounted retinas and visually guided behaviors in freely moving animals [44]. Ongoing work is aimed at optimizing this approach in the central nervous system of awake, behaving mice using optical fiber-based manipulation.

2 Materials

2.1 Cell Culture

1. HEK 293 or 293T cells. *See* **Note 1** for alternative cellular models.
2. Biosafety cabinet (laminar flow hood).
3. 37 °C Incubator with 5% CO_2 and >50% relative humidity.
4. 37 °C Water bath.
5. Table top centrifuge.
6. Inverted light microscope with 10× or 20× objective to visualize cells.
7. "T25" Flasks: 25 cm^2 sterile polystyrene cell culture flasks with filter cap.
8. Polystyrene 12-well tissue culture-treated plates.
9. 15 or 18 mm diameter round glass coverslips (thickness ~0.1 mm).
10. Thin metal forceps for handling coverslips.
11. Sterile distilled water.
12. Culture Media: Dulbecco's Modified Eagle's Medium (DMEM) with 5% Fetal Bovine Serum. Filter after making 500 mL stock and store at 4 °C.
13. 0.25% Trypsin with EDTA

14. Borate Buffer: For a 500 mL solution, use 1.55 g Boric Acid (50.14 mM), 2.375 g Sodium Borate (23.61 mM), 500 mL dH_2O, pH 8.5, store at 4 °C.
15. 20× Poly-L-Lysine (PLL) Stock Solution (10 mg/mL) diluted in water.
16. Dulbecco's buffered saline (DPBS) without Ca^{2+} and Mg^{2+} (Optional).

2.2 DNA and Transfection

1. Lipofectamine (2000 or 3000) or other lipid-based transfection systems. *See* **Note 2** for alternatives transfection options for other cell lines.
2. DNA Construct for SNAP-tagged GPCR of interest (in this protocol, SNAP-mGluR2). *See* **Notes 3** and **4** for alternatives and other considerations.
3. DNA construct for reporter of GPCR activation (in this protocol, GIRK1). *See* **Notes 3** and **5** for alternatives and other considerations.
4. DNA Construct for reporter of transfection efficiency (in this protocol, tdTomato). *See* **Note 3** for considerations.

2.3 Solutions for Labeling and Electrophysiology

1. Extracellular Solution ("EX"): 135 mM NaCl, 5.4 mM KCl, 10 mM HEPES, 2 mM $CaCl_2$, 1 mM $MgCl_2$. Adjust pH to 7.4 with NaOH. Filter and store at room temperature. *See* **Note 6** for additional usages for these solutions.
2. High Potassium Solution ("HK_{120}"): 120 mM KCl, 25 mM NaCl, 10 mM HEPES, 2 mM $CaCl_2$, 1 mM $MgCl_2$. Adjust pH to 7.4 with KOH. Filter and store for up to 1 week at room temperature. *See* **Note 7** for usage.
3. Intracellular Solution: 140 mM KCl, 10 mM HEPES, 5 mM EGTA, 3 mM $MgCl_2$, 3 mM Na_2ATP, 0.2 mM Na_2GTP (Na_2ATP and Na_2GTP are added last and maintained on ice). Adjust pH to 7.4 with KOH. Filter, make 1–2 mL aliquots, and store at −20 °C.
4. Glutamate Stock Solution: 100 mM high purity L-glutamic acid in HK_{120} Solution. Store at 4 °C. *See* **Note 8** for details and alternatives.

2.4 PORTLs and Related Compounds

1. Materials for Synthesis (*see* ref. 30). Semi-preparative or preparative reverse-phase C18 HPLC is of vital importance.
2. Spectrophotometer. *See* **Note 9** for usage.
3. PORTLs (available compounds are shown in Table 1, *see* **Note 10**): Powder stocks are solubilized in a minimal volume of dry DMSO. PORTL aliquots are stored at 10–100 mM in DMSO at −20 °C in dark tubes with desiccant. *See* **Note 11** for storage

recommendations. *See* **Note 12** for determining PORTL concentration.

4. Soluble test compounds (*see* **Note 13**).

2.5 Patch Clamp Electrophysiology

1. Vibration isolation table attached to house air.
2. Inverted Microscope (Ex. Olympus IX73) that is suitable for basic fluorescence imaging. 20× or 40× objectives with large working distances are ideal for patch clamp experiments.
3. Filter cubes/Filters for fluorescence imaging and optical manipulation. *See* **Note 14** for considerations.
4. Micromanipulator (Ex. Scientifica PatchStar Micromanipulator).
5. Faraday Cage.
6. Amplifier (Ex. Axopatch 200B from Molecular Devices).
7. Digitizer (Ex. Axon Digidata 1550B from Molecular Devices).
8. Silver wire (>99% purity) for microelectrode and ground electrodes, respectively. Silver wire is chlorinated by incubating it in bleach for 30–60 min.
9. Microelectrode holder and headstage (Ex. Axopatch CV-203BU).
10. Illumination system. *See* **Note 15** for options.
11. Borosilicate, thin wall capillary glass with filament for patch pipettes (Ex. G150TF-3 from Warner Instruments).
12. Micropipette Puller (Ex. Sutter Instruments Model P-97).
13. 1 mL Syringe to fill pipette with Intracellular Solution. Syringe tips for filling micropipettes (Ex. Microfil 28 AWG, MF28G67-5, World Precision Instruments).
14. Tubing. *See* **Note 16** for how to set up tubing system for controlling pipette pressure.
15. Salt bridges made from bent glass capillaries and filled with 3 M KCl in 1% agarose and stored at 4 °C in 3 M KCl.
16. Desktop computer with Clampex (or similar) software for voltage clamp control and current recording.
17. Gravity-Driven Perfusion System mounted on a support ring stand containing syringe holders, 60 mL syringes, a stopcock for each syringe, tygon tubing (Ex. Outer diameter: 1/8 in; Inner diameter: 1/16 in) and a manifold (Ex. Warner MP series with ≤8 inputs). A flow regulator may be placed after the manifold to allow for modulation of the flow rate for all channels.
18. Vacuum line connecting suction reservoir of chamber to large flask for collection of perfusion output throughout experiment.

19. Recording/Imaging Chamber (Ex. Warner Instruments RC-25F) and Platform (Ex. Warner Instruments PM-4). *See* **Note 17** for considerations.

3 Methods

For synthetic schemes of PORTLs we refer the reader to Broichhagen et al. [30] and Levitz et al. [20]. *See* **Note 18** on CuAAC, SpAAC, peptide coupling, and NHS-ester handling.

3.1 Coverslip Preparation

1. To prepare the 1× PPL Solution, combine 1.5 mL of PLL 20× stock to 30 mL Borate Buffer (final working concentration of PLL is 0.5 mg/mL). For a 12-well plate, add ~2 mL to each well; add coverslips to well and ensure it is completely covered by the PLL solution by pushing down with metal forceps.
2. Leave coverslips in PLL solution ≥12 h prior to seeding cells. PLL solution is re-usable.

3.2 Cell Preparation and Seeding (Day 1)

1. Place coverslips in sterile 12-well plate with ~2 mL H_2O in each well to wash poly-L-lysine. Thoroughly aspirate dH_2O (optional: wash a second time with dH_2O) and let coverslips air dry in the biosafety cabinet. Once dry (10–15 min), add 1.5 mL culture media to each well and place in incubator.
2. Aspirate media from T25 flask of confluent cells and add 3 mL pre-warmed trypsin and place in incubator for 3 min.
3. Harvest cells and spin in table-top centrifuge at 1500 × *g* for 3 min to produce a small pellet. Aspirate supernatant, add 6 mL media and resuspend cells by pipetting up and down.
4. Seed cells on coverslips (Fig. 2). Density should be sparse to have single cells or small clusters (≤3 cells) for electrophysiology recordings 36–72 h later. To achieve this density, ~50 μL of cell suspension are added to each well. The amount of cells added can be adjusted empirically taking into account the density of the parent flask, the rate of cell growth, and the timing of seeding and subsequent experiments. Seeding at higher densities can be used for imaging experiments when using optical readouts. *See* **Note 19** for additional considerations.

3.3 Transfection (Day 2)

1. Transfect cells ~12 h after seeding according to the manufacturer's protocol (Fig. 2). For Lipofectamine transfection in a 12-well plate, 0.75 μg SNAP-tagged GPCR, 0.75 μg GIRK1-F137S, and 0.1 μg tdTomato are used per well. Less DNA is typically used for optical reporters, such as GCaMP, because soluble proteins express more easily than membrane proteins.

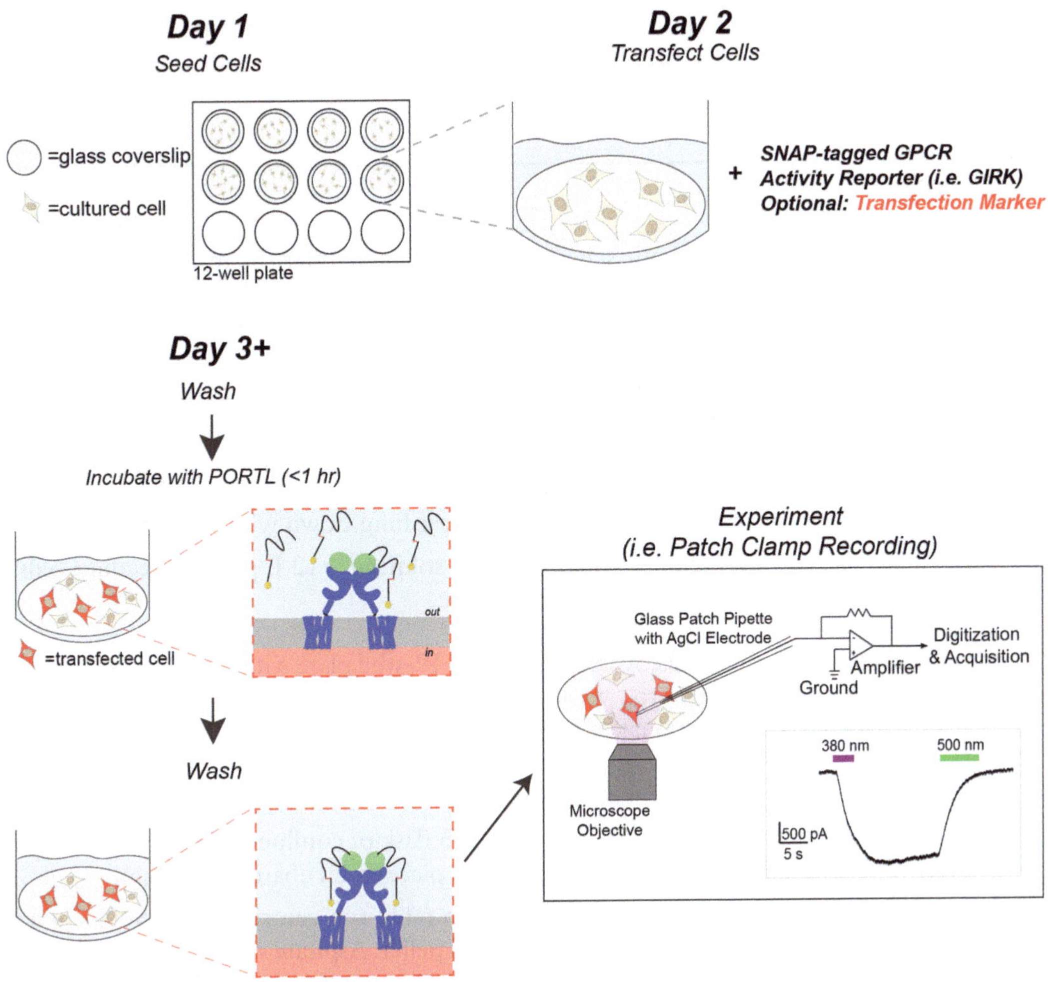

Fig. 2 Workflow for PORTL-mediated optical control of SNAP-tagged GPCRs in cultured cells. On day 1 cells are cultured in a 12-well plate and are transfected with relevant DNA on day 2. After 24–48 h of expression, cells are labeled with PORTL, washed and mounted on an inverted microscope for electrophysiology. Light is transmitted through the objective to illuminate the sample and cells are patch-clamped in whole cell mode using a glass patch micropipette with an electrode connected to digitization and acquisition equipment. The inset gray box shows a representative result, where UV light (~380 nm) is used to drive photoactivation and visible light (>480 nm) is used to reverse the photoswitch and terminate photoactivation

2. Allow 24–48 h for expression or longer for primary cells (3–5 days). *See* **Note 20** for optional strategies to ensure cell health. *See* **Note 21** for strategies to circumvent low expression issues.

3.4 PORTL Labeling (Day 3)

1. Prepare dilution of PORTLs for labeling cells in EX solution (Fig. 2). 1–10 μM is typically sufficient to provide saturating labeling at 37 °C for 45–50 min [30]. *See* **Note 22** for determining optimal labeling conditions.

2. Following a brief wash in 1–2 mL of EX, coverslips are labeled in 400 μL of PORTL solution (i.e. BGAG) in a 12-well plate (Fig. 2). Due to the stability of the BG moiety in aqueous buffer, the same dilution of BGAG can be re-used during a full day of experiments with multiple coverslips.

3.5 Whole Cell Patch Clamp Recording (Day 3)

1. While a coverslip is being labeled with PORTLs, thaw an Intracellular Solution aliquot on ice.
2. Load perfusion system with HK_{120} and relevant compounds based on the particular experiment (*see* **Note 23**).
3. Pull pipettes using the parameters necessary to achieve a resistance of 3–7 MΩ when pipette is placed in the bath.
4. Turn on equipment on patch electrophysiology rig: microscope illumination, amplifier, digidata, computer, and micromanipulator. Using a Faraday cage will reduce electrical noise, and draping a black sheet over it will prevent excess light from reaching the sample. Additionally, a vibration isolation table attached to house air will minimize vibration-induced noise in patch clamp experiments.
5. Following 45–50 min of labeling, wash cells with EX solution. An efficient way to do this is to move the entire coverslip to a new well within the 12-well plate containing 1–2 mL EX. Repeat this 2–3 times.
6. Prepare the cells for recording by setting the coverslip in the recording chamber and attaching it to the perfusion system. Set up perfusion system to achieve a flow rate of ~1–2 mL/min (using flow regulators) and position end of vacuum line in chamber to maintain a constant flow of HK_{120} from perfusion inlet port, through chamber, and out the suction reservoir. Ensure that the ground electrode, or salt bridges connected to a batch containing HK_{120}, is also in the suction reservoir connected to the bath.
7. Select a healthy cell (i.e. spread, flat, isolated, not round) that is expressing the fluorescent reporter construct and adjust stage/objective so that it is in the center of the field of view (Fig. 2). It is important to be aware that the light used to excite the fluorescent reporter might impact the photoswitch. Ideally, the excitation light used for the fluorescent reporter should not fall under the spectral profile for the azobenzene state that activates the receptor. For example, tdTomato can be excited at 550 nm, which will maintain $BGAG_n$ in the inactive *trans* state for SNAG-mGluR2. In the absence of an orthogonal fluorescent reporter that falls outside of the emission range of the active configuration of the PORTL, illuminating the sample for the minimal amount of time is crucial to prevent excessive receptor activation which can lead to desensitization. For

example, when using tdTomato in conjunction with group III SNAG-mGluRs (where *trans*-agonism is observed) brief (<10 s) 550 nm illumination can be used to visualize tdTomato and a short pulse at 380 nm can be used to switch the PORTL back to the inactive *cis* state.

8. Fill micropipette with Intracellular Solution using a syringe (be sure to remove all bubbles from micropipette by flicking it with fingers) and attach the micropipette to the pipette holder. Make sure that the AgCl electrode is in contact with intracellular solution and that the pipette is not filled more than is necessary to make a connection with the electrode.
9. Bring micropipette tip near the cell, apply and lock positive pressure into the pipette tip, and bring pipette offset to 0 pA while voltage clamping. Resistance of the pipette in the bath should be 3–7 MΩ as calculated in software via a 10 ms, −5 mV test pulse.
10. Approach cell slowly until resistance increases by ~0.1–0.3 MΩ, which indicates that the glass pipette has touched the cell. Immediately release the positive pressure and apply gentle suction. Ramp voltage down to ~−20 mV to facilitate seal formation. Allow resistance to increase until a >1 GΩ seal is formed. This may be quick for healthy cells with optimized conditions (<10 s) or may take a few minutes otherwise.
11. While in the cell-attached mode, turn perfusion and vacuum on, allowing HK_{120} to perfuse through the chamber to maintain a steady level. Because patch stability is higher in the cell-attached mode than in whole-cell mode, adjust filters to the desired settings for the subsequent experiment before breaking in (*see* **step 9**).
12. Once a stable GΩ seal is formed in cell-attached mode, apply brief suction to break the cell membrane at the tip of the pipette to enter whole-cell mode. The test pulse should show an increase in the time constant of the exponential decay due to the increased capacitance due to gaining electrical access to the entire cell. A decrease in resistance is also expected.
13. Begin recording in the gap-free mode with a holding potential of −60 mV at a sampling rate of 10 kHz. *See* **Note 24** for more details on voltage-dependence.
14. Record for 30–60 s to achieve a steady baseline and allow cell dialysis with intracellular solution to occur. At a holding potential of −60 mV, the current level should be in the −100 to −500 pA range. Basal currents levels are due to basal GIRK activity and basal receptor activity as well as the leakiness of the patch. Generally, as long as the baseline is less than 1 nA and is stable, one can proceed with the stimulation protocol.

15. Turn on light to drive photoisomerization of the azobenzene to photoactivate receptors. For $BGAG_n$ with SNAP-mGluR2, the optimal wavelength to drive the *cis* state for photoagonism is ~380 nm. Either brief pulses or constant illumination can be used to induce receptor photoactivation to the same level. Photoactivation of SNAP-mGluR2 with $BGAG_n$ typically produces anywhere from 100 pA to 1–2 nA. *See* **Note 25** for details on timing and wavelength of illumination for different PORTLs. Photoactivation of the receptor will produce an inward current from K^+ ion influx which is induced downstream of G protein activation via binding of $G_{\beta\gamma}$ to the GIRK channel. Once current has reached a steady-state plateau (typically within 5–10 s), turn on light to drive photoisomerization back to the *trans* state (>480 nm for BGAGn PORTLs). Current readout will return to baseline within 10–20 s.
16. Repeat photoactivation multiple times (at least 2–3) to assess repeatability and then apply a saturating agonist to allow for quantification of photoswitch efficiency (*see* Fig. 3). *See* **Note 26** for consideration about desensitization. Many variations of this protocol with variable illumination timing and frequency can be devised depending on the precise application. *See* **Note 27** for further controls and characterization. *See* **Note 28** for using optical readouts and **Note 29** for consideration when imaging and photoswitching simultaneously. Generally, the variability in photoswitch efficiency relative to saturating agonist will be around 10–15%, where at least four cells on at least two separate coverslips should be patched to calculate reliable photoswitching efficiencies.

4 Notes

1. HEK293T cells are well-suited to these experiments because of the ease of handling them for splitting, seeding, transfecting, and imaging or patch clamping and their lack of endogenous metabotropic glutamate receptors and other class C GPCRs (with the exception of $GABA_{B1}R$) [45] and low levels of endogenous ion channels. HEK 293T cells, which contain the SV40 large T-antigen, generally show higher levels of expression than HEK 293. Other mammalian cell lines also work for similar experiments, including COS-7, CHO, and HeLa cells. Primary cells, such as hippocampal or cortical neurons or glia that tolerate transfection or viral infection may also be used for experiments aimed at studying class C GPCRs in a more physiological context.

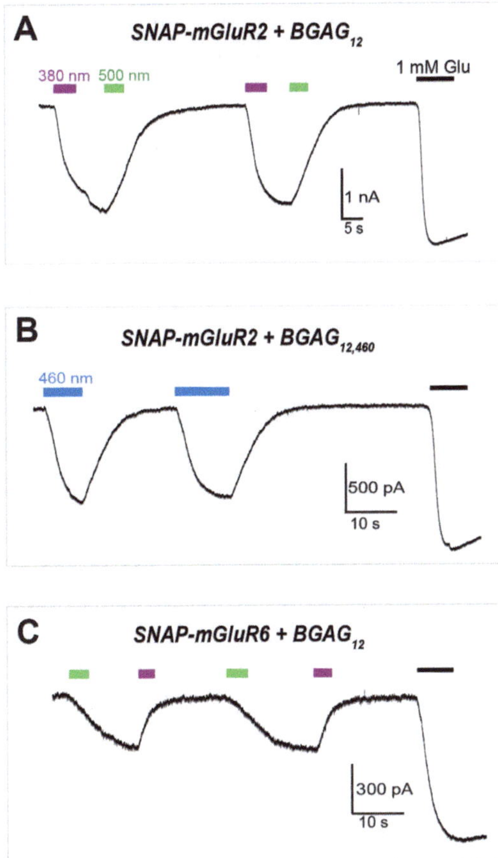

Fig. 3 Photoactivation of SNAP-tagged mGluRs with various PORTLs. (**a**) Representative trace showing SNAP-mGluR2 photoactivation via $BGAG_{12}$ in HEK 293T cells co-expressing GIRK channels. This combination averages a ~60% photoswitching efficiency relative to saturating glutamate upon photoactivation by 380 nm light and termination of photoactivation by 500 nm light. (**b**) Representative trace showing SNAP-mGluR2 photoactivation via $BGAG_{12,460}$. This combination averages ~40% photoswitching efficiency upon photoactivation by 460 nm light and termination of photoactivation in the dark. (**c**) Representative trace of SNAP-mGluR6 activated by $BGAG_{12}$. This combination averages a ~45% photoswitching efficiency upon photoactivation by >480 nm light and termination of photoactivation by ~380 nm light

2. For primary cells, calcium phosphate-based transfection is generally more effective and better tolerated.
3. DNA is extracted and purified from competent *E. coli* cells using a miniprep kit. The concentration of the DNA is determined via absorbance at 260 nm using a spectrophotomer or nanodrop. Concentrations over 600 ng/μL are desirable. Purity of the DNA constructs is confirmed by measuring the ratio of the absorbance at 260 nm to the absorbance at 280 nm

(A_{260}/A_{280}), where ratios between 1.7 and 2.0 are acceptable. DNA constructs should be fully sequenced to confirm their identity and identify any mutations. For HEK cell experiments, constructs with a CMV promoter to drive strong expression of relevant proteins in mammalian cells are used. Other DNA constructs can also be used with this protocol.

4. Any cDNA with a SNAP- or CLIP-tagged GPCR of interest can be used. N-terminally SNAP-tagged mGluRs were developed by Doumazane et al. [46]. All existing work on PORTLs has been done using N-terminal SNAP- or CLIP-tagged GPCRs that contain a minimal linker ("TR" in the case of SNAP-mGluR2) between the tag and the start of the GPCR [46]. For GPCRs with defined signal sequences, the SNAP tag should be placed following this. For those without defined signal sequence, the tag placement will have to be empirically optimized, but the N-terminus is probably a good starting point. Assessing the prior use of GFP (or other fluorescent protein) tags should guide SNAP or CLIP attachment site as positions that tolerate GFP are likely to tolerate SNAP or CLIP, which are ~70% as large. In principle, tags may be placed at either the N-terminus (extracellular for GPCRs) or C-terminus (intracellular for GPCRs) or, potentially, at intramolecular sites. However, for every new construct, labeling and receptor function should be carefully characterized to assess efficiency and orthogonality. In addition, to SNAP and CLIP, other protein tags should be amenable to the PORTL approach, including Halo tags [32].
5. The reporter of GPCR activation needs to be altered for the GPCR of interest. For *$G_{i/o}$-coupled receptors* (Ex. Group II/III mGluRs and $GABA_BRs$), a good reporter is G protein-coupled Inward Rectifying Potassium Channel (GIRK1). The GIRK1-F137S homo-tetramerization mutant is used to reduce the number of constructs needed to simplify transfection and expression [20]. For *G_q-coupled receptors* (Ex. Group I mGluRs): to detect receptor activation with GIRK channels a chimeric $G\alpha_{i/o}$ protein can be co-transected. This chimera contains the three C-terminal acids of $G\alpha_q$ [47] which allows a G_q-coupled receptor to initiate $G\alpha_{i/o}$-type signaling (i.e. GIRK activation). An alternative to electrophysiology is to use calcium sensors, such as genetically encoded GCaMP6 [48] or synthetic dyes [49], as an optical readout. *See* **Note 28** for other potential fluorescent reporters that may be used in conjunction with PORTLs. While GIRK channels offer a simple, robust readout of $G_{i/o}$-coupled receptors and other G protein pathways via chimeric G_α proteins, other ion channels may be used to assess receptor activation and provide an alternative vantage point compared to GIRK since unique signaling

mechanisms are employed for each. For example, inhibition of voltage-gated calcium channels [50] or potentiation of TREK channels [51] may be used for $G_{i/o}$-coupled receptors and activation of TrpC3 channels [52] may be used for G_q-coupled receptors. Different voltage protocols and ionic conditions are needed depended on the channel reporter of choice (*see* referenced papers for details). In primary neurons, a GPCR activation reporter is not necessary since native effectors exist. A number of GPCR-sensitive ion channels are expressed in neurons, providing a native-like complex response that can be detected electrophysiologically. For $G_{i/o}$-coupled receptors activation typically results in hyperpolarization, while G_q-coupled receptors typically produce depolarization.

6. EX Solution is used for labeling cells with PORTLs or fluorophores and can be used for imaging studies or electrophysiology with activity reporters other than GIRK channels.
7. High Potassium Solution ("HK_{120}") is used as a recording solution when performing GIRK reporter experiments. The high extracellular potassium allows large inward currents to be evoked at negative holding potentials.
8. Glutamate stock (100 mM) should be diluted to the desired working concentration in fresh HK_{120} for experiment. For other GPCRs, a suitable stock of an agonist is required.
9. Spectrophotometer is used for determining PORTL concentration via absorbance measurements.
10. It is worth noting that there are a variety of PORTLs available and their photopharmacological profiles vary based on the receptor they are attached to. Table 1 shows a current list of PORTLs and how they function in various receptor combinations (*see* Table 2 for chemical structures). $BGAG_{12}$ has been modified to photoisomerize at red-shifted wavelength, resulting in $BGAG_{12,460}$ (due to *cis*-photoisomerization occurring with 460 nm light, instead of 380 nm). This PORTL maintains the agonism profile of $BGAG_{12}$ (*cis* state is an agonist, *trans* state is not an agonist) on group II mGluRs. However, the pharmacological unit that activates a given receptor can be inverted, such as with group III mGluRs. For these receptors, the photopharmacology of $BGAG_{12}$ is inverted (*cis* state is not an agonist, *trans* state is an agonist). Although this is still unclear, the current thought is that the precise shape of the linker-azobenzene moiety plays a role in whether the glutamate is able to fit into the LBD of a specific receptor.
11. For PORTL solubilization, 0.75 mL DMSO-d6 from ampules is recommended—do not use DMSO that has been standing on the lab bench for extended periods. The concentration of PORTL in DMSO may be determined using a dilution of the

Table 3
Extinction coefficient (ε) and peak absorption wavelength (λ) for BGAG PORTLs

Compound	ε (M^{-1} cm^{-1})	λ (nm)
$BGAG_n$	19,600	378
$BGAG_{n,460}$	24,600	420

PORTL in EX Solution. The absorbance of the dilution can be measured using a spectrophotometer, and the concentration can be calculated using the extinction coefficient and peak wavelength given below (*see* Table 3 and **Note 12**). The BG moiety is extremely stable if stored properly. Individual batches have shown robust labeling and photoswitching for years following many rounds of freeze/thaw cycles.

12. The Beer–Lambert Law ($A = \ell * \varepsilon * c$) is used for the calculation of concentration where A is absorbance, ℓ is path length in cm, ε is extinction coefficient in M^{-1} cm^{-1}, and c is concentration in *M*. *See* Table 3 for values used for BGAG PORTLs.

13. When developing PORTLs a simple, iterative approach can be used to design an effective compound. First, a screen of the pharmacological literature can provide valuable information about what attachments are tolerated in the parent ligand. Such a literature search led to the observation that 4'D attachment to L-glutamate is tolerated in group II and III mGluRs [20], a key first step toward the development of D-MAG PTLs and BGAG PORTLs. Test compounds may be synthesized and tested that contain a substituted parent ligand and the first ring of the azobenzene (*see* "Tether Models" in Volgraf et al. [24] and Levitz et al. [20]). If these test compounds are functional, soluble photochromic ligands (PCL) can be developed. PCLs contain all parts of the PORTL except for the attachment moiety, allowing them to be tested as soluble drugs in both *cis* and *trans*, by alternating the illumination wavelength, on the receptor of interest. For example, AG_{12} (Table 2) serves as a *cis*-agonist on mGluR2 [30]. This approach can serve as a way to screen photoswitchable compounds and ensure their photoagonist activity before incorporating attachment with a PORTL, which will require additional considerations outside the pharmacology of the compound, such as linker length and flexibility. In all cases tested so far, the directionality of photoswitching of a PCL is predictive of a PORTL based on its design. In addition, weak PCL photoswitching does not preclude high-efficiency PORTL photoswitching since the enhanced effective concentration following ligand attachment

can drastically improve this. For example, AG_{12} shows <20% photoactivation but $BGAG_{12}$ shows >60% on mGluR2 [30].

14. When selecting a filter cube for assessing expression markers, standard GFP or RFP cubes may be used to visualize GFP or tdTomato or other fluorophores. For photoswitching a filter cube with either a mirror or far-red (~650 nm) high-pass dichroic filter should be used to maximize the illumination that reaches the sample. No excitation or emission filters are needed. *See* **Note 29** for considerations needed for imaging experiments.
15. Various illumination systems can work well. LED-based systems or Xenon lamp-based systems allow for rapid switching between different wavelengths via TTL pulses from digitizer. Laser illumination in full field modes or scanning can be used as well for more targeted applications.
16. Tubing should be connected from the pipette holder to a syringe or mouth tip with a valve for opening and closing. This will allow the application of negative and positive pressure to the sample. A pressure meter may be attached for assessing positive and negative pressure values.
17. The recording chamber must have an inlet for perfusion and ideally an outlet or separate reservoir for ground electrodes and vacuum. A glass-bottom petri dish is the simplest option and can work fine, but has limitations in terms of perfusion speed and stability.
18. In the original BGAG design, a modular approach was chosen for the connection of attachment site (BG), linker, and azobenzene-containing pharmacophore. Accordingly, Cu(I)-catalyzed alkyne-azide click chemistry (CuAAC) was envisioned to serve as a late-stage modification to introduce the attachment molecule in an orthogonal way. However, in our initial studies [30] we realized that the reaction needs high temperatures to proceed together with high, over-stoichiometric catalyst loadings. In addition, the CuAAC on the red-shifted $BGAG_{12,460}$ did not proceed at all, and was therefore circumvented by the use of a strain-promoted alkyne-azide click (SpAAC) using DBCO as an alkyne. Albeit resulting in the formation of two stereoisomers (with neglectable influence on tether length), this approach was further expanded in a follow-up study [31], where BCAGs were made using SpAAC chemistry. As a first example, alkyne-azide click chemistry was completely abolished by making $BGAG_{28}$, as standard Fmoc-amine deprotection and subsequent peptide coupling employing TSTU proved to be beneficial with respect to being less labor intensive, less toxic, enabling shorter reaction times and facilitating a more straightforward purification. Furthermore,

acids activated by TSTU to their NHS-ester counterparts can be chromatographed and isolated by means of reverse-phase high-performance liquid chromatography (RP-HPLC) when the mobile phase is buffered with an acid (usually MeCN/H_2O with 0.1% formic acid or TFA). Given these experiences, we recommend focusing on peptide couplings for building PORTLs rather than designing click partners.

19. The specific timing and methodology will differ for different cell lines or for primary cells. For a protocol on cultured neuron preparation, *see* Ref. 53.

20. To ensure cell health, ~8 h after transfection, cell media may be changed. Some GPCRs are toxic to cells and this can be alleviated by adding an antagonist at this stage. For example, 5 μM LY341495, a pan-mGluR competitive antagonist, can help maintain the health of mGluR-transfected cells. It is crucial that the reversibility and the effects of an antagonist on receptor expression are thoroughly characterized before using.

21. If the DNA of interest requires longer expression time, transfected cells may be re-seeded 24–48 h following transfection. For re-seeding, media is aspirated and 0.5 mL calcium-free DPBS solution is added to each well. Following a 20-min incubation, 100–200 μL (amount determined empirically) is added to freshly washed coverslips in 1.5 mL culture media in a new 12-well plate. Typically, at least 12 h are needed following re-seeding for cells to re-adhere and develop a normal, spread-out morphology.

22. Previous studies of SNAP-mGluRs have shown saturating labeling for $BGAG_n$ and $BCAG_n$ at concentrations as low as 1 μM for 45 min at 37 °C [30, 31]. However, sufficient labeling concentrations may depend on the PORTL, the receptor itself, the expression system, and may be sensitive to the state of the PORTL. To ensure sufficient labeling, the surface labeling efficiency of PORTLs may be tested using a competition-based assay with cell-impermeable BG-conjugated fluorophores (such as Surface SNAP-Alexa-546 from New England Biolabs). By initially labeling cells with the PORTL at a given concentration and then labeling with saturating amounts of the cell-impermeable BG-conjugated fluorophore, the residual fluorescence can be compared to cells incubated only with the fluorophore to calculate the PORTL labeling efficiency. These results can be used to determine the saturating concentration of a PORTL needed for photoswitching experiments and can give a reasonable estimate of the overall labeling efficiency. *See* Levitz et al. [31] for an example of this experiment where $BGAG_n$ labeling of SNAP-mGluR2 was estimated to be 80–90%.

23. For any quantification of photoswitch efficiency, a saturating concentration of an agonist is applied for normalization. For mGluR PORTL experiments, 1 mM glutamate (or 10 mM for mGluR7) is used as a standard saturating concentration. In addition to assessing PORTL efficiency as part of the tool engineering and characterization processes, this quantification can also be used for testing the effects of different mutations, especially in the context of ligand occupancy-dependent cooperativity (*see* Ref. 26).

24. While native currents from GIRK channels [54] are typically outward, when using GIRK as an exogenous reporter of GPCR activity the inward rectification of these channels allows larger currents to be obtained by measuring inward currents. To permit inward currents at physiological membrane potentials (~−60 mV), a high potassium extracellular solution is used (HK_{120}). To confirm that potassium-selective GIRK currents are being recorded, a current–voltage (*I–V*) relationship can be calculated by measuring the current evoked following steps to different voltages (Ex. 500 ms steps from a baseline of −60 ranging from +40 to −100). Subtracting a baseline *I–V* curve from the *I–V* curve obtained in the presence of ligand or PORTL photoactivation should reveal an inward rectifying current with a reversal potential that matches what is expected for potassium in the ionic conditions of the experiment (~0 mV). Alternatively, a blocker can be used to confirm that currents are mediated by GIRK channels (*see* **Note 27g**).

25. Light requirements vary depending on whether the PORTL being used contains the standard (i.e. $BGAG_n$) or red-shifted (i.e. $BGAG_{n,460}$) azobenzene core. Data on intensity and time requirements of light have been characterized in HEK 293T cells for $BGAG_{12}$ and $BGAG_{12,460}$ on SNAP-mGluR2 [44]. Similar results are expected for any PORTL with either a standard or red-shifted azobenzene core. $BGAG_{12}$ and $BGAG_{12,460}$ vary in their photophysical profile and, thus, require very different light exposures to drive and reverse photoisomerization. $BGAG_n$ is bistable, achieving a maximal photostationary state under ~380 nm light (*cis* state) or >480 nm light (*trans* state), without reverting to *trans* spontaneously in the dark [11]. $BGAG_{n,460}$ photoisomerizes to the *cis* state under ~460 nm illumination (*see* Ref. 55 for full spectral profile of a related azobenzene PTL), but this red-shift results in a loss of bistability so that red-shifted photoswitches revert to *trans* spontaneously in the dark with a time constant of ~700 ms [55]. The bistability of standard azobenzenes, and the subsequent lack of relaxation from *cis* to *trans* on the time scale of receptor activation, allows $BGAG_{12}$ to require much lower light intensities than $BGAG_{12,460}$.

$BGAG_{12}$ produces near maximal photocurrents with light intensity levels as low as ~10 mW/cm^2, while $BGAG_{12,460}$ requires ~1000 mW/cm^2 for maximal photocurrents [44]. Light intensities should be measured post-objective at the sample's location with a power meter (Ex. ThorLabs PM100D) and should be monitored over time. For standard azobenzene cores, decreased light intensity does not decrease photoswitching efficiency but only alters the time it takes to reach the steady state with less intense light producing slower currents. With typical light intensities of ~1000 mW/cm^2, ~200–300 ms of 380 nm illumination are sufficient for maximal photoswitching amplitude. For red-shifted azobenzene cores, the intensity of light determines both the amplitude and speed of the photoactivation. For SNAP-mGluR2 + $BGAG_{12,460}$ in HEK 293T cells with GIRK as a readout, ~5 s of 470 nm illumination at 1000 mW/cm^2 are required for a maximal response.

26. Although mGluR2 is resistant to desensitization [56] other GPCRs will need to be characterized for their desensitization kinetics. Typically, photoswitch efficiency is stable form cell to cell due to minimal desensitization typically observed with mGluR2, but this may vary from receptor to receptor.

27. To properly characterize the photoswitching of PORTL/SNAP-GPCR combinations, there are some key controls and further experiments that should be done:
    (a) *Labeling Specificity Control*: An important negative control is to perform the experiment either without application of the PORTL or without transfecting cells with the SNAP- or CLIP-tagged receptor. Photocurrent should be completely dependent on the presence of both the PORTL and the tagged receptor. This condition is especially important in neurons and tissue, where native receptors are present and PORTL washout may be more complicated.
    (b) *Competitive antagonist*: A saturating concentration of a competitive antagonist may be used to block photoactivation of the receptor. For example, $BGAG_{12}$ photoactivation is completely blocked by LY341495, a competitive mGluR antagonist [30]. This blockage confirms that the PORTL ligand is binding at the orthosteric site. Additionally, this experiment can be used to calculate basal activity due to potential *trans* (or *cis* in the case of a *trans*-agonist, i.e. group III mGluRs) state activation of the receptor. This can be observed using electrophysiology by recording a baseline and then adding the antagonist to block basal activity; if there is any basal activity, an

outward current is induced due to the termination of GIRK activation. The measured basal activity can be due to either spontaneous basal activity of the receptor or *trans* activation from the photoswitch. To separate these, basal activity of the receptor with or without the PORTL attached can be measured to determine the contribution of each.

(c) *Action Spectra*: The wavelength-dependence of the PORTL can be described by measuring photoactivation and de-activation at various wavelengths. As previously described using azobenzene-glutamate PTLs [11], maximal photoisomerization of BGAG$_n$ occurs at 380 nm (to *cis* state) and 500 nm (to *trans* state) and deviations from this results in sub-optimal photoswitching. Similar characterization experiments can be performed to determine an action spectra for any PORTL by simply modifying the wavelengths used for activation or de-activation (*see* Ref. 11 or Ref. 55 for examples with PTLs). It is important to use the same bandwidth for each wavelength and to make sure that long enough illumination times are used to reach steady state since sub-optimal wavelengths may take longer to photoactivate. While the general action spectra can be predicted based on PORTL structure, subtle effects may be produced by different ligands or binding to its target [57] and further engineering of the azobenzene core to tweak photophysical properties is an active area of research [58, 59].

(d) *Bistability*: As described above, PORTLs with a standard azobenzene core (Ex. BGAG$_n$) will display bistability. Following photoisomerization to the *cis* state they will maintain the *cis* configuration in the dark with a time constant that is typically >20 min [11]. Exposure to a second wavelength of light (i.e. ~500 nm) is required to drive its relaxation to the *trans* state. In contrast, PORTLs with a red-shifted azobenzene core (ex. BGAG$_{n,460}$) spontaneously relax in the dark to the *trans* state. Thus, constant illumination is required to drive such PORTLs. To determine the stability profile of a particular PORTL, shine the light expected to drive photoisomerization to the *cis* state for a short amount of time (typically <5 s to produce maximal population of the *cis* state) and return to the dark; after reaching a stable level of activation, shine light to drive photoisomerization to the *trans* state. Repeat photoactivation but without returning to dark between the two light exposures. Compare these two results; if a compound is bistable throughout the time of the recording, the results should be identical. If a

compound is not bistable, there will be a full or partial return to baseline conditions while the sample is in the dark. Varying the dark time will allow for the relaxation rate to be estimated. Crucially, following full photoactivation with 380 nm light and incubation in the dark, a second pulse of 380 nm can be used to see if significant relaxation back to the *trans* state has occurred. If relaxation has occurred, the second 380 nm pulse should increase the inward current. For $BGAG_n$ PORTLs, following complete photoactivation at least a minute or two in the dark is required before any response to 380 nm can be observed.

(e) *Effective concentration*: The effective local concentration of the PORTL ligand can be probed using low-affinity receptor mutants. Depending on the effective ligand concentration, lowering the affinity of the receptor will either reduce photoswitch efficiency or have no effect. For example, the low-affinity R57A mGluR2 mutation, which reduces glutamate affinity by ~30×, abolishes photoactivation of SNAG-mGluR2 [31]. However, with $BGAG_0$, but not $BGAG_{12}$, some photocurrent is rescued in the presence of glutamate, indicating that the glutamate moiety of $BGAG_0$ is in a higher effective concentration than in $BGAG_{12}$ and that there is intersubunit cooperativity in mGluR2. In contrast, the same mutation in LimGluR2 (mGluR2-300C, R57A + D-MAG-0) does not alter photoswitch efficiency [26], indicating that PTLs are in a much higher local concentration than PORTLs. Comparing photoswitch efficiency to a dose–response curve of a soluble compound that models the ligand moiety of the PORTL, as was done for LiGluR with different PTLs [11], can allow an estimate of the actual effective concentration. Based on such experiments it is generally estimated that azobenzene-glutamate PTLs are in a >10 mM local concentration while PORTLs are in the hundreds of μM to single digit mM range.

(f) *Partial versus Full Agonism*: an important aspect of PORTL characterization is to gain a full understanding of the basis of the efficiency of the photoswitching. In even the best cases, PORTL photoswitching is well below 100% relative to saturating concentrations of the native ligand. For SNAP-mGluR2 with $BGAG_n$ this incomplete photoswitching is largely explained by incomplete labeling, incomplete *cis* occupancy in the photostationary state and receptor cooperativity (*see* discussion in Ref. 31). However, one alternative possibility is that the PORTL ligand can serve as a partial agonist. To test this possibility,

use the protocol described in this chapter to set up your experimental setup. Use the perfusion system to apply a saturating amount of a full agonist (be sure to not be oversaturating) and once the response has stabilized, use light to photoactivate the receptor with the PORTL ligand. The competition between the full agonist and the PORTL ligand will result in some LBDs being bound by the PORTL ligand instead of the agonist. If the PORTL ligand is a partial agonist of sufficient effective concentration, this will result in a decrease in readout response (i.e. apparent antagonism). Generally, photoswitching in the presence of a wide-range of agonist concentrations may be a valuable way to characterize the system and the agonism profile of the PORTL ligand.

(g) *Channel Block*: To confirm that the observed photocurrent is due to GIRK channel opening, either barium [60] or tertiapin-q [61] blockers may be applied. Typically, one round of photoactivation is performed, then the drug is applied and optical stimulation should be occluded. Following washout, the photocurrent should return. For alternative ion channel readouts (*see* **Note 5**), different blockers may be used to confirm the basis of the light response.

(h) *Light Timing*: Playing with the timing of illumination can be a valuable way to learn both about the PORTL system and the GPCR itself, especially in the context of comparing related receptors. Optical control with PORTLs allows the ON and OFF kinetics, frequency-dependence and minimal activation and de-activation times to be tested. For ion channel-based readouts of GPCRs it is important to keep in mind that the photocurrent kinetics are not a direct readout of receptor activity but are also dependent on a number of downstream processes, such as G protein activation, GTP exchange, and G protein-effector binding.

(i) *PORTL Multiplexing*: The labeling specificity of SNAP and CLIP tags for O^6-benzylguanine and O^2-benzylcytosine, respectively, allows for separate control of SNAP- and CLIP-tagged receptors. It is critical for such experiments that the two PORTLs used have spectral differences between them to allow for separate control. $BGAG_{12}$, $BGAG_{12,460}$, and the related $BCAG_{12}$ and $BCAG_{12,460}$ have previously been shown to have the necessary spectral differences due to the red-shifted azobenzene in the 460 variants [31]. This allows for co-expression of a SNAP-tagged receptor and a CLIP-tagged receptor, labeling each with a BG-based PORTL

or a BC-based PORTL, respectively, each controlled by different light exposures. *See* Fig. 8 in ref. 31 for an example. Light intensities and wavelengths must be determined empirically for a give set up, but in the previously reported experiments low intensity 380 and 590 nm light was used to activate and de-activate standard azobenzene core photoswitches (i.e. $BGAG_n$ or $BCAG_n$) without co-activating red-shifted azobenzene core photoswitches (i.e. $BGAG_{n,460}$ or $BCAG_{n,460}$).

(j) *Agonist Titration*: Dose–response curves for the native ligand are a useful test to see if either SNAP (or other protein tags) attachment or PORTL conjugation alter receptor function. Simply adding in and washing out a given concentration of a native ligand (ex. glutamate) and quantifying the ensuing current amplitude relative to a saturating concentration of the same ligand allows a dose–response curve to be constructed. For a new PORTL/GPCR combination, dose–response curves should be produced for the wild-type receptor, the SNAP-tagged receptor, and the PORTL-labeled receptor (in the inactive state of the PORTL). Any shift induced by just the SNAP tag may motivate further modification to the construct and any shift induced by the PORTL attachment indicates some effect of the less active form of the ligand. For instance, it is easy to imagine that a PORTL may serve as an agonist in the *cis* state but as an antagonist in the *trans* state (or vice-versa) which would produce a right-shifted dose–response curve.

28. In addition to sensors for calcium, other fluorescent sensors can also be used to observe GPCR activity. Both $G_{i/o}$ and G_s result in a change in cAMP concentration, making a cAMP sensor [62–64] a potential optical readout for these receptors. G_q activation can also be sensed using PH domain-based phosphatidylinositol sensors, which can detect the phospholipase C-mediated hydrolysis of phosphatidylinositol 4,5-bisphosphate (PIP_2) into diacylglycerol (DAG) and inositol 1,4,5-trisphosphate (IP_3) [65]. Other pathways for which optical sensors exist have also been implicated in GPCR-induced G protein and ß-arrestin-mediated signaling, including ERK [66, 67], mTOR [68], and CaMKII [69]. These sensors may be ideal in native contexts (i.e. cultured neurons or glia) to dissect the spatial and temporal signaling patterns induced by subcellular activation of GPCRs with PORTLs. *See* **Note 29** for considerations when designing dual photoswitching/imaging experiments.

29. Similar to the important considerations needed when choosing a fluorescent reporter, it is crucial to account for the effects of

imaging light on the *cis–trans* isomerization of PORTLs. In this note we focus on imaging while using a photoswitch that reverts to the *trans* state upon application of visible light (>480 nm), such as $BGAG_n$. When trying to photoactivate *cis*-active PORTLs while imaging, the major concern is that the imaging light will revert the photoswitch to *trans*, preventing sufficient population of the active *cis* state. To circumvent this, there are a few options including spatially separating the imaging and photoactivation illumination areas, which is possible in scanning confocal microscopy, or lowering the frequency and intensity of imaging light to allow the photoswitching light to dominate. For instance, in wide-field epifluorescence imaging, G_q-coupled calcium signals can often easily be imaged at frequencies less than 1 Hz with low illumination intensities and short exposures. For example, 50 ms exposure at 1 Hz can be used to get a baseline (with the PORTL in *trans*) and photoactivation can be produced by constant illumination at ~380 nm. The constant UV should be able to maintain the PORTL in *cis* despite the competing effects of the imaging light. Alternatively, to minimize photon exposure a sufficient ~380 nm pulse can be given just after the imaging light is applied to ensure that the *cis* state is maintained for the entire period in the dark until the next image is captured. It is ideal to use a sensor, such as GCaMP6, that requires low levels of excitation light and gives a large $\Delta F/F$ response upon activation. Red-shifted sensors which are excited at wavelengths longer than 580 nm should be used when available since standard azobenzene cores will only weakly absorb in this range. It is important to confirm that the optical sensor being used is not sensitive to photostimulation light, as has been reported for fluorescent protein-based sensors [70], which can introduce artifacts that confound interpretation of experiments. Finally, red-shifted BGAG variants (i.e. $BGAG_{12,460}$) require additional considerations given the broad spectra of this azobenzene [55] which makes it difficult to image without directly isomerizing the PORTL unless very dim or red-shifted light is used.

Acknowledgements

We thank the Levitz and Broichhagen labs for useful discussion, and all previous and current members of the Isacoff and Trauner labs for their contributions to the development and application of photoswitchable ligands and receptors. J.B. thanks Kai Johnsson for constant support. J.L. is supported by an R35 grant from the National Institute of General Medical Science (1 R35 GM124731).

References

1. Lagerstrom MC, Schioth HB (2008) Structural diversity of G protein-coupled receptors and significance for drug discovery. Nat Rev Drug Discov 7(4):339–357
2. Hilger D, Masureel M, Kobilka BK (2018) Structure and dynamics of GPCR signaling complexes. Nat Struct Mol Biol 25(1):4–12
3. Niswender CM, Conn PJ (2010) Metabotropic glutamate receptors: physiology, pharmacology, and disease. Annu Rev Pharmacol Toxicol 50:295–322
4. Pin JP, Bettler B (2016) Organization and functions of mGlu and GABAB receptor complexes. Nature 540(7631):60–68
5. Riccardi D, Kemp PJ (2012) The calcium-sensing receptor beyond extracellular calcium homeostasis: conception, development, adult physiology, and disease. Annu Rev Physiol 74:271–297
6. Clemmensen C, Smajilovic S, Wellendorph P, Bräuner-Osborne H (2014) The GPCR, class C, group 6, subtype A (GPRC6A) receptor: from cloning to physiological function. Br J Pharmacol 171(5):1129–1141
7. Sutton LP, Orlandi C, Song C, Oh WC, Muntean BS, Xie K, Filippini A, Xie X, Satterfield R, Yaeger JDW, Renner KJ, Young SM Jr, Xu B, Kwon H, Martemyanov KA (2018) Orphan receptor GPR158 controls stress-induced depression. Elife 7. https://doi.org/10.7554/eLife.33273
8. Spangler SM, Bruchas MR (2017) Optogenetic approaches for dissecting neuromodulation and GPCR signaling in neural circuits. Curr Opin Pharmacol 32:56–70
9. Callaway EM, Katz LC (1993) Photostimulation using caged glutamate reveals functional circuitry in living brain slices. Proc Natl Acad Sci U S A 90(16):7661–7665
10. Reiner A, Levitz J, Isacoff EY (2015) Controlling ionotropic and metabotropic glutamate receptors with light: principles and potential. Curr Opin Pharmacol 20:135–143
11. Gorostiza P, Volgraf M, Numano R, Szobota S, Trauner D, Isacoff EY (2007) Mechanisms of photoswitch conjugation and light activation of an ionotropic glutamate receptor. Proc Natl Acad Sci U S A 104:10865–10870
12. Roth BL (2016) DREADDs for neuroscientists. Neuron 89(4):683–694
13. Armbruster BN, Li X, Pausch MH, Herlitze S, Roth BL (2007) Evolving the lock to fit the key to create a family of G protein-coupled receptors potently activated by an inert ligand. Proc Natl Acad Sci U S A 104(12):5163–5168
14. Gomez JL, Bonaventura J, Lesniak W, Mathews WB, Sysa-Shah P, Rodriguez LA, Ellis RJ, Richie CT, Harvey BK, Dannals RF, Pomper MG, Bonci A, Michaelides M (2017) Chemogenetics revealed: DREADD occupancy and activation via converted clozapine. Science 357(6350):503–507
15. Rost BR, Schneider-Warme F, Schmitz D, Hegemann P (2017) Optogenetic tools for subcellular applications in neuroscience. Neuron 96(3):572–603
16. Morri M, Sanchez-Romero I, Tichy AM, Kainrath S, Gerrard EJ, Hirschfeld PP, Schwarz J, Janovjak H (2018) Optical functionalization of human Class A orphan G-protein-coupled receptors. Nat Commun 9(1):1950
17. Airan RD, Thompson KR, Fenno LE, Bernstein H, Deisseroth K (2009) Temporally precise in vivo control of intracellular signalling. Nature 458(7241):1025–1029
18. Siuda ER, Copits BA, Schmidt MJ, Baird MA, Al-Hasani R, Planer WJ, Funderburk SC, McCall JG, Gereau RWt, Bruchas MR (2015) Spatiotemporal control of opioid signaling and behavior. Neuron 86(4):923–935
19. Oh E, Maejima T, Liu C, Deneris E, Herlitze S (2010) Substitution of 5-HT1A receptor signaling by a light-activated G protein-coupled receptor. J Biol Chem 285(40):30825–30836
20. Levitz J, Pantoja C, Gaub B, Janovjak H, Reiner A, Hoagland A, Schoppik D, Kane B, Stawski P, Schier AF, Trauner D, Isacoff EY (2013) Optical control of metabotropic glutamate receptors. Nat Neurosci 16(4):507–516
21. Masseck OA, Spoida K, Dalkara D, Maejima T, Rubelowski JM, Wallhorn L, Deneris ES, Herlitze S (2014) Vertebrate cone opsins enable sustained and highly sensitive rapid control of Gi/o signaling in anxiety circuitry. Neuron 81(6):1263–1273
22. Kramer RH, Mourot A, Adesnik H (2013) Optogenetic pharmacology for control of native neuronal signaling proteins. Nat Neurosci 16(7):816–823
23. Banghart M, Borges K, Isacoff E, Trauner D, Kramer RH (2004) Light-activated ion channels for remote control of neuronal firing. Nat Neurosci 7(12):1381–1386
24. Volgraf M, Gorostiza P, Numano R, Kramer RH, Isacoff EY, Trauner D (2006) Allosteric control of an ionotropic glutamate receptor with an optical switch. Nat Chem Biol 2(1):47–52
25. Levitz J, Popescu AT, Reiner A, Isacoff EY (2016) A toolkit for orthogonal and in vivo

optical manipulation of ionotropic glutamate receptors. Front Mol Neurosci 9:2

26. Levitz J, Habrian C, Bharill S, Fu Z, Vafabakhsh R, Isacoff EY (2016) Mechanism of assembly and cooperativity of homomeric and heteromeric metabotropic glutamate receptors. Neuron 92(1):143–159
27. Carroll EC, Berlin S, Levitz J, Kienzler MA, Yuan Z, Madsen D, Larsen DS, Isacoff EY (2015) Two-photon brightness of azobenzene photoswitches designed for glutamate receptor optogenetics. Proc Natl Acad Sci U S A 112(7): E776–E785
28. Li D, Herault K, Zylbersztejn K, Lauterbach MA, Guillon M, Oheim M, Ropert N (2015) Astrocyte VAMP3 vesicles undergo Ca2+-independent cycling and modulate glutamate transporter trafficking. J Physiol 593 (13):2807–2832
29. DuBay KH, Iwan K, Osorio-Planes L, Geissler PL, Groll M, Trauner D, Broichhagen J (2018) A predictive approach for the optical control of carbonic anhydrase II activity. ACS Chem Biol 13(3):793–800
30. Broichhagen J, Damijonaitis A, Levitz J, Sokol KR, Leippe P, Konrad D, Isacoff EY, Trauner D (2015) Orthogonal optical control of a G protein-coupled receptor with a SNAP-tethered photochromic ligand. ACS Cent Sci 1(7):383–393
31. Levitz J, Broichhagen J, Leippe P, Konrad D, Trauner D, Isacoff EY (2017) Dual optical control and mechanistic insights into photoswitchable group II and III metabotropic glutamate receptors. Proc Natl Acad Sci U S A 114 (17):e3546–e3554
32. Gautier A, Juillerat A, Heinis C, Correa IR Jr, Kindermann M, Beaufils F, Johnsson K (2008) An engineered protein tag for multiprotein labeling in living cells. Chem Biol 15 (2):128–136
33. Schonberger M, Trauner D (2014) A photochromic agonist for mu-opioid receptors. Angew Chem Int Ed Engl 53(12):3264–3267
34. Donthamsetti PC, Winter N, Schonberger M, Levitz J, Stanley C, Javitch JA, Isacoff EY, Trauner D (2017) Optical control of dopamine receptors using a photoswitchable tethered inverse agonist. J Am Chem Soc 139 (51):18522–18535
35. Broichhagen J, Podewin T, Meyer-Berg H, von Ohlen Y, Johnston NR, Jones BJ, Bloom SR, Rutter GA, Hoffmann-Roder A, Hodson DJ, Trauner D (2015) Optical control of insulin secretion using an incretin switch. Angew Chem Int Ed Engl 54(51):15565–15569
36. Broichhagen J, Johnston NR, von Ohlen Y, Meyer-Berg H, Jones BJ, Bloom SR, Rutter GA, Trauner D, Hodson DJ (2016) Allosteric optical control of a class B G-protein-coupled receptor. Angew Chem Int Ed Engl 55 (19):5865–5868
37. Agnetta L, Kauk M, Canizal MCA, Messerer R, Holzgrabe U, Hoffmann C, Decker M (2017) A photoswitchable dualsteric ligand controlling receptor efficacy. Angew Chem Int Ed Engl 56(25):7282–7287
38. Westphal MV, Schafroth MA, Sarott RC, Imhof MA, Bold CP, Leippe P, Dhopeshwarkar A, Grandner JM, Katritch V, Mackie K, Trauner D, Carreira EM, Frank JA (2017) Synthesis of photoswitchable delta(9)-tetrahydrocannabinol derivatives enables optical control of cannabinoid receptor 1 signaling. J Am Chem Soc 139(50):18206–18212
39. Hauwert NJ, Mocking TAM, Da Costa Pereira D, Kooistra AJ, Wijnen LM, Vreeker GCM, Verweij EWE, De Boer AH, Smit MJ, De Graaf C, Vischer HF, de Esch IJP, Wijtmans M, Leurs R (2018) Synthesis and characterization of a bidirectional photoswitchable antagonist toolbox for real-time GPCR photopharmacology. J Am Chem Soc 140 (12):4232–4243
40. Podewin T, Ast J, Broichhagen J, Fine NHF, Nasteska D, Leippe P, Gailer M, Buenaventura T, Kanda N, Jones BJ, M'Kadmi C, Baneres JL, Marie J, Tomas A, Trauner D, Hoffmann-Roder A, Hodson DJ (2018) Conditional and reversible activation of class A and B G protein-coupled receptors using tethered pharmacology. ACS Cent Sci 4 (2):166–179
41. Shields BC, Kahuno E, Kim C, Apostolides PF, Brown J, Lindo S, Mensh BD, Dudman JT, Lavis LD, Tadross MR (2017) Deconstructing behavioral neuropharmacology with cellular specificity. Science 356(6333). https://doi.org/10.1126/science.aaj2161
42. Nishiyama J, Mikuni T, Yasuda R (2017) Virus-mediated genome editing via homology-directed repair in mitotic and post-mitotic cells in mammalian brain. Neuron 96 (4):755–768.e5
43. Farrants H, Acosta-Ruiz A, Gutzeit VA, Trauner D, Johnsson K, Levitz J, Broichhagen J (2018) SNAP-tagged nanobodies enable reversible optical control of a G protein-coupled receptor via a remotely tethered photoswitchable ligand. ACS Chem Biol 13 (9):2682–2688
44. Berry MH, Holt A, Levitz J, Broichhagen J, Gaub BM, Visel M, Stanley C, Aghi K, Kim YJ, Cao K, Kramer RH, Trauner D, Flannery J,

Isacoff EY (2017) Restoration of patterned vision with an engineered photoactivatable G protein-coupled receptor. Nat Commun 8 (1):1862

45. Atwood BK, Lopez J, Wager-Miller J, Mackie K, Straiker A (2011) Expression of G protein-coupled receptors and related proteins in HEK293, AtT20, BV2, and N18 cell lines as revealed by microarray analysis. BMC Genomics 12:14
46. Doumazane E, Scholler P, Zwier JM, Trinquet E, Rondard P, Pin JP (2011) A new approach to analyze cell surface protein complexes reveals specific heterodimeric metabotropic glutamate receptors. FASEB J 25 (1):66–77
47. Conklin BR, Farfel Z, Lustig KD, Julius D, Bourne HR (1993) Substitution of three amino acids switches receptor specificity of Gq alpha to that of Gi alpha. Nature 363 (6426):274–276
48. Chen TW, Wardill TJ, Sun Y, Pulver SR, Renninger SL, Baohan A, Schreiter ER, Kerr RA, Orger MB, Jayaraman V, Looger LL, Svoboda K, Kim DS (2013) Ultrasensitive fluorescent proteins for imaging neuronal activity. Nature 499(7458):295–300
49. Deo C, Lavis LD (2018) Synthetic and genetically encoded fluorescent neural activity indicators. Curr Opin Neurobiol 50:101–108
50. Li X, Gutierrez DV, Hanson MG, Han J, Mark MD, Chiel H, Hegemann P, Landmesser LT, Herlitze S (2005) Fast noninvasive activation and inhibition of neural and network activity by vertebrate rhodopsin and green algae channelrhodopsin. Proc Natl Acad Sci U S A 102 (49):17816–17821
51. Sandoz G, Levitz J, Kramer RH, Isacoff EY (2012) Optical control of endogenous proteins with a photoswitchable conditional subunit reveals a role for TREK1 in GABA (B) signaling. Neuron 74(6):1005–1014
52. Hofmann T, Obukhov AG, Schaefer M, Harteneck C, Gudermann T, Schultz G (1999) Direct activation of human TRPC6 and TRPC3 channels by diacylglycerol. Nature 397(6716):259–263
53. Peng Y, Xiong WC, Mei L (2013) Culture of dissociated hippocampal neurons. Methods Mol Biol 1018:39–47
54. Lüscher C, Slesinger PA (2010) Emerging concepts for G protein-gated inwardly rectifying potassium (GIRK) channels in health and disease. Nat Rev Neurosci 11(5):301–315
55. Kienzler MA, Reiner A, Trautman E, Yoo S, Trauner D, Isacoff EY (2013) A red-shifted, fast-relaxing azobenzene photoswitch for visible light control of an ionotropic glutamate receptor. J Am Chem Soc 135 (47):17683–17686
56. Iacovelli L, Molinaro G, Battaglia G, Motolese M, Di Menna L, Alfiero M, Blahos J, Matrisciano F, Corsi M, Corti C, Bruno V, De Blasi A, Nicoletti F (2009) Regulation of group II metabotropic glutamate receptors by G protein-coupled receptor kinases: mGlu2 receptors are resistant to homologous desensitization. Mol Pharmacol 75(4):991–1003
57. Barber DM, Liu SA, Gottschling K, Sumser M, Hollmann M, Trauner D (2017) Optical control of AMPA receptors using a photoswitchable quinoxaline-2,3-dione antagonist. Chem Sci 8(1):611–615
58. Beharry AA, Woolley GA (2011) Azobenzene photoswitches for biomolecules. Chem Soc Rev 40(8):4422–4437
59. Dong M, Babalhavaeji A, Samanta S, Beharry AA, Woolley GA (2015) Red-shifting azobenzene photoswitches for in vivo use. Acc Chem Res 48(10):2662–2670
60. Alagem N, Dvir M, Reuveny E (2001) Mechanism of Ba2+ block of a mouse inwardly rectifying K+ channel: differential contribution by two discrete residues. J Physiol 534 (Pt 2):381–393
61. Jin W, Lu Z (1998) A novel high-affinity inhibitor for inward-rectifier K+ channels. Biochemistry 37(38):13291–13299
62. Hackley CR, Mazzoni EO, Blau J (2018) cAMPr: a single-wavelength fluorescent sensor for cyclic AMP. Sci Signal 11(520). https://doi.org/10.1126/scisignal.aah3738
63. Klarenbeek J, Goedhart J, van Batenburg A, Groenewald D, Jalink K (2015) Fourth-generation epac-based FRET sensors for cAMP feature exceptional brightness, photostability and dynamic range: characterization of dedicated sensors for FLIM, for ratiometry and with high affinity. PLoS One 10(4): e0122513
64. Harada K, Ito M, Wang X, Tanaka M, Wongso D, Konno A, Hirai H, Hirase H, Tsuboi T, Kitaguchi T (2017) Red fluorescent protein-based cAMP indicator applicable to optogenetics and in vivo imaging. Sci Rep 7 (1):7351
65. Stauffer TP, Ahn S, Meyer T (1998) Receptor-induced transient reduction in plasma

membrane PtdIns(4,5)P2 concentration monitored in living cells. Curr Biol 8(6):343–346

66. Harvey CD, Ehrhardt AG, Cellurale C, Zhong H, Yasuda R, Davis RJ, Svoboda K (2008) A genetically encoded fluorescent sensor of ERK activity. Proc Natl Acad Sci U S A 105(49):19264–19269
67. de la Cova C, Townley R, Regot S, Greenwald I (2017) A real-time biosensor for ERK activity reveals signaling dynamics during C. elegans cell fate specification. Dev Cell 42 (5):542–553.e4
68. Zhou X, Clister TL, Lowry PR, Seldin MM, Wong GW, Zhang J (2015) Dynamic visualization of mTORC1 activity in living cells. Cell Rep. https://doi.org/10.1016/j.celrep.2015.02.031
69. Lee SJR, Escobedo-Lozoya Y, Szatmari EM, Yasuda R (2009) Activation of CaMKII in single dendritic spines during long-term potentiation. Nature 458(7236):299–304
70. Ai M, Mills H, Kanai M, Lai J, Deng J, Schreiter E, Looger L, Neubert T, Suh G (2015) Green-to-red photoconversion of GCaMP. PLoS One 10(9):e0138127

Chapter 7

Chemoselective Acylation of Hydrazinopeptides to Access Fluorescent Probes for Time-Resolved FRET Assays on GPCRs

Sridévi M. Ramanoudjame, Lucie Esteoulle, Stéphanie Riché, Jean-François Margathe, Thierry Durroux, Iuliia A. Karpenko, and Dominique Bonnet

Abstract

Fluorescence techniques represent a powerful tool to investigate dynamic and functional architecture of GPCRs. Thus, fluorescent GPCR ligands have found various applications in cellular imaging, in the development of binding assays as replacements for radioligands in the study of ligand–receptor but also in receptor–receptor interactions at the cell surface or in native tissues. To extend the applicability of these techniques, the design and the synthesis of fluorescent probes are critical steps. As there are numerous peptide receptors in the GPCR family, fluorescent peptide-based probes are of importance. Herein, we present a convenient method to facilitate the solution-phase fluorescent labeling of peptides which is based on the chemoselective acylation of α-hydrazinopeptides. This approach combines the advantages to use commercially available amine-reactive dyes and very mild conditions that are fully compatible with the chemical sensitivity of the dyes. It gives a rapid access to fluorescent peptidic probes compatible with the time-resolved fluorescence resonance energy transfer (TR-FRET) techniques.

Key words Chemical ligation, Chemoselective acylation, Hydrazinopeptides, Fluorescent probes, GPCR, TR-FRET

1 Introduction

G protein-coupled receptors (GPCRs), the largest family of cell-surface membrane proteins encoded by the human genome, represent the target of about 30% of prescription drugs on the market [1]. These receptors are responsible for a wide variety of signaling responses in diverse cell types. Despite the recent breakthrough in structural biology [2], fluorescence techniques remain a powerful tool to investigate dynamic and functional architecture of GPCRs

---

Sridévi M. Ramanoudjame and Lucie Esteoulle contributed equally to this work.

Mario Tiberi (ed.), *G Protein-Coupled Receptor Signaling: Methods and Protocols*, Methods in Molecular Biology, vol. 1947, https://doi.org/10.1007/978-1-4939-9121-1_7,

[3]. Thus, fluorescent GPCR ligands have found various applications in cellular imaging [4, 5], in the development of binding assays as replacements for radioligands [6] in the study of ligand-receptor but also in receptor-receptor interactions at the cell surface or in native tissues [7, 8]. To extend the applicability of these techniques, the design and the synthesis of fluorescent probes is a critical step. As there are numerous peptide receptors in the GPCR family, fluorescent peptide-based probes are of importance [3].

The attachment of a fluorophore to the peptides is generally performed through a spacer to limit the impact of the bulky dyes on the pharmacological properties of the native ligands. The dye can be attached to the peptide on solid phase following a classical Fmoc/*t*-butyl strategy by a simple condensation reaction or by chemical ligation [9, 10]. However, these approaches are not compatible with the incorporation of acid-sensitive and/or expensive fluorophores. In these cases, the labeling can be performed in solution on a fully deprotected peptide by amine-reactive dyes [11]. Widely used for the fluorescent labeling of peptides and proteins, this approach suffers from a lack of selectivity when several nucleophilic amino groups are present on the peptide. To overcome this limitation, bioorthogonal techniques have been developed, among which the oxime and the hydrazone/semicarbazone chemical ligation, the maleimide-based conjugation, or the "click" chemistry [12, 13].

1.1 Principle of the Chemoselective Acylation Method

To facilitate the labeling of GPCR ligands derived from peptides [14], we have developed a convenient and mild method for the solution-phase labeling of peptides by chemoselective acylation of fully deprotected α-hydrazinopeptides [15]. This approach displays the major advantage over the existing biorthogonal approaches to use commercially available dyes activated by succinimidyl ester groups (Fig. 1).

The acylation has been developed to allow the chemoselective introduction of dyes into peptides owning in their sequence other nucleophiles such as the ε-amino group of lysine. Thereby, the reaction carried out at pH 5.2 enables the protection of the ε-amino group ($pK_a \sim 10$) through protonation and the discrimination between the ε-amino and the α-hydrazino ($pK_a \sim 6$) groups. The acylation of the ε-amino group is thus very limited at pH 5.2 whereas it greatly increases at pH 6.5 and 8.0 leading to a mixture of monoacylpeptide and diacylpeptide (Fig. 2) [15].

1.2 Synthesis of α-Hydrazinopeptides

Hydrazinopeptides are synthesized following a solid-phase approach. Peptide sequences are first elaborated from a polystyrene resin following a Fmoc/*t*-butyl strategy either manually or by using a peptide synthesizer (Fig. 3). After peptide chain elongation, the obtained resin is *N*-acylated with tri-Boc-hydrazinoacetic acid ($(Boc)_3$aza-Gly) [16] either directly (Method A) or through an aminoundecanoic acyl spacer (Method B). Deprotection and

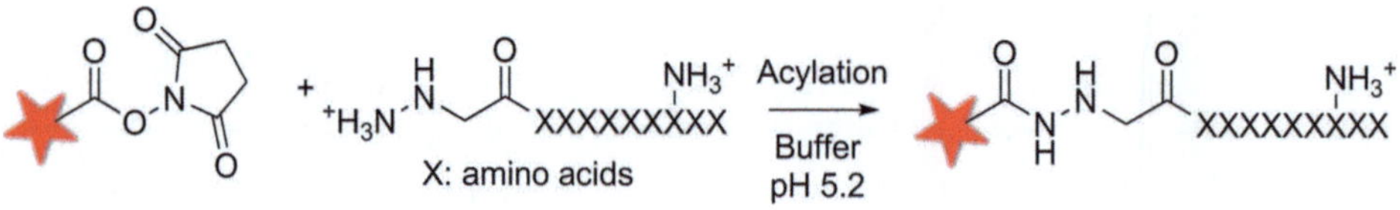

Fig. 1 General strategy for the synthesis of fluorescent probes by chemoselective acylation of α-hydrazinopeptides

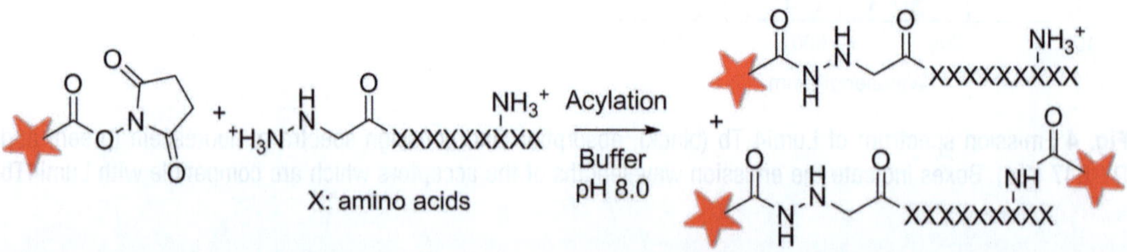

Fig. 2 Acylation performed at pH 8.0 in the presence of a dye activated by succinimidyl ester group leading to a mixture of mono and diacylpeptides

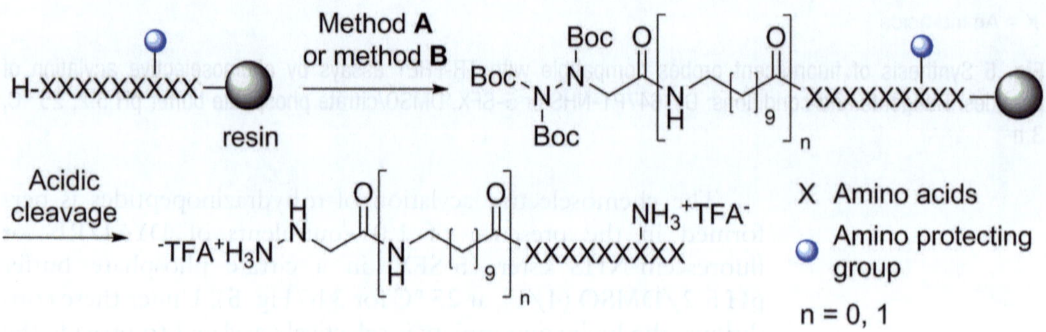

Fig. 3 Solid-phase synthesis approach for the synthesis of α-hydrazinopeptides

cleavage from solid phase under acidic conditions lead to *N*-terminal acetylated hydrazinopeptides which are purified by RP-HPLC. The isolated hydrazinopeptides are then freeze-dried. Their purity and identity are determined by analytical RP-HPLC and ESI-MS analysis.

1.3 Labeling of α-Hydrazinopeptides with Dyes Compatible with TR-FRET

TR-FRET, utilizing rare-earth lanthanides with long emission half-lives as donor fluorophores, combines standard FRET with the time-resolved fluorescence detection. Terbium cryptate (Lumi4-Tb) is of particular interest, thanks to its FRET compatibility with green (fluorescein) and red (DY647) fluorescent dyes, which can therefore be used as acceptors (Fig. 4). The high sensitivity of TR-FRET, associated with efficient fluorescent chemical probes, provides an excellent opportunity to investigate ligand–protein and protein–protein interactions.

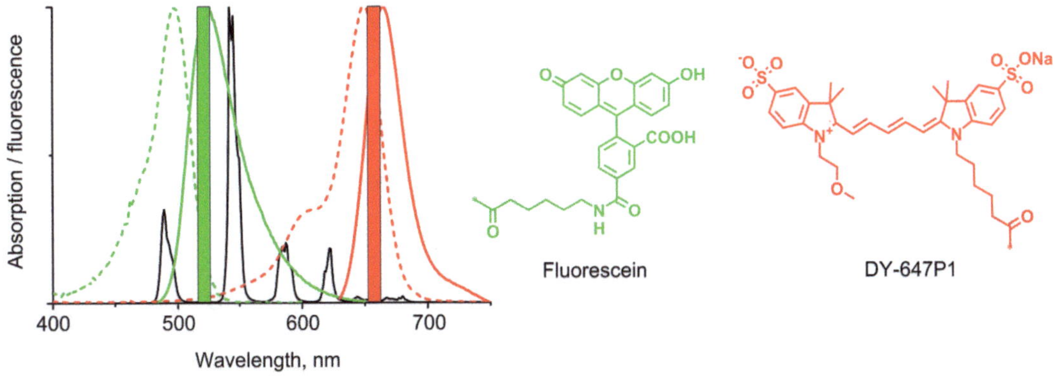

Fig. 4 Emission spectrum of Lumi4 Tb (black), absorption and emission spectra of fluorescein (green) and DY-647 (red). Boxes indicate the emission wavelengths of the acceptors which are compatible with Lumi4Tb

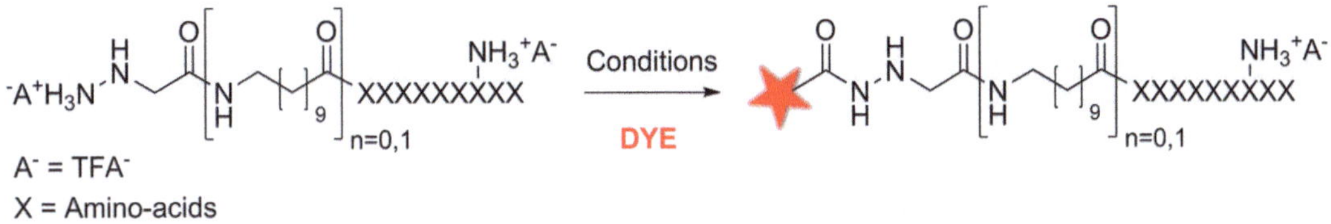

Fig. 5 Synthesis of fluorescent probes compatible with TR-FRET assays by chemoselective acylation of peptides. Reagents and conditions: DY-647P1-NHS or 5-SFX, DMSO/citrate phosphate buffer pH 5.2, 25 °C, 3 h

The chemoselective acylation of α-hydrazinopeptides is performed in the presence of 1.0 equivalents of DY647P1- or fluorescein-NHS ester (5-SFX) in a citrate phosphate buffer pH 5.2/DMSO (4/1), at 25 °C for 3 h (Fig. 5). Under these conditions, the hydrazino moiety is selectively acylated to provide the expected fluorescently labeled peptide, which is further purified by semi-preparative RP-HPLC. Noteworthy, the very mild conditions used for the acylation are fully compatible with the sensitivity of dyes and α-hydrazinopeptides even those which contain *oxidation-sensitive* amino acids such as methionine. Hence, as expected, the reaction carried out at pH 5.2 enables the protection of the ε-amino group of lysine through protonation and the discrimination between the α-amino and the α-hydrazino groups.

2 Materials

2.1 Solid-Phase Synthesis of α-Hydrazinopeptides

1. Peptide synthesis manual reaction vessels and accessories are commercially available from Sigma-Aldrich (Supelco Analytical).
2. Reagents and solvents: *N,N*-dimethylformamide (DMF) anhydrous, dichloromethane (DCM), diethyl ether (Et_2O), piperidine, *N,N*-diisopropylethylamine (DIPEA), *o*-(benzotriazol-1-

yl)-*N*,*N*,*N*′,*N*′-tetramethyluronium hexafluorophosphate (HBTU), 1-hydroxybenzotriazole hydrate (HOBt) and benzotriazole-1-yl-oxy-tris-pyrrolidino-phosphonium hexafluorophosphate (PyBOP).

3. Rink Amide AM or pre-loaded Wang resin (*see* **Note 1**).
4. Fmoc-amino acids.
5. Tri-Boc-hydrazinoacetic acid ((Boc)$_3$aza-Gly) and 11-(Fmoc-amino)undecanoic acid (Fmoc-11-Aun-OH).
6. Solutions for Kaiser test: (a) 1 g of ninhydrin in 20 mL of ethanol, (b) 64 g of phenol in 16 mL of ethanol, and (c) 2 mL of 1 mM water solution of KCN in 98 mL of pyridine. Can be stored from 6 to 9 months at 4 °C.
7. 2,4,6-Trinitrobenzenesulfonic acid (TNBS) and a solution of 10% DIPEA in DMF (v/v). Can be stored from 6 to 9 months at 4 °C.
8. Solutions for Chloranil test: (a) 2% acetaldehyde in DMF (v/v), (b) 2% chloranil (2,3,5,6-tetrachloro-1,4-benzoquinone) in DMF (w/v). Can be stored from 6 to 9 months at −20 °C.
9. 20% piperidine in DMF (v/v). Can be stored for 3 months at room temperature.
10. Cleavage cocktail (*see* **Note 2**): 95% trifluoroacetic acid (TFA), 2.5% triisopropylsilane (TIS) and 2.5% H_2O (v/v) or Reagent K (82.5% TFA (v/v), 5% phenol (w/v), 5% thioanisole (v/v), 2.5% 1,2-ethanedithiol (v/v) and 5% H_2O (v/v)). Should be freshly prepared and can be stored for 24 h at 4 °C.
11. Reverse-phase semi-preparative high-performance liquid chromatography (RP-HPLC) system for α-hydrazinopeptide purification.
12. Freeze-dryer.
13. Liquid chromatography-high resolution mass spectrometry (LC-HRMS) system for purity and identity analysis.

2.2 Chemoselective Acylation of α-Hydrazinopeptides with Fluorescent Probes

1. Purified α-hydrazinopeptides (>95% purity determined by RP-HPLC).
2. DY-647P1-NHS ester (Dyomics), 5-SFX (6-(fluorescein-5-carboxamido)hexanoic acid NHS ester, ThermoFisher Scientific), or any other commercially available dye activated by succinimidyl (NHS) ester group. Store at −20 °C.
3. Citrate phosphate buffer pH 5.2 (*see* **Note 3**).
4. pH-meter.
5. RP-HPLC system for purification.
6. Freeze-dryer.
7. LC-HRMS system for purity and identity analysis.

3 Methods

3.1 Solid-Phase Synthesis of α-Hydrazinopeptides

α-Hydrazino-derivatives of desired peptides are synthesized on solid phase following a classical Fmoc/*t*-butyl peptide synthesis approach. Here, we describe a general protocol for the manual assembly of the desired peptide sequence (**steps 1–10**). For the direct introduction of the α-hydrazino group onto the N-terminus of the peptide, follow Method A. For the introduction of the α-hydrazino group through the aminoundecanoic acyl spacer, follow Method B.

1. Swell the dry resin by shaking it in a peptide synthesis reactor in three-time bed volume of DMF for 30 min.
2. Remove the Fmoc protecting group by treating the resin twice with 20% piperidine in DMF for 15 min.
3. Wash the resin successively with DMF (3×) and DCM (3×).
4. Confirm the removal of the Fmoc group by colorimetric tests: Kaiser test and TNBS test for all the amino acids except proline, Chloranil test for proline (*see* **Note 4**). For Kaiser test, transfer few resin beads to a small glass tube and add one drop of the ninhydrin solution, one drop of the phenol solution, and three drops of KCN in pyridine. Mix and heat well using heat gun. Dark blue resin beads indicate the presence of resin-attached primary amines (*see* **Note 5**). For TNBS test, transfer few resin beads to a small glass tube and add 3 μL of TNBS and 5 μL of 10% DIPEA in DMF. Wait for 5 min. Red or deep-orange resin beads indicate the presence of resin-attached primary amines. For Chloranil test, transfer few resin beads to a small glass tube and add one drop of the acetaldehyde solution and one drop of the chloranil solution. Mix and wait for 10 min. Blue resin beads indicate the presence of resin-attached proline (*see* **Note 6**).
5. Wash the resin with anhydrous DMF.
6. Weight out into a dry glass vial Fmoc-amino acid (4 eq), HBTU (3.8 eq), and HOBt (4 eq). Dissolve in dry DMF (typically 2.5 mL for 0.1 mmol of resin). Add DIPEA (12 eq) and mix thoroughly (*see* **Note** 7).
7. Add the solution to the *N*-deprotected resin and agitate at room temperature for 45 min.
8. Wash the resin successively with DMF (3×) and DCM (3×).
9. Confirm the coupling by Kaiser and TNBS tests (*see* **step 4**).
10. Repeat **steps 2–9** to assemble the desired peptide sequence, then follow Method A or Method B to *N*-acylate the peptide with tri-Boc-hydrazinoacetic acid.

3.1.1 Method A

1. Remove the Fmoc group following **steps 2–5** of Subheading 3.1.
2. Weight out into a dry glass vial tri-Boc-hydrazinoacetic acid (5 eq) and PyBOP (5 eq). Dissolve in dry DMF (typically 2.5 mL for 0.1 mmol of resin). Add the solution to the *N*-deprotected resin. Add DIPEA (12 eq) and mix thoroughly. Agitate at room temperature for 1 h.
3. Wash the resin successively with DMF (3×) and DCM (3×).
4. Confirm the coupling by Kaiser and TNBS tests (*see* Subheading 3.1, **step 4**) (*see* **Note 8**).
5. Wash the resin with Et_2O.
6. Dry the resin under high vacuum for at least 3 h.
7. Add cleavage cocktail (TFA/TIS/water or Reagent K, typically 7.5 mL for 0.1 mmol of resin) and agitate at room temperature for 3 h (*see* **Note 2**).
8. Collect the solution by filtration, wash the resin twice with 0.5 mL of TFA.
9. Add the filtrate dropwise to 8 volumes of cold Et_2O (kept at −20 °C for 1 h). Centrifuge the mixture for 5 min at 1811 × *g* at 4 °C and carefully remove the solvent by decantation. To the residue add the same volume of cold Et_2O, triturate thoroughly, centrifuge for 5 min at 1811 × *g* at 4 °C and carefully remove the solvent (*see* **Note 9**).
10. Dry the residue under high vacuum for at least 1 h to remove the traces of solvents, then solubilize in a small amount of acetonitrile/water 50/50 and freeze-dry.
11. Purify the crude product by semi-preparative RP-HPLC (*see* **Note 10**). Confirm the purity and the identity by LC-HRMS.

3.1.2 Method B

1. Remove the Fmoc group following **steps 2–5** of Subheading 3.1.
2. Weight out into a dry glass vial Fmoc-11-amino-undecanoic acid (5 eq), HBTU (4.95 eq) and HOBt (5 eq). Dissolve in dry DMF (typically 2.5 mL for 0.1 mmol of resin). Add DIPEA (12 eq) and mix thoroughly.
3. Add the solution to the *N*-deprotected resin and agitate at room temperature for 1 h.
4. Wash the resin successively with DMF (3×) and DCM (3×).
5. Confirm the coupling by Kaiser and TNBS tests (*see* Subheading 3.1, **step 4**).
6. Follow Method A (**steps 1–11** of Subheading 3.1.1) to *N*-acylate the peptide with tri-Boc-hydrazinoacetic acid.

3.2 Chemoselective Acylation of α-Hydrazinopeptides with Fluorescent Probes

1. Dissolve the desired fluorophore NHS ester in dry DMSO at 10 mM (*see* **Note 11**).
2. In a 1.5 mL microcentrifuge tube, dissolve the desired amount of α-hydrazinopeptide (1 eq) in citrate phosphate buffer pH 5.2 to get 1.8 mM solution.
3. Add the required amount of the DMSO solution of the fluorophore (1.1 eq) to the solution of α-hydrazinopeptide.
4. Add DMSO to adjust the final ratio citrate phosphate buffer/DMSO to 4/1 (v/v) (*see* **Note 12**).
5. Cover the tube with an aluminum foil to protect from light. Stir or shake the reaction mixture at 25 °C for 3 h. The completion of the reaction may be monitored by analytical RP-HPLC.
6. Purify the crude product by semi-preparative RP-HPLC (*see* **Note 13**). Confirm the purity and the identity by LC-HRMS.

4 Notes

1. Typical loading ranges for resins are between 0.4 and 1 mmol/g. Sometimes, using low loading resins (0.1–0.4 mmol/g) can improve the yields of difficult (long, hydrophobic, or sterically hindered) peptides.
2. Usually peptides are cleaved with a mixture of 95% TFA (v/v), 2.5% TIS (v/v), and 2.5% H_2O (v/v). For cysteine and methionine-containing peptides, we advise to use Reagent K: 82.5% TFA (v/v), 5% phenol (w/v), 5% thioanisole (v/v), 2.5% 1,2-ethanedithiol (v/v), and 5% H_2O (v/v).
3. Dissolve 21.0 g of citric acid monohydrate in 1 L of deionized water to get 0.1 M solution. Dissolve 28.4 g of disodium hydrogen phosphate (Na_2HPO_4) in 1 L of deionized water to get 0.2 M solution. Mix 23.3 mL of 0.1 M solution of citric acid with 26.7 mL of 0.2 M solution of Na_2HPO_4 and add water to a total of 100 mL. Measure pH and adjust if necessary to 5.2 by adding few drops of either 0.1 M solution of citric acid or 0.2 M solution of Na_2HPO_4. Store at 4 °C.
4. Proline being secondary amine does not react in Kaiser test. The deprotection of proline and the coupling of the next amino acid should be monitored with the Chloranil test.
5. Some deprotected amino acids may not show the expected dark blue color typical for resin-attached primary amines. That was reported for serine, asparagine, and aspartic acid (*see* ref. 17). We find that *N*-deprotected Cys(Mmt) gives red color of beads in Kaiser test. We thus recommend always performing both Kaiser and TNBS tests.

6. Primary amines will also react in Chloranil test, however in this case the beads will turn brown-red.
7. Although routinely we use HBTU/HOBt activation, other activating reagents such as HATU or PyBOP (in situ carboxyl activation) can be used to improve the efficiency of difficult sterically hindered couplings (*see* ref. 18).
8. If at least one of the tests turns out positive, we advise to repeat **steps 2–4** of Subheading 3.1.1 to ensure the completion of the coupling.
9. We find that highly hydrophobic peptides have increased solubility in Et_2O and thus their precipitation may not be complete. If the presence of unprecipitated peptide is confirmed in the Et_2O fraction, we recommend to evaporate the solvent, combine all the peptide residues and repeat **step 9**, Subheading 3.1.1, using a mixture of Et_2O/pentane 50/50 (v/v) as solvent.
10. For RP-HPLC purification we use a linear gradient of solvent A (acetonitrile with 0.1% TFA) in solvent B (water with 0.1% TFA), usually from 5% to 60% in 30 min, and a flow rate of 20 mL/min. The injection is performed in a minimum amount of DMSO (usually 0.5 mL). Although it was never observed in our laboratory, it has been reported that some α-hydrazinopeptides may be instable when stored in the lyophilized form following RP-HPLC purification with acetonitrile as the mobile phase (*see* ref. 19). In this case the replacement of acetonitrile by 2-propanol is advised.
11. To avoid the hydrolysis of NHS esters, dry fresh DMSO should be used. We find it useful to dissolve NHS esters in deuterated DMSO (DMSO-*d6*), which can be obtained in single use 0.5 mL glass ampoules from Aldrich. The DMSO solution of NHS esters of dyes can be stored at −20 °C for at least 1 year.
12. In a typical example, we used 0.36 μmol of peptide, which was dissolved in 200 μL of citrate phosphate buffer to get 1.8 mM solution. To the obtained solution, 0.40 μmol (1.1 eq) of NHS ester of the fluorescent dye was added as 40 μL of 10 mM solution. Finally, 10 μL of DMSO were added to adjust the final ratio citrate phosphate buffer/DMSO to 4/1 (v/v).
13. *N*-Acylated α-hydrazinopeptides are purified by RP-HPLC using a linear gradient of solvent A (acetonitrile with 0.1% TFA) in solvent B (water with 0.1% TFA), usually from 10% to 100% in 40 min, and a flow rate of 20 mL/min. The injection is performed in a minimum amount of DMSO (usually 0.5 mL). *N*-Acylated α-hydrazinopeptides were found to be perfectly stable in 0.1% TFA in acetonitrile/water used for the purification and during the long-term storage in lyophilized form at −20 °C.

Acknowledgements

The development of the method was supported by the Fonds Unique Interministériel (Cell2lead program n° F1005035J/ ATFUAA00LB/AAP9), the Centre National de la Recherche Scientifique and the Université de Strasbourg. Sridévi M. Ramanoudjame and Lucie Esteoulle contributed equally to this work.

References

1. Ma Z, Du LP, Li MY (2014) Toward fluorescent probes for G-Protein-Coupled Receptors (GPCRs). J Med Chem 57(20):8187–8203. https://doi.org/10.1021/jm401823z
2. Congreve M, Langmead CJ, Mason JS, Marshall FH (2011) Progress in structure based drug design for G protein-coupled receptors. J Med Chem 54(13):4283–4311. https://doi.org/10.1021/jm200371q
3. Sridharan R, Zuber J, Connelly SM, Mathew E, Dumont ME (2014) Fluorescent approaches for understanding interactions of ligands with G protein coupled receptors. Biochim Biophys Acta 1838(1):15–33. https://doi.org/10.1016/j.bbamern.2013.09.005
4. Karpenko IA, Klymchenko AS, Gioria S, Kreder R, Shulov I, Villa P, Mely Y, Hibert M, Bonnet D (2015) Squaraine as a bright, stable and environment-sensitive far-red label for receptor-specific cellular imaging. Chem Commun 51(14):2960–2963. https://doi.org/10.1039/c4cc09113b
5. Karpenko IA, Margathe JF, Rodriguez T, Pflimlin E, Dupuis E, Hibert M, Durroux T, Bonnet D (2015) Selective nonpeptidic fluorescent ligands for oxytocin receptor: design, synthesis, and application to time-resolved FRET binding assay. J Med Chem 58 (5):2547–2552. https://doi.org/10.1021/jm501395b
6. Zwier JM, Roux T, Cottet M, Durroux T, Douzon S, Bdioui S, Gregor N, Bourrier E, Oueslati N, Nicolas L, Tinel N, Boisseau C, Yverneau P, Charrier-Savournin F, Fink M, Trinquet E (2010) A fluorescent ligand-binding alternative using Tag-lite (R) technology. J Biomol Screen 15 (10):1248–1259. https://doi.org/10.1177/1087057110384611
7. Loison S, Cottet M, Orcel H, Adihou H, Rahmeh R, Lamarque L, Trinquet E, Kellenberger E, Hibert M, Durroux T, Mouillac B, Bonnet D (2012) Selective fluorescent nonpeptidic antagonists for vasopressin V-2 GPCR: application to ligand screening and oligomerization assays. J Med Chem 55 (20):8588–8602. https://doi.org/10.1021/jm30061461
8. Albizu L, Cottet M, Kralikova M, Stoev S, Seyer R, Brabet I, Roux T, Bazin H, Bourrier E, Lamarque L, Breton C, Rives ML, Newman A, Javitch J, Trinquet E, Manning M, Pin JP, Mouillac B, Durroux T (2010) Time-resolved FRET between GPCR ligands reveals oligomers in native tissues. Nat Chem Biol 6(8):587–594. https://doi.org/10.1038/Nchembio.396
9. Tornoe CW, Christensen C, Meldal M (2002) Peptidotriazoles on solid phase: [1,2,3]-triazoles by regiospecific copper(I)-catalyzed 1,3-dipolar cycloadditions of terminal alkynes to azides. J Org Chem 67(9):3057–3064. https://doi.org/10.1021/jo011148j
10. Bonnet D, Riche S, Loison S, Dagher R, Frantz MC, Boudier L, Rahmeh R, Mouillac B, Haiech J, Hibert M (2008) Solid-phase organic tagging resins for labeling biomolecules by 1,3-dipolar cycloaddition: application to the synthesis of a fluorescent non-peptidic vasopressin receptor ligand. Chem Eur J 14 (20):6247–6254. https://doi.org/10.1002/chem.200800273
11. Terrillon S, Cheng LL, Stoev S, Mouillac B, Barberis C, Manning M, Durroux T (2002) Synthesis and characterization of fluorescent antagonists and agonists for human oxytocin and vasopressin V-1a receptors. J Med Chem 45(12):2579–2588. https://doi.org/10.1021/jm010526+
12. King M, Wagner A (2014) Developments in the field of bioorthogonal bond forming reactions—past and present trends. Bioconjug Chem 25(5):825–839. https://doi.org/10.1021/bc500028d
13. Kolb HC, Finn MG, Sharpless KB (2001) Click chemistry: diverse chemical function from a few good reactions. Angew Chem Int Ed Engl 11:2056–2075. https://doi.org/10.

1002/1521-3773(20010601)40:11<2004::Aid-Anie2004>3.0.Co;2-5

14. Bonnet D, Ilien B, Galzi JL, Riche S, Antheaune C, Hibert M (2006) A rapid and versatile method to label receptor ligands using “click” chemistry: validation with the muscarinic M1 antagonist pirenzepine. Bioconjug Chem 17(6):1618–1623. https://doi.org/10.1021/bc060140j
15. Margathe JF, Iturrioz X, Regenass P, Karpenko IA, Humbert N, de Rocquigny H, Hibert M, Llorens-Cortes C, Bonnet D (2016) Convenient access to fluorescent probes by chemoselective acylation of hydrazinopeptides: application to the synthesis of the first far-red ligand for apelin receptor imaging. Chem Eur J 22(4):1399–1405. https://doi.org/10.1002/chem.201503630
16. Bonnet D, Ollivier N, Gras-Masse H, Melnyk O (2001) Chemoselective acylation of fully deprotected hydrazino acetyl peptides. Application to the synthesis of lipopeptides. J Org Chem 66(2):443–449. https://doi.org/10.1021/Jo0010577
17. Chan W, White P (2000) Fmoc solid phase peptide synthesis. A practical approach. Oxford University Press, Oxford
18. Montalbetti CAGN, Falque V (2005) Amide bond formation and peptide coupling. Tetrahedron 61(46):10827–10852. https://doi.org/10.1016/j.tet.2005.08.031
19. Bonnet D, Grandjean C, Rousselot-Pailley P, Joly P, Bourel-Bonnet L, Santraine V, Gras-Masse H, Melnyk O (2003) Solid-phase functionalization of peptides by an alpha-hydrazinoacetyl group. J Org Chem 68(18):7033–7040. https://doi.org/10.1021/jo0343432

Part III

Bioluminescence and Fluorescence Resonance Transfer Energy Approaches

Chapter 8

Time-Resolved FRET-Based Assays to Characterize G Protein-Coupled Receptor Hetero-oligomer Pharmacology

Joyce Heuninck, Candide Hounsou, Elodie Dupuis, Eric Trinquet, Bernard Mouillac, Jean-Philippe Pin, Dominique Bonnet, and Thierry Durroux

Abstract

Although G protein-coupled receptor (GPCR) oligomerization is a matter of debate, it has been shown that the nature of the GPCR partners within the oligomers can influence the pharmacological properties of the receptors. Therefore, finding specific ligands for homo- or hetero-oligomers opens new perspectives for drug discovery. However, no efficient experimental strategy to screen for such ligands existed yet. Indeed, conventional binding strategies do not discriminate ligand binding on GPCR monomers, homo- or hetero-oligomers. To address this issue, we recently developed a new assay based on a time-resolved FRET method that is easy to implement and that can focus on ligand binding specifically on the hetero-oligomer.

Key words Tag-lite® screening, G protein-coupled receptor, Fluorescent ligand, HTRF®, Time-resolved FRET, Lanthanide, Terbium, Self-labeling enzyme, Binding experiment, Homo- and hetero-oligomerization

1 Introduction

G protein-coupled receptors (GPCRs) constitute the largest family of membrane proteins and they participate to the regulation of all physiological functions. They are the major drug targets for development of therapeutics with about 30% of the drugs currently on the market [1]. However, there are still several reasons for screening new drugs targeting GPCRs. First, only few receptors are targets of drugs. Second, because full agonists display side effects, there is a need to screen for biased agonists devoid of any side effects. Lastly, the concept of GPCR oligomerization, although still a matter of debate, opens new perspectives. Indeed, GPCRS have the propensity to oligomerize forming homo-oligomers (made of identical receptors) or hetero-oligomers (made of different receptors). Various studies have shown that the binding and/or coupling

Mario Tiberi (ed.), *G Protein-Coupled Receptor Signaling: Methods and Protocols*, Methods in Molecular Biology, vol. 1947, https://doi.org/10.1007/978-1-4939-9121-1_8, © Springer Science+Business Media, LLC, part of Springer Nature 2019

properties of a given receptor can be modified upon oligomer formation with another GPCR. Therefore, finding ligands selective for oligomers opens tremendous perspectives in terms of drug development.

However, until recently, no experimental strategy exists to screen for ligands selective for oligomers. The method [2] described below, derived from the Tag-lite® assays based on a time-resolved strategy [3–9], is therefore promising to find oligomer-specific ligands. The method is compatible with high-throughput screening, easy to implement and constitutes an attractive strategy for both academic laboratories and big pharmaceutical companies.

1.1 Principle of the Method

The principle of the strategy is based on a Förster (or Fluorescence) resonance energy transfer (FRET) [10] between a receptor tagged with a fluorophore and a fluorescent ligand bound to a second receptor (Fig. 1a). As for all other FRET-based strategies, the non-radiative transfer occurs between a fluorophore donor and an acceptor. Three parameters are crucial to make the non-radiative transfer efficient: (1) the energy compatibility between the fluorophores. The greater the overlap between the emission spectrum of the donor and the excitation spectrum of the acceptor, the better the FRET; (2) the relative orientation of the donor and the acceptor (*see* **Note 1**): the FRET is maximal when the dipole transition moments of the donor and the acceptor are parallel; (3) the distance between the donor and the acceptor. For every pair of fluorophores, a distance called the Forster distance (R_0) can be determined. It corresponds to the distance that gives half maximal FRET efficiency. Because the efficiency of the transfer varies as a function of the inverse of the distance to the sixth power, a distance between the fluorophores superior to 1.5 times R_0 or inferior to 0.5 R_0 results in the absence of FRET or a maximal efficiency, respectively. Classic pairs of fluorophores usually have a R_0 between 35 and 80 Å.

FRET techniques do not usually exhibit a high signal-to-noise ratio because of various contaminating signals due to (1) the direct excitation of the acceptor fluorophore at the excitation wavelength of the donor; (2) the emission of the donor at the emission wavelength of the acceptor; or (3) the background fluorescence of the biological samples or medium. Therefore, various acquisition protocols have been developed to eliminate this contamination from the FRET signal. They consist of measuring the fluorescence of the donor or the acceptor after photobleaching of the acceptor or the donor, respectively. These methods require complex data analysis such as normalization by the expression of the donor or the acceptor or the subtraction of the background fluorescence. All these steps can again in turn generate artifacts.

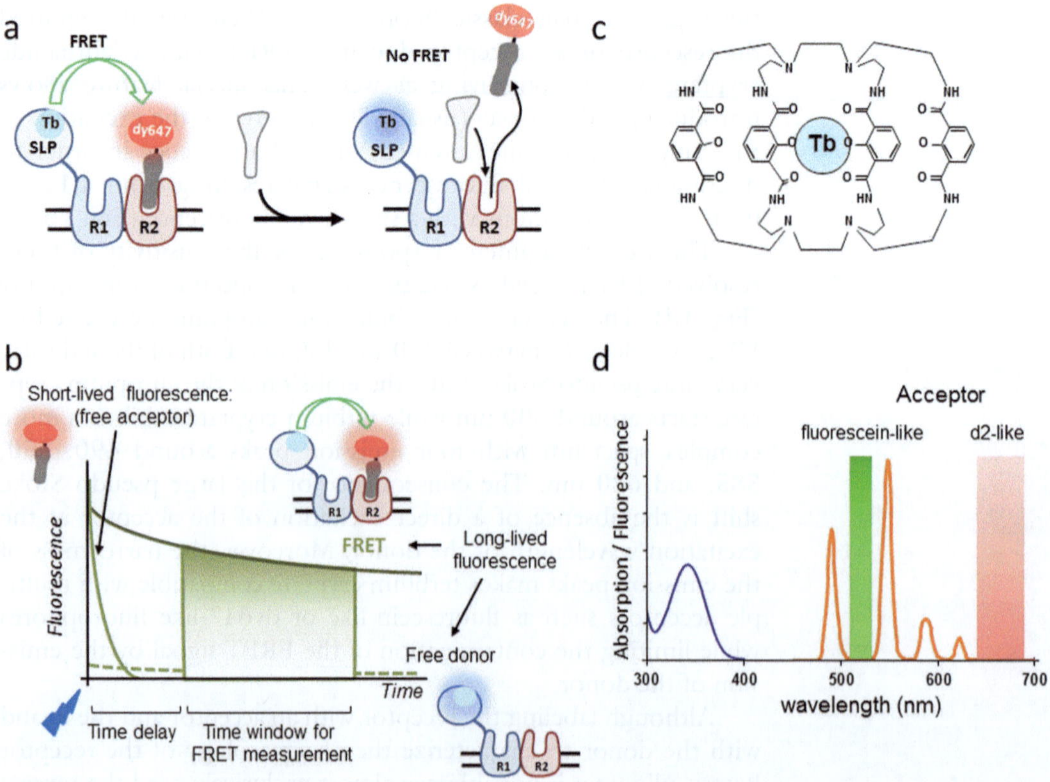

Fig. 1 Principle of the time-resolved FRET technique with HTRF® technology. (**a**) Principle of the assay to characterize hetero-oligomer pharmacology. (**b**) Temporal selectivity in HTRF® technology: TR-FRET signal is measured 50 μs after the excitation, in a time window of 400–500 μs. During the delay, all short-lived fluorescence is decreased to zero and only long-lived time fluorescence FRET and free donor fluorescence can be measured. Because the fluorescence emission of the donor is weak at 665 nm (emission wavelength of the acceptor), the contamination caused by the free donor fluorescence is often negligible. (**d**) Spectral compatibility with HTRF® donor and acceptors: absorption (dark blue) and emission (orange) spectra of Lumi4®-Tb. Green and red box indicate the emission wavelengths of the acceptors which are compatible with Lumi4®-Tb. (**c**) Structure of the Lumi4®-Tb. *SLP* self-labeling protein

An alternative strategy consists in implementing time-resolved FRET, a strategy derived from the HTRF® technology, which displays a higher signal-to-noise ratio due to two properties: the temporal selectivity and the spectral compatibility [11]. The temporal selectivity is provided by the use of cryptates of lanthanide, more specifically cryptate of europium and terbium, as donor (Fig. 1b). The cryptate moiety (Fig. 1c) is essential [12]: (1) it allows the linkage of the lanthanide on the molecule of interest (ligand or receptor); (2) it is an antenna increasing the quantum yield of the donor; (3) it protects lanthanides from quenching, especially quenching by water. Although lanthanides are luminescent but not fluorescent molecules, their resonance energy transfer on a fluorophore is following the same rules. Lanthanides exhibit a long-lasting emission, greater than 1 ms, i.e. more than 100,000

times greater than classic fluorophores. Therefore, the emitted fluorescence of an acceptor due to a FRET with a lanthanide cryptate will be long-living as well. This special feature allows introducing a time delay (usually 50 μs) between the excitation of the sample and the measurement of the fluorescence in order to discriminate the real fluorescence signal resulting from a FRET from all contaminations with a short-live fluorescence (Fig. 1b).

The second parameter responsible for the sensitivity of time-resolved FRET-based strategies is the spectral compatibility (Fig. 1d). The cryptates of terbium and europium are excited by UV at wavelengths between 330 and 350 nm. Both of them display very large pseudo Stoke shifts: the emission of the europium cryptate starts around 600 nm while terbium cryptate exhibits a more complex spectrum with four emission peaks around 490, 550, 585, and 620 nm. The consequence of this large pseudo Stoke shift is the absence of a direct excitation of the acceptor at the excitation wavelength of the donor. Moreover, the narrowness of the emission peaks makes terbium cryptate compatible with multiple acceptors such as fluorescein-like or dy647-like fluorophores while limiting the contamination of the FRET signal by the emission of the donor.

Although labeling the receptor with an acceptor and the ligand with the donor to characterize the pharmacology of the receptor hetero-oligomer is possible, we almost exclusively used the reverse labeling, i.e. the labeling of the receptor with the donor and the ligand with the acceptor. This choice is based on at least two reasons: first, most fluorescent ligands have already been derivatized with classic organic fluorescent acceptors compatible with the cryptates of lanthanide to perform TR-FRET experiments but almost never with cryptates of lanthanides; and second, the signal-to-noise ratio when labeling the receptor with the donor and the ligand with the acceptor is higher than for the reverse labeling.

1.2 Labeling of G Protein-Coupled Receptors

Various strategies can be used to label GPCRs. A first set of strategies is based on a non-covalent labeling of the receptor with fluorescent antibodies against either the receptor itself or a tag fused to the receptor N-terminus. Although positive results have been obtained, it is noteworthy that antibodies are large molecules (150 kDa) and induce steric hindrance [13]. One alternative approach consists in using fluorescently labeled nanobodies which are much smaller. A second drawback is linked to the non-covalent binding itself, which results in an equilibrium between unlabeled and labeled GPCRs. Since an excess of antibodies is usually used to get the largest fraction of labeled GPCRs, washing steps are necessary to eliminate free antibodies which can be the source of a non-specific FRET signal. The kinetics of dissociation of the antibodies/receptor complex should therefore be slow to keep the receptor labeled after the washing steps.

The second set of strategies is based on the covalent labeling of the receptors. This can be reached by the fusion of a self-labeling protein (also called suicide enzyme) to the N-terminus of the receptor. The self-labeling proteins catalyze the transfer of a chemical group from a substrate to themselves. Providing a fluorescent substrate to the self-labeling protein will therefore result in the transfer of the fluorescent group onto the self-labeling protein/receptor chimera. Interestingly, different self-labeling proteins with each their specific substrates have been developed allowing the labeling of multiple receptors at the same time. One of the most commonly self-labeling protein used is the SNAP-tag® (20 kDa) that has been derived from O^6-alkylguanine-DNA alkyl transferase (AGT) [14–17], a DNA-repair enzyme. Optimization of the structure of the wild-type enzyme increased its enzymatic activity and reduced its size. Addition of fluorescent benzylguanine (BG) substrates results in the transfer of the fluorescent benzyl group to the protein. More recently, also other self-labeling proteins such as CLIP-Tag® (21 kDa) [18] and Halo-Tag® (33 kDa) [19] have been developed to label receptors.

These strategies have two main advantages: (1) because of the high enzymatic activity of these proteins, 100% of the receptors can be labeled [20] and the labeling is not sensitive to washing steps and (2) the steric hindrance is reduced since these proteins are much smaller than antibodies. The main drawback is that the fusion of the self-labeling protein to the receptor can influence the pharmacological properties of the receptor, although it has not yet been reported.

1.3 Ligand Labeling

As mentioned above, various acceptors are compatible with a lanthanide cryptate and particularly with the terbium cryptate because of its multiple emission peaks. These acceptors can be organic fluorophores or fluorescent proteins such as Green Fluorescent Protein (GFP) or its derivatives. The choice is mainly driven by the easiness to synthesize the fluorescent ligands. Organic fluorophores can often be easily linked to small ligands during their synthesis [4, 9, 13, 21–24]. By contrast, labeling large ligands such as proteins is more difficult and expensive and fusing them to fluorescent proteins can be a good alternative if the fluorescent partner does not induce a large steric hindrance. Moreover, before starting, you have to be sure that your device is compatible with green and red FRET measurements and thus equipped with the appropriate filters.

However, there are no obvious general rules for ligand labeling. It seems that each ligand is unique. We have observed that the linkage of fluorescein dramatically alters the affinity of some ligands for their cognate receptors while linkage of a d2 acceptor does not, but we also observed the contrary suggesting that all physicochemical parameters (size, polarity, hydrophobicity/hydrophilicity, etc.),

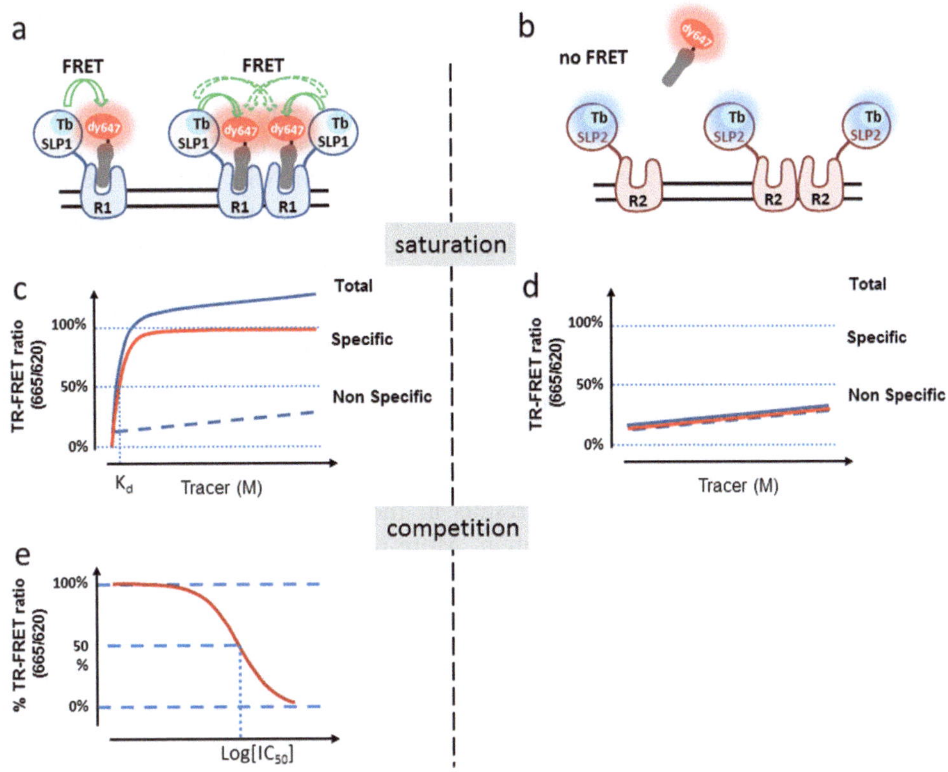

Fig. 2 Time-resolved FRET-based binding assay with HTRF® technology (**a**, **b**). Saturation (**c**, **d**) and competition (**e**) experiments can be performed. Two steps are required to develop a time-resolved FRET-based assay. The tracer should exhibit a high affinity and a high selectivity for one of the receptors

the length of the spacer if any, and the position of the derivatization on the ligands seem to be important. Although analyses of structure–activity relationship of the ligand–receptor complex are very helpful, our experience shows that the impact of ligand labeling on its affinity is not always predictable.

1.4 Overview of the Assay Implementation

The development of an assay to investigate hetero-oligomer binding properties requires various steps:

1. A full characterization of the pharmacological properties of the tracer, the fluorescent ligand, should be performed on each receptor expressed alone (Fig. 2a, b) to determine the affinity of the tracer for each receptor (Fig. 2c, d). The tracer should display a high affinity and selectivity for one of the two receptors. Competition experiments can be carried out to validate the method with the tracer if it exhibits high affinity for the receptor (Fig. 2e).
2. The implementation of the assays on hetero-oligomers (Fig. 3): the expression of both receptors should be optimized (Fig. 3a) in order to get the amount of hetero-oligomers compatible

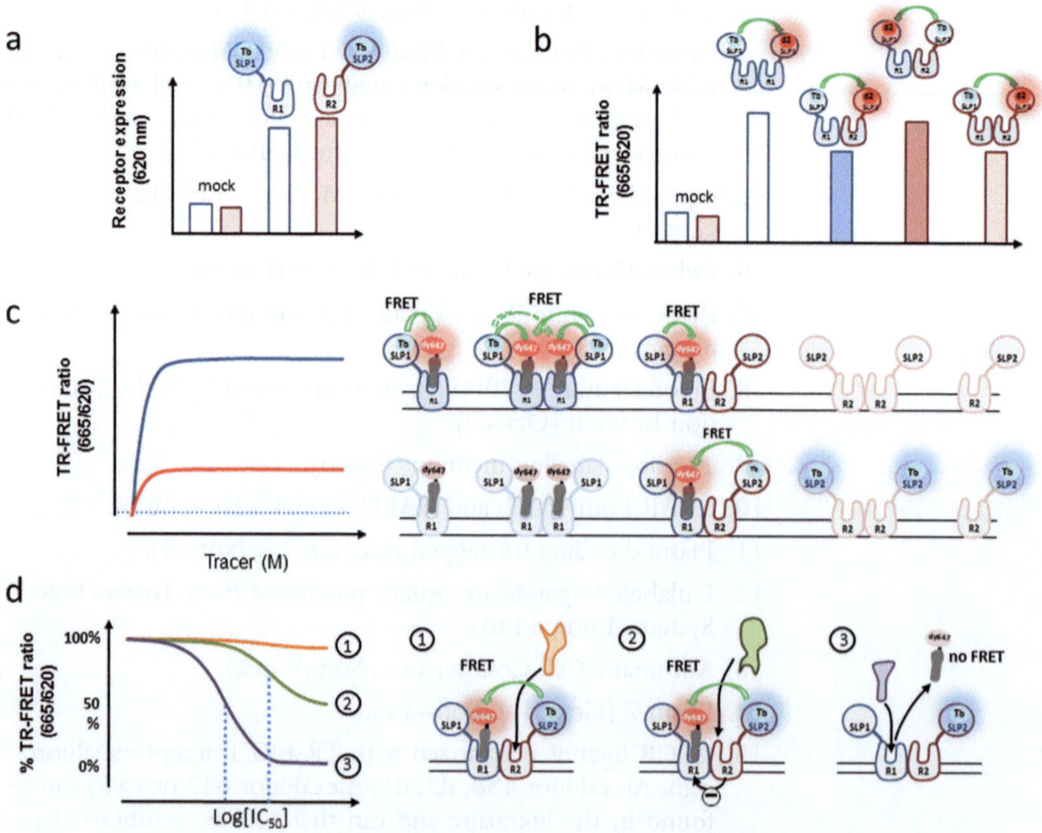

Fig. 3 Time-resolved FRET-based binding assay when two receptors are co-expressed with HTRF® technology. Estimation of the expression of each receptor (**a**) and of the formation of both homo- or hetero-oligomers (**b**), saturation (**c**), and competition (**d**) curve

with the sensitivity of the technique (Fig. 3b). Saturation experiments should be carried out while labeling receptor 1 or receptor 2 (Fig. 3c). Finally, competition experiments can be implemented with reference compounds for each receptor (Fig. 3d).

The affinity of the tracer and of the competitors for each receptor can be influenced by the receptor hetero-oligomerization because of cross-cooperativity between the two protomers. In this assay, the cooperativity phenomena can be investigated by comparing the binding data obtained on receptors expressed alone or co-expressed.

2 Materials

1. HEK293 or CHO cells (*see* **Note 2**).
2. Phosphate buffered saline (PBS): 0.9% NaCl, 10 mM sodium phosphate pH 7.

3. Trypsin-EDTA solution: 0.05% Trypsin-EDTA.
4. Complete Dulbecco's Modified Eagle's Medium (DMEM): DMEM medium supplemented with 10% Fetal Calf Serum and 1% penicillin/streptomycin antibiotics and MEM Non-essential amino acids (only for CHO cells).
5. Opti-MEM® Reduced Serum Medium (Life Technologies, Gibco).
6. Poly-L-Ornithine Solution: 0.1 mg/mL in PBS.
7. Black- or white 96-well plate (e.g. Greiner Cell Star 96-well plate).
8. Lipofectamine™ 2000 (Invitrogen) or jetPEI DNA Transfection Reagent (Ozyme).
9. Tag-lite® labeling medium (Cisbio).
10. SNAP-Lumi4®-Tb and HALO-Lumi4®-Tb (Cisbio).
11. Plasmid coding for tagged receptors (*see* **Note 3**).
12. Unlabeled ligands are usually purchased from Tocris (R & D Systems Europe Ltd).
13. Automatic Cell Counter Eve (NanoEntek).
14. Prism-7 (GraphPad Software).
15. GPCR ligands derivatized with TR-FRET acceptors (fluorescein, AlexaFluor 488, d2, d1, AlexaFluor 647 or Cy5) can be found in the literature and can therefore be synthesized (*see* **Note 4**) and prepared ideally at a stock concentration of about 300 μM (*see* **Note 5**).
16. Microplate readers compatible with Time-resolved FRET (*see* https://www.cisbio.com/drug-discovery/htrf-microplate-readers) (*see* **Note 6**).

3 Methods

3.1 Expression of G Protein-Coupled Receptors in HEK293 Cells

The receptor expression in cell lines should be optimized for each G protein-coupled receptor when co-expressed or expressed alone.

1. HEK293 cells are grown in culture in an atmosphere of 95% air and 5% CO_2 at 37 °C in complete DMEM. Split cells before they reach confluence with trypsin.
2. Transfect cells using manufacturer's Lipofectamine 2000® Transfection protocol. Coat black- or white-96-well plates (flat bottom) with Poly-L-Ornithine diluted at 0.1 mg/mL in sterile PBS (50 μL/well) during 30 min at 37 °C.
3. Wash plates with 100 μL sterile PBS per well.
4. Dilute Lipofectamine® 2000 Transfection Reagent (0.5 μL/well) in Opti-MEM (25 μL /well) (5 min at room

temperature), and then add to Opti-MEM mixture containing plasmid coding for the GPCR of interest and non-coding DNA (25–200 ng/well in 25 μL Opti-MEM per well). Incubate the mixture 20 min at room temperature. Add 50 μL of the DNA: lipofectamine mixture to each well (*see* **Note** 7).

5. Harvest HEK293 cells when they are at 80% of confluence, count cells in Automatic Cell Counter Eve, resuspend cells in Opti-MEM® medium at a density of 250,000 or 500,000 cells/mL to follow 24 or 48 h after transfection respectively. Plate 100 μL per well on top of the DNA mixture for 24 or 48 h.
6. Binding experiments can be carried out 24 or 48 h after transfection depending on the level of expression of the receptors (*see* **Note** 7).

3.2 Expression of G Protein-Coupled Receptors in CHO Cells

1. CHO cells are grown in an atmosphere of 95% air and 5% CO_2 at 37 °C in complete DMEM. Split cells before they reach confluence.
2. Transfect cells using manufacturer's jetPEI transfection protocol.
3. The day before the transfection, harvest CHO cells when they are at 80% of confluence, count in Automatic cell counter EVE, resuspend cells in complete DMEM at a density of 100,000–170,000 cells/mL and plate (100 μL/well).
4. On the day of transfection, dilute jetPEI Transfection Reagent (0.5 μL/well) in 25 μL 150 mM NaCl.
5. Prepare DNA mix (plasmids coding for the receptors of interest and non-coding DNA) in 150 mM NaCl (200 ng DNA/25 μL NaCl/well) in a separate tube (*see* **Note** 7). Vortex gently.
6. Add the jetPEI solution to the DNA mix and vortex immediately.
7. Keep the jetPEI/DNA mix at room temperature for 30 min.
8. During that time, remove the 100 μL DMEM from the plate with cells and add 50 μL of fresh complete DMEM per well.
9. After 30 min incubation, add 50 μL of the jetPEI/DNA mix per well.
10. Binding experiments can be carried out 24 or 48 h after transfection (*see* **Note** 7).

3.3 Labeling of G Protein-Coupled Receptors Expressed at the Cell Surface

1. Incubate cells expressing SNAP-tag® or Halo-Tag® receptors in the presence of their cognate substrates, i.e. SNAP-Lumi4®-Tb or Halo-Lumi4®-Tb.
2. Dilute SNAP-Lumi4®-Tb or Halo-Lumi4®-Tb substrates in Tag-lite® labeling medium to get a final concentration of 100 nM (*see* **Note 8**).

3. Remove medium from cells.
4. Wash cells once with 100 μL Tag-lite® labeling medium.
5. Dispense 100 μL/well of 100 nM SNAP-Lumi4®-Tb and/or 100 nM HALO-Lumi4®-Tb solution (*see* **Note 9**).
6. Incubate cells for 1–2 h at 37 °C.
7. Remove the medium and proceed to 4 washes with Tag-lite® labeling medium.

3.4 Optimization of the Condition to Detect Receptor Hetero-oligomerization

It is important to define experimental conditions which optimize receptor hetero-oligomerization. Transfections should be carried out with various ratios of plasmids coding for the two receptors containing a different tag (in this example SNAP for receptor 1 and Halo for receptor 2). For each transfection, we suggested to label homo- and hetero-oligomers (Fig. 3). For the labeling of homo-oligomers, both substrates should be mixed before and then added to the cells in order to get the same proportion of receptor labeled with the donor and with the acceptor. This is not crucial for hetero-oligomer labeling since the substrates do not cross react with the other self-labeling protein.

To this end:

1. Prepare mixes of substrates as follows:
   - 100 nM SNAP-Lumi4®-Tb and 300 nM SNAP-redor 100 nM SNAP-green (receptor 1 homo-oligomer).
   - 100 nM Halo-Lumi4®-Tb and 300 nM Halo-red (receptor 2 homo-oligomer).
   - 100 nM SNAP-Lumi4®-Tb and 300 nM Halo-red (receptor 1/receptor 2 hetero-oligomer).
   - 100 nM Halo-Lumi4®-Tb and 300 nM SNAP-red or 300 nM SNAP-green (receptor 1/receptor 2 hetero-oligomer) (*see* **Note 10**).
2. Remove medium from cells.
3. Wash cells once with 100 μL with Tag-lite® labeling medium.
4. Dispense 100 μL/well of the different mixes prepared in **step 1**.
5. Incubate cells for 1–2 h at 37 °C.
6. Remove the medium and proceed to four washes with Tag-lite® labeling medium.
7. Measure Donor fluorescent signal at 620 nm (fluorescence of the donor) and FRET signal either at 520 nm (for green acceptor) or at 665 nm (for red acceptor) in a time-resolved mode (*see* **Notes 11** and **12**).

3.5 Saturation Binding Experiments

As mentioned above, saturation experiments have to be performed on each receptor expressed alone and on co-expressed receptors.

These experiments should allow determining the concentration of the fluorescent ligand that should be used for the competition experiments. One prerequisite to carry out saturation experiments is that ligands should be in excess with respect to the receptor expression (*see* **Note 13**).

1. After the washing steps (*see* Subheading 3.3), add 50 μL Tag-lite® labeling medium per well.
2. Prepare a serial dilution of the fluorescent ligands in Tag-lite® labeling medium. At this point, prepare all the ligands at four times the desired final concentrations.
3. Add 25 μL of fluorescent ligands per well.
4. Add 25 μL of Tag-lite® labeling medium or 25 μL of unlabeled ligand in excess in each well to determine total binding or non-specific binding, respectively (*see* **Note 14**).
5. Measure the fluorescent signal of the donor at 620 nm and FRET signal either at 520 nm (for green acceptor) or at 665 nm (for red acceptor) in a time-resolved mode.
6. The dissociation constant can only be determined when equilibrium is reached. Because the assay is performed in homogeneous conditions, meaning the fluorescent ligand is not washed away, the evolution of the TR-FRET signal can be followed over time. When the signal reaches a plateau for all the concentrations of the tracer, one can consider that the equilibrium is reached (*see* **Note 15**).
7. Analyze data as described in Subheading 3.7.

3.6 Competition Binding Experiments

One prerequisite to carry out competition experiments is that ligands (tracer and competitors) should be in excess with respect to the receptor expression (*see* **Note 13**).

1. Prepare the tracers at concentration four times the K_d to use them at a final concentration close to K_d and make dilutions in Tag-lite® labeling medium.
2. Perform serial dilutions of the competitors in Tag-lite® labeling medium and prepare ligands at four times the desired final concentration.
3. After the washing steps of the labeling procedure, dispense 50 μL of Tag-lite® labeling medium per well.
4. Add 25 μL of tracer prepared in **step 1** to all the wells and 25 μL per well of one of the various competitor solutions from the serial dilution prepared in **step 2**.
5. Two controls should be included in the experiment: the “total binding” and “non-specific binding” conditions. Substitute the 25 μL of competitor solution with 25 μL of Tag-lite® labeling medium for the total binding and with 25 μL of

unlabeled ligand at high concentration for the non-specific binding (*see* **Note 14**).

6. Measure the fluorescent signal of the donor at 620 nm and the FRET signal either at 520 nm (for green acceptor) or at 665 nm (for red acceptor) in a time-resolved mode.
7. For the first experiments, read fluorescent signals at different times to define the incubation time required to reach equilibrium (Fig. 2b). If long incubation is required to reach equilibrium, perform overnight incubation at 4 °C.

3.7 Analysis of Saturation and Competition Curves

1. Two methods are generally considered to analyze the FRET signals: (1) the first method takes only in account the raw FRET signal measured at the emission wavelength of the acceptor, 520 nm or at 665 nm for green or red acceptor, respectively. This method is the simplest one but is relevant only if the expression of the receptor is kept constant; (2) the second method normalizes the FRET signal by the receptor expression. For experiments performed on a single receptor, receptor expression is generally estimated by the intensity of the signal at 620 nm although this peak is not the most intense. Therefore, depending on the acceptor, the ratios 520/620 or 665/620 can be used (*see* https://www.cisbio.com/drug-discovery/htrf-ratio-and-data-reduction). This normalization can be performed well by well. In case of the hetero-oligomer, normalization is more complicated since the donor complex density is dependent on the expression of both receptors but only one receptor can be labeled with the donor at a time. Therefore, the double normalization cannot be performed well by well. The expression of the second receptor can only be estimated by a mean value after the labeling of the second receptor in separate wells. This second normalization is required if one wants to compare experiments performed at various receptor densities (*see* **Note 16**).
2. For saturation curves, when equilibrium is reached, i.e. the TR-FRET signal is stable at all the ligand concentrations tested, the TR-FRET signal or the 520/620 or 665/620 ratio can be plotted as a function of the tracer concentration to get a saturation curve. The specific TR-FRET signal is obtained after subtracting the non-specific signal from the total TR-FRET signal. The specific TR-FRET signal should reach a plateau proving that equilibrium has been reached. It is noteworthy that the non-specific signal is no longer linear at the highest concentrations of the tracer because of a dynamic FRET between free labeled ligand and the labeled receptor.

   The following equation is used to fit the data corresponding to the saturation curve:

$$F = F_{max}{}^{*}[\text{tracer}]/(K_d + [\text{tracer}])$$

in which F is the TR-FRET signal, F_{max} the maximal TR-FRET signal (at the plateau of the saturation curve), [tracer] the concentration of tracer and K_d the constant of dissociation of the tracer for the receptor. It is noteworthy that for some receptors, more complex equations considering two or more binding sites or the Hill equation have to be considered to get good fits for the experimental data.

3. For competition curves, when equilibrium is reached, the TR-FRET signal can be plotted as a function of the concentration of the competitor. It is worthy to note that the time needed to reach equilibrium in not necessarily the same as for the saturation experiments because it depends on the binding kinetics of the tracer and the competitors of the different receptors.

The curve can be fitted with the following equation:

$$F = F_{min} + (F_{max} - F_{min})/\left(1 + 10^{(\text{Log}([\text{competitor}]) - \text{Log}(\text{IC}_{50}))}\right)$$

in which F is the TR-FRET signal, F_{max} the maximal TR-FRET signal (obtained in the absence of competitor), F_{min} the minimal TR-FRET signal (obtained in the non-specific conditions), [competitor] the concentration of competitor and IC_{50} the concentration of competitor leading to the half-maximal TR-FRET signal. The inhibition constant can be deduced from the IC_{50} value when using the Cheng–Prusoff equation:

$$K_i = \text{IC}_{50}/(1 + [\text{Tracer}]/K_d)$$

in which K_i is the inhibition constant of the competitor, IC_{50} the competitor concentration leading to a half-maximum TR-FRET signal, and K_d the dissociation constant of the tracer. As for saturation experiments, more complex models such as two binding sites models for example can be considered to get better fits.

4 Notes

1. The relative orientation of the donor and acceptor is a crucial parameter for classic FRET- or BRET-based experiments but it is less important in Time-resolved FRET based because of the use of a cryptate of lanthanides [11].
2. Experiments can in theory be performed in any cell line that can be efficiently transfected. We chose HEK293 and CHO cells since these models are often used in literature. CHO cells adhere very well and hence no poly-D-ornithine is needed,

while HEK293 cells are easily washed away, so coating of the 96-well plate is crucial.

3. Various tags have been developed. They are either self-labeling proteins (also called suicide enzymes) such as SNAP-tag®, CLIP-tag® or HaloTag® or substrates for an enzyme (ACP--tag®). For all self-labeling proteins, specific substrates have been developed. The plasmids can be homemade plasmids or purchased from different manufacturers. A large collection of these plasmids is now commercially available (Cisbio) (*see* https://www.cisbio.com/drug-discovery/receptor-binding-assays). Tags are generally fused to the N-terminus of the receptor and the fusion has been shown not to impact receptor functioning. However, this needs to be verified for all receptors. Insertion of a SNAP-tag® or a CLIP-tag® inside extracellular loops is not recommended since it generally affects receptor conformation and modifies receptor binding and functioning properties. By contrast, ACP-tag® or Halo-tag® insertions in extracellular loops are generally more tolerated.
4. A large collection of these ligands are now commercially available (Cisbio) since they are used in Tag-lite® binding assays (*see* https://www.cisbio.com/drug-discovery/receptor-binding-assays).
5. Ligands are dissolved in 10% DMSO in case of a peptide or protein ligand or in 100% DMSO for an organic ligand. The concentration of the stock solution is determined by using the Beer–Lambert relationship: $A = \varepsilon l C$ in which A is the absorbance, ε is the molar extinction coefficient, l the width of the cuvette, and C the concentration of the solution.

   Regarding red acceptor-derivatized ligands, the following values for the molar extinction coefficient (ε) (L/mol/cm) at 649 nm were used: d1: 250,000; d2: 225,000; Bodipy: 80,000. The ratio of the absorption measured at 649 and 604 nm was systematically defined and should be around 3.3. A lower value can reflect a degradation of the ligand or some difficulties to dissolve it.

   Regarding green acceptor-derivatized ligands, aliquots for measuring absorption were generally diluted in a 100 mM carbonate buffer, pH 9 and the value of the molar extinction coefficient is 75,000 (68,000 at pH 7.4) at 495 nm.
6. We read the TR-FRET signal on a Pherastar reader (BMG Labtechnologies) and on a Tecan Infinite F500 microplate reader (Tecan) but numerous readers are compatible with TR-FRET-based strategies.
7. This step can need optimization. First, the final concentration of DNA is dependent on the method used for the transfection. Second, the plasmid including the coding sequences for the

receptors of interest should generally be mixed with non-coding plasmid. The ratio of each plasmid should be optimized to get receptor expressions compatible with binding. Third, the level of expression is also dependent on the time between the transfection and the binding experiments (24 or 48 h). It is noteworthy that a longer time does not necessarily result in a higher receptor expression. A too low expression will prevent a reliable measure of the FRET signal; a too high expression will induce ligand depletion.

8. The concentration of substrates is dependent on the self-labeling protein and the substrate itself. The duration of the incubation and the concentration of the substrate solution are defined in such a way to get almost 100% of the receptor labeled. However, these parameters can be modified to get a faster labeling (shorter incubation and higher substrate concentration) or to use less substrates (longer incubation and smaller substrate concentration). The concentrations are indicated in the Table below.

Tag	Name of the substrate	Final concentration (nM)
SNAP-tag®	SNAP-Lumi4®-Tb	100
	SNAP-green	300
	SNAP-red	300
CLIP-tag®	CLIP-Lumi4®-Tb	400
	CLIP-green	300
	CLIP-red	100
HALOTag®	HALO-Lumi4®-Tb	100
	HALO-red	300

9. To perform simultaneous labeling of receptors, ensure that the substrates are specific for one self-labeling protein and cannot cross react with a second one. It is noteworthy that CLIP substrates can react on the SNAP tag. Therefore, it is important to add an excess of unlabeled SNAP substrate to label a CLIP tag in the presence of a SNAP-tagged protein.

10. For the labeling of homo-oligomers, a mix of both substrates should be prepared in advance and then added to the cells in order to get the same proportion of receptor labeled with the donor and with the acceptor. This is not crucial for hetero-oligomer labeling since the substrates do not cross react with the other self-labeling protein.

11. In the time-resolved mode, the FRET signal measurement is delayed with respect to the excitation. The parameters have

been optimized on the various plate readers. The delay and the time window for the FRET measurement are usually 50 and 400 μs on the Pherastar device and 150 and 500 μs on Infinite 500 device. The wavelength of excitation is 337 nm on the Pherastar device and 340 nm on the Infinite 500 device.

12. Amplitude of the signals obtained with the different mixes should not be compared since the FRET signal will depend not only on the number of oligomers but also on the distance between the donor and the acceptor and, to a lesser extent, on their relative orientation. The distance and orientation can vary from one complex to another. By contrast, for a given complex, the amplitude of the TR-FRET ratio resulting from different transfections can be compared.
13. In general, the amount of free ligand should be at least ten times greater than the amount of the bound fraction. Because the assays are generally performed in small volumes (100 μL in 96-well plates as described here, but smaller volumes may be involved when using 384-well plates), the experimenter has to be sure that ligands are in excess. One method to verify this is to remove the medium containing the fluorescent ligand after the equilibrium is reached, and to compare the fluorescence remaining in the medium to the fluorescence bound to the cell.
14. Unlabeled ligand is added in excess to determine the non-specific binding. This ligand can be the unlabeled homologue of the tracer but it can also be a well-characterized ligand for the GPCR of interest. This ligand needs to bind on the same site as the tracer. It should be used in such a way that the probability of binding of the unlabeled ligand is at least 100 times greater than for the tracer. Importantly, the non-specific signal depends not only on the non-specific binding of the ligand but also on the dynamic FRET resulting from the random collisions of the free fluorescent ligand molecule and the tagged receptor. The dynamic FRET is often negligible when fluorescent ligands are used at concentration inferior to 10 nM but can be very important for concentrations superior to 100 nM. This dynamic FRET does not increase linearly in function of the tracer concentration and could not easily be extrapolated. Therefore, it is important to determine the non-specific signal for each tracer concentration.
15. Depending on the fluorescent ligands, great variations in the incubation time can be observed. Equilibrium can be reached within 1 h, but sometimes only after 8 h. For long incubation times, overnight incubation can be performed at 4 °C. Moreover, when ligands are or are presumed to be agonists, incubation can be done at a temperature lower than 16 °C to prevent receptor internalization and recycling.

16. Close attention should be given when calculating the 520/620 or 665/620 ratios. If variations of the signal at 620 nm are small (<5%), calculation of the ratios can be carried out. By contrast larger decrease or increase (>50%) can be observed in saturation or competition experiments respectively, when tracer or competitor concentration increases. These variations are probably due to a high FRET efficiency between donor and acceptor. In such conditions, two alternative strategies can then be used to calculate the 520/620 or 665/620 ratios: the first and the most relevant strategy consists in the determination of the signal at 620 nm before adding the fluorescent tracer. The second method consists in considering an average value of the signal at 620 nm determined only from non-specific binding conditions. With the latter method, potential variations in cell density or receptor expression between wells will not be considered.

Acknowledgements

The development of the technique has been supported by the Fonds Unique Interministériel and OSEO in a collaborative program named "CELL2Lead". Thanks are due to the plateforme Arpège of the Institut de Génomique Fonctionnelle. This work has been funded by the European Consortium Oncornet (HORIZON 2020 MSCA–ITN–2014–ETN–Project 641833 ONCORNET) (to J.H., and J.-P.P. and T.D.).

References

1. Overington JP, Al-Lazikani B, Hopkins AL (2006) How many drug targets are there? Nat Rev Drug Discov 5:993–996
2. Hounsou C, Margathe J-F, Oueslati N, Belhocine A, Dupuis E, Thomas C, Mann A, Ilien B, Rognan D, Trinquet E, Hibert M, Pin J-P, Bonnet D, Durroux T (2015) Time-resolved FRET binding assay to investigate hetero-oligomer binding properties: proof of concept with dopamine D1/D3 heterodimer. ACS Chem Biol 10:466–474
3. Zwier JM, Roux T, Cottet M, Durroux T, Douzon S, Bdioui S, Gregor N, Bourrier E, Oueslati N, Nicolas L, Tinel N, Boisseau C, Yverneau P, Charrier-Savournin F, Fink M, Trinquet E (2010) A fluorescent ligand-binding alternative using Tag-lite® technology. J Biomol Screen 15:1248–1259
4. Loison S, Cottet M, Orcel H, Adihou H, Rahmeh R, Lamarque L, Trinquet E, Kellenberger E, Hibert M, Durroux T, Mouillac B, Bonnet D (2012) Selective fluorescent nonpeptidic antagonists for vasopressin V_2 GPCR: application to ligand screening and oligomerization assays. J Med Chem 55:8588–8602
5. Cottet M, Faklaris O, Falco A, Trinquet E, Pin J-P, Mouillac B, Durroux T (2013) Fluorescent ligands to investigate GPCR binding properties and oligomerization. Biochem Soc Trans 41:148–153
6. Emami-Nemini A, Roux T, Leblay M, Bourrier E, Lamarque L, Trinquet E, Lohse MJ (2013) Time-resolved fluorescence ligand binding for G protein-coupled receptors. Nat Protoc 8:1307–1320
7. Oueslati N, Hounsou C, Belhocine A, Rodriguez T, Dupuis E, Zwier JM, Trinquet E, Pin J-P, Durroux T (2015) Time-resolved FRET strategy to screen GPCR ligand library. Methods Mol Biol 1272:23–36

8. Valencia C, Dujet C, Margathe J-F, Iturrioz X, Roux T, Trinquet E, Villa P, Hibert M, Dupuis E, Llorens-Cortes C, Bonnet D (2017) A time-resolved FRET cell-based binding assay for the apelin receptor. ChemMedChem 12:925–931
9. Hounsou C, Baehr C, Gasparik V, Alili D, Belhocine A, Rodriguez T, Dupuis E, Roux T, Mann A, Heissler D, Pin J-P, Durroux T, Bonnet D, Hibert M (2018) From the promiscuous asenapine to potent fluorescent ligands acting at a series of aminergic G-protein-coupled receptors. J Med Chem 61:174–188
10. Vogel SS, Thaler C, Koushik SV (2006) Fanciful FRET. Sci STKE 2006:re2
11. Selvin PR (2002) Principles and biophysical applications of lanthanide-based probes. Annu Rev Biophys Biomol Struct 31:275–302
12. Mathis G, Bazin H (2011) Stable luminescent chelates and macrocyclic compounds. In: Hänninen P, Härmä HD (eds) In lanthanide luminescence. Springer, Berlin, pp 47–88
13. Albizu L, Teppaz G, Seyer R, Bazin H, Ansanay H, Manning M, Mouillac B, Durroux T (2007) Toward efficient drug screening by homogeneous assays based on the development of new fluorescent vasopressin and oxytocin receptor ligands. J Med Chem 50:4976–4985
14. Keppler A, Gendreizig S, Gronemeyer T, Pick H, Vogel H, Johnsson K (2003) A general method for the covalent labeling of fusion proteins with small molecules in vivo. Nat Biotechnol 21:86–89
15. Juillerat A, Gronemeyer T, Keppler A, Gendreizig S, Pick H, Vogel H, Johnsson K (2003) Directed evolution of O6-alkylguanine-DNA alkyltransferase for efficient labeling of fusion proteins with small molecules in vivo. Chem Biol 10:313–317
16. Juillerat A, Heinis C, Sielaff I, Barnikow J, Jaccard H, Kunz B, Terskikh A, Johnsson K (2005) Engineering substrate specificity of O6-alkylguanine-DNA alkyltransferase for specific protein labeling in living cells. Chembiochem 6:1263–1269
17. Gronemeyer T, Chidley C, Juillerat A, Heinis C, Johnsson K (2006) Directed evolution of O6-alkylguanine-DNA alkyltransferase for applications in protein labeling. Protein Eng Des Sel 19:309–316
18. Gautier A, Juillerat A, Heinis C, Corrêa IR, Kindermann M, Beaufils F, Johnsson K (2008) An engineered protein tag for multiprotein labeling in living cells. Chem Biol 15:128–136
19. Zhang Y, So M-K, Loening AM, Yao H, Gambhir SS, Rao J (2006) HaloTag protein-mediated site-specific conjugation of bioluminescent proteins to quantum dots. Angew Chem Int Ed Engl 45:4936–4940
20. Maurel D, Comps-Agrar L, Brock C, Rives M-L, Bourrier E, Ayoub MA, Bazin H, Tinel N, Durroux T, Prézeau L, Trinquet E, Pin J-P (2008) Cell-surface protein-protein interaction analysis with time-resolved FRET and snap-tag technologies: application to GPCR oligomerization. Nat Methods 5:561–567
21. Durroux T, Peter M, Turcatti G, Chollet A, Balestre MN, Barberis C, Seyer R (1999) Fluorescent pseudo-peptide linear vasopressin antagonists: design, synthesis, and applications. J Med Chem 42:1312–1319
22. Terrillon S, Cheng LL, Stoev S, Mouillac B, Barberis C, Manning M, Durroux T (2002) Synthesis and characterization of fluorescent antagonists and agonists for human oxytocin and vasopressin V(1)(a) receptors. J Med Chem 45:2579–2588
23. Albizu L, Cottet M, Kralikova M, Stoev S, Seyer R, Brabet I, Roux T, Bazin H, Bourrier E, Lamarque L, Breton C, Rives M-L, Newman A, Javitch J, Trinquet E, Manning M, Pin J-P, Mouillac B, Durroux T (2010) Time-resolved FRET between GPCR ligands reveals oligomers in native tissues. Nat Chem Biol 6:587–594
24. Margathe J-F, Iturrioz X, Regenass P, Karpenko IA, Humbert N, de Rocquigny H, Hibert M, Llorens-Cortes C, Bonnet D (2016) Convenient access to fluorescent probes by chemoselective acylation of hydrazinopeptides: application to the synthesis of the first far-red ligand for apelin receptor imaging. Chem Eur J 22:1399–1405

Chapter 9

Combining Conformational Profiling of GPCRs with CRISPR/Cas9 Gene Editing Approaches

Kyla Bourque, Dominic Devost, Asuka Inoue, and Terence E. Hébert

Abstract

Ligand-biased signaling could have a significant impact on drug discovery programs. As such, many approaches to screening now target a larger section of the signaling responses downstream of an individual G protein-coupled receptor (GPCR). Biosensor-based platforms have been developed to capture signaling signatures. Despite the ability to use such signaling signatures, they may still be particular to an individual cell type and thus such platforms may not be portable from cell to cell, necessitating further cell-specific biosensor development. We have developed a complementary strategy based on capturing receptor-proximal conformational profiles using intra-molecular BRET-based sensors composed of a *Renilla* luciferase donor engineered into the carboxy-terminus and CCPGCC motifs which bind fluorescent hairpin biarsenical dyes engineered into different positions into the receptor primary structure. Here, we discuss how these experiments can be conducted and combined with CRISPR/Cas9 genome editing to assess specific G protein-dependent and -independent events.

Key words GPCRs, BRET, Genome-editing, Conformation-sensitive biosensors, CRISPR, Drug discovery

1 Introduction

Despite the broad physiological and clinical significance of G protein-coupled receptors (GPCRs), their signaling properties in their native cellular contexts remains incompletely understood, which presents a critical impediment to drug discovery. Individual cells express numerous GPCRs which can couple to diverse signaling pathways with convergent effects on several downstream processes such as cell division, cell growth, and other phenotypic parameters. While profiling of signaling events has been shown to be of significant utility in designing and performing drug screens [1–4], it is likely that a "one size fits all" approach will not work as we move from one cell type to another. The ability to measure GPCR activity directly at the level of the receptor allows us to

Mario Tiberi (ed.), *G Protein-Coupled Receptor Signaling: Methods and Protocols*, Methods in Molecular Biology, vol. 1947, https://doi.org/10.1007/978-1-4939-9121-1_9,

complement such screens and may help in decisions about whether or not to expand panels of signaling biosensors [5–8].

We have engineered FlAsH-BRET (bioluminescence resonance energy transfer) conformational biosensors into several GPCRs [5–8] and have used them to characterize specific conformational changes that occur in response to ligand or protein interactions. Specific conformational states are likely associated with the activation of different downstream signaling pathways and these state transitions can be detected in real-time in live cells using the FlAsH-BRET method described below. Such profiles do not require knowledge of signaling pathways a priori. Briefly, this technique relies on BRET between a bioluminescent enzyme, *Renilla* luciferase II (RlucII) and a small molecule dye called FlAsH which can bind a short peptide sequence CCPGCC [9, 10]. As the efficiency of energy transfer from the donor (RlucII) to the acceptor (FlAsH) is a function of distance and orientation of the donor and acceptor moieties, it is sensitive to conformational rearrangements within the receptor. FlAsH only fluoresces when bound selectively to the tetracysteine tag above, which can be introduced at multiple locations within the receptor by insertional mutagenesis. By fusing RlucII to the receptor C-tail and inserting FlAsH binding motifs throughout different intracellular loops using insertional mutagenesis, panels of biosensors can be generated to capture different conformational "perspectives" of a receptor of interest in response to ligand or allosteric modulators.

Biosensors designed in our lab for the prostaglandin F (FP), angiotensin II type 1 (AT_1R), and β_2-adrenergic receptors show this approach to be robust and reproducible [5, 6, 8]. Such biosensors could differentiate between balanced or biased ligands with different profiles demonstrating that these assays can be used to guide drug discovery efforts [6]. We also used the technique to study allosteric crosstalk between receptor dimers and with intracellular protein partners [7]. We feel that the FlAsH-BRET approach offers simple and robust insights into the conformational dynamics of GPCRs in a live cell context. Such insights can help understand the complex mechanisms linking receptor conformations and functional outputs as measured using signaling outputs.

1.1 Generation of FlAsH-BRET-Based Conformation-Sensitive Biosensors

The precise strategies to generate FlAsH-BRET-based, conformation-sensitive biosensors within the angiotensin II type I receptor (AT_1R) and β_2-adrenergic receptor (β_2AR) have previously been published [5–8]. To summarize here, the wild-type AT_1R sequence was used as template for overlap PCR in order to introduce a signal peptide and a FLAG-tag at the 5′ end of the receptor and to insert the FlAsH-binding tetracysteine tag, CCPGCC at various different positions within the coding sequence of the receptor. The resulting PCR product was ligated in a restriction enzyme compatible pIRESH plasmid which contained a *Renilla* luciferase

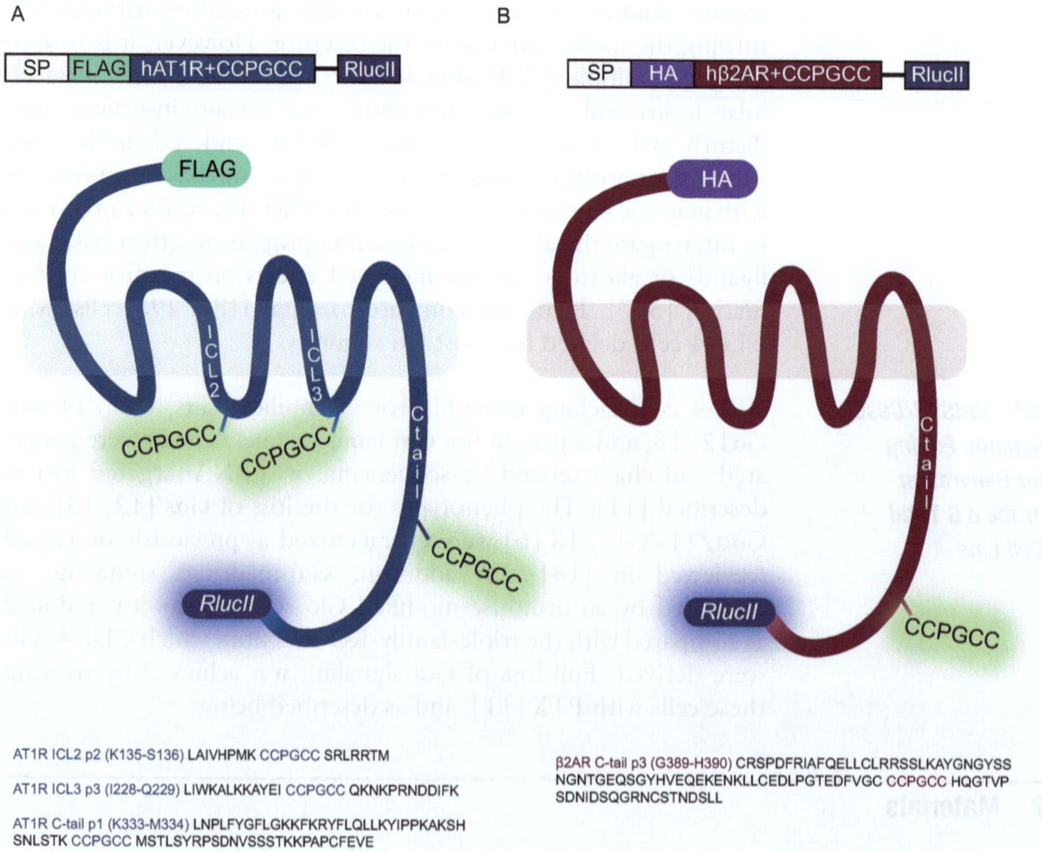

Fig. 1 Cartoon depiction of FlAsH BRET-based conformation sensitive biosensors. Both (**a**) AT_1R and (**b**) β_2AR constructs contain an upstream signal peptide for cell surface targeting as well as a FLAG- or HA-tag for confirming membrane localization. The *Renilla* luciferase donor was introduced onto the C-termini of both the AT_1R and the β_2AR. Within the coding sequence of the AT_1R there is one FlAsH acceptor insertion site located in the second intracellular loop (ICL2 p2), one within the third loop (ICL3 p3) as well as one in the C-tail (C-tail p1). For the β_2AR, the tetracysteine motif was introduced in the distal portion of the C-tail (C-tail p3). This BRET-based biosensor configuration allows to detect relative changes in receptor conformation upon agonist stimulation

construct fused at the 3′ end. Sanger sequencing was performed to confirm the sequence of the resulting plasmids containing SP-Flag-hAT_1R-CCPGCC-RlucII in pIRESH backbone vector (Genome Québec). A similar strategy was used to construct FlAsH-BRET sensors in the β_2AR. The sensors used in this paper are shown schematically in Fig. 1.

This biosensor configuration is ideal to capture the conformational flexibility within a receptor. Introducing the tetracysteine tag at positions distributed within different loops of the receptor is favorable in order to have multiple conformational perspectives to sample from. The small size of the FlAsH motif enables us to

capture conformational dynamics within a receptor without perturbing the native function of the receptor. However, it is wise to generate multiple FlAsH insertion sites within the distinct intracellular loops and carboxyl terminus since certain insertions may disturb wild-type function and will be rendered useless for subsequent analysis. Combining such biosensor-based approaches with genome editing techniques such as CRISPR/Cas9 can be used to interrogate the ability of different G proteins to effect basal and ligand- or allosteric-partner-mediated effects on receptor conformation [5–7]. Here, we compare parental HEK 293 cells, with CL4-4 cells deleted for most Gα subunits.

1.2 CRISPR/Cas9 Genome Editing for Generation of the Δ G Total Cell Line

CL4-4 cells lacking three G protein families (Gαs, Gαq/11 and Gα12/13, and a part of the Gαi family, namely Gαi2) were generated and characterized by sequencing of sgRNA-targeted loci as described [11]. The phenotypes for the loss of Gαs [12, 13] and Gαq/11/G12/13 [6] were characterized as previously described (reviewed in [14]). In addition, Gαi-mediated signaling, as measured by an in-house-modified Glo-sensor assay was reduced as compared with the triple-family-KO cells from which CL4-4 cells were derived. Full loss of Gαi signaling was achieved by treating these cells with PTX [11], and as described below.

2 Materials

2.1 Cell Culture

1. Tissue culture incubator.
2. Biosafety cabinet.
3. T75 flasks.
4. 6-Well dishes.
5. White flat bottom 96-well plates.
6. Dulbecco's Modified Eagle Medium, DMEM.
7. Fetal bovine serum, FBS (Wisent, QC).
8. 0.25% Trypsin-EDTA.
9. Penicillin-streptomycin.
10. Lipofectamine 2000 (Invitrogen).
11. Poly-L-ornithine.
12. Pertussis toxin (Sigma-Aldrich).

2.2 FlAsH Labeling Procedure

1. Chemical Fume Hood.
2. 37 °C incubator.
3. 1.5 mL Eppendorf tubes.
4. 15 mL conical tubes.

5. 50 mL conical tubes.
6. Multi-channel or repeater pipette.
7. Pipette wash basin.
8. Aluminum foil.
9. 1, 2-Ethanedithiol (EDT, Sigma-Aldrich) (*see* **Note 1**).
10. Dimethylsulfoxide (DMSO, Sigma-Aldrich).
11. FlAsH reagent (Invitrogen). Store at −20 °C protected from light.
12. Hank's balanced salt solution 1× (HBSS) without phenyl red, with sodium bicarbonate, calcium, and magnesium. Store at 4 °C and keep sterile (Wisent, 311-513-CL).
13. 25 mM 2,3-dimercapto-1-propanol (BAL wash buffer, Invitrogen). Store at 4 °C (*see* **Note 2**).
14. Kreb's buffer: 146 mM NaCl, 4.2 mM KCl, 0.5 mM $MgCl_2$, 1 mM $CaCl_2$, 10 mM HEPES pH 7.4, 0.1% glucose. Store at room temperature, in an environment protected from light.
15. 1× Phosphate Buffered Saline: 137 mM NaCl, 2.7 mM Potassium Chloride, 8 mM Sodium Phosphate dibasic, 2 mM Potassium Phosphate monobasic, pH 7.4.

2.3 BRET Assay

1. A filter-based detection plate-reader that can detect light sequentially at two wavelengths equipped with an injector unit.
2. Timer.
3. Coelenterazine h (powder form, NanoLight Technologies) Reconstitute in 100% ethanol at a 1 mM stock concentration. Aliquot the solution and keep aliquots at −80 °C for long-term storage. It is preferable to store aliquots in dark screw cap tubes to protect the substrate from light (*see* **Note 3**).
4. Ligands of interest.

3 Methods

3.1 Cell Culture

1. *Day 0*—Grow stock flask of parental HEK 293 cells [6] and CL4-4 cells to ~80% confluency in DMEM with 5% fetal bovine serum (vol./vol.) at 37 °C in a humidified atmosphere with 95% air and 5% CO_2 (*see* **Note 4**).
2. *Day 1*—Plate 200,000 parental cells and 800,000 CL4-4 cells in 2 mL DMEM + 5% FBS in a 6-well plate (*see* **Note 5**).
3. *Day 2 AM*—Transfect cells with 1.5 μg total plasmid DNA per well using Lipofectamine 2000 according to the manufacturer's instructions. Briefly, for one well, combine 1 μg of the FlAsH BRET-based conformation-sensitive biosensor plasmid

DNA, in a tube, with 0.5 μg empty vector pcDNA3.1 and complete to 100 μL with DMEM. The Lipofectamine to plasmid DNA ratio should be maintained at 2:1. In a separate tube, add 3 μL Lipofectamine 2000 with 97 μL of DMEM. Wait 5 min and then mix the DNA tube with the Lipofectamine tube for a total volume of 200 μL. Wait 30 min for the lipid–DNA complexes to form. Add this transfection solution drop-wise onto the cells (*see* **Note 6**).

4. *Day 2 PM*—Change media to 2 mL DMEM + 5% FBS 5 h post transfection.
5. *Day 3 AM*—Replate 30,000 parental and CL4-4 cells per well in a poly-L-ornithine coated white 96-well plate (*see* **Note 7**).
6. *Day 3 PM*—Wait approximately 5–6 h for the cells to adhere to the 96-well plate. For conditions where cells must be pre-treated with PTX, prepare a solution of 100 ng/mL PTX in DMEM + 5% FBS. Aspirate the media and add 100 μL of this 100 ng/mL PTX solution per well. Incubate cells overnight with PTX for at least 16 h in a 37 °C incubator (*see* **Note 8**).
7. *Day 4 AM*—Follow the FlAsH labeling procedure (*see* below), which takes roughly 1.5–2 h to complete.

3.2 FlAsH Labeling Procedure

1. In a water bath, heat the HBSS to reach a working temperature of 37 °C.
2. Make a 25 mM solution of 1,2-ethanedithiol (EDT). To avoid pipetting small volumes, first prepare a 1 M solution of EDT by diluting it in DMSO as follows; in an Eppendorf containing 55 μL DMSO, add 5 μL of 12 M EDT.
3. Mix by vortex.
4. Further dilute the 1 M EDT solution in DMSO to make a 25 mM solution of EDT. As such, pipette 12.5 μL of 1 M EDT into 487.5 μL DMSO.
5. Mix by vortex (*see* **Note 9**).
6. Make the FlAsH-EDT_2 solution. Add one volume of FlAsH reagent (2 mM) to two volumes of 25 mM EDT to make a 667 μM FlAsH-EDT_2 solution (*see* **Note 10**).
7. Incubate the FlAsH-EDT_2 solution for 10 min at room temperature.
8. Then, add 100 μL of HBSS to the 667 μM FlAsH-EDT_2 solution.
9. Continue the incubation for 5 min at room temperature.
10. Complete the volume with HBSS to make a solution with final concentration of 750 nM FlAsH-EDT_2.
11. Mix by vortex.

12. Prior to the FlAsH labeling, wash cells with 150 μL of HBSS (*see* **Note 11**).
13. Aspirate the HBSS from each well.
14. Add 50–60 μL of the 750 nM FlAsH-EDT_2 solution and incubate for 60 min at 37 °C, protected from any source of direct light. Due to the unpleasant FlAsH-EDT odor, prior to incubation at 37 °C, wrap the 96-well plate with parafilm. This will also prevent accidental spills.
15. Wash cells with 2,3-dimercapto-1-propanol (BAL). Make 100 μM BAL wash buffer by pipetting 40 μL of 25 mM BAL solution in 10 mL HBSS. You may adjust the final volume of BAL wash buffer depending on the number of wells labeled on your assay plate.
16. Mix by vortex.
17. Aspirate the FlAsH-EDT_2 labeling solution.
18. Wash cells twice with 100 μL of a 100 μM BAL wash buffer diluted in HBSS buffer. Incubate the first wash for 10 min at 37 °C. Perform the second wash without incubation.
19. Aspirate the BAL-wash solution.
20. Wash cells once with 150 μL of Kreb's assay buffer.
21. Let cells sit in 80 μL of Kreb's buffer for 2 h at room temperature, in an environment protected from light, prior t o the BRET assay (*see* **Notes 12** and **13**).
22. Proper and safe disposal procedures must be taken when discarding the FlAsH and EDT particularly since FlAsH is complexed with arsenic. All solutions containing FlAsH-EDT_2 must be re-collected and disposed of according to the regulations in place by waste management facilities at your institution.

3.3 Setting Up the Assay Protocol

3.3.1 Preparation of the Stock Ligand Concentrations

For angiotensin II (Sigma-Aldrich), dissolve it in ddH_2O to make a stock concentration at 10 mM. For stability purposes, aliquot and store at −20 °C. For isoproterenol (Sigma-Aldrich), dissolve it in 100 mM ascorbic acid to make a stock concentration at 100 mM. For stability purposes, aliquot and store at −20 °C. We recommend using fresh aliquots of isoproterenol, as it is not always stable even when frozen.

3.3.2 Preparation of the Working Ligand Concentrations

Dilute the concentrated ligand stock to make a saturating concentration of the agonist (*see* **Note 14**). For angiotensin II, use at 1 μM final concentration. For isoproterenol, use at 10 μM final concentration.

3.3.3 Preparing the Plate Reader

A filter-based detection plate-reader that can detect light sequentially at two wavelengths (F485, F530) with a relatively rapid read speed is optimal for kinetic-based measurements, however single point assays also work. For ideal performance, the plate reader should be equipped with injector units as they are essential for capturing kinetic readings pre- and post-ligand injection. A temperature control unit is also necessary to maintain the temperature constant for the duration of the assay. In BRET-based assays, we use white bottom microplates to maximize the amount of light collected by the detector considering that white plates reflect the light to increase the signal. In our lab, we use the TriStar2 multimode plate reader from Berthold Technologies and the Victor X Light from Perkin Elmer.

1. Set the temperature of the plate reader between 25 and 28 °C (*see* **Note 15**).
2. Wash the injectors once with Kreb's assay buffer.
3. Prime the injectors with the appropriate assay vehicle and agonist.
4. Set up the kinetic protocol.

3.3.4 Preparing the BRET Substrate

The donor moiety of the FlAsH BRET conformation-sensitive biosensors is a mutated version of the bioluminescent enzyme *Renilla* luciferase (RlucII). The production of light thus requires the oxidation of the substrate, coelenterazine h.

1. Prepare the working solution of the substrate. Dilute coelenterazine h in Kreb's assay buffer to attain a 20 μM working concentration (*see* **Note 16**).
2. Add 10 μL of the 20 μM coelenterazine h solution in 6–18 wells maximum (*see* **Note 17**).
3. Incubate 5 min at room temperature (*see* **Note 18**).
4. Start the BRET protocol on the plate reader (*see* **Note 19**).

Our basic experimental protocol is depicted in Fig. 2. Using this kinetic protocol, basal BRET readings are first collected followed by vehicle/agonist injection onto the cells where BRET is continuously monitored thereafter. The BRET ratio can subsequently be calculated by computing the F530/F485 ratio. The change in BRET as a response to the addition of the vehicle/agonist can also be computed by subtracting the pre-injection BRET from the post-injection BRET. The pre-injection or basal BRET is defined by the average of the first ten measurements and the post-injection BRET can be calculated by averaging the last ten measurements. Such changes in BRET can then give us insight on relative conformational changes within the structure of the receptor as a response to agonist. An increase or decrease in BRET after

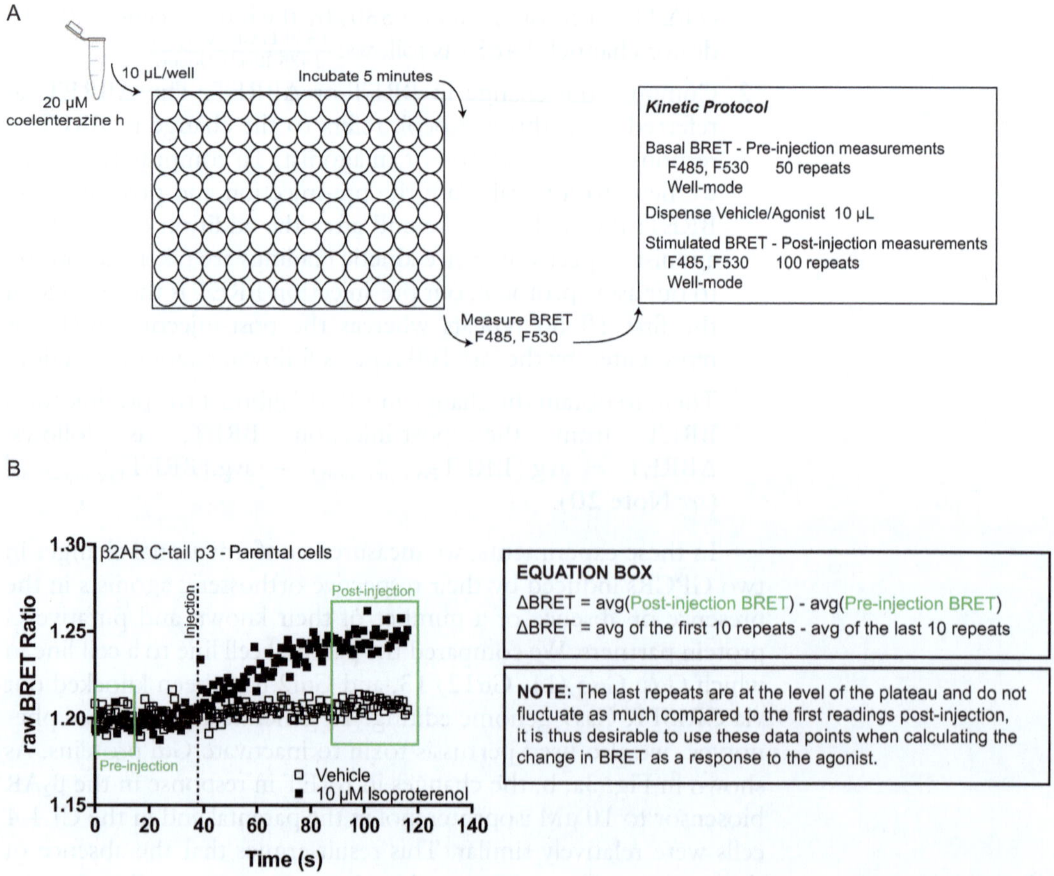

Fig. 2 Overview of the experimental protocol when performing and analyzing FlAsH BRET-based conformation-sensitive biosensor experiments. (**a**) Upon substrate addition (coelenterazine h), the bioluminescent enzyme RlucII emits light at 485 nm. If the acceptor moiety is in close enough proximity (within 10 nm), this leads to the transfer of resonance energy and the subsequent emission of light at 530 nm. (**b**) Kinetic measurements can be taken both pre- and post-stimulation by vehicle or agonist. The BRET ratio can be calculated by dividing the acceptor emission (F530) by the signal from the luminescent donor (F485). The equation box contains details on how to calculate the corresponding change in BRET as a result to vehicle/agonist stimulation as exemplified above

agonist stimulation is indicative of distinct conformational changes. Conformational profiles can then be generated by comparing the conformation at different FlAsH insertion sites giving us insights on the determinants of receptor conformation. The expression of our various biosensors in different CRISPR lines can also give us insight on the requirements of different G proteins or β-arrestins in regard to agonist-induced receptor conformation.

3.4 Data Analysis

1. Compute the BRET ratio. The software driving many plate readers collates the raw data in an excel file (.xls). The BRET ratio can then be computed by dividing the fluorescence

(FlAsH, acceptor channel, F530) by the luminescence (RlucII, donor channel, F485) as follows: $\frac{F530\ \text{FlAsH Acceptor}}{F485\ \text{RlucII Donor.}}$

2. Compute the change in BRET or ΔBRET. The ΔBRET, as referred to in this protocol, refers to the change in BRET in response to the addition of an agonist. To compute this value, we need to first calculate the pre-injection and post-injection BRET; the BRET before and after the addition of vehicle or agonist respectively. Since kinetic readings are taken, according to our assay protocol, our pre-injection BRET is the average of the first 10–50 repeats whereas the post-injection BRET is represented by the last 100 repeats following agonist injection.
3. Then, to obtain the change in BRET, subtract the pre-injection BRET from the post-injection BRET, as follows: $\Delta\text{BRET} = \text{avg.}[\text{BRET}_{\text{Post-injection}}] - \text{avg.}[\text{BRET}_{\text{Pre-injection}}]$ (*see* **Note 20**).

In these experiments, we measured conformational changes in two GPCRs induced by their respective orthosteric agonists in the presence or absence of a number of their known and putative G protein partners. We compared the parental cell line to a cell line in which Gαs, Gαq/11, Gα12/13, and Gαi2 have been knocked out via CRISPR-Cas9 genome editing. To generate a "total" Gαi phenotype, we also used pertussis toxin to inactivate Gαi proteins. As shown in Fig. 3a, b, the changes in BRET in response in the β_2AR biosensor to 10 μM isoproterenol in the parental and in the CL4-4 cells were relatively similar. This result argues that the absence of the various Gα subunits does not seem to influence the

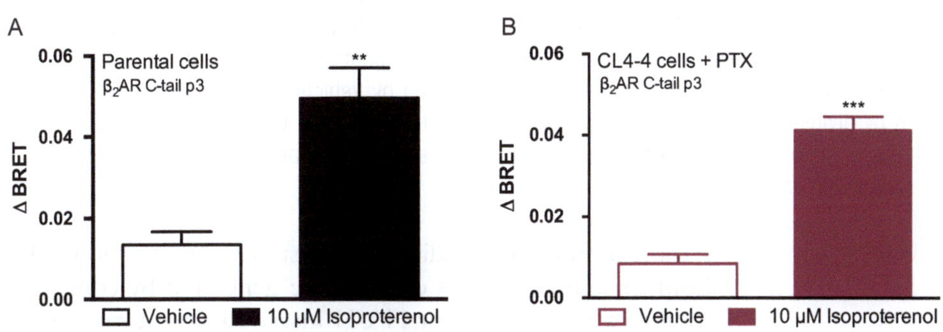

Fig. 3 Isoproterenol-driven conformational rearrangements in the β_2AR expressed in two different genetic backgrounds; (**a**) parental and (**b**) in CRISPR-mediated absence of Gαs, Gαq/11, Gα12/13, and Gαi2 (CL4-4) cells (with additional treatment with PTX) measured by the FlAsH BRET method. Parental cells and CL4-4 cells were plated and transient transfected with HA-hβ_2AR-CCPGCC-Ctail-p3-RlucII biosensor. CL4-4 cells were pre-treated for 16 h with 100 ng/mL PTX. Parental and CL4-4 cells were labeled with the FlAsH dye and stimulated with vehicle or 10 μM isoproterenol followed by kinetic BRET measurements. Data represent four independent experiments; error bars represent mean ± SEM. *T*-tests were performed analyzing the vehicle with the agonist response, asterisks represent $^{**}p \leq 0.01$, $^{***}p \leq 0.001$ Graphs were plotted using GraphPad Prism software

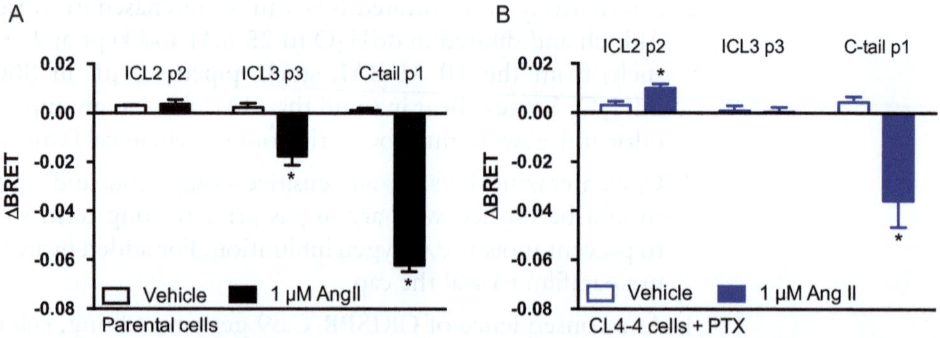

Fig. 4 Angiotensin II-induced BRET changes from the three AT_1R sensors in (**a**) parental and (**b**) CL4-4 cells treated with PTX. Both cell types were transiently transfected with either ICL2 p2, ICL3 p3, and C-tail p1 AT1 FlAsH-based sensors for 48 h followed by FlAsH labeling and BRET assessment; 16 h before the end of the transfection procedure, the CL4-4 cells were treated with 100 ng/mL PTX. Data represent averaged values from 3 (parental) or 4 (CL4-4 cells) independent experiments. Error bars represent mean ± SEM. *T*-test with correction for multiple comparisons were performed between vehicle and Ang II-treated groups using the GraphPad Prism software; $^{*}p < 0.05$

isoproterenol-induced conformational rearrangement of the β_2AR at the C-tail p3 position. Figure 4 shows results from the three conformation-sensitive AT_1R biosensors [6] expressed in both the control parental cell line (Fig. 4a) and in the CL4-4 KO cells treated with PTX (Fig. 4b). In contrast to measurements with the β_2AR, when comparing angiotensin II-induced BRET changes (ΔBRET) from the 3 AT_1R sensors, we noticed a different response profile between the two cell lines. For ICL2 p2, we observed a statistically significant Ang II response (when compared to vehicle treatment) in the CL4-4 KO cell line which was absent in the parental cells while this cell-specific response profile was inverted for ICL3 position 3 sensor. The lack of most of the Gα proteins in the CL4-4 KO cells led to an abrogated Ang II-response from the C-tail p1 sensor when compared to the response measured in the parental cell line. Finally, although all experiments on CL4-4 KO cells were done in presence of PTX to assure complete Gαi inactivation, similar results were observed in absence of the toxin (data not shown). Taken together, our data indicate that these biosensors can capture both G protein-dependent and -independent conformational changes induced by orthosteric agonists.

4 Notes

1. EDT has a distinct unpleasant odor which necessitates its use in a chemical fume hood. The procedure should be performed under dim lighting where possible given that the FlAsH reagent is light-sensitive.

2. Alternatively, concentrated BAL can be purchased from Sigma-Aldrich and diluted in ddH_2O to 25 mM and kept at 4 °C. As such, from the 10 M BAL stock, pipette 2 μL in 800 μL ddH_2O. Vortex. Bear in mind that BAL also has an unpleasant odor and all work must be carried out in a chemical fume hood.
3. Coelenterazine h is an air-sensitive compound and aliquots should be flushed with argon gas prior to long-term storage to prevent moisture/oxygen infiltration. For added protection, use parafilm to seal the cap.
4. As a consequence of CRISPR-Cas9 genome editing, cell morphology as well as growth rate can occasionally be affected. In this particular case, the removal of the various Gα subunits in CL4-4 cells has significantly slowed down cell growth and the cells are also smaller in size. Therefore, it is wise to observe and monitor the cells for a week or more prior to conducting experiments.
5. Growth rate is an important factor when determining the seeding density prior to transfection. Since the CL4-4 cells grow at a significantly slower rate compared to the parental line, we plated four times more cells. Confluency of the two cell types must be relatively similar as such, optimization is required at this step.
6. If culturing cells in presence of antibiotics, it is important to remove them during transfection, as they will reduce the transfection efficiency.
7. Prior to plating the cells in flat bottom white 96-well plates, it is essential to coat the wells with poly-L-ornithine. Briefly, pipette 50 μL of 1× poly-L-ornithine solution in phosphate-buffered saline at 0.1 mg/mL per well and incubate for 10 min. Aspirate and wash once with sterile water. Keep the plate opened in the cell culture hood for any remaining liquid to dry off. This step is vital otherwise the cells will most likely lift off during the FlAsH labeling protocol due to the multiple washing steps.
8. Caution must be taken when working with pertussis toxin (PTX) as it is a highly dangerous toxin that comes from the bacterium *Bordetella pertussis*. Extreme care must be taken when handling this product. Quick spin tubes prior to opening. You will need to work under Biosafety containment level 2. You should also bleach all surfaces and microtips that have come in contact with PTX. Contact the Environmental Health & Safety Office at your institution for approval of your standard operating procedure prior to working with this toxin.
9. We recommend changing gloves regularly as they will be contaminated with EDT (i.e. the unpleasant odor).

10. Considering that the volumes are small, first pipette the EDT as a small droplet on the side of the 15 mL conical tube and then add the required FlAsH volume to the same droplet.
11. If cells have been pre-treated with pertussis toxin, it is best to remove the media in a biosafety cabinet and subsequently bleach contaminated media and other solutions as required. The same safety precautions apply when removing the HBSS prior to the FlAsH labeling as the HBSS may also contain traces of the toxin.
12. The final assay volume depends on the assay to be performed. Typically, in our assays, 80 μL is standard since we add the BRET substrate (10 μL) and then the ligand (10 μL) for a final volume of 100 μL. If pre-treatments are required, the volume can be altered.
13. The ideal location to incubate the assay microplate is in the plate reader to be used. This will allow time for the cells to equilibrate to the assay temperature. This is possible for the Victor X Light but not for the Tristar2 multimode plate reader. If using the latter, incubate the plate in the same room as the instrument, and protected from any source of direct light.
14. Consider using a tenfold higher concentration because we inject a small volume of agonist.
15. Program the temperature unit approximately 45 min earlier than the time you envision starting the experiment. This will allow sufficient time for the targeted temperature to be reached and stabilize.
16. When thawing a new aliquot of coelenterazine h, keep the tube at room temperature for approximately 20 min prior to opening the tube.
17. Due to the high surface tension in the wells of the 96-well plate, it is important to carefully pipette the substrate solution to the bottom of the well. Gently shaking the 96-well plate will also help evenly distribute the substrate throughout the well.
18. Considering that the luminescence signal degrades over time, the 20 μM coelenterazine h solution should be prepared only when needed. For example, prepare a 100 μL solution to stimulate 6 wells. Likewise, reading 6–18 wells at a time ensures that the luminescence output remains stable during the experiment. For this reason, it is important to prepare the 20 μM coelenterazine h solution right before its addition onto the cells. After the addition of the substrate into the well, wait 5 min prior to measuring the luminescence signal. This lag time will ensure that all the readings will be comparable in intensity considering that after the 5 min the luminescence signal is reasonably stable.

19. If kinetic measurements are not required, single point measurements are also feasible. However, optimization may be required. We also recommend generating stable cell lines for more consistent results when considering collecting data using single point measurements.
20. The equation provided as to how to compute the ΔBRET is only our recommendation. The change in BRET can be calculated by taking the average of the last 10–50 measurements and subtracting them by the first 10–50 measurements. For the post-injection BRET, taking the last ten measurements is recommended, as it is the plateau region which is relatively stable. Plotting the BRET ratio as a function of time is also a way to gain a better comprehension of how the BRET ratio is changing over time.

Acknowledgements

The work was funded by a grant to TEH from the Canadian Institutes for Health Research (MOP-130309). KB was supported by an internal Faculty of Medicine Studentship from McGill University.

References

1. Sauliere A et al (2012) Deciphering biased-agonism complexity reveals a new active AT1 receptor entity. Nat Chem Biol 8(7):622–630
2. Beautrait A et al (2017) A new inhibitor of the β-arrestin/AP2 endocytic complex reveals interplay between GPCR internalization and signalling. Nat Commun 8:15054
3. Namkung Y et al (2016) Monitoring G protein-coupled receptor and β-arrestin trafficking in live cells using enhanced bystander BRET. Nat Commun 7:12178
4. Besserer-Offroy E et al (2017) The signaling signature of the neurotensin type 1 receptor with endogenous ligands. Eur J Pharmacol 805:1–13
5. Bourque K et al (2017) Distinct conformational dynamics of three G protein-coupled receptors measured using FlAsH-BRET biosensors. Front Endocrinol (Lausanne) 8:61
6. Devost D et al (2017) Conformational profiling of the AT1 angiotensin II receptor reflects biased agonism, G protein coupling, and cellular context. J Biol Chem 292 (13):5443–5456
7. Sleno R et al (2017) Conformational biosensors reveal allosteric interactions between heterodimeric AT1 angiotensin and prostaglandin F2α receptors. J Biol Chem 292 (29):12139–12152
8. Sleno R et al (2016) Designing BRET-based conformational biosensors for G protein-coupled receptors. Methods 92:11–18
9. Machleidt T, Robers M, Hanson GT (2007) Protein labeling with FlAsH and ReAsH. Methods Mol Biol 356:209–220
10. Hoffmann C et al (2010) Fluorescent labeling of tetracysteine-tagged proteins in intact cells. Nat Protoc 5(10):1666–1677
11. Grundmann M et al (2018) Lack of β-arrestin signaling in the absence of active G proteins. Nat Commun 9(1):341
12. Stallaert W et al (2017) Purinergic receptor transactivation by the β_2-adrenergic receptor increases intracellular $Ca^{(2+)}$ in nonexcitable cells. Mol Pharmacol 91(5):533–544
13. O'Hayre M et al (2017) Genetic evidence that β-arrestins are dispensable for the initiation of β_2-adrenergic receptor signaling to ERK. Sci Signal 10(484). https://doi.org/10.1126/scisignal.aal3395
14. Milligan G, Inoue A (2018) Genome editing provides new insights into receptor-controlled signalling pathways. Trends Pharmacol Sci 39 (5):481–493

Chapter 10

Measuring GPCR Stoichiometry Using Types-1, -2, and -3 Bioluminescence Resonance Energy Transfer-Based Assays

James H. Felce, John R. James, and Simon J. Davis

Abstract

How G protein-coupled receptors are assembled is a matter of considerable interest owing in large part to their remarkable pharmacological importance. For determining receptor stoichiometry, resonance energy transfer-based methods offer considerable advantages insofar as they provide the necessary spatial resolution, and because measurements can be made in situ, relatively easily. This chapter describes three complementary stoichiometric assays that rely on measurements of bioluminescence resonance energy transfer. These quantitative approaches make it possible to identify true protein–protein interactions from non-specific associations that inevitably result from constraining proteins in cellular membranes. In our experience, concordant data obtained in two or more of these assays, benchmarked with suitable controls, strongly predict receptor stoichiometry.

Key words G protein-coupled receptors, Membrane proteins, Bioluminescence resonance energy transfer (BRET), Stoichiometry, Oligomerization, Luciferase, GFP

1 Introduction

The stoichiometries of G protein-coupled receptors (GPCRs) (i.e. whether they comprise monomers, dimers, or oligomers), and how they may change on ligand-binding, are central to understanding their functions and pharmacology, and as such have been both intensively studied and debated for more than 20 years. Bioluminescence resonance energy transfer (BRET) allows the detection of protein–protein interactions via the measurement of non-radiative energy transfer from a donor luciferase (Rluc) molecule to an acceptor green fluorescent protein (GFP), which shifts the emission wavelength from that of the donor to that of the acceptor. Energy transfer is only possible when the two molecules are separated by less than ~10 nm, however, which confers spatial information. A major advantage of BRET over other forms of RET technology is the very high signal:noise ratio gained from luminescence detection, which is also unaffected by photobleaching or

Mario Tiberi (ed.), *G Protein-Coupled Receptor Signaling: Methods and Protocols*, Methods in Molecular Biology, vol. 1947, https://doi.org/10.1007/978-1-4939-9121-1_10,

other optical effects. This allows protein interactions to be detected at physiological levels of expression. Using BRET to determine the stoichiometry of transmembrane proteins such as GPCRs is not necessarily straightforward, however. The major complicating factor is that surprisingly high levels of BRET can occur between non-specifically interacting proteins when they are over-expressed in the crowded membrane environment. It is therefore necessary to distinguish BRET signals due to bona fide cognate interactions, from those arising non-specifically.

In suitable assay formats monomers and dimers give qualitatively different data, which greatly aids in the BRET-based assignment of receptor stoichiometry. Here, we describe in detail three types of BRET assays of receptor stoichiometry that we have developed [1, 2], the utility of which has been confirmed by their ability to correctly distinguish between known monomeric and dimeric control proteins, i.e. CD86 (a monomer), and CD28 and GABAB1 (both homodimers). We have also shown with each of these assays that both the β2 adrenergic receptor, whose stoichiometry was contentious, and a mouse cannabinoid receptor mCannR, were exemplars of monomeric behavior. These assignments were further strengthened using single molecule fluorescence-based imaging [3], and two of the assays have subsequently been used to determine the stoichiometric "signatures" of >60 human GPCRs [4]. We have called the three approaches types-1, -2, and -3 BRET assays (Fig. 1). In the type-1 assay [1], the relationship between BRET efficiency ($BRET_{eff}$) and acceptor:donor ratio (A:D) differs systematically when total receptor density is kept constant, in contrast to a BRET saturation assay [5]. For type-2 assays [1] $BRET_{eff}$, measurements are made at increasing receptor concentrations, allowing extrapolation to conditions in which non-specific energy transfer does not occur. Finally, the type-3 assay [2] relies on the disruption of BRET-productive interactions within oligomers using unlabeled competitors, measured at different expression levels. The type-1 and -3 assays are especially complementary insofar as type-1 assays rarely give false-dimer artifacts but are susceptible to false-monomer results when there is higher-order oligomerization or very weak dimerization, whereas type-3 assays avoid false-monomer results but can generate false dimer signals when, for example, the competitor causes previously clustered tagged receptors to be relaxed, reducing the effective receptor density.

It is important to emphasize that these assays are best suited to the identification of monomers and homodimers. This is because each assay relies on the expression of the tagged and unlabeled receptors being equivalent, which is unlikely generally to be the case for the subunits of heterodimeric complexes. The use of confocal fluorescence microscopy and, where possible, flow cytometry

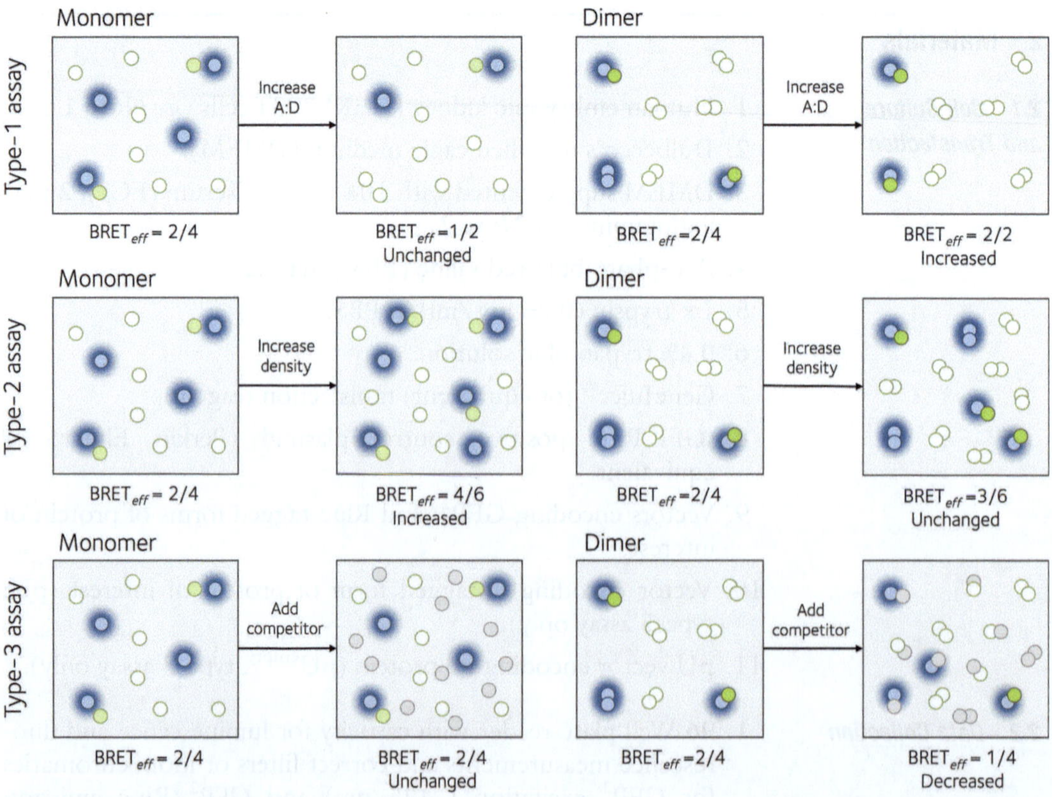

Fig. 1 Principles of the types-1, -2, and -3 BRET assays. In a type-1 assay (top), total density is kept constant while the ratio of donor (blue circles) to acceptor (fluorescing = green circles, non-fluorescing = white circles) is varied. This effectively replaces donor molecules with acceptors as A:D increases. For a monomer, this does not affect $BRET_{eff}$ above an A:D ratio of ~2 as the mean distance of the remaining donors to their nearest acceptor is unchanged. For a dimer, increasing A:D results in fewer donor–donor pairs and thus the remaining donors are more likely to be BRET-productive, meaning $BRET_{eff}$ increases with A:D in a hyperbolic manner. In a type-2 assay (middle), A:D is kept constant and total density is varied. BRET between monomers is highly density-dependent and so $BRET_{eff}$ is zero when extrapolated to very low densities. By contrast, BRET within dimers is independent of total density (as long as dimer stability is not concentration-dependent) and hence $BRET_{eff}$ is non-zero at low densities. For a type-3 assay (bottom), $BRET_{eff}$ is measured at a constant A:D in the presence or absence of untagged 'competitor' molecules (gray circles). $BRET_{eff}$ between monomers is unaffected by competitors, whereas dimers exhibit reduced $BRET_{eff}$ in the presence of untagged proteins due to competition for positions within BRET-productive dimers. $BRET_{eff}$ is measured over a range of total receptor densities to account for competitor-dependent reductions in tagged protein expression, which could otherwise lead to false identification of dimers

to confirm the intracellular localization of tagged GPCRs should also be considered to rule out aggregation-induced oligomerization. We refer the reader to the cited papers for full explanations of the assay principles and their relative utility.

2 Materials

2.1 Cell Culture and Transfection

1. Human embryonic kidney (HEK) 293T cells (*see* **Note 1**).
2. Dulbecco's modified eagle medium (DMEM).
3. DMEM supplemented with 10% fetal calf serum (FCS), 2 mM L-glutamine (*see* **Note 2**).
4. Phosphate-buffered saline (PBS), pH 7.2.
5. 1× trypsin (0.25 mg/ml) in PBS.
6. 0.4% trypan blue solution.
7. GeneJuice® (or equivalent) transfection reagent.
8. sGFP-Rluc positive control plasmid (Perkin Elmer) or equivalent.
9. Vectors encoding GFP^2- and Rluc-tagged forms of protein of interest.
10. Vector encoding untagged form of protein of interest (pU; type-3 assay only).
11. pU vector encoding no protein (pU^{empty}, type-3 assay only).

2.2 Data Collection

1. 96-Well plate-reader with capacity for luminescence and fluorescence measurements and correct filters or monochromatics for GFP^2 excitation (~425 nm) and GFP^2/Rluc emission (~410 and 515 nm).
2. Optiplate-96 well plate (Perkin Elmer) or similar.
3. 1 mM stock solution Coelenterazine-400A in ethanol, store protected from light at −20 °C.
4. 1 mM stock solution Coelenterazine-h in ethanol, store protected from light at −20 °C.
5. PBS, pH 7.2.

3 Methods

All procedures should be performed at room temperature unless otherwise specified.

3.1 Cell Plating

1. HEK-293T cells should be maintained in supplemented DMEM, 37 °C, 5% CO_2 at 50–80% confluency, passaging every 3–4 days. Do not allow cells to become overgrown and do not use cells that have recently been over/under-confluent for transfection as this may affect efficiency. For most applications cells should be cultured in a 75 cm^2 tissue culture flask; the volumes referred to here correspond to a flask of this size.

Table 1
Number and format of wells required for each BRET assay

Assay	Wells per gene	Format	Control wells
Type-1	8	On as few plates as possible	3
Type-2	5	Each well on a separate plate	3
Type-3	10	5 plates at 2 wells per plate	3

2. To plate cells for transfection, remove growth medium, wash with 10 ml sterile PBS, and cover with 5 ml trypsin solution. Leave for 5 min then detach cells by gently tapping the side of the flask (*see* **Note 3**). Add 10 ml of supplemented DMEM to inhibit trypsin activity.
3. Move cell suspension into a 50 ml conical tube, removing a small volume for cell counting, and centrifuge the remainder at $500 \times g$ for 5 min. Count the cells using a hemocytometer or equivalent using an appropriate dead-cell stain, such as trypan blue. It is not advisable to use cells with a large dead-cell population, as this will negatively affect transfection.
4. Remove the supernatant from the centrifuged cells and resuspend in an appropriate volume of pre-warmed supplemented DMEM to give 3×10^5 cells/ml. Add cells to 6-well plates at 2 ml/well. The number and format of wells vary depending on the assay and are given in Table 1 and Fig. 1 (*see* **Note 4**). Take care not to swirl the plates, as this will cause the cells to accumulate in the center of the well.
5. Prepare three additional wells for control transfections. These can be on the same plate.
6. Incubate cells at 37 °C, 5% CO_2 for 24 h.

3.2 Transfection

This transfection protocol is optimized for use with the GeneJuice® transfection system, however other transient transfection systems are also suitable. The DNA concentrations and amounts given assume the use of the pGFP^2, pRluc BRET^2 vectors and modified pU vector [2] or an equivalent transient expression vector under cytomegalovirus immediate-early promoter control. Other expression strategies will require a re-optimized transfection protocol.

1. Dilute pGFP^2 and pRluc vectors to 50 ng/μl in dH_2O. The total volume required depends on the assay (Table 2). For the type-3 assay, also prepare dilutions of pU and pU^{empty} at the same concentration.
2. Prepare final DNA mixes in the ratios given in Table 2. We recommend using an 8-tube PCR strip for the type-1 assay mixes, and 1.5 ml tubes for types-2 and -3. For the type-3

Table 2
Transfection ratios required for each assay

Assay	pRluc:pGFP^2	pGFP^2 (μl)	pRluc (μl)	pU (μl)	pU^{empty} (μl)
Type-1	2:1	6.7	13.3	–	–
	1:1	10.0	10.0	–	–
	1:3	15.0	5.0	–	–
	1:6	17.1	2.9	–	–
	1:9	18.0	2.0	–	–
	1:12	18.5	1.5	–	–
	1:15	18.75	1.25	–	–
	1:39	19.5	0.5	–	–
Type-2	1:12	92	8	–	–
Type-3	1:12	46	4	100	–
	1:12	46	4	–	100

assay, it is recommended that a single mix of 92 μl pGFP^2 and 8 μl pRluc is first prepared then separated into two tubes of 50 μl before adding pU/pU^{empty}, as this will minimize variation between the pU- and pU^{empty}-containing samples.

3. Prepare positive and negative controls: 1 μg sGFP-Rluc (*see* **Note 5**) in 20 μl dH_2O, and 20 μl dH_2O without DNA, respectively.
4. Prepare a luciferase-only control: 1 μg pRluc in 20 μl dH_2O. If assaying multiple proteins, a single luciferase-only control is sufficient for all samples—it is not necessary to prepare one per gene.
5. Make a 3% GeneJuice® solution in non-supplemented DMEM. Allow 100 μl of transfection solution per well to be transfected (i.e. 3 μl Genejuice), plus excess for accurate pipetting. Mix thoroughly and leave for 5 min to equilibrate.
6. Add transfection solution to DNA preparations—100 μl/tube for the type-1 assay, 500 μl/tube for types-2 and -3. Add 100 μl transfection solution to each control.
7. Incubate samples for 20 min.
8. Add 120 μl DNA-GeneJuice® mix to each well of cells. For the type-1 assay this is simply one well per pRluc:pGFP^2 ratio. For the types-2 and -3 assays, add the same transfection mix to five wells. For the type-3 assay, transfect one well of each plate with the pU-containing sample and one with the pU^{empty}-containing. Each control should be added to a single well. To achieve

even mixing of the DNA within the growth medium, tilt the plate and slowly add the transfection mix in a drop-wise manner to the pooled medium (not onto exposed cells). Do not swirl the plate as this may concentrate DNA near the center.

9. Incubate at 37 °C, 5% CO_2 for the appropriate time (*see* Subheading 3.3).

3.3 Sample Preparation

Measurements for each BRET assay are performed in an identical manner (*see* Subheading 3.3.4), however the incubation time between transfection and measurement varies by assay.

3.3.1 Type-1 Assay

Incubation for 24 h is optimal for most proteins. If protein expression is too weak to allow accurate measurement at this stage cells can be left for up to 48 h post-transfection before measurement. All wells, including controls, are measured at the same time.

3.3.2 Type-2 Assay

Measurements are made at 2 h intervals to obtain a range of expression levels. Measuring the first well 7 h post-transfection is optimal for most proteins but requires optimization in each case. Incubation time should be such that the first measurement is performed near the lower limits of the detector's sensitivity, as this will allow the most precise determination of the *y*-intercept of $BRET_{eff}$ vs. expression. Controls should be measured at the middle time point.

3.3.3 Type-3 Assay

Measurements are made at 2 h intervals to obtain a range of expression levels. At each time point, one pU- and one pU^{empty}-transfected well is measured. Performing the first measurement 16 h post-transfection is optimal for most proteins but requires optimization in each case. Unlike the type-2 assay, the first read does not need to be near the detector's lower sensitivity limit, however incubating cells for too long will lead to marginal increases in expression between time points. Optimal results are obtained with a broad range of expression levels. Controls should be measured at the middle time point.

3.3.4 General Sample Preparation Protocol

1. Remove medium from transfected cells and wash with 1 ml PBS, taking care not to dislodge cells at this step. Remove PBS.
2. Use 1 ml PBS to dislodge cells from the plate by pipetting gently up and down three or four times. Use a P1000 pipette rather than a stripette or smaller volume pipette. Be as consistent as possible between wells as the number of cells collected should be as similar as possible between samples.
3. Transfer each well of dislodged cells into a 1.5 ml centrifuge tube and spin at 700 × *g* for 3 min.

4. Remove supernatant and resuspend cells in 200 μl PBS. For the positive control, resuspend cells in 1 ml (*see* **Note 6**).
5. Split each sample into two wells of an Optiplate-96 well plate or similar opaque plate; 100 μl/well. One well will be used to measure GFP^2 and Rluc expression, the other for $BRET_{eff}$ (Fig. 2). Arrange samples such that there is at least one-well separation between wells used for each assay in order to minimize bleed-through from the Rluc measurement into the less bright $BRET_{eff}$ measurement.
6. Perform measurements as described below as soon as possible to minimize cell stress.

3.4 BRET$_{eff}$ Measurement

1. Prepare $BRET_{eff}$ measurement settings on a 96-well plate-reader. These should measure luminescence at the donor (410 nm) and acceptor (515 nm) wavelengths with appropriate bandwidths to allow sufficient sensitivity (e.g. 410 ± 40 and 515 ± 15 nm). Voltage and gain settings must be optimized for each instrument. Typically, settings should ensure signals from samples are at approximately 50% of the saturation intensity of the instrument in order to allow greatest range of sensitivity. The same settings should be used for all experiments. The instrument should integrate three measurements of each channel over 1 s in order to increase precision.
2. Freshly prepare a 100 μM solution of coelenterazine-400A in PBS from the stock solution. 10 μl will be required for each sample, plus excess for accurate pipetting. Keep on ice protected from light when not in use.
3. Perform $BRET_{eff}$ measurement on 1 well from each sample and control: Add 10 μl coelenterazine-400A solution to the 100 μl cell suspension and begin $BRET_{eff}$ measurement immediately (*see* **Note 7**).
4. Use the values from the donor and acceptor channels to determine $BRET_{eff}$ for each sample:

$$BRET_{eff} = \frac{\left(BRET_A - BRET_A^{negative}\right) - \left(\left(BRET_D - BRET_D^{negative}\right) \times cf\right)}{BRET_D - BRET_D^{negative}}$$

where:

$BRET_D$ = relative luminescence in the BRET assay donor channel.

$BRET_A$ = relative luminescence in the BRET assay acceptor channel.

$BRET^{negative}$ = values measured for the negative control.

cf = correction factor for Rluc emission in GFP channel

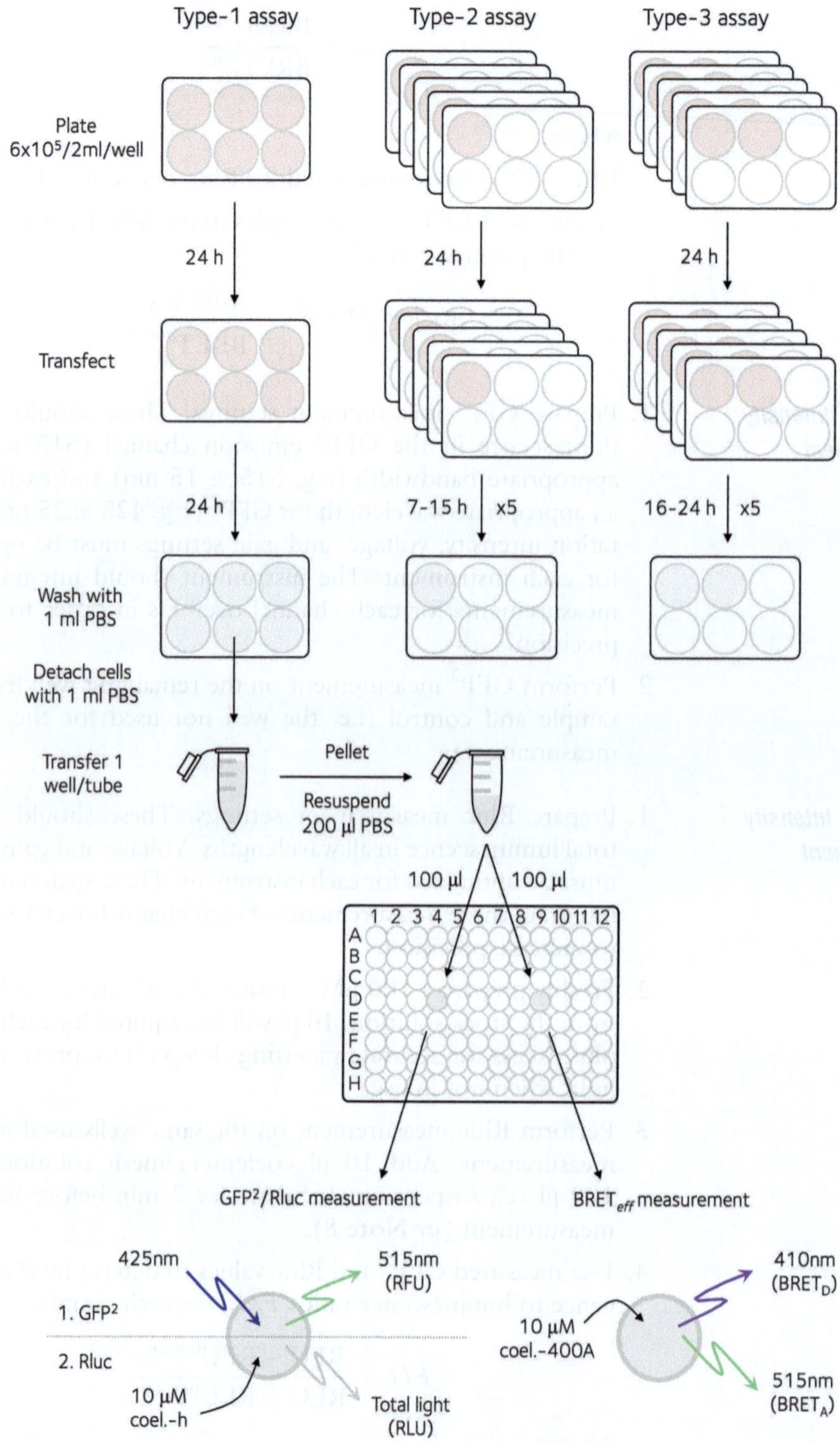

Fig. 2 Schematic workflow for the types-1, -2, and -3 BRET assays. Cells are plated in the format shown, transfected after 24 h, and then removed and assayed at different time points depending on the assay. Each well of transfected cells is split into two wells of a suitable 96-well plate for separate measurement of GFP^2 intensity, Rluc intensity, and $BRET_{eff}$

$$cf = \frac{BRET_A^{Rluc}}{BRET_D^{Rluc}}$$

where:

$BRET^{Rluc}$ = values measured for Rluc-only control.

Normalize $BRET_{eff}$ for all samples to the $BRET_{eff}$ measured for the positive control:

$$BRET_{eff}^{normalized} = \frac{BRET_{eff}}{BRET_{eff}^{positive}}$$

3.5 GFP² Intensity Measurement

1. Prepare GFP^2 measurement settings. These should measure fluorescence in the GFP^2 emission channel (515 nm) with appropriate bandwidth (e.g. 515 ± 15 nm) and excitation at an appropriate wavelength for GFP^2 (e.g. 425 ± 25 nm). Excitation intensity, voltage, and gain settings must be optimized for each instrument. The instrument should integrate three measurements for each channel over 1 s in order to increase precision.
2. Perform GFP^2 measurement on the remaining well from each sample and control (i.e. the well not used for the $BRET_{eff}$ measurement).

3.6 Rluc Intensity Measurement

1. Prepare Rluc measurement settings. These should measure total luminescence in all wavelengths. Voltage and gain settings must be optimized for each instrument. The instrument should integrate three measurements of each channel over 1 s in order to increase precision.
2. Freshly prepare a 100 μM solution of coelenterazine-h in PBS from the stock solution. 10 μl will be required for each sample, plus excess for accurate pipetting. Keep on ice protected from light when not in use.
3. Perform Rluc measurement on the same wells used for GFP^2 measurement. Add 10 μl coelenterazine-h solution to the 100 μl cell suspension. Incubate for 2 min before beginning measurement (*see* **Note 8**).
4. Use measured GFP^2 and Rluc values to determine the fluorescence to luminescence ratio (F/L) for each sample:

$$F/L = \frac{RFU - RFU^{negative}}{RLU - RLU^{negative}}$$

where:

RFU = relative fluorescence units.

RLU = relative luminescence units.

$RU^{negative}$ = values measured for negative control.

Convert F/L to the molar GFP:Rluc ratio by normalizing to the F/L for the positive control (which has an inherent GFP:Rluc of 1):

$$[\text{GFP}]/[\text{Rluc}] = \frac{F/L}{F/L^{\text{positive}}}$$

For the type-3 assay, also convert RLU to total expression using the GFP:Rluc ratio (*see* **Note 9**):

$$\text{Expression} = \left(\text{RLU} - \text{RLU}^{\text{negative}}\right) \times (1 + [\text{GFP}]/[\text{Rluc}])$$

3.7 Data Analysis and Interpretation

3.7.1 Type-1 Assay

1. Plot [GFP]/[Rluc] vs. normalized BRET_{eff}.
2. Using appropriate analytical software, fit all points within [GFP]/[Rluc] values of 2 and 15 (*see* **Note 10**) to the model:

$$\text{BRET}_{eff} = \left(1 - \frac{1}{(1+f)^{n-1}}\right)\text{BRET}_{\max}$$

where:

f = [GFP]/[Rluc].

n = stoichiometry (i.e. 2 for dimer).

($\text{BRET}_{\max}$ is a constant calculated from the data by fitting to the model rather than the measured $\text{BRET}_{\max}$).

3. Determine the coefficient of determination (R^2) value for the fitted model to the data.
4. A negative R^2 value indicates that the data have a poorer goodness-of-fit to the model than to a model of $\text{BRET}_{eff} = \text{BRET}_{\max}$, as predicted for a monomer, and hence the protein of interest is more likely to be monomeric than the modeled stoichiometry. A positive R^2 indicates better goodness-of-fit, suggesting that the protein of interest corresponds to the modeled stoichiometry. Greater deviation of R^2 from zero indicates greater confidence in the conclusion (Fig. 3).

3.7.2 Type-2 Assay

1. Plot Rluc emission vs. normalized BRET_{eff} (*see* **Note 11**).
2. Model the data to a linear least squares regression using appropriate analytical software.
3. Determine the 95% confidence limits of the *y*-intercept of the linear regression.
4. A *y*-intercept significantly greater than zero indicates dimerization or oligomerization of the protein of interest. A *y*-intercept that does not significantly deviate from zero is reflective of monomeric behavior (*see* **Note 12**; Fig. 3).

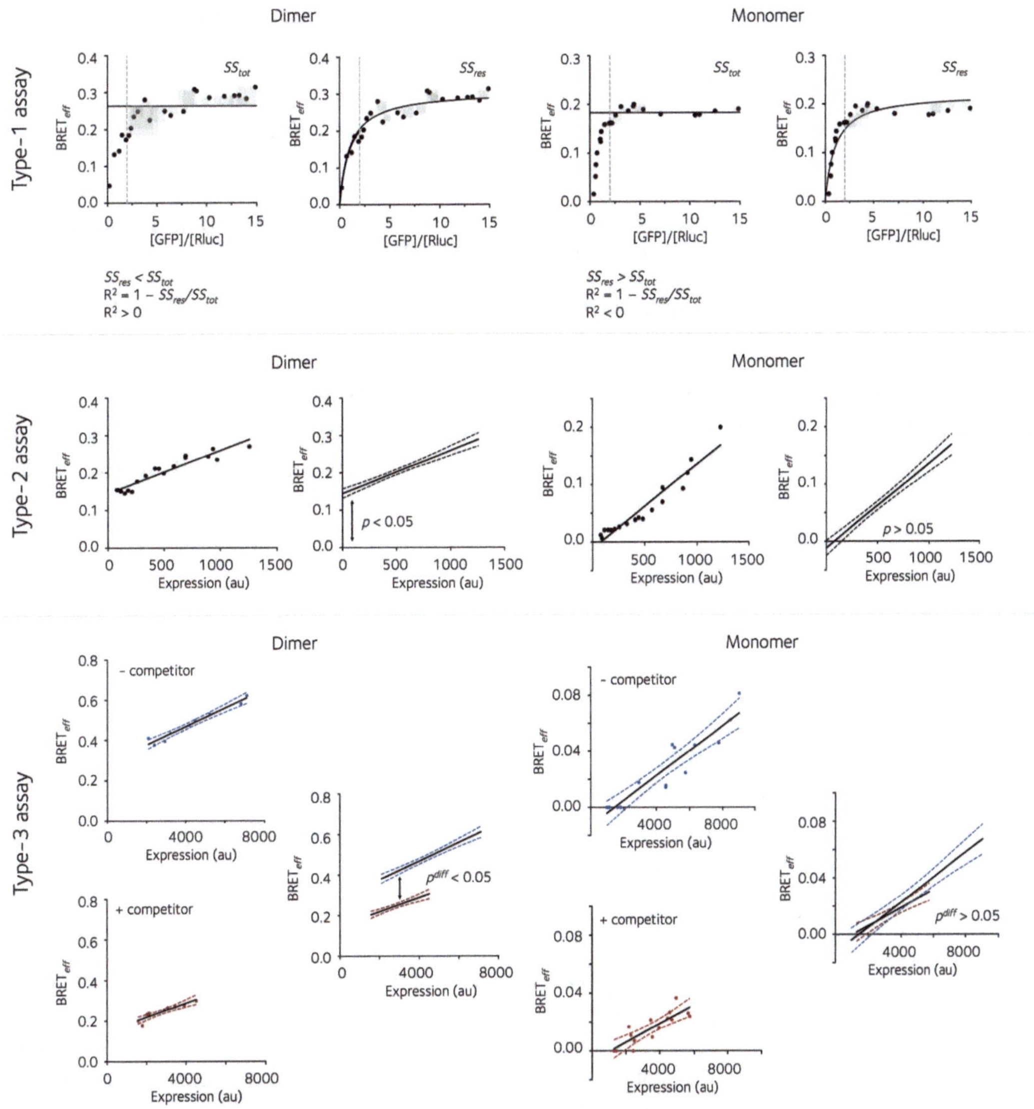

Fig. 3 Summary of data analysis approaches for the types-1, -2, and -3 BRET assays. In the type-1 assay (top), $BRET_{eff}$ is plotted against [GFP]/[Rluc] and fitted to a dimer model (hyperbolic; right) for [GFP]/[Rluc] values between 2 and 15. The R^2 value of this fit is calculated to determine if this is a better fit than a monomer model (constant; left). This compares the residual sum of squares for the dimer model fit (SS_{res}) to that of a flat line (i.e. the monomer model; SS_{tot}). R^2 is determined as 1—SS_{res}/SS_{tot}. For a dimer, SS_{res} is smaller than SS_{tot}, so R^2 is positive. For a monomer, SS_{res} is larger than SS_{tot}, so R^2 is negative. Type-2 assay data (middle) are plotted as $BRET_{eff}$ vs. expression (Rluc) and fitted to a linear least-squares regression model. Stoichiometry is inferred from the 95% confidence limits of this fit at $y = 0$. A significantly non-zero y intercept indicates dimerization or oligomerization, whereas monomers exhibit an intercept that does not significantly deviate from zero. For type-3 BRET assays (bottom), both datasets (i.e. with and without competitor) are fitted to a linear regression least squares model. The difference in the elevation of the two fits is tested using a t test to determine the probability that the two fits came from samples with identical t distributions. A significant difference in the linear regression models results in a p^{diff} value below 0.05, whereas a p^{diff} over 0.05 indicates no significant difference. The existence of a significant difference between datasets indicates the presence of dimers or oligomers, whereas no difference is observed for monomers

3.7.3 Type-3 Assay

1. Plot total expression vs. normalized $BRET_{eff}$. Plot data collected with competitor (pU) and without competitor (pU^{empty}) as separate datasets.
2. Model each dataset to separate linear least squares regressions using appropriate analytical software.
3. Use a 2-tailed *t* test to determine the probability (p^{diff}) that the two datasets fit significantly different linear regressions (*see* **Note 13**).
4. If $p^{diff} > 0.05$ the two datasets are not significantly different, suggesting that the protein of interest is a monomer. If $p^{diff} < 0.05$ the two datasets are significantly different, suggesting the protein of interest is a dimer or oligomer (*see* **Note 14**; Fig. 3).

4 Notes

1. While these assays have been optimized for use in HEK-293T cells, they can in principle be performed in any cell line. If alternative cell lines are to be used the means of transfection may differ from that described above, and the amounts of DNA and incubation times must be re-optimized.
2. Antibiotics may also be added to growth medium if desired and will not affect the assay.
3. Firm hitting of tissue culture flasks during cell detachment can cause cells to clump together, which will ultimately lead to reduced transfection efficiency and lower reproducibility between experiments. Other detachment strategies can be used (e.g. 1 mM EDTA), however care should always be taken to avoid cell aggregation.
4. The reason for using separate plates for the type-2 and -3 assays is because the BRET assay will be performed at different times for each well and so this avoids disruption of wells not being measured. If assaying several proteins, it is often most efficient to do multiples of 6 (type-2 assay) or 3 (type-3 assay) targets per experiment as this allows all wells of each plate to be used.
5. Any positive BRET control consisting of a GFP-Rluc fusion is suitable.
6. As a cytoplasmic protein, the positive control is typically far more highly expressed than transmembrane proteins of interest and hence requires additional dilution to avoid saturation of the detector. If an alternative BRET positive control is used this may need re-optimizing. Similarly, if a sample protein has extremely high expression then cells can also be diluted more thoroughly, however in this instance a reduced incubation time

post-transfection may be preferable as it will minimize non-specific BRET due to high protein density.

7. Rluc luminescence measured using coelenterazine 400A as substrate decreases rapidly with time and hence the signal should be measured as soon as possible once substrate is added. Brighter variants of Rluc, e.g. Rluc8, should still be measured immediately, although the window allowing precise measurements is longer.
8. Always perform the Rluc measurement after the GFP^2 measurement to avoid complications of Rluc emission in the GFP^2 channel. Emission does not decay as rapidly with coelenterazine-h as with coelenterazine-400A and hence measurement is not performed immediately. Nonetheless, care should be taken to ensure equivalent incubation times across samples to allow accurate comparisons. This is most easily achieved using a plate-reader with automated fluidics. However, the substrate can also be added at the rate the instrument moves between wells during measurement to ensure a consistent incubation time.
9. Determining total expression in this way helps to minimize the effect of slight variations in [GFP]/[Rluc] between pU and pU^{empty} samples. Nonetheless, [GFP]/[Rluc] should be broadly similar between the two datasets. If it deviates substantially, confirm the concentrations of the DNA preparations used in order to avoid variable transfection with pU and pU^{empty}.
10. The set of 2–15 [GFP]/[Rluc] values is used because this represents the range over which models of different stoichiometries are most easily distinguished. Below a [GFP]/[Rluc] of 2, the independence of $BRET_{eff}$ on [GFP]/[Rluc] for a monomer breaks down and $BRET_{eff}$ converges on zero. Above a [GFP]/[Rluc] of 15, the sigmoidal relationship for dimers and oligomers flattens to the point that it is not possible to distinguish a dimer from a monomer model given the precision of the assay. We typically collect and plot data at [GFP]/[Rluc] values between 0 and 2 to confirm that $BRET_{eff}$ reduces to zero as expected, however these measurements are not used for data analysis.
11. Expression is determined as Rluc emission alone for the type-2 assay, rather than as a function of RLU and [GFP]/[Rluc] as in the type-3 assay. This is because the signal-to-noise of Rluc measurements is far better than that of GFP^2 measurements due to the lack of substantial luminescence background vs. fluorescence background. Hence, at early time points, RFU values are less precise than RLU and so would generate unreliable expression values if used as in the

type-3 assay. Nonetheless, we recommend collecting RFU values for all points in order to confirm that [GFP]/[Rluc] does not vary substantially across time.

12. If the earliest data points are collected at too high an expression level in the type-2 assay the confidence limits of the *y*-intercept will be extremely broad and hence may miss-identify an oligomer as a monomer. This is why early time points should be measured while expression is near the lower sensitivity limit of the detector used.
13. In order to allow confident comparison of the two datasets in the type-3 assay, there must be at least a partial overlap of the expression levels. Expression is typically lower in the pU-transfected samples, however later time points for these samples should overlap with the expression measured for earlier time points of pU^{empty}-transfected samples. If this is not the case, optimize incubation time post-transfection to achieve overlap.
14. This assumes that $BRET_{eff}$ at a given expression level is lower in the presence of competitor (pU) than its absence (pU^{empty}). We have never experienced $BRET_{eff}$ increasing in the presence of competitor, but in principle this could occur for proteins with complex factors influencing stoichiometry and/or organization (e.g. if competitor expression leads to internalization of the tagged protein). In such cases, the type-3 assay is not suitable for determining stoichiometry.

Acknowledgements

This work was supported by the Wellcome Trust (098274/Z/12/Z to S.J.D., 107375/Z/15/Z to J.H.F.).

References

1. James JR, Oliveira MI, Carmo AM, Iaboni A, Davis SJ (2006) A rigorous experimental framework for detecting protein oligomerization using bioluminescence resonance energy transfer. Nat Methods 3(12):1001–1006. https://doi.org/10.1038/nmeth978
2. Felce JH, Knox RG, Davis SJ (2014) Type-3 BRET, an improved competition-based bioluminescence resonance energy transfer assay. Biophys J 106(12):L41–L43. https://doi.org/10.1016/j.bpj.2014.04.061
3. Latty SL, Felce JH, Weimann L, Lee SF, Davis SJ, Klenerman D (2015) Referenced single-molecule measurements differentiate between GPCR oligomerization states. Biophys J 109(9):1798–1806. https://doi.org/10.1016/j.bpj.2015.09.004
4. Felce JH, Latty SL, Knox RG, Mattick SR, Lui Y, Lee SF, Klenerman D, Davis SJ (2017) Receptor quaternary organization explains G protein-coupled receptor family structure. Cell Rep 20(11):2654–2665
5. Felce JH, Davis SJ (2012) Unraveling receptor stoichiometry using bret. Front Endocrinol 3:86–86. https://doi.org/10.3389/fendo.2012.00086

Chapter 11

Combining SRET^2 and BiFC to Study GPCR Heteromerization and Protein–Protein Interactions

Amina M. Bagher, Melanie E. M. Kelly, and Eileen M. Denovan-Wright

Abstract

G protein-coupled receptors (GPCRs) are the target for many drugs. Evidence continues to accumulate demonstrating that multiple receptors form homo- and heteromeric complexes, which in turn dynamically couple with G proteins, and other interacting proteins. Here, we describe a method to simultaneously determine the identity of up to four distinct constituents of GPCR complexes using a combination of sequential bioluminescence resonance energy transfer 2—fluorescence resonance energy transfer (SRET^2) with bimolecular fluorescence complementation (BiFC). The method is amenable to moderate throughput screening of changes in response to ligands and time-course analysis of protein–protein oligomerization.

Key words BiFC, Bimolecular fluorescence complementation, BRET^2, Bioluminescence energy transfer 2, FRET, Fluorescence resonance energy transfer, SRET^2, Sequential BRET^2-FRET

1 Introduction

Class A G protein-coupled receptors (GPCRs) physically associate to form homomers and heteromers when expressed in heterologous expression systems. In addition, the existence of GPCR heteromers in native tissues and animal models has been demonstrated (reviewed in [1–7]). Oligomerization of class A GPCRs can affect nearly every aspect of GPCR functions including biosynthesis, trafficking, ligand pharmacology, signal transduction, and internalization [4–6, 8]. GPCR oligomerization is considered a dynamic state and the composition and number of individual proteins influences the function of the GPCR complex. As such, there is considerable interest to determine the number and identify of GPCR subunits involved in the formation of oligomeric complexes and examine change in complexes over time in the presence and absence of specific ligands. Changes in the absolute levels of receptors during development or pathophysiological processes are likely to influence the relative proportion and distribution of distinct GPCR oligomeric units [4–6, 8].

Mario Tiberi (ed.), *G Protein-Coupled Receptor Signaling: Methods and Protocols*, Methods in Molecular Biology, vol. 1947, https://doi.org/10.1007/978-1-4939-9121-1_11, © Springer Science+Business Media, LLC, part of Springer Nature 2019

While techniques such as co-immunoprecipitation and immunohistochemical co-localization have provided evidence of protein–protein interactions, the field of GPCR oligomerization developed rapidly after the development of techniques such as bioluminescence resonance energy transfer 2 ($BRET^2$) and Förster resonance energy transfer (FRET) that allowed for the detection and quantification of protein–protein interactions in intact cells [9–11]. $BRET^2$ employs engineered expression vectors that produce recombinant proteins expressed from cDNAs of proteins of interest in frame with recombinant luciferase (Rluc) or green fluorescent protein 2 (GFP^2) cDNA. Rluc can hydrolyze Coelenterazine 400a and the product emits light between 290 and 405 nM. In $BRET^2$, if the Rluc tagged partner is in close proximity (approximately 50–100 Å) to the GFP^2 tagged partner and Coelenterazine 400a is present in saturating amounts, there will be a non-radiative resonance energy transfer from the Rluc donor to the GFP^2 acceptor, which leads to a subsequent fluorescent emission at 505–510 nM [9, 10] (Fig. 1). The efficiency of energy transfer is dependent on the relative distance between the donor and acceptor molecules and the relative orientation of the fluorophores. To study the interaction between three proteins, $BRET^2$ was combined with FRET in a technique called sequential $BRET^2$-FRET ($SRET^2$). In $SRET^2$, an Rluc donor can excite a GFP^2 acceptor, which in turn acts as a donor to enhanced yellow fluorescent protein (EYFP) (Fig. 2) [12, 13]. $SRET^2$ was originally used to demonstrate that the adenosine receptor A_{2A} (A_{2A}), the dopamine receptor type 2 long form (D_{2L}), and the cannabinoid receptor type 1 (CB_1) form heterotrimers [12, 13]. Protein complementation approaches such as bimolecular fluorescence complementation (BiFC) have also been used to study interacting pairs of proteins. For BiFC, proteins of interest are engineered as non-fluorescent subfragments of a fluorescent protein such as EYFP Venus [14, 15]. If the proteins of interest are stably associated in close proximity, the non-fluorescent fragments can physically interact to reconstitute fluorescent Venus with emission at 530 nm (Fig. 3). The Venus acceptor protein in these analyses is composed of the protein of interest fused to the Venus N-terminal hemiprotein (VN) and a second potentially interacting protein of interest fused to the Venus C-terminal hemiprotein (VC) [16] (Fig. 3).

In our work [17], we combined $SRET^2$ and BiFC to simultaneously detect the interaction of four GPCRs (Fig. 4). Here we describe the combined $SRET^2$-BiFC method to identify interacting GPCRs and to quantify interactions between the GPCRs and interacting ancillary proteins including G proteins and other allosteric modulators of G protein complex function. The method is amenable to determination of the relative composition of complexes in the presence of defined doses of ligand. This approach extends the range of molecular pharmacological tools to fully understand

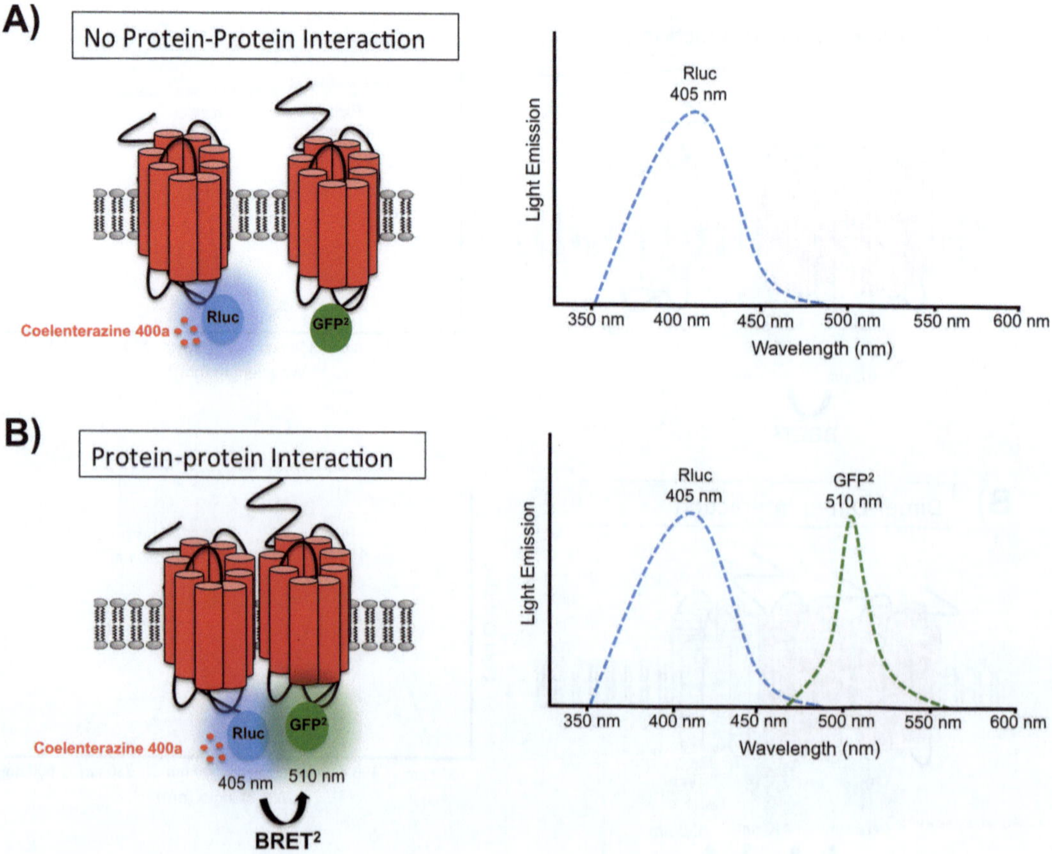

Fig. 1 Bioluminescence Resonance Energy Transfer 2 (BRET^2). GPCRs are tagged at their carboxy-termini with either Rluc or GFP^2. (**a**) The left panel illustrates when the tagged GPCRs are not interacting. Following the addition of the Rluc substrate Coelenterazine 400a it is oxidized by Rluc, causing Rluc to emit blue light at ~405 nm, but no energy is transferred to the acceptor GFP^2, and therefore no green light is emitted. The right panel illustrates the emission spectra for co-expressed Rluc and GFP^2 in the presence of Coelenterazine 400a when the Rluc and GFP^2 are not in close proximity. When Rluc and GFP^2 are not in close proximity resonance energy transfer does not occur; resulting in a peak at 405 nm from Rluc emission. (**B**) When the tagged GPCRs are interacting, the oxidation of Coelenterazine 400a by Rluc emits blue light, which is transferred to the acceptor GFP^2 when it is in close enough proximity to Rluc. This allows resonance energy transfer to occur, causing GFP^2 excitation, resulting in the emission of green light at ~510 nm. The right panel shows the emission spectra for co-expressed Rluc and GFP^2 in the presence of Coelenterazine 400a when the Rluc and GFP^2 are sufficiently close to allow for resonance energy transfer to occur. This results in two peaks in the emission spectra; one at ~405 nm and one at ~510 nm. BRET^2 signals are measured as the ratio of the 510 nm to the 405 nm peaks

GPCR protein complex structure and function. We provide examples based on our study of CB_1 and D_{2L} heterotetrameric complexes in HEK293A cells [17]. Our technique was based on the techniques described in [12, 13].

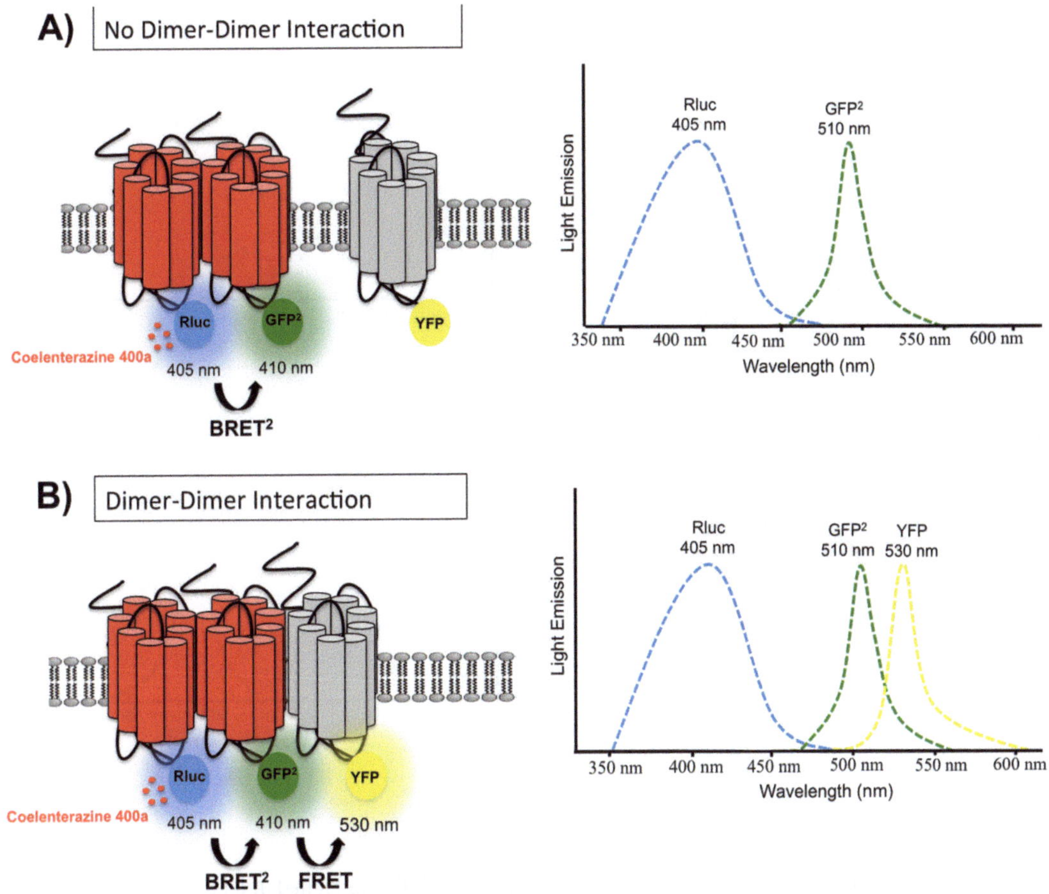

Fig. 2 Sequential Resonance Energy Transfer 2 (SRET^2). GPCRs are tagged at their carboxy-termini with Rluc, GFP^2, or EYFP. (**a**) The left panel shows Rluc and GFP^2 tagged GPCRs interacting. Thus, on the addition of the Rluc substrate, Coelenterazine 400a, the oxidation of Coelenterazine 400a by Rluc tagged GPCRs triggers acceptor excitation of GFP^2 tagged GPCRs by BRET^2. Since Venus tagged GPCR does not interact with Rluc and the GFP^2 tagged GPCRs, no energy transfer occurs from GFP^2 tagged GPCRs to Venus tagged GPCR by FRET. In the right panel, emission spectra for co-expressed Rluc and GFP^2 in the presence of Coelenterazine 400a when the Rluc and GFP^2 are in close proximity and resonance energy transfer can occur. There are only peaks at 405 and 510 nm. (**b**) The left panel shows Rluc, GFP^2, or Venus tagged GPCRs interacting. In the left panel, as a result of this, on the addition of Coelenterazine 400a, the oxidation of Coelenterazine 400a by Rluc emits blue light and triggers the excitation of the acceptor GFP^2 by BRET^2, which emits green light. Since Venus-tagged GPCR is now in close enough proximity to GFP^2−tagged GPCRs, resonance energy transfer does occur to the acceptor Venus by FRET. In the right panel, emission spectra for co-expressed Rluc, GFP^2, and Venus tagged GPCR^2 in the presence of Coelenterazine 400a, when the Rluc, GFP^2 and Venus-tagged GPCR^2 are sufficiently close to allow resonance energy transfer to occur by BRET^2 and FRET. There will be three peaks at 405 nm resulting from Rluc emission, at 510 nm resulting from GFP^2 emission, and 530 nm resulting from Venus emission. Net SRET^2 signals are measured as the ratio of the 530 nm to the 405 nm peaks

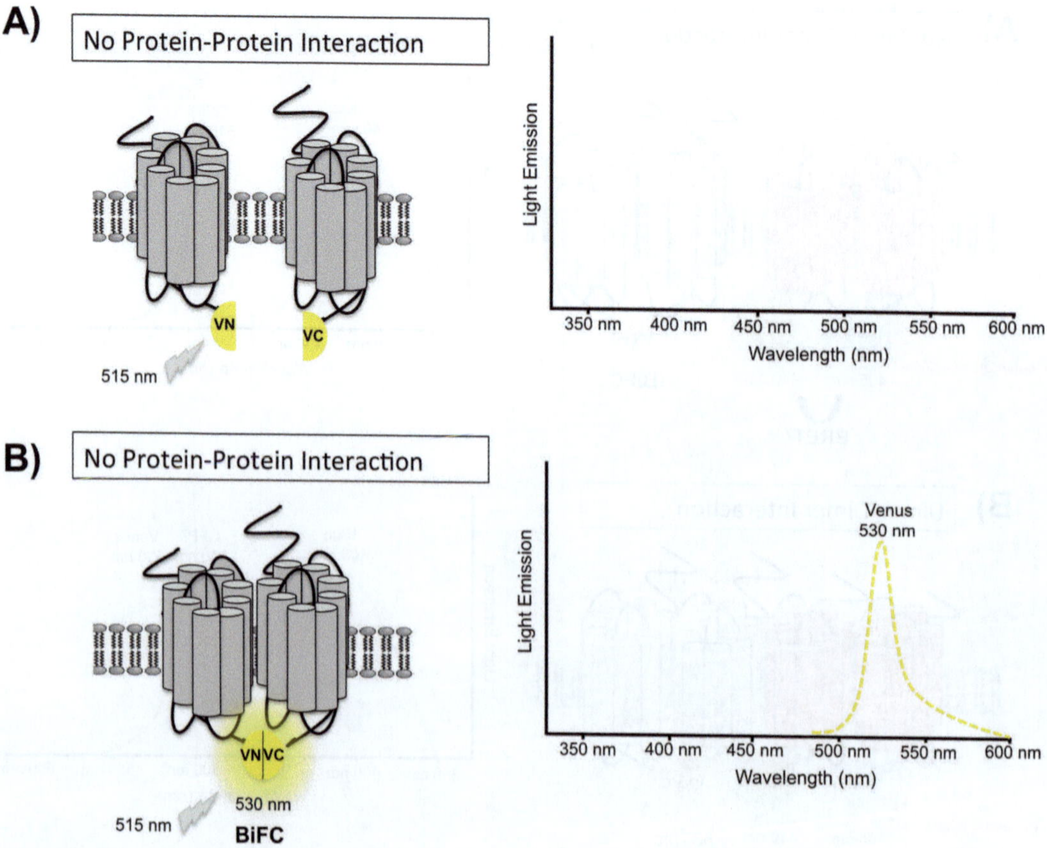

Fig. 3 Bimolecular Fluorescence Complementation (BiFC). GPCRs are tagged at their carboxy-termini with non-fluorescent protein fragments of the enhanced yellow fluorescent protein (EYFP) Venus, the EYFP Venus N-terminal (VN) or EYFP Venus C-terminal (VC). (**a**) The left panel illustrates tagged GPCRs that are not interacting. In this case, the two Venus fragments do not come into close proximity and there is no fluorescence. The right panel illustrates the emission spectra for co-expressed Venus-VN and Venus-VC tagged GPCRs when the Venus-VN and Venus-VC tagged are not in close proximity resulting in no detectable signal using an excitation filter of 515 nm and an emission filter of 530 nm. (**b**) The left panel illustrates tagged GPCRs interactions. As a result of the interaction, the two Venus fragments associate and refold allowing fluorescence to occur. The right panel illustrates the emission spectra for co-expressed Venus-VN and Venus-VC tagged GPCRs when the Venus-VN and Venus-VC are in close proximity allowing the two fragments associate, resulting in a detectable signal using an excitation filter of 515 nm and an emission filter of 530 nm

2 Materials

Prepare all solutions using ultrapure water and analytical grade reagents. Prepare and store all reagents at room temperature (unless indicated otherwise).

2.1 Reagents

1. cDNA for proteins of interest cloned into pcDNA3.1 (+) plasmid.
2. cDNA for negative control non-interacting receptor cloned into pcDNA3.1 (+) plasmid (*see* **Note 1**).

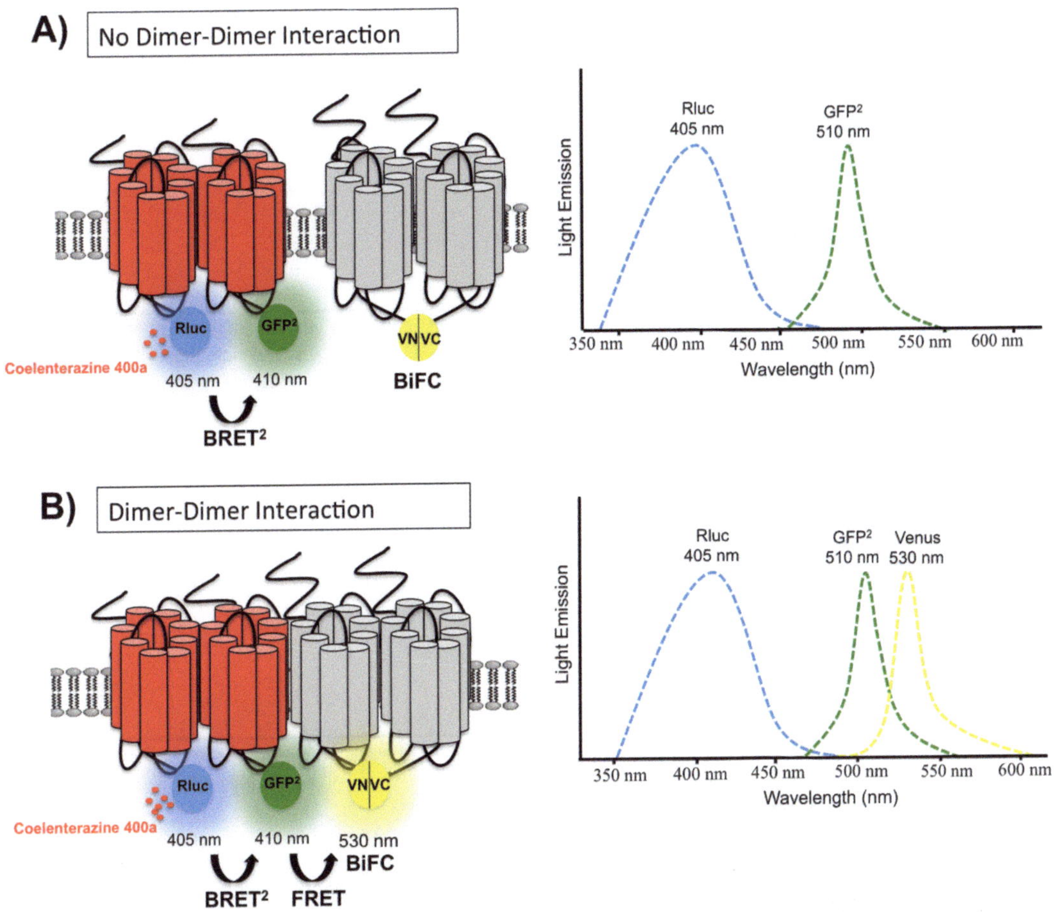

Fig. 4 Sequential Resonance Energy Transfer 2 ($SRET^2$) Combined with Bimolecular Fluorescence Complementation (BiFC). GPCRs are tagged at their carboxy-termini with Rluc, GFP^2, or Venus fragments (Venus-VN and Venus-VC). (**a**) The left panel shows Rluc- and GFP^2-tagged GPCRs interacting. Thus, on the addition of the Rluc substrate, Coelenterazine 400a, the oxidation of Coelenterazine 400a by Rluc tagged GPCRs triggers acceptor excitation of GFP^2 tagged GPCRs by $BRET^2$. Since Venus-VN and Venus-VC tagged GPCRs interact together, but not with Rluc and GFP^2 tagged GPCRs, no energy transfer occurs from GFP^2 tagged GPCRs to Venus tagged GPCRs by FRET. In the right panel, emission spectra for co-expressed Rluc and GFP^2 in the presence of Coelenterazine 400a when the Rluc and GFP^2 are in close proximity and resonance energy transfer can occur. There is only a peak at 405 and 510 nm. (**b**) The left panel shows Rluc, GFP^2, or Venus-tagged GPCRs interacting. In the left panel, as a result of this, on the addition of Coelenterazine 400a, the oxidation of Coelenterazine 400a by Rluc emits blue light and triggers the excitation of the acceptor GFP^2 by $BRET^2$, which emits green light. Since Venus-tagged GPCRs are now in close enough proximity to GFP^2-tagged GPCRs, resonance energy transfer does occur to the acceptor Venus by FRET. In the right panel, emission spectra for co-expressed Rluc, GFP^2 and Venus-tagged $GPCR^2$ in the presence of Coelenterazine 400a, when the Rluc, GFP^2, and Venus-tagged $GPCR^2$ are sufficiently close to allow resonance energy transfer to occur by $BRET^2$ and FRET. There will be three peaks at 405 nm resulting from Rluc emission, at 510 nm resulting from GFP^2 emission, and 530 nm resulting from Venus emission. Net $SRET^2$ signals are measured as the ratio of the 530 nm to the 405 nm peaks

3. pRluc-N1 vector (Rluc expressing vector, PerkinElmer).
4. pGFP^2-N3 vector (GFP^2 expressing vector, PerkinElmer).
5. GFP^2-Rluc vector (GFP^2-Rluc fusion construct, PerkinElmer).
6. pEYFP vector (Venus expression vector, Addgene).
7. Venus N-terminal hemiprotein (VN) vector pBiFC-VN173 (Addgene), Venus C-terminal hemiprotein (CN) vector pBiFC-VC155 (Addgene).
8. pcDNA3.1 (+) empty vector (Thermo Fisher Scientific).
9. One Shot® TOP10 Chemically Competent *E. coli* (Thermo Fisher Scientific).
10. GenElute™ Plasma Miniprep Kit (Sigma-Aldrich).
11. Human embryonic Kidney 293A (HEK293A) cells.
12. Lipofectamine® 2000 reagent (Thermo Fisher Scientific). Store at 4 °C.
13. Opti-MEM® Reduced-Serum Medium (Thermo Fisher Scientific). Store at 4 °C.
14. Coelenterazine 400a (Biotium) in anhydrous ethanol as a stock solution. Store aliquot at −20 °C protected from light.
15. 6-Well plate for cell culture and transfection.
16. 96-Well white microplates for SRET^2.

2.2 Buffers

1. Complete Dulbecco's Modified Eagle's Medium (DMEM) media: Add to 445 ml of DMEM 50 ml of 10% FBS and 5 ml of 10^4 U/ml penicillin/streptomycin. Store at 4 °C.
2. Serum-free DMEM media: Add to 495 ml DMEM 5 ml of 10^4 U/ml penicillin/streptomycin (exclude for transfections). Store at 4 °C.
3. 10× phosphate buffered saline (PBS). Store at room temperature until dilution into the working solution (1× PBS). Sterilize by autoclaving.
4. Assay buffer: 1× PBS supplemented with glucose (1 mg/ml), benzamidine (10 mg/ml), leupeptin (5 mg/ml), and a trypsin inhibitor (5 mg/ml) [9].

2.3 Equipment

1. Cell culture incubator at 37 °C with 5% CO_2.
2. Fluorescence microscope (Olympus IX81) with excitation filters at 488 nm, and emission filter at 530 nm for Venus visualization.
3. FLx800 fluorescence plate reader (BioTek Instruments Inc.) with excitation filters at 485 and 515 nm and emission filters at 510 and 530 nm.

4. Luminoskan Ascent plate reader (Thermo Scientific) with detection filter 405 and 530 nm to measure $SRET^2$ signals.
5. GraphPad Prism v. 6 or higher.

3 Methods

Carry out all procedures at room temperature unless otherwise specified.

3.1 Generation of Fusion Constructs

1. Generate fusion constructs containing cDNAs of the proteins of interest cloned in-frame with the cDNA of the bioluminescent or fluorescent donor or acceptor molecules using Rluc-N1, GFP^2-N3, or Venus (pBiFC-VN173 and pBiFC-VC155) expression vectors (*see* **Note 2**).
2. Amplify proteins of interest using polymerase chain reaction (PCR) without their stop codon utilizing a high-fidelity DNA polymerase and using appropriate forward and reverse primers (*see* **Note 3**).
3. Fractionate the PCR products on an agarose gel. Extract bands of the expected size from the agarose gel, isolate the PCR product from the gel using any standard method and digest with appropriate restriction enzymes. Perform the same restriction enzyme digestions on the vectors. Inactivate the restriction enzymes and ligate PCR fragments into the vectors using a T4 DNA ligase overnight at 4 °C.
4. Transform the ligation mix into One Shot® TOP10 Chemically Competent *E. coli* and plate on agar plates containing either kanamycin (30 μg/ml) for selection of protein-Rluc, zeocin (25 μg/ml) for selection of protein-GFP_2, or ampicillin (100 μg/ml) for selection of protein-VN and protein-VC constructs. Incubate plates overnight at 37 °C to allow individual colonies to form.
5. Isolate positive colonies and allow growing overnight in 2 ml Luria-Bertani (LB) broth containing appropriate antibiotics. Extract plasmids using a GenElute™ Plasma Miniprep Kit, and identify clones containing appropriate inserts by restriction digestions of each individual DNA sample followed by gel electrophoresis. Subject a clone containing appropriate sized insert to bidirectional sequencing using universal primers to confirm their full cDNA sequence and reading-frame.

3.2 Transient Cell Transfection

1. Maintain HEK293A cells in complete DMEM at 37 °C and 5% CO_2.
2. Plate cells to be transformed in 6-well plate (10 cm^2/ml) with complete DMEM for 24–48 h, until cells reached 90% confluence.

3. Transfect HEK293A cells using Lipofectamine® 2000 reagent following the manufacturer's protocol. Briefly, each well of the 6-well plate receives 400 μg of the required plasmid(s) diluted in 250 μl Opti-MEM® Reduced-Serum Medium. Keep the total amount of DNA/well constant by using a pcDNA3.1+ empty vector as required. Mix diluted plasmid DNA with 250 μl Opti-MEM® Reduced-Serum Medium containing 5 μl of Lipofectamine® 2000 reagent. Incubate the solution at room temperature for 20 min and then add it to a well of the 6-well plate containing fresh serum-free DMEM media. Cells are cultured for 48 h.
4. Prior to performing experiments to determine protein–protein interactions, validate that fusion constructs can be expressed and are appropriately localized in HEK293A cells (*see* **Note 4**). For the protein-Rluc construct, validation of luminescence activity can be conducted by measuring luminescence following the addition of Coelenterazine 400a, while protein localization is determined using confocal microscopy using an antibody against the specific receptor. For the protein-GFP^2 construct, fluorescence can be measured using FLx800 fluorescence plate reader (BioTek Instruments) with excitation and emission filters at 485 and 510 nm, respectively. Protein-GFP^2-expressing constructs localization can be detected using confocal microscopy. Rluc and GFP^2 controls and empty vector can serve as positive and negative controls. For protein-VN and protein-VC constructs, fluorescence and localization can be detected in cells co-expressing both constructs using a fluorescence microscope (Olympus IX81) with excitation filters at 488 nm and emission filter at 530 nm.
5. Use functional assays of expressed proteins to validate the activity of the fused protein of interest (*see* **Note 5**).

3.3 BRET^2 Analysis

For BRET^2 experiments to test the protein–protein interactions, cells are co-transfected with one protein of interest tagged with Rluc and another protein of interest tagged with GFP^2. To determine if the observed BRET^2 signal is the result of specific protein--protein interaction or random collision due to high-level expression of membrane-bound proteins, BRET^2 saturation analysis for each BRET^2 pair should be completed.

1. Transfect HEK293A cells grown in 6-well plates with a fixed amount of protein of interest-Rluc BRET^2 donor and increasing amounts of protein of interest-GFP^2 BRET^2 acceptor plasmid. Correct total plasmid concentrations per transfection by adding pcDNA3.1 (+) empty vector (*see* **Note 6**).
2. Forty-eight hours after transfection, wash cells twice with cold 1× PBS and resuspend the cells in approximately 1 ml of freshly

prepared assay buffer to achieve a density of 0.20×10^6 cells per ml [9]. Keep this tube on ice during experimental manipulations.

3. Dispense 90 μl of cell suspension (approximately 18,000 cells) into a white 96-well plate.
4. Treat cells with 1 μl of vehicle or ligand if applicable for the experimentally defined period of time.
5. Add 10 μl of 50 μM Coelenterazine 400a and measure emission of Rluc and GFP^2 at 405 and 510 nm, respectively. The 510/405 ratio is converted to $BRET_{Eff}$ by determining the ratio for each sample and subtracting the minimum 510/405 ratio from cells expressing only the Rluc-N1 construct, then dividing the maximum 510/405 ratio obtained from cells expressing the GFP^2-Rluc fusion construct.
6. To generate $BRET^2$ saturation curve, $BRET_{Eff}$ is plotted against the ratio of GFP^2 fluorescence (measured in the absence of Coelenterazine 400a substrate, and with direct excitation of GFP^2) to Rluc bioluminescence emission.
7. Fit curves to a nonlinear regression equation (Hyperbola), assuming a single phase using Graph-Pad Prism software (*see* **Note** 7).
8. $BRET_{Max}$ (B_{Max}) and $BRET_{50}$ (K_d) can be obtained from the $BRET^2$ saturation curve. The $BRET_{50}$ value can serve as an estimate of the affinity of the donor and acceptor molecules. The $BRET_{Max}$ value reflects the distance, relative orientation and expression levels of the donor and acceptor molecules (Fig. 5).

3.4 BiFC

To test the interaction of a separate set of interacting proteins, the proteins of interest are engineered to express the Venus N-terminal (VN) (protein-VN) and Venus C-terminal (VC) (protein-VC) constructs.

1. Transiently transfect HEK293A cells grown in 6-well plates with intact pEYFP construct, protein-VN construct alone, and protein-VC construct. Co-transfect cells with protein-VN and protein-VC constructs at 1:1 ratio. The intact pEYFP construct serves as a positive control while protein-VN alone and protein-VC alone serve as negative controls.
2. Forty-eight hours after transfection, wash cells twice with cold 1× PBS buffer and suspended in 90 μl of assay buffer.
3. Dispense cells into white 96-well plates and Venus fluorescence is measured using the FLx800 fluorescence plate reader with excitation at 515 nm and emission measured at 530 nm.
4. Vary the absolute amount of 1:1 protein-VN and protein-VC constructs mix to determine the maximum Venus levels relative

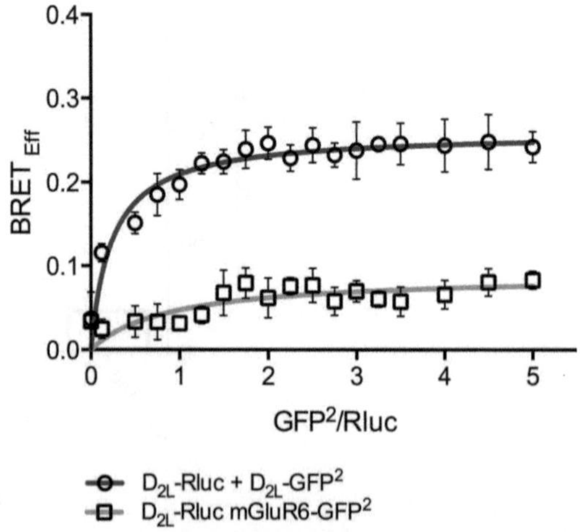

Fig. 5 The BRET^2 saturation curve is used to determine if the interaction between Rluc and GFP^2 tagged proteins are specific and saturable. BRET^2 saturation curves obtained from cells transiently transfected with constant amounts of protein-Rluc-tagged BRET^2 donor (D_{2L}-GFP^2) and increasing amounts of protein-GFP^2-tagged BRET^2 acceptor (D_{2L} -GFP^2) or a negative control (mGLuR6-GFP^2). $BRET_{Eff}$ was plotted against the ratio of GFP^2 fluorescence and Rluc emission. The data were fit to a rectangular hyperbola. $BRET_{Max}$ (B_{Max}) and $BRET_{50}$ (K_d) are obtained from the BRET^2 saturation curve. The D_{2L} saturation curve resulted in $BRET_{50}$ value of 0.28 ± 0.04 and 0.26 ± 0.02

to the intact pEYFP construct. Both construct pairs should be tested to ensure efficient reconstitution of Venus from the VN: VC fragments (Fig. 6).

3.5 SRET^2 Combined with BiFC

Once specific interaction between Rluc- and GFP^2-tagged proteins have been confirmed using BRET^2 saturation curves and the interaction between VC- and VN-tagged proteins has been confirmed using BiFC, SRET^2 combined with BiFC can be conducted to test if the four proteins of interest simultaneously interact. In SRET^2 experiments, the oxidation of an Rluc substrate triggers GFP^2 excitation by BRET^2 and energy transfer occurs to the FRET acceptor Venus. In this case, the Venus acceptor results from BiFC interaction (Fig. 4). It is critical to determine and transfect the optimal ratio of cDNA (*see* **Note 8**).

1. Transiently transfect HEK293A cells grown in 6-well plates with various amounts of plasmid encoding fusion proteins (protein-Rluc, protein-GFP^2, protein-NV, and protein-CV). In our example, cells were transfected with D_{2L}-Rluc, D_{2L}-GFP^2, CB_1-VN, and CB_1-VC or the corresponding negative controls as shown in Fig. 7a.
2. Forty-eight hours after transfection, wash cells for each sample twice with cold 1× PBS and resuspended in freshly prepared

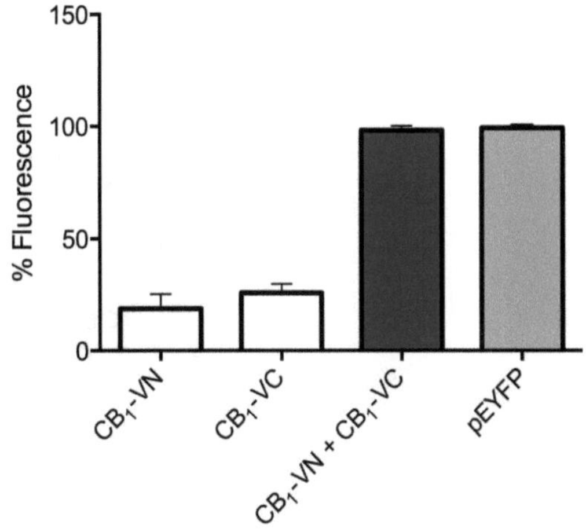

Fig. 6 BiFC is used to confirm protein - protein interaction. HEK 293A cells were transfected with CB_1-VC, CB_1-VN, pEYFP or CB_1-VC and CB_1-VN. CB_1 can form homodimers when expressed in HEK293A cells as measured using BiFC

assay buffer to achieve a density of 0.20×10^6 cells per ml. Keep this tube on ice during experimental manipulations.

3. Divide each cell suspension into aliquots of 90 μl (approximately 18,000 cells).
4. It is important to quantify the expression levels of protein-GFP^2, protein-VN, and protein-VC prior to the $SRET^2$ experiments to ensure that the expression levels of all tagged proteins are similar in all the samples.
5. For quantification of protein-GFP^2 expression level, measure GFP^2 fluorescence emission. Dispense the first 90 μl aliquot of cell suspension into a white 96-well plate and measure GFP^2 emission using the FLx800 fluorescence plate reader with excitation at 410 nm and emission measured at 510 nm. Determine protein-Venus expression by measuring fluorescence emission using an excitation filter of 485 nm and emission measured at 530 nm.
6. Given the close spectral emission of the GFP^2 and Venus, it is important to determine the exact fluorescence emission of protein-GFP^2 and protein-Venus for $SRET^2$ experiments. To determine exact emission of GFP^2 and Venus, measure fluorescence emissions from cells expressing only protein-GFP^2 or protein-Venus using both channels (510 and 530 nm), then normalize the fluorescence signals to the sum of the fluorescence signals using the two detection channels (510 and 530 nm).
7. Use 90 μl aliquot of cells suspension to analyze the transfer of energy from the $BRET^2$ pair donor (Rluc) to the $SRET^2$ acceptor (Venus). ($SRET^2$). In this case, the Venus acceptor

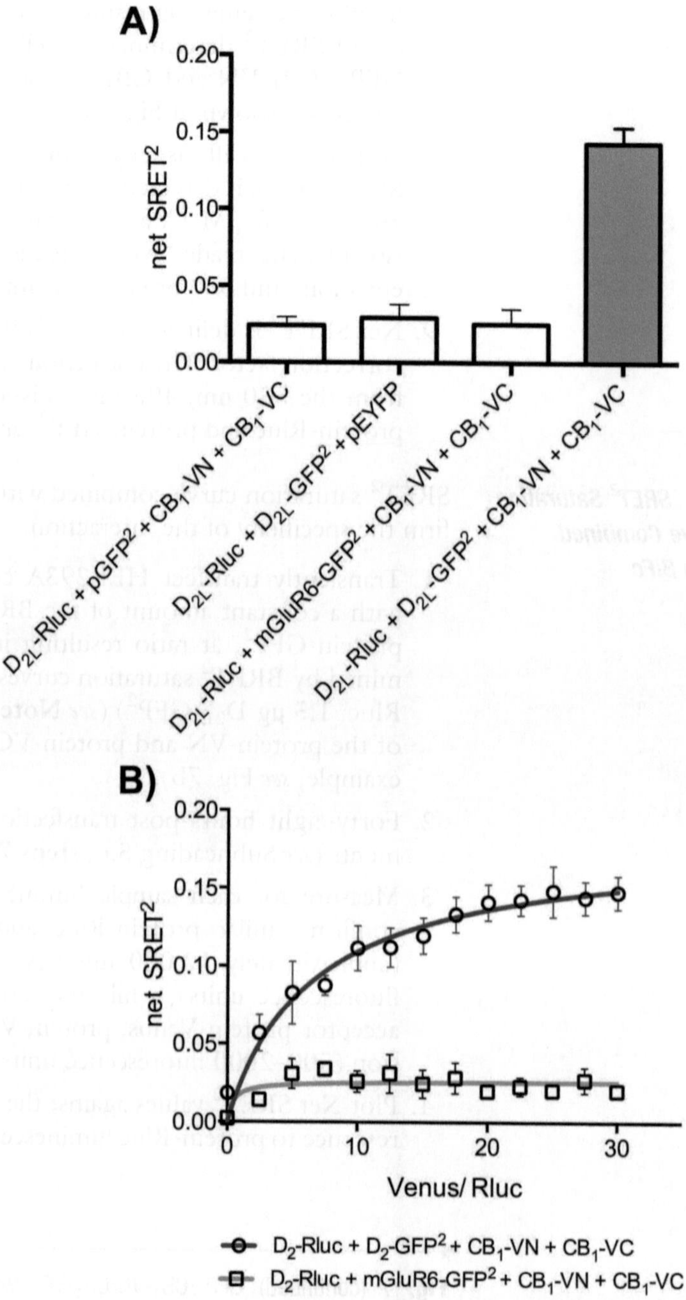

Fig. 7 SRET^2 Combined with BiFC for CB_1 and D_{2L} Receptors. (**a**) SRET^2 combined with BiFC assays were performed 48 h after transfection. Cells were transfected with D_{2L}-Rluc (1 μg of cDNA; 12,000 luminescence units), D_{2L}-GFP^2 (1.5 μg of cDNA; 3000 fluorescence units), CB_1-VN, and CB_1-VC (4 μg of cDNA; 2000 fluorescence units, ratio of 1:1). Net SRET^2 was obtained by measuring the Venus fluorescence emission after Coelenterazine 400a addition, with subtraction of the fluorescence value obtained from cells transfected the same amount of D_{2L}-Rluc and D_{2L}-GFP^2. Significant net SRET^2 was detected for D_{2L}-Rluc/D_{2L}-

fluorescence emission results from BiFC interaction. An example of $SRET^2$ determined in cells expressing D_{2L}-Rluc, D_{2L}-GFP^2, CB_1-VN and CB_1-VC or the corresponding negative controls is shown in Fig. 7a.

8. Dispense the cell suspension into a white 96-well plate and the $SRET^2$ signal is detected immediately following the addition of 10 μl of 50 μM Coelenterazine 400a using a Luminoskan Ascent plate reader with detection filters for 405 nm (Rluc emission) and 530 nm (Venus emission).
9. Net $SRET^2$ is defined as the [(530 emission)/(405 emission)] correction factor. The correction factor is the value determined from the 530 nm/405 nm emission for cells expressing only protein-Rluc and protein-GFP^2 (*see* **Note 9**).

3.6 $SRET^2$ Saturation Curve Combined with BiFc

$SRET^2$ saturation curve combined with BiFC is generated to confirm the specificity of the interaction.

1. Transiently transfect HEK293A cells grown in 6-well plates with a constant amount of the $BRET^2$ pair, protein-Rluc and protein-GFP^2, at ratio resulting in $BRET_{50}$ values as determined by $BRET^2$ saturation curves (in our example 1 μg D_{2L}-Rluc: 1.5 μg D_{2L}-GFP^2) (*see* **Note 8**) and increasing amounts of the protein-VN and protein-VC constructs (1:1 ratio) (for example, *see* Fig. 7b).
2. Forty-eight hours post transfection, record $SRET^2$ measurements (*see* Subheading 3.6, **steps 7–9**).
3. Measure for each sample luminescence and fluorescence to confirm similar protein-Rluc and protein-GFP^2 expression (approximately 12,000 luminescence units and 3000 GFP^2 fluorescence units), while increasing the fluorescence of the acceptor protein-Venus, protein-VN, and protein-VC expression (500–2000 fluorescence units).
4. Plot Net $SRET^2$ values against the ratio of protein-Venus fluorescence to protein-Rluc luminescence emission. Fit curves to a

Fig. 7 (continued) GFP^2/CB_1-VN/CB_1-VC, while negligible net $SRET^2$ was obtained from cells expressing equivalent amounts of D_{2L}-Rluc, CB_1-VN, CB_1-VC and empty pGFP^2-N3 or the negative control mGluR6-GFP^2, or D_{2L}-Rluc/D_{2L}-GFP^2 and pEYFP. (**b**) $SRET^2$ saturation curves were obtained using HEK293A cells transfected with a constant amount of D_{2L}-Rluc and D_{2L}-GFP^2 (1 μg:1.5 μg) and increasing amounts of Venus-tagged CB_1 (CB_1-VN and CB_1-VC at 1:1 ratio). Net $SRET^2$ was plotted against the ratio of Venus fluorescence and Rluc emission. As a negative control, cells were transfected with equivalent amounts of D_{2L}-Rluc + mGluR6-GFP^2, and increasing amounts of Venus-tagged CB_1 (CB_1-VN and CB_1-VC at 1:1 ratio)

nonlinear regression equation (Hyperbola), assuming a single phase using for instance GraphPad Prism software.

5. From the $SRET^2$ saturation curves, determine $SRET_{Max}$ and $SRET_{50}$ values as described for $BRET^2$ saturation assays. A $SRET^2$ saturation curve obtained by increasing CB_1-VN and CB_1-VC expression (the ratio of the CB_1-VN and CB_1-VC was kept constant at a ratio of 1:1) while maintaining the same D_{2L}-Rluc and D_{2L}-GFP^2 ratio is shown in Fig. 7b. In our example, we determined that, for the tetramer D_{2L}-Rluc/D_{2L}-GFP^2/CB_1-VN:CB_1-VC that the $SRET_{Max}$ value was 0.18 ± 0.01 and the $SRET_{50}$ value was 0.13 ± 0.02 (Fig. 7b).

4 Notes

1. The negative control should be considered within the experimental context. Appropriate negative controls include pEYFP, GFP^3-N3, Rluc-N1, and a protein that is known not to interact with your protein of interest. For our work [17], mGluR6 (GenBank accession number: NC_000005.10), which is a GPCR known to be expressed on the cell membranes but not to interact with D_{2L} or CB_1, was used as a negative control.
2. By way of example, we describe the combined $SRET^2$-BiFC method to identify interaction between D_{2L} and CB_1 receptors [17]. The D_{2L} cDNA was cloned into Rluc-N1 and GFP^2-N3 vectors to generate D_{2L}-Rluc and D_{2L}-GFP^2 constructs. The CB_1 was cloned into Venus vector pBiFC-VN173 (VN) and pBiFC-VC155 (VC) to generate CB_1-VN and CB_1-VC constructs. In our example, the human D_{2L} and CB_1 were amplified from D_{2L}-pcDNA3.1 and CB_1-pcDNA3.1, respectively.
3. PCR primers were chosen such that the stop codon and the resulting product could be ligated in-frame with the reporter protein. In our example, the appropriate forward primer possessed an *EcoR1* restriction site and the reverse primer possessed a *Kpn1* restriction site.
4. The amount of fusion proteins transfected should be near the physiological ranges. While heterologous cells are often used for $BRET^2$, BiFC and $SRET^2$, other cell types can be transfected and analyzed using these techniques. In some cases, it may be preferable to test protein–protein interactions in cells that more closely model the endogenous cells of interest. In this case, special consideration must be given for cell lines that express the same protein of interests at high levels as endogenous unlabeled proteins of interest are likely to interact with the transfected labeled protein of interest and influence the quantification of protein interaction.

5. For example, measuring extracellular signal-regulated kinases (ERK) phosphorylation following the addition of GPCR agonists that couple to $G_{\alpha i}$ protein could be used to demonstrate that a given receptor known to couple to $G_{\alpha i}$ is expressed and is functioning in response to ligand as expected. Detailed procedures for functional analyses can be found in [18].
6. Both orientations of the proteins of interest tagged with Rluc and GFP^2 should be tested, as one orientation may be more efficient than the other due to steric constraints.
7. For $BRET^2$ saturation curve, interactions that are specific and saturable, within the range of plasmid ratios tested, fit a rectangular hyperbola, whereas combinations of non-interacting proteins that undergo random collision are described by a linear increase without a plateau.
8. For $SRET^2$ assays, it is important to select the protein of interest-Rluc and protein of interest-GFP^2 cDNA ratio that gives a $BRET_{Eff}$ signal approaching $BRET_{50}$ for $SRET^2$ assays. Higher ratio will lead to excessive fluorescence background, which overlaps with Venus emission resulting in poor $SRET^2$ signal. If the ratio is low, the acceptor protein (Venus) will not be excited [12].
9. A positive control is recommended for $SRET^2$ experiments. The positive control consists of a vector encoding Rluc-GFP^2-EYFP fusion protein. The construction of Rluc-GFP^2-EYFP was described previously [13].

References

1. Rios CD, Jordan BA, Gomes I et al (2001) G-protein-coupled receptor dimerization: Modulation of receptor function. Pharmacol Ther 92:71–87
2. Milligan G (2004) G protein-coupled receptor dimerization: function and ligand pharmacology. Mol Pharmacol 66:1–7
3. Milligan G (2009) G protein-coupled receptor hetero-dimerization: contribution to pharmacology and function. Br J Pharmacol 158:5–14
4. Ferré S, Casadó V, Devi LA et al (2014) G protein-coupled receptor oligomerization revisited: functional and pharmacological perspectives. Pharmacol Rev 66:413–434
5. Ferré S (2015) The GPCR heterotetramer: challenging classical pharmacology. Trends Pharmacol Sci 36:145–152
6. Franco R, Martínez-Pinilla E, Lanciego JL et al (2016) Basic pharmacological and structural evidence for class A G-protein-coupled receptor heteromerization. Front Pharmacol 7:76
7. Gaitonde SA, González-Maeso J (2017) Contribution of heteromerization to G protein-coupled receptor function. Curr Opin Pharmacol 32:23–31
8. Gomes I, Ayoub MA, Fujita W et al (2016) G protein-coupled receptor heteromers. Annu Rev Pharmacol Toxicol 56:403–425
9. James JR, Oliveira MI, Carmo AM et al (2006) A rigorous experimental framework for detecting protein oligomerization using bioluminescence resonance energy transfer. Nat Methods 3:1001–1006
10. Pfleger KD, Seeber RM, Eidne KA (2006) Bioluminescence resonance energy transfer (BRET) for the real-time detection of protein-protein interactions. Nat Protoc 1:337–345
11. Marullo S, Bouvier M (2007) Resonance energy transfer approaches in molecular pharmacology and beyond. Trends Pharmacol Sci 28:362–365

12. Carriba P, Navarro G, Ciruela F et al (2008) Detection of heteromerization of more than two proteins by sequential BRET-FRET. Nat Methods 5:727–733
13. Navarro G, McCormick PJ, Mallol J et al (2013) Detection of receptor heteromers involving dopamine receptors by the sequential BRET-FRET technology. Methods Mol Biol 964:95–105
14. Hu CD, Chinenov Y, Kerppola TK (2002) Visualization of interactions among bZIP and Rel family proteins in living cells using bimolecular fluorescence complementation. Mol Cell 9:789–798
15. Vidi PA, Przybyla JA, Hu CD et al (2010) Visualization of G protein-coupled receptor (GPCR) interactions in living cells using bimolecular fluorescence. Curr Protoc Neurosci. Chapter 5:Unit 5.29
16. Shyu YJ, Liu H, Deng X et al (2006) Identification of new fluorescent protein fragments for bimolecular fluorescence complementation analysis under physiological conditions. Biotech 40:61–66
17. Bagher AM, Laprairie RB, Toguri J et al (2017) Bidirectional allosteric interactions between cannabinoid receptor type 1 (CB_1) and the dopamine receptor type 2 (D_2) agonists mediated through CB_1/D_{2L} heterotetramers (2017). Eur J Pharmacol 813:66–83
18. Bagher AM, Laprairie RB, Kelly MEM et al (2018) Methods to quantify cell signaling and GPCR receptor ligand bias: characterization of drugs that target the endocannabinoid receptors in Huntington's disease. Methods Mol Biol 1780:549–571

Chapter 12

Quantification and Comparison of Signals Generated by Different FRET-Based cAMP Reporters

Andreas Koschinski and Manuela Zaccolo

Abstract

A variety of FRET-based biosensors are currently in use for real-time monitoring of dynamic changes of intracellular cAMP. Due to differences in sensor properties, unique features of the cell type under examination and diverse specifications of the imaging setups in different laboratories, data generated using these sensors may not be immediately comparable within the same study or across studies. To facilitate comparison, often FRET data are normalized and expressed as fractional change of the maximal FRET response at sensor saturation. However, this approach may lead to misinterpretation of the underlying cAMP change. In this chapter, we provide examples of the problems that may arise when using normalized FRET data and present a method based on the conversion of FRET ratio changes into actual cAMP concentrations that mitigates these issues.

Key words Fluorescence resonance energy transfer, FRET, Biosensors, cAMP, Protein kinase A, Real time imaging, Intracellular signaling

1 Introduction

The phenomenon known as Förster Resonance Energy Transfer (FRET) was first described nearly 70 years ago [1]. This non-radiative transfer of energy by resonance between a donor and an acceptor fluorophore strictly depends on the distance and angle between the two fluorophores [2], a feature that has been exploited in various ways to detect changes in distance between protein domains and has been implemented in the development of biosensors to measure chemical compounds or chemical reactions within intact cells. One application of FRET-based biosensors is real-time monitoring of the dynamic changes of the intracellular second messenger cAMP [3–7]. Since their first description about 20 years ago, these sensors have been constantly improved and refined. Currently, a multitude of different variants of these probes are available for detection of cAMP. Most of these sensors share a similar design that includes a protein domain which interacts with

Mario Tiberi (ed.), *G Protein-Coupled Receptor Signaling: Methods and Protocols*, Methods in Molecular Biology, vol. 1947,
https://doi.org/10.1007/978-1-4939-9121-1_12,

cAMP and is fused to two spectral variants of the green fluorescent protein, typically CFP acting as the donor fluorophore and YFP acting as the acceptor fluorophore. Upon binding of cAMP, the sensor changes its conformation so that the distance and/or the angle between CFP and YFP changes, resulting in an increase ("gain of FRET" sensors) or a decrease ("loss of FRET" sensors) in the ratio between acceptor and donor fluorescence emissions intensity (*see* **Note 1**).

One recent application of FRET-based reporters has been their use to define the differences in amplitude and kinetics of cAMP signaling events that occur at distinct subcellular sites. This is typically achieved by targeting the sensor to defined subcellular structures via fusion to a targeting domain [8–10]. A limitation of these applications is that data obtained using different reporters often cannot be directly compared. Most published data are expressed as percent increase of the acceptor/donor emission value relative to the acceptor/donor emission value at baseline, which does not take into account the fact that the baseline itself might be different under different conditions. In addition, changes relative to the baseline are influenced by at least two major factors: (1) the dynamic range of the sensor, and therefore the extent of the FRET change at a certain concentration of cAMP, may vary depending on the cellular environment and (2) each FRET measuring setup will affect the measure by a constant that is setup-specific (*see* **Note 2**).

As these two factors will only influence the absolute value of the FRET change and not the general response of the sensor to cAMP, normalization to the maximal FRET change at saturation measured at a given setup can provide values that are more comparable (*see* **Note 3**). Therefore, normalization is commonly applied to compare the responses of different sensors and to express the changes of an individual sensor [11–14].

Nonetheless, such normalization inherits three major drawbacks.

First, it does not take into account the non-linearity between FRET change and change in cAMP concentration. Second, it does not account for different basal intracellular levels of cAMP. Third, it does not account for different EC_{50} or Hill coefficient values that different sensors may exhibit.

The most accurate method to analyze and compare results obtained with FRET sensors is to calibrate the sensors and express the changes as real cAMP changes. This, however, is a laborious and time-consuming process. As calibration data are now available for a number of cAMP FRET sensors [10, 15, 16], an alternative, at least for some commonly used cell systems, is to use these published values and the tools presented in this chapter to convert measured FRET changes into the underlying cAMP concentration changes.

In this chapter, we provide a detailed description of how to convert FRET ratio changes into cAMP concentration changes to allow a more accurate comparison of data. We also illustrate the drawbacks of normalizing FRET values to maximal FRET change and how the ostensible linearity of the sensor response expressed in percent may lead to misinterpretation of the results. We use data obtained with cAMP-FRET sensors as specific examples, although the same principles apply to any FRET-based reporter.

2 Materials

1. Dataset (own original FRET data set or mean values ± SD/SEM obtained from published data).
2. Maximal FRET change of the sensor starting from basal cAMP concentration (R_{max}) (*see* **Note 4**).
3. The negative FRET change starting from basal cAMP and going to zero cAMP (R_{min}) or, alternatively, information about the basal cAMP concentration (x at R_0) in the specific cell type under consideration (*see* **Note 5**) (some values for basal cAMP are provided in Table 1).
4. The cAMP concentration leading to half-maximal FRET change (EC_{50}) and the Hill coefficient (n) obtained from the concentration-dependency curve for the specific sensor (some information provided in Table 2).
5. Standard spreadsheet calculation program (Excel, Open Office Calc, or similar).
6. Analysis program such as GraphPad Prism or similar. Alternative methods for the interpolation of data are also briefly mentioned.

Table 1
Basal intracellular cAMP-concentration (x at R_0) in selected cell types

Cell type/line	Basal cAMP (μM)	References
Adult cardiomyocte (guinea pig)	1.2	Iancu et al. [17]
Adult cardiomyocte (mouse)	1.3	Boerner et al. [15]
Adult cardiomyocte (rat)	1.3	Our observation
Chinese Hamster ovary cells (CHO)	1.0	Koschinski and Zaccolo [16]
GH_3B_6 (pituitary gland neoplasm, Rat)	0.3	Wachten et al. [18]
Thyroid cells (mouse)	0.4	Boerner et al. [15]

Table 2
EC_{50} and Hill coefficient values for selected cAMP FRET-based sensors

Sensor	EC_{50} (μM)	Hill coefficient *n*	Determination method	References
AKAP18δ-CUTie	7.143	1.045	In cell	Surdo et al. [10]
AKAP79-CUTie	7.179	1.230	In cell	Surdo et al. [10]
AKAR3[a]	2.086	2.971	In cell	Koschinski and Zaccolo [16]
CUTie cytosolic	7.382	1.071	In cell	Surdo et al. [10]
EPAC1-camps	2.5	0.74	In vitro	Boerner et al. [15]
EPAC S-H187	7.406	1.750	In cell	Koschinski and Zaccolo [16]
H9C6 (PKA-based sensor)	5.231	2.191	In cell	Koschinski and Zaccolo [16]
TPNI-CUTie	7.243	1.152	In cell	Surdo et al. [10]

[a]The EC_{50} for cAMP of AKAR3 might vary in different cell systems depending on the activity of PKA and phosphatases

3 Methods

All analysis and conversions discussed below are based on the availability of a calibration curve for the specific sensor. GraphPad Prism or similar biostatistics programs allow for easy generation of a concentration-dependency curve either from original experimental data or using essential parameters obtained from the literature. From this curve the program can then extrapolate values of interest. The program also provides a graph of the curve which is useful to visualize the conversion and can be used for a graphical conversion approach or to control the plausibility of calculated results. The essential equations to calculate the relevant values manually or in any spreadsheet program (e.g. Excel, Open Office, etc.) are also provided.

The method illustrated below focuses on the generation of a curve and the conversion of FRET change values into cAMP concentrations when only the minimal set of values (basal cAMP concentration at R_0, R_{max}, EC_{50}, and Hill coefficient n) are available (*see* **Note 6**). This is the typical case when data from the literature are compared, though the same approach can be used to calculate a curve from existing original data or to convert own FRET data into cAMP concentrations.

3.1 Creating a cAMP-FRET Change Concentration-Dependency Curve with GraphPad Prism

The method that follows describes the essential steps when using the actual GraphPad Prism version 7. As older versions of the software differ slightly, but are still widely in use, we have also included an adapted tutorial for GraphPad Prism version 5.

In the following example, the aim is to generate a [cAMP]-FRET change concentration-dependency curve for the "gain of FRET" AKAP79-CUTie cAMP sensor in mouse cardiac myocytes using parameters from the literature [10, 16] (key values are reported in Tables 1 and 2). All settings (range, scaling) for the graph to be created are optimized to encompass the curve for this sensor, but can easily be adapted for any other sensor.

3.1.1 Procedure for GraphPad Prism 7

1. Open the program. If the "Welcome" window does not open automatically, select "File" from the menu, then "New", then "New Project File".
2. In the "Welcome" window select "XY"-table and graph, X should be "Numbers", Y should be "Enter and plot a single Y value for each point". By clicking the "create"-button an empty table will appear.
3. Enter the normalized values (compare Table 1) for the basal cAMP-concentration R_0 (enter 1.3 μM into the X-column), the respective FRET change value (enter 0 % into the first Y-column) and the maximal FRET change at a saturating cAMP-concentration R_{max} (enter 1000 μM into the X column, enter 100 % into the first Y column) into the table (*see* **Note 7**). Enter only the values, without the units. Naming the columns in the "Title" fields will automatically generate axis labels for the graph that will be created later. Figure 1 shows an example of how to enter the data into the GraphPad Prism Data table.
4. From the "Analysis" menu tab select "Analyze". In the pop-up window open the "Transform, Normalize, . . ." point and select "Transform Concentrations [X]" in the submenu. Click "OK".
5. In the following pop-up window ("Parameters: Transform Concentrations [X]") select "Transform to logarithms". Keep all other parameters to their defaults.
6. Clicking "OK" leads to the next table, now with the concentrations expressed as logarithms. With this table active (open) select again "Analyze". This time select the point "XY analyses", and here "Nonlinear Regression (Curve Fit)". Confirm with "OK".
7. Now the "Parameters: Nonlinear Regression" window will pop up. It can also be opened by clicking on "Nonlin Fit of transform X of Data 1" in the navigation tree (sidebar) and then "Change"—"Analysis Parameters" in the menu bar.

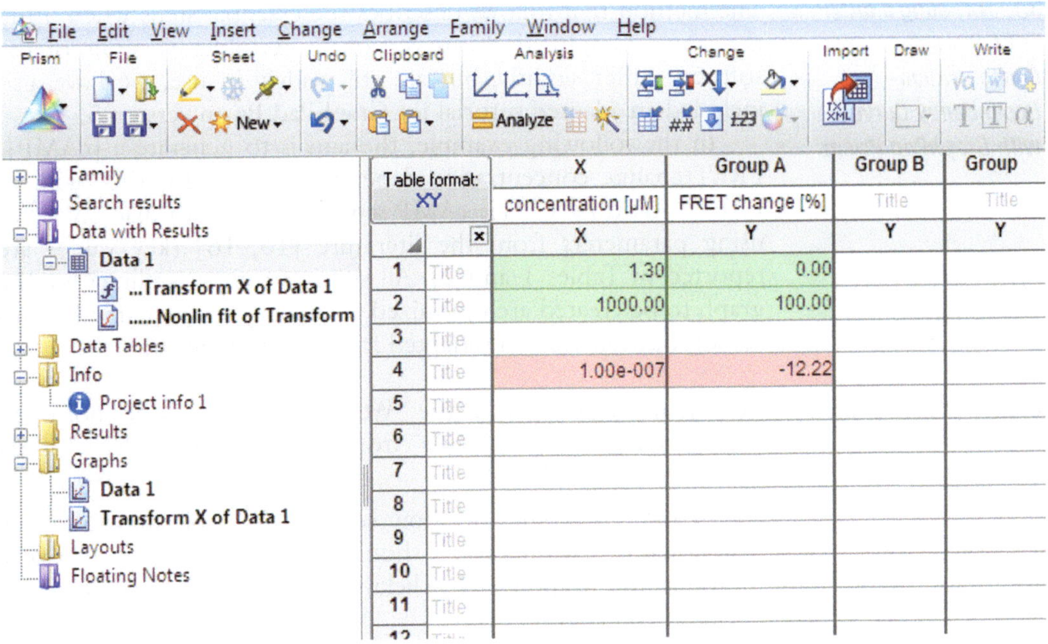

Fig. 1 Screenshot of the final data table created in GraphPad Prism 7. Green highlighted are the data that have to be entered initially: $R_0 = 0$ %, x at $R_0 = 1.3$ µM, $R_{max} = 100$ %, x at $R_{max} = 1000$ µM. Red highlighted is the value calculated by Prism for $R_{min} = -12.22$ and x at R_{min}, which should be zero, but for a specific reason has to be in this case 10^{-7} µM (*see* **Note 8**). These two values have to be entered manually in a second step after they have been calculated. At the left part of the screenshot the sidebar with the navigation tree is shown. Here all data tables, calculations, or graphs can be accessed directly

8. If not already active, select the "Fit" tab. Here select "Dose-Response - Stimulation" and in the submenu "log[agonist] vs response -- Variable slope (four parameters)". Tick the box "Interpolate unknowns from standard curve". Leave all other options to their defaults.
9. In the same pop-up window now select the "Constrain"-tab. Fix the "Top" of the curve (R_{max}) by selecting "constant or equal to" from the drop-down menu and entering "100" manually in the corresponding field. Do the same for Log EC_{50} (constant or equal to 0.8561) and Hill slope (constant or equal to 1.230) (*see* **Note 9**). The "bottom" (R_{min}) should not be constrained.
10. "OK" opens the "Results" table ("Nonlin fit of Transforms of Data 1") which can be found in the "Results"-folder in the navigation tree at the left sidebar of the program window.

    GraphPad Prism automatically calculates the "bottom" of the curve (R_{min}, here −12.22%) (*see* **Note 10**). It also automatically calculates a curve based on the entered parameters and creates a graph which will need some refinement:

11. Open the "Graphs"-folder in the navigation tree (left sidebar) and select the "Transform X of data 1"-graph. Double-click directly on an axis (not on an axis-label) to open the axis-dialogue pop-up window.
12. On the "X axis" tab untick "Automatically determine the range and interval". Now enter the minimum range (= −2) (*see* **Note 11**) and maximum range (= 3). Select "Major tick interval" (= 1), "Starting at X" (= −2). Tick the "log" box and select nine minor intervals. As "Number format" select "Antilog". All other available options can be kept by defaults.
13. On the "Left Y axis" tab untick the automatic determination and format the axis to encompass all values of the whole curve (here −20 to 100). "Major tick interval" should be set to 20, "starting point" at −20. "Minor tick interval" depends on the size and purpose of the graph. Here 4 or 10 minor intervals should be fine (*see* **Note 11**). All other values can stay at their defaults. Do not tick the "log" box for the *Y*-axis.
14. By pressing OK, the graph now shows the positive part of the curve.
15. To also show the negative range enter the calculated "bottom" value (R_{min}, −12.22, see above) manually into the first table "Data 1" (in the navigation tree, left side bar). Do not enter "0" as concentration, but a concentration sufficiently low for the respective sensor. In the case of the AKAP79-CUTie sensor a concentration of 0.0000001 (10^{-7}) μM would be significantly below the detection limit and therefore suitable for the purpose of creating a concentration point close to the calculated R_{min} at zero cAMP (*see* **Note 8**).

3.1.2 Procedure for GraphPad Prism 5

1. Open the program. If the "Welcome" window does not open automatically, select "File" from the menu, then "New", then "New Project File".
2. In the "Welcome" window choose "XY" from the "New Table & Graph" group. Tick "Start with an empty XY-data table". The graph-type is not important, keep the default. For "X" choose "Enter and plot a single Y-value for each point". By pressing the "create"-button an empty table will appear.
3. Enter the normalized values (compare Table 1) for the basal cAMP-concentration R_0 (X-column, enter 1.3 μM), the respective FRET change value (first Y-column, enter 0 %) and the maximal FRET change at a saturating cAMP-concentration R_{max} (X-column, enter 1000 μM, first Y-column, enter 100 %) into the table (*see* **Note 7**). Enter only the values, without the units. Naming the columns will automatically generate axis labels for the graph that will be created later. An example of the table with the entered values is shown in Fig. 1.

4. From the "Analysis" menu tab select "Analyze" and in the pop-up window select "transform" in the submenu of the point "Transform, Normalize...", then click "OK."
5. In the following pop-up window select "Transform X values using" "X = Log(X)" from the drop-down menu. Make sure that the "Create a new graph of the results" box is ticked.
6. Clicking "OK" leads to the next table, now with the concentrations expressed as logarithms. With this table ("Transform of Data 1", accessible via the navigation tree in the left sidebar of the program window) active (open) select again "Analyze". This time select the point "XY analyses" and here "Nonlinear Regression (Curve Fit)".
7. "OK" leads to the "Parameters: Nonlinear Regression" window. It can also be opened by clicking on "Nonlin Fit of transform X of Data 1" in the navigation tree (sidebar) and then "Change"—"Analysis Parameters" in the menu bar.
8. In the "Fit"-tab of this window select either "Dose-Response - Stimulation" and in the sub-menu "log [agonist] vs. response -- Variable slope (four parameters)" or "Classic Equations" and in the sub-menu "Sigmoidal Dose-Response (Variable Slope)". The latter option is identical to the option in Prism 3. Also tick the box "Interpolate unknowns from standard curve."
9. In the "Constraints"-tab fix the top of the curve (R_{max}) by selecting "constant equal to" from the drop-down menu and enter "100" manually in the field after. Do the same for Log EC_{50} (0.8561) and Hill slope (1.230) (*see* **Note 9**). "OK" opens the results table.

   Prism automatically calculates the "bottom" of the curve (R_{min}, here −12.22 %) (*see* **Note 10**).
   It also automatically calculates a curve based on the entered parameters and creates a graph which will need some refinement.
10. Open the "Graphs"-folder in the navigation tree at the left sidebar of the program window and select the "Transform of Data 1"-graph by double-click. Double-click directly on an axis (not the axis-label) of the shown graph opens the axis-dialogue pop-up window.
11. On the "X axis" tab untick "Automatically determine the range and interval." Now enter the minimum range = −2 (*see* **Note 11**) and maximum range = 3. Select "Major tic interval" = 1, "Starting at X" = −2. Tick the "log" box and select "9 minor intervals". "Number format" should be "Antilog". All other available options can be kept at their defaults.
12. On the "Left Y axis" tab untick the automatic determination and format the axis to encompass all values of the whole curve

(here −20 to 100). "Major ticks interval" should be set to 20, "starting point" at −20. "Minor tics interval" depends on the size and purpose of the graph (*see* **Note 11**). Four or ten tics are sensible choices. All other available options can be kept at their defaults. Do not tick the log box for the *Y*-axis. The "Transform of Data 1" graph now shows the positive part of the curve.

13. To show also the negative range enter the calculated "bottom" (R_{min}, −12.22, *see* above) manually into the first data table "Data 1" (in the navigation tree in the left sidebar). As concentration do not enter "0," but a concentration sufficiently low for the respective sensor (*see* **Note 8**). In the case of the AKAP79-CUTie sensor a concentration of 0.0000001 (10^{-7}) μM would be significantly below the detection limit and therefore suitable for the purpose of creating a point near zero.

The "Transform of Data 1" graph (selectable in the navigation tree in the left sidebar) will now also show the negative part of the curve.

Once generated, the curve can be used for a graphical determination of cAMP concentrations from FRET change values and to visualize such conversions (Figs. 3, 4, 5, and 6). The curve can also be used as the standard curve to directly interpolate unknown *X* or *Y*-values using the program (*see* **Note 12**).

3.2 Calculating cAMP Concentration Changes from FRET Changes Manually or Using Spreadsheet Programs

Creating a sigmoidal dose–response curve in other commonly used spreadsheet programs is not as simple, but these programs can still be used to facilitate the calculation of unknown values using the equations shown below.

1. If unknown, R_{min} must be calculated according to the following Eq. 1:

$$\Delta R_{min}\,[\%] = \frac{R_{max} - R_0}{10^{((\log EC_{50} - \log x)\times n)}} + R_0 \times (-1) \qquad (1)$$

Here, ΔR_{min} is the FRET change value from the basal cAMP concentration x to undetectable cAMP concentrations ("zero"), R_0 is the FRET change value at the basal cAMP concentration, R_{max} is the maximal FRET change at saturation for the given sensor, and x the cAMP-concentration (μmol/L) at R_0 (basal cAMP level). n is the Hill coefficient and EC_{50} the cAMP concentration that leads to half-maximal FRET change.

Due to the normalization of the FRET change values, the FRET change at basal cAMP-levels R_0 is defined to be zero. The general equation can therefore be simplified as shown in Eq. 2:

$$\Delta R_{\min}[\%] = \frac{R_{\max}}{10^{\left((\log \mathrm{EC}_{50} - \log x) \times n\right)}} \times (-1) \quad (2)$$

Equations 1 and 2 are derived from the equation used in GraphPad Prism to extrapolate unknowns from a standard curve, with the nomenclature used by GraphPad Prism adapted to fit better to standard equations. The multiplicator (-1) is introduced to yield a negative FRET change for "gain of FRET sensors". It has to be omitted for "loss of FRET" sensors, if their curves are not inverted (*see* **Note 1**).

2. $R_{\min}$ is then entered into the following Eq. 3 to calculate cAMP changes from FRET change values.

$$x[\mu\mathrm{M}] = 10^{-\left(\frac{\log\left(\frac{R_{\max}-R_{\min}}{\Delta R - R_{\min}} - 1\right)}{n} - \log\mathrm{EC}_{50}\right)} \quad (3)$$

This equation can be adapted to be used in any spreadsheet program (e.g. Excel, Open Office Calc) as shown in Eq. 4 below:

$$x[\mu\mathrm{M}] = 10^{\left(\left(\left(\mathrm{(LOG}\left(\left(\left((R_{\max}-R_{\min})/(\Delta R - R_{\min}))-1\right);10\right)\right)/n)-\mathrm{LOG}(\mathrm{EC}_{50};10)\right)\times(-)\right)} \quad (4)$$

Here, ΔR is the FRET change value for which the cAMP concentration needs to be determined, $R_{\min}$ is the negative FRET change from basal cAMP levels to "zero" or undetectable levels, $R_{\max}$ the maximal FRET change and x is the unknown concentration in μmol/L. n is the Hill coefficient and EC_{50} the cAMP concentration that leads to half-maximal FRET change. If a spreadsheet is used the units and symbols in Eq. 4 must be replaced with the column and row number of the cell in the worksheet in which the corresponding numerical values are placed (*see* input box of the spreadsheets shown in Fig. 2).

To effectively work with these equations, worksheets can be created as shown in Fig. 2 (*see* **Note 13**).

In the example shown in Fig. 2a, the mean FRET change on application of the stimulus is 70.0 ± 5.8 % (SEM, $n = 3$). This is converted into a cAMP concentration of $16.3 + 4.2/-3.0$ μM. The asymmetric error is inherent to the method and can only be avoided if the FRET change values are converted first individually into concentrations and the mean concentration is calculated then as shown in Fig. 2b (*see* **Notes 13** and **14**).

Figure 3 demonstrates graphically why the initially symmetric error is asymmetric after conversion and how to determine the error values manually.

a

G11 f_x =10^((((LOG((((C$7-C$6)/(C11-C$6))-1);10))/C$9)-LOG(C$8;10))*(-1))

	A	B	C	D	E	F	G	H	I	J	K
1											
2		**Calculation of cAMP concentrations from mean FRET changes**									
3		(Values normalized to respective saturating stimulus)									
4											
5											
6		**R_{min} (for basal cAMP of 1.3 μM)**	-12.225								
7		**R_{max}**	100.000				**Calculated**				
8		**EC_{50}**	7.1790				**cAMP conc.**				
9		**n (Hill-coeff.)**	1.2300				**[μM]**				
10											
11		**Mean of FRET-changes ± SEM**	70.00	+/-	5.774		16.2954				
12		Mean + SEM	75.77				20.4884				
13		Mean - SEM	64.23				13.3109				
14											
15											
16		**Concentration derived from mean FRET change +/- SEM**					16.30 +		4.19 / -		2.98
17											

b

G11 f_x =10^((((LOG((((C$7-C$6)/(C11-C$6))-1);10))/C$9)-LOG(C$8;10))*(-1))

	A	B	C	D	E	F	G	H	I	J	K
1											
2		**Calculation of cAMP concentrations from measured FRET changes**									
3		(Values normalized to respective saturating stimulus)									
4											
5											
6		**R_{min} (for basal cAMP of 1.3 μM)**	-12.225								
7		**R_{max}**	100.000				**Calculated**				
8		**EC_{50}**	7.1790				**cAMP conc.**				
9		**n (Hill-coeff.)**	1.2300				**[μM]**				
10											
11		**Delta R (measured)**	60.00				11.6066				
12			70.00				16.2954				
13			80.00				24.8745				
14											
15											
16		**Mean of single converted concentrations +/- SEM**					17.59 +/-		3.89		
17											

Fig. 2 (**a**) Example spreadsheet to calculate cAMP concentrations, e.g. from literature FRET changes. Shown is the conversion of a mean FRET change value ± SEM (highlighted in green) into cAMP concentrations (highlighted in red) according to Eq. 4. This equation is also shown in the input box of the spreadsheet program. Because of the conversion via the curve function, the error of the mean FRET change (column E, line 11) cannot be converted directly, but it has to be added or subtracted from the mean value (column C, line 11). After conversion of these two new values (column C, lines 12 and 13), the differences to the mean concentration have to be calculated to yield the positive and negative error range. (**b**) Example of a spreadsheet to calculate cAMP concentrations, e.g. from measured FRET changes. Shown is the conversion of the single FRET change values (highlighted in green) into cAMP concentrations (highlighted in red) according to Eq. 4 and the calculated mean ± SEM of the single concentration values. Here the dataset is restricted to the three values used for the example calculations, but it can be easily extended to any needed number of values. For curve parameters and basal cAMP refer to Tables 1 and 2. R_{min} was calculated according to Eq. 1

a

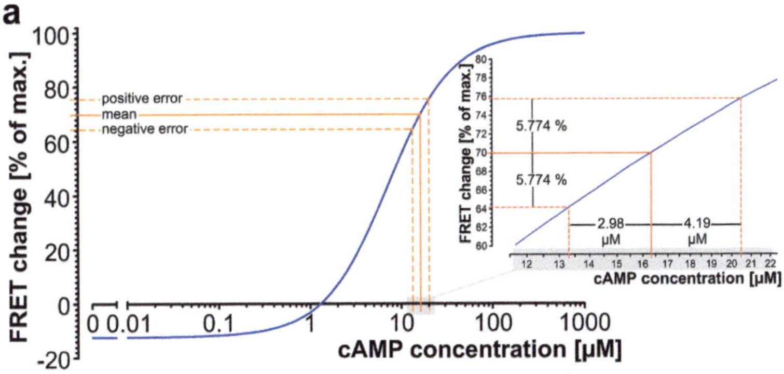

b mean concentration :

$$x_{[\mu M]} = 10^{-\left(\frac{\log\left(\frac{100-(-12.225)}{(70.0)-(-12.225)}-1\right)}{1.230} - \log 7.179\right)} = 16.2954\ \mu\text{M}\ (16.3\ \mu\text{M})$$

c positive error :

$$x_{[\mu M]} = 10^{-\left(\frac{\log\left(\frac{100-(-12.225)}{(70.0+5.774)-(-12.225)}-1\right)}{1.230} - \log 7.179\right)} = 20.4884\ \mu\text{M},$$

$$\Rightarrow +\ 4.2\ \mu\text{M}$$

d negative error :

$$x_{[\mu M]} = 10^{-\left(\frac{\log\left(\frac{100-(-12.225)}{(70.0-5.774)-(-12.225)}-1\right)}{1.230} - \log 7.179\right)} = 13.3109\ \mu\text{M},$$

$$\Rightarrow -\ 3.0\ \mu\text{M}$$

Fig. 3 Conversion of mean FRET changes into cAMP concentration changes. (**a**) cAMP concentrations determined with the graphical approach. Orange horizontal line projects the normalized FRET changes onto the curve, vertical line projects to the corresponding cAMP concentration. Dashed lines indicate error limits (here SEM). Insert shows a magnified part of the graph for better visualization of the transformation of the initially symmetric percentage errors into asymmetric errors after conversion into concentrations. Error values are shown in the insert. The curve shown is the calculated concentration-dependency curve for AKAP79-CUTie. Parameters were: $R_0 = 0$ % at a basal cAMP concentration of 1.3 μM, $R_{max} = 100$ %, n (Hill coefficient) $= 1.230$, $EC_{50} = 7.179$ μM. R_{max}, n and EC_{50} were fixed to the respective values, R_{min}, the theoretical FRET change from basal cAMP to zero cAMP, was extrapolated by Prism to be −12.2 % (*see* **Note 15**). (**b**) Calculation of the mean cAMP concentration from the mean FRET change with the mathematical approach following Eq. 3, (**c**, **d**) Error limits (SEM) calculated with the same approach. Note that because of the logarithmic relationship between concentration and FRET change the errors have to be calculated separately by adding or subtracting the respective values to the mean concentration value (highlighted in green and red in **c** and **d**, respectively). This nonlinear relation also causes the asymmetric error. It can be avoided by first converting each single FRET response into a concentration and then calculating the mean (*see* **Notes 13** and **14**)

In the following section, we will use the calibration curves and calculations illustrated above to show the importance of converting FRET changes into the underlying cAMP concentrations for data interpretation. We provide examples of how expressing FRET change values as relative to maximal FRET change may lead to data misinterpretation.

3.3 Examples

3.3.1 Impact of Sensor Non-linearity

A normalized FRET change from 5 % to 25 %, e.g. by blocking phosphodiesterases (PDEs) after a previous treatment with a low dose of Isoproterenol (ISO) might be considered as similar to a FRET change from 75 % to 95 % (blocking PDEs after treatment with prostaglandin, PGE_2). As in both cases 20 % increase in FRET is observed (Fig. 4, left panel), one may conclude that the contribution of PDEs in regulation of the cAMP response is similar. However, if the FRET change values are converted into cAMP concentrations as shown here using the graphical approach (Fig.4, middle panel), it is clear that the cAMP change resulting from PDE inhibition after application of ISO is 2.3 μM (from 1.8 to 4.1 μM) whereas PDE inhibition after application of PGE_2 results in 67 μM increase in cAMP (from 19.8 to 86.8 μM), a change that is about 30 times higher than in the ISO experiment (compare Fig. 4, right panel).

3.3.2 Impact of Different Basal Intracellular cAMP Concentrations

The baseline typically used in FRET experiments to calculate the FRET change is not equal to zero cAMP, as at baseline (unstimulated cells) a certain level of basal cAMP already is present in the cell. This should be taken into consideration when comparing cAMP changes in different cell types. If for example the effect of an agonist in thyroid cells (basal cAMP = 0.4 μM) is compared to

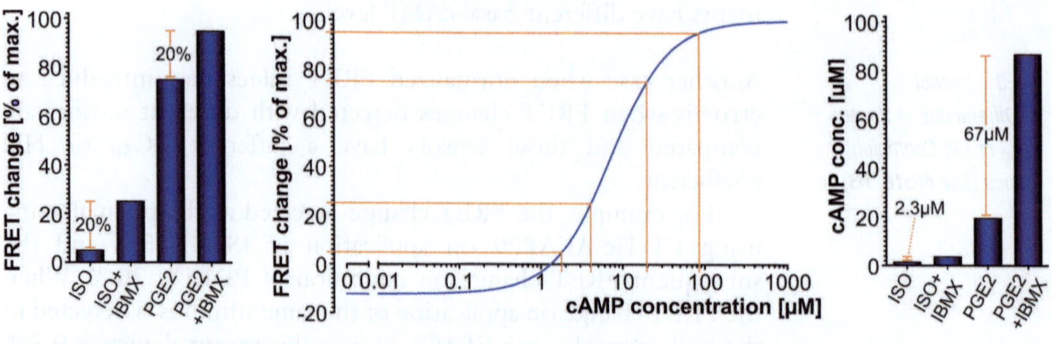

Fig. 4 Comparison of the results from an hypothetical experiment when using normalized FRET values (left) or FRET values converted into cAMP concentrations (right). The curve in the middle panel shows the graphical conversion of the FRET change values into cAMP concentrations using the calculated dose–response curve for the cAMP-dependent FRET change of the AKAP79-CUTie-sensor. Parameters used are: $R_0 = 0$ % at a basal cAMP concentration of 1.3 μM, $R_{max} = 100$ %, n (Hill coefficient) = 1.230, $EC_{50} = 7.179$ μM. R_{max}, n and EC_{50} were fixed to the respective values, R_{min}, the theoretical FRET change from basal cAMP to zero cAMP, was extrapolated by Prism to be −12.2 %

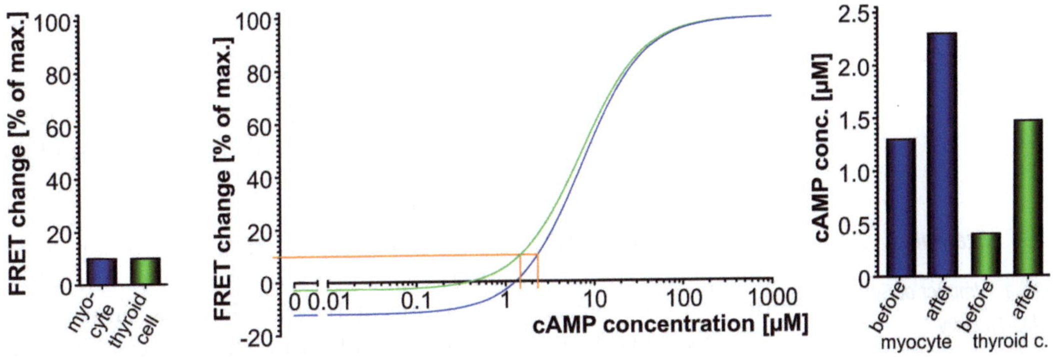

Fig. 5 Comparison of the hypothetical response to an agonist in myocytes (blue bars) or thyroid cells (green bars) when normalized (left) or when converted in concentrations (right). The normalized presentation suggests the same effect in both cell types, whereas the conversion into concentrations (right) clearly shows that both cell types react with different increases in their cAMP levels. "Before" refers to the basal cAMP level before the response to the agonist. The curves (middle) show the graphical conversion of the FRET change values into cAMP concentrations. Curve parameters for the AKAP79-CUTie sensor in thyroid cells (green curve) were the same as in adult mouse myocytes (blue curve) except for the lower basal cAMP concentration (0.4 µM) and therefore a lower R_{min} (−2.867 %). Basal cAMP level of thyroid cells is according to [15]

the effect of the same dose of agonist in cardiac myocytes (basal cAMP = 1.3 µM) and agonist application induces in both cell types a 10 % FRET change, the conclusion that the effect of agonist stimulation generates the same response in the two cell types is incorrect. As illustrated in Fig. 5, in thyroid cells cAMP increases about fourfold (from 0.4 to 1.5 µM), whereas in myocytes it only approximately doubles (from 1.3 to 2.3 µM). Comparison of FRET changes detected with sensors targeted to different subcellular compartments could lead to a similar error if the different compartments have different basal cAMP levels.

*3.3.3 Impact of Differences in Sensor EC_{50} or Hill Coefficient Values (See **Note 16**)*

Another case when normalized FRET values may introduce an error is when FRET changes detected with different sensors are compared and these sensors have a different EC_{50} or Hill coefficient.

For example, the FRET change detected at the plasmalemma using CUTie-AKAP79 on application of ISO is 5 % and the subsequent FRET change on inhibition of PDEs is 20 %. When the FRET change on application of the same stimulus is detected in the bulk cytosol using EPAC1-camps, this sensor detects a 9.3 % and 24.3 % signal increase, respectively (Fig. 6, left panel). These values may be misinterpreted to suggest that both the cAMP response on ISO application and the contribution of PDEs in limiting the cAMP increase are higher in the bulk cytosol compared to the plasma membrane. However, when the FRET change values are transformed into cAMP concentrations (Fig. 6, right panel)

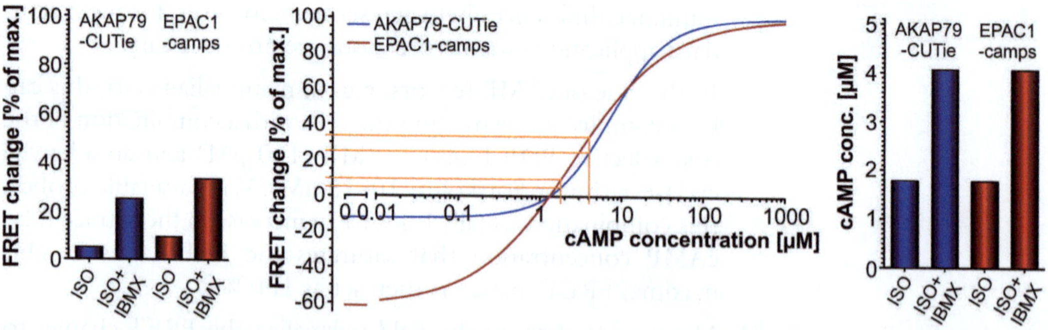

Fig. 6 Comparison of the hypothetical responses of the membrane bound AKAP79-CUTie FRET-sensor (blue bars) and the cytosolic EPAC1-camps (red bars) when normalized (left) or when converted into concentrations (right). The normalized presentation suggests higher cAMP-levels in the cytosol, whereas the conversion into concentration clearly shows that both sensors register the same concentrations. The curves (middle) show the graphical conversion of the FRET change values into cAMP concentrations for AKAP79-CUTie (blue) and EPAC1-camps (red). Curve parameters for the AKAP79-CUTie sensor were the same as shown before. Parameters for the EPAC1-camps sensor are: $R_0 = 0$ % at a basal cAMP concentration of 1.3 μM, $R_{max} = 100$ % at an estimated concentration of 10^4 μM, n (Hill coefficient) = 0.74, $EC_{50} = 2.5$ μM. EC_{50} and n were taken from [15], R_{min} (−61.5%) was calculated by GraphPad Prism. For the comparison, the FRET ratio of the "loss of FRET" sensor EPAC1-camps was inverted (*see* **Note 1**)

using the concentration-dependency curve for each sensor (Fig. 6, middle), it is clear that the response is identical in the two compartments.

4 Notes

1. FRET changes are typically measured as sensitized emission ratio, that is fluorescence intensity at the acceptor emission wavelength divided by fluorescence intensity at the donor emission wavelength (emission at 535 nm/emission at 480 nm, respectively for YFP and CFP) on excitation of the sample at the donor excitation wavelength (430 nm for CFP). For "loss of FRET" sensors it may be convenient to express the FRET changes as donor emission intensity/acceptor emission intensity. In this way an increase in cAMP translates in an increase in the ratio value. This may be particularly useful when comparing concentration-dependency curves of sensors with opposite mechanisms ("loss of FRET" versus "gain of FRET"), as shown in the example in Fig. 6.
2. The setup-specific constant is influenced by the specific filter set, dichroic, light source and coating and composition of the lenses used in the acquisition system. Correction for "bleed through" fluorescence (emission of one fluorophore that can be seen in the other fluorophore channel) and cross-excitation of the acceptor at the excitation wavelength of the donor can

minimize differences between setups. However, for most standard applications such corrections are not necessary.

3. In the case of cAMP sensors, e.g. in mammalian cells, this can be accomplished by treating the cells with a combination of the non-selective PDE-blocker IBMX (100 μM) and an adenylyl cyclase activator Forskolin (10–25 μM). With available probes, this combination usually leads to an increase in the intracellular cAMP concentration that saturates the FRET sensor. This maximal FRET change is then set as 100 %.
4. Many researchers in the field normalize the FRET change to the maximal FRET change at saturation (e.g. with Forskolin and IBMX). Therefore, R_{max} will probably be known and already set to 100 %. In case the available data are not yet normalized, but a value for FRET change at a saturating stimulus is provided, this value can be set as a 100 % and used to normalize the values of interest.
5. Information about the FRET change from the basal intracellular cAMP-concentration to undetectable levels ("zero cAMP") is difficult to find. If possible, this value should be determined experimentally in the specific cellular system of interest.

   One method to determine the basal concentration of cAMP is to use microinfusion [19], which requires a combined patch-clamp and FRET-imaging setup. In brief, cells expressing the FRET sensor are microinfused using a glass patch pipette containing known concentrations of cAMP going from zero cAMP to saturating concentrations, e.g. 1 mM cAMP. The resulting FRET change at each cAMP concentration is recorded and the values are computed into a concentration–response curve using GraphPad Prism. An alternative method is based on permeabilization of the cells with saponin or triton X and "wash-out" of intracellular cAMP. This approach only requires a FRET-imaging setup. For both methods, the composition of the cAMP-containing solution (either in the pipette or applied to the bath) is crucial for the correctness of the results. This solution must match the intracellular pH and ionic composition accurately as even small differences can lead to artificial conformational changes of the sensor or affect fluorescence emission, which then will interfere with the real FRET change due to the wash-out of intracellular cAMP. One way to verify if the composition of the buffer systems used is appropriate is to test it using a commonly used cell line where the basal cAMP concentration has been determined before. For example, the basal cAMP concentration in CHO cells has been reported to be around 1 μM (1.03 ± 0.12 μM, SEM, $n = 4$, calculated from single values of four different cAMP sensors published in [16]). The correct buffer is not expected to generate any change in FRET in CHO cells when 1 μM cAMP is added to it.

For some cell types there are published values for basal intracellular cAMP-concentrations (*see* Table 1). If using any of these cells the existing values can be used.

Another, apparently simple method to acquire information regarding the FRET ratio at zero cAMP is to block adenylyl cyclases, for example with the inhibitor MDL 12,330A [15]. This method, however, is dependent on robust activity of intracellular phosphodiesterases, the enzymes that degrade cAMP. If the activity of the phosphodiesterases is low, it will take several minutes up to half an hour until cAMP is degraded to undetectable concentrations. In our experience, many cell types die during that time span. HEK293 and CHO cells show significant blebbing shortly after application of MDL 12,330A and show unexpected changes in fluorescence that can be erroneously interpreted as FRET changes (e.g. an apparent significant cAMP increase instead of the expected decrease) before they die. Adult rat ventricular myocytes are more robust and survive the treatment for longer, but also show partly inexplicable effects on the FRET sensors. Therefore, the results obtained with this method must be interpreted critically.

6. Most data available in the literature report FRET changes as relative to the basal FRET value, which is set as $R_0 = 0$ % FRET change. However, in cells R_0 is the FRET value that corresponds to a basal concentration of cAMP which is typically higher than zero. If the cell is depleted of cAMP (for example by microdialysis through a glass pipette containing no cAMP), the sensor is expected to show a negative FRET change until it reaches its minimal FRET ratio, R_{min}, at cAMP concentrations lower than it's detection limit. This is true for "gain of FRET" sensors. Non-inverted "loss of FRET" sensors would show a further increase of the ratio until they reach the maximum.
7. If no information is available for the concentration at which the sensor is saturated, there is a simple way to estimate it. As a general rule of thumb for dose–response curves with a Hill slope around one, a hundred-fold higher concentration than the EC_{50} will yield about 99 % of the maximal response, thousand-fold the EC_{50} will yield 99.9 % of the maximal FRET change. Higher Hill slopes will lead to saturation already at lower concentrations and vice versa. As a single measured curve value (e.g. R_0), together with the EC_{50} and the Hill coefficient are the main determinants of the curve shape, a small aberration from the absolute concentration value for saturation does not significantly influence the curve. Concentrations between 100- and 1000-fold the EC_{50} will typically lead to good estimates of saturation values.
8. To estimate the "zero" FRET value in principle the same considerations apply as for the maximal FRET change (*see* **Note 7**).

At a concentration 1000-fold lower than the EC_{50} the FRET change is only 0.1 %. In most cases this is already below the actual detection limit of the sensor. Any concentration around or below this concentration will yield a good estimate of the (negative) FRET value at non-detectable cAMP levels (R_{min}). Avoid entering "zero" ("0"), as "zero" (as well as "100 %") in theory are only achievable at infinite low or high concentrations. In some equations a "zero" value might lead to errors. Although Prism calculates R_{min} at zero cAMP correctly, it will not show the negative part of the curve if R_{min} is entered manually at a concentration of zero cAMP.

9. If necessary, all these parameters can be changed again later by selecting "Change analysis parameters" from the "Analysis" menu tab.

10. In case original data for the FRET changes at different cAMP concentrations are available, these can be used to create a measured concentration-dependency curve by entering the cAMP concentrations (μM) in the X column and the respective FRET change values in the Y column. If there are replicates, the column format can be changed to accommodate the respective maximal number of replicates with the "Format data table" pop-up in the "Change" menu. In this case the parameters for the analysis of the sigmoidal dose–response curve should not be constrained to yield the best possible fit.

11. If for clarity reasons *X* and *Y*-axes should have an "offset" (*X* and *Y* axis not crossing each other), choose a value slightly more negative for the *X*-axis (e.g. −2.001). This is a way to work around a program bug which results in the first tick on the axis not being shown when offsetted. In this case, after customizing the *Y*-axis, also open the "Frame and Origin" tab, set the "Origin" to "Lower left" or "Custom", and the "Frame Style" to "Offset X & Y axes". In case you selected Custom, set "Y intersects the X axis at X=" to −2.001 and "X intersects the Y axis at X=" to 0.

    In Prism 5 the "Origin" option "Automatically" combined with "Offset X & Y axes" will directly lead to the same appearance of the graph.

    All other fields can stay at their defaults. However, if the graph is to be used for graphical conversions, an increased size allows more intermediate ticks and increases accuracy. For publication purposes other features may be more important.

12. For the interpolation of data in Prism either enter the concentration for which a FRET change value is wanted in the initial Table (*X*-value column) and leave the FRET change (*Y*-value column) open, or enter FRET change values and leave the concentration open. Make sure that the "Interpolate

unknowns from standard curve" box in the fitting menu is ticked. The interpolated values can be found in the "Results" folder in the navigation menu in the left sidebar of the program window. In Prism the interpolated concentrations are expressed as the logarithm of the concentration. They can be converted by using Eq. 5 below:

$$\text{Concentration } x(\mu\text{M}) = 10^{\log x} \tag{5}$$

13. As well as to convert means ± SD or SEM, the Excel spreadsheet shown in Fig. 2a can also be used to convert a complete set of FRET change data, e.g. own original measurements (Fig. 2b). The latter case—if the original data are available—is preferable over the conversion of the mean ± SD or SEM, as it will yield the real mean concentration and a normal, symmetric error (*see* **Note 14**).
14. There will be slightly different results if the single FRET data are averaged first and then converted into the cAMP change or concentration or if the FRET data are converted separately first and then averaged. This is due to the fact that the FRET change is represented on a linear scale whereas the cAMP concentration is represented on a logarithmic scale. A simple example will demonstrate this: for a dataset consisting of FRET changes of 60 %, 70 %, and 80 % as shown in Fig. 2, the average FRET change would be 70 ± 5.8 % and the calculated cAMP concentration would be 16.3 + 4.2/−3.0 μM. However, converting the single values first would result in the single concentration values of 11.6, 16.3, and 24.9 μM, which would result in a mean of 17.6 ± 3.9 μM (Fig. 2b). The difference in the mean values is more pronounced the more the data vary. In addition, converting the SD or SEM error values of the averaged FRET change will yield an asymmetric error for the concentration, while converting the single FRET values before averaging will lead to a normal, symmetric error. Therefore it is recommend, when possible, to first convert the single FRET change values into concentrations and then average the concentrations.
15. If a curve is calculated from only the minimal possible parameter and is fixed for these values it does not show a confidence interval. To have an impression of typical confidence intervals and of the potential error that might be introduced by the conversion *see* [16]. When analyzing for significance, this error would also have to be taken into account and all errors would sum up. Therefore, we would recommend testing for significance already at the level of the FRET change values, as here the potential error of the calibration curve is not yet included. However, this will only be possible if the sensors share similar properties (EC_{50}, Hill coefficient).

16. Be aware that this is not only an issue when comparing different sensors. Fusion of targeting sequences might alter the characteristics of parental sensors unpredictably (for examples *see* [10]). Therefore, curve parameters determined for a parental sensor should not be applied to targeted versions unless the characteristics of their concentration-dependency curves have been verified.

Acknowledgments

This work was supported by the British Heart Foundation (PG/10/75/28537 and RG/17/6/32944) and the BHF Centre of Research Excellence, Oxford (RE/13/1/30181).

References

1. Förster T (1948) Zwischenmolekulare energiewanderung und fluoreszenz. Ann Phys 437:55–57. https://doi.org/10.1002/andp.19484370105
2. Stryer L (1978) Fluorescence energy transfer as a spectroscopic ruler. Annu Rev Biochem 47:819–846. https://doi.org/10.1146/annurev.bi.47.070178.004131
3. Adams SR, Harootunian AT, Buechler YJ et al (1991) Fluorescence ratio imaging of cyclic AMP in single cells. Nature 349:694–697. https://doi.org/10.1038/349694a0
4. Zaccolo M, De Giorgi F, Cho CY et al (2000) A genetically encoded fluorescent indicator for cyclic AMP in living cells. Nat Cell Biol 2:25–29. https://doi.org/10.1038/71345
5. DiPilato LM, Cheng X, Zhang J (2004) Fluorescent indicators of cAMP and Epac activation reveal differential dynamics of cAMP signaling within discrete subcellular compartments. Proc Natl Acad Sci U S A 101(47):16513–16518. https://doi.org/10.1073/pnas.0405973101
6. Nikolaev VO, Bünemann M, Hein L et al (2004) Novel single chain cAMP sensors for receptor-induced signal propagation. J Biol Chem 279:37215–37218. https://doi.org/10.1074/jbc.C400302200
7. Klarenbeek J, Goedhart J, van Batenburg A et al (2015) Fourth-generation epac-based FRET sensors for cAMP feature exceptional brightness, photostability and dynamic range: characterization of dedicated sensors for FLIM, for ratiometry and with high affinity. PLoS One 10(4):e0122513. https://doi.org/10.1371/journal.pone.0122513
8. Di Benedetto G, Zoccarato A, Lissandron V et al (2008) Protein kinase A type I and type II define distinct intracellular signaling compartments. Circ Res 103(8):836–844. https://doi.org/10.1161/CIRKRESAHA.108.174813
9. Sprenger JU, Perera RK, Steinbrecher JH et al (2015) In vivo model with targeted cAMP biosensor reveals changes in receptor-microdomain communication in cardiac disease. Nat Commun 6:6965. https://doi.org/10.1038/ncomms7965
10. Surdo NC, Berrera M, Koschinski A et al (2017) FRET biosensor uncovers cAMP nano-domains at β-adrenergic targets that dictate precise tuning of cardiac contractility. Nat Commun 8:15031. https://doi.org/10.1038/ncomms15031
11. Nikolaev VO, Bünemann M, Schmitteckert E et al (2006) Cyclic AMP imaging in adult cardiac myocytes reveals far-reaching β_1-adrenergic but locally confined β_2-adrenergic receptor-mediated signaling. Circ Res 99:1084–1091. https://doi.org/10.1161/01.RES.0000250046.69918.d5
12. Herget S, Lohse MJ, Nikolaev VO (2008) Real-time monitoring of phosphodiesterase inhibition in intact cells. Cell Signal 20:1423–1431. https://doi.org/10.1016/j.cellsig.2008.03.011
13. Halls ML, Cooper DMF (2010) Sub-picomolar relaxin signalling by a pre-assembled RXFP1, AKAP79, AC2, β-arrestin 2, PDE4D3 complex. EMBO J 29:2772–2787. https://doi.org/10.1038/emboj.2010.168
14. Agarwal SR, Yang PC, Rice M et al (2014) Role of membrane microdomains in compartmentation of cAMP signaling. PLoS One 9(4):

e95835. https://doi.org/10.1371/journal.pone.0095835

15. Boerner S, Schwede F, Schlipp A et al (2011) FRET measurements of intracellular cAMP concentrations and cAMP analog permeability in intact cells. Nat Protoc 6:427–438. https://doi.org/10.1038/nprot.2010.198
16. Koschinski A, Zaccolo M (2017) Activation of PKA in cell requires higher concentration of cAMP than in vitro: implications for compartmentalization of cAMP signalling. Sci Rep 7 (1):14090. https://doi.org/10.1038/s41598-017-13021-y
17. Iancu RV, Ramamurthy G, Warrier S et al (2008) Cytoplasmic cAMP concentrations in intact cardiac myocytes. Am J Physiol Cell Physiol 295(2):C414–C422. https://doi.org/10.1152/ajpcell.00038.2008
18. Wachten S, Masada N, Ayling LJ et al (2010) Distinct pools of cAMP centre on different isoforms of adenylyl cyclase in pituitary-derived. GH3B6 Cells 123(Pt 1):95–106. https://doi.org/10.1242/jcs.058594
19. Koschinski A, Zaccolo M (2015) A novel approach combining real-time imaging and the patch-clamp technique to calibrate FRET-based reporters for cAMP in their cellular microenvironment. Methods Mol Biol 1294:25–40. https://doi.org/10.1007/978-1-4939-2537-7_3

Part IV

Exploring GPCR Signaling Properties

Chapter 13

Measuring GPCR-Induced Activation of Protein Tyrosine Phosphatases (PTP) Using In-Gel and Colorimetric PTP Assays

Geneviève Hamel-Côté, Fanny Lapointe, and Jana Stankova

Abstract

Given the increasing amount of data showing the importance of protein tyrosine phosphatases (PTPs) in G protein-coupled receptor (GPCR) signaling pathways, the modulation of this enzyme family by that type of receptor can become an important experimental question. Here, we describe two different methods, an in-gel and a colorimetric PTP assay, to evaluate the modulation of PTP activity after stimulation with GPCR agonists.

Key words Protein-tyrosine phosphatase, In-gel phosphatase assay, Colorimetric phosphatase assay, GPCR, Phosphatase activity

1 Introduction

Post-translational modification of proteins is one of the mechanisms by which cells modulate protein interaction, enzymatic activity, and cell localization. Among post-translational modifications, phosphorylation is the most studied [1]. This interest could be due to the fact that many phosphorylation sites can be found on the same protein, allowing the integration of different signaling pathways and, maybe the most important point, this kind of post-translational modification is reversible, thus enabling intermediate phosphorylation states or recovery of unphosphorylated protein, allowing a fine-tuning of protein activity [1]. Whereas histidine phosphorylation is well characterized in eubacteria, plants, and fungi [1, 2], it is less known in vertebrates (only two enzymes, both isoforms of the nucleoside diphosphate kinase had been identified [2]), whereas serine/threonine and tyrosine phosphorylation is well documented [1]. In addition to the modulation of phosphorylation levels by regulation of kinase activity, phosphatases are an important part of the equation. As there are three main

Mario Tiberi (ed.), *G Protein-Coupled Receptor Signaling: Methods and Protocols*, Methods in Molecular Biology, vol. 1947, https://doi.org/10.1007/978-1-4939-9121-1_13, © Springer Science+Business Media, LLC, part of Springer Nature 2019

phosphorylation types, three major phosphatase categories exist: Serine/threonine phosphatases, histidine phosphatases, and protein tyrosine phosphatases (PTPs) [3].

PTPs form an enzyme family which comprises 107 members, subdivided into four different classes (I–IV). Whereas class IV PTPs, which include the Eyes Absent (EyA) tyrosine-specific phosphatases and Haloacid dehalogenase (HAD) family, initiate the enzymatic reaction by a nucleophilic attack dependent on an aspartate residue in their catalytic pocket, the nucleophilic attack depends on a conserved cysteine in the catalytic pocket for class I–III PTPs [4, 5]. The PTPs in these three classes have a consensus motif [I/V]HCXXGXXR[S/T] involved in the first step of the reaction where the sulfur of a thiol group on the conserved cysteine [6], found in the thiolate form at pH higher than 6, reacts with the phosphate group of the phospho-tyrosine on the substrate, forming a covalent link between the latter and the PTP [6]. Next, as described for PTP1B, a class I PTP prototype, sequential acid/base reactions take place involving the conserved aspartate and glycine residues, found in the catalytic pocket, allowing the release of first, the substrate, then, the phosphate and the recovery of the cysteine thiolate group (for an extensive description of the reaction steps *see* [6]).

Whereas the class II PTPs contain only the Low-molecular-mass PTPs (LMW PTP) and the class III only the Cdc25 protein family, the class I accounts for the majority of the PTPs known to date [5, 7]. This latter class is also subdivided into two different categories, with their own characteristics. Some members of this class have been shown to be regulated by, or involved in, G-protein coupled receptor (GPCR) signaling. The class I subcategory with the most members is the Dual-specificity phosphatases (DUSP) with at least 43 members divided in three classes according to their sequence homology [8]. In addition to Tyr residues, these PTPs can dephosphorylate phospho-Ser/Thr residues or phospholipids. The best known members of this category are the DUSPs with a MAP Kinase Phosphatase (MKPs) activity that dephosphorylate MAP kinases on the Thr and Tyr residues, found in their activation loops. In addition, Phosphatase and TENsin homolog (PTEN) is best known in GPCR signaling pathways for its ability to dephosphorylate phosphatidylinositol 3,4,5-triphosphate at the 3′ position, although it has functions that are independent of its lipid-phosphatase activity [8–11].

Many GPCR agonists modulate expression of DUSPs, such as the platelet-activating factor (PAF) which increases DUSP-1 (MKP-1) and DUSP5 expression [12], protease-activated receptor (PAR)-2 agonists which increase DUSP 6 expression [13] or the prostaglandin E2 (PGE2) which increases DUSP-1 [14]. Modulation of DUSPs levels by GPCRs can impact signaling pathways induced by subsequent stimulation as illustrated by data showing

that the PGE2-induced DUSP-1 expression results in a decreased in interleukin-1ß-induced p38 MAPK activation, an important pro-inflammatory pathway [14]. Furthermore, DUSPs are also directly involved in the modulation of signaling pathways of some GPCRs such as the angiotensin type 1a receptor where DUSP-1 and -16 (MKP-7) down-regulate STAT1 and JNK3 activity, respectively [15, 16].

The second subcategory of the class I PTPs, the "classical" PTPs, includes PTPs targeting only phospho-tyrosines. The classical PTPs are also divided into two groups: the receptor and the non-receptor PTPs. With its 21 members, the receptor PTP (RPTP) family is the larger of the two subcategories [17]. These PTPs are differentiated by a highly variable extracellular N-terminal domain, a single transmembrane domain and an intracellular domain characterized, for the majority of RPTPs, by two phosphatase domains in tandem. The membrane proximal phosphatase domain (D1) is the principal catalytic domain whereas the membrane distal phosphatase domain (D2) seems to be a pseudo-phosphatase domain, mostly involved in modulation of RPTP activity [17]. In addition to oxidization, the dimerization ability which is shared by many of RPTPs is another mechanism which could modulate phosphatase activity levels [18]. Data from the literature show that several RPTP family members modulate GPCR signaling pathways and cellular responses. For example, in mice, CD45 (encoded by *Ptprc and expressed exclusively by hematopoietic cells*) and CD148 (encoded by *Ptprj*) had been shown to have opposite effects on formyl-Met-Leu-Phe (fMLF)-induced Akt phosphorylation and calcium and chemotactic responses in neutrophils; CD148 negatively and CD45 positively modulate these cell responses [19]. Moreover, GPCR agonists such as gintonin, via the lysophosphatidic acid (LPA) receptor [20], 5-hydroxytryptamine (5-HT) via the 5-HT2C receptors [21] and carbachol, via the M1 muscarinic acetylcholine receptor [22] inhibit voltage-gated potassium (Kv) channels in a protein tyrosine phosphatase α (RPTPα)-dependent manner.

Finally, the second group of classical PTPs, the non-receptor PTPs, has 17 members and, among them, the prototypical PTP, PTP1B. Phosphatases of this subfamily usually have only one PTP domain and at least one regulatory domain involved in substrate/enzyme interactions and regulation of cellular localization, given that these PTPs can be found in different cellular compartments such as the cytoplasm, nucleus or at the cytoplasmic face of the endoplasmic reticulum [18]. The involvement of this PTP family in GPCR signaling is well documented. For example, both SHP-1 (encoded by *PTPN6*) and SHP-2 (encoded by *PTPN11*) bind to CXCR4 and modulate SDF-1α-induced signaling pathways and chemotaxis [23, 24]. Moreover, in cells stimulated with ghrelin-1, an agonist of the growth hormone secretagogue receptor type 1a

(GHSR1a), SHP-1 is recruited to the arrestin/receptor complex and activated by Src, which leads to a decrease of Akt activity [25, 26]. Finally, in monocyte-derived dendritic cells (Mo-DCs) PAF increases the PTP activity levels of PTP1B which positively modulates PAF-induced IL-6 mRNA levels [27].

Given the growing amount of data demonstrating the importance of PTPs in GPCR signaling pathways and cell responses, specific tools to monitor the effect of GPCR on PTP activity are needed. Here, we described two different methods providing a measure of PTP activity after cell stimulation. The first one to be described, in-gel PTP assays, is a sensitive method that can be useful in detecting the PTP activity levels of different isoforms of a given PTP or revealing PTP activity levels according to molecular weight, which can guide the identification of PTPs that are modulated by a given stimulus. This method can also be useful in detecting isoforms or PTPs associated with ones' protein of interest [28, 29] and, with some modifications, which will not be discussed here (*see* [30]), can be used as a reliable approach to detect modulations of activity due to oxidation. However, it is a longer and more expensive method than the p-nitrophenyl phosphate (pNPP) PTP assay. Moreover, in-gel PTP assays do not allow for the determination of enzymatic parameters such as enzymatic rates and the denaturation/renaturation steps, which make this method less sensitive for some RPTP such as RPTPα (the protein quantity needed is at the microgram scale as compared to 100 pg for DUSP, for example [31]) or totally ineffective for other RPTPs, such as CD45 [31], making this method a PTP assay biased toward non-receptor PTPs.

For their part, colorimetric PTP assays based on pNPP are a faster and less expensive method than in-gel assays and allow for evaluation of enzymatic activity for a given PTP and the comparison between different conditions. Moreover, given that there is no denaturation step, this method allows for detection of RPTP activity. On the other hand, this method cannot discriminate between isoforms and, given the use of a reducing agent, the presence of PTPs in their oxidized state. Finally, this method gives an incomplete overview of the catalytic activity. Indeed, one should keep in mind that the measured activity levels with this approach do not necessarily mirror what happens in vivo since protein/protein interactions between enzyme and substrate can affect the hydrolysis rate.

It should also be noted that since PTPs can be "activated" (i.e. that their capacity to dephosphorylate substrate is increased), by changing their subcellular localization to reach their targets, an unchanged enzymatic activity ex vivo is not a proof of unchanged activation of a particular enzyme in vivo.

2 Materials

2.1 General Consideration

1. Prepare fresh solutions with nanopure grade water and keep them at room temperature unless otherwise specified. For dithiothreitol (DTT) and pNPP, which are hydroscopic powders, store them at −20 °C, away from humidity, and allow them to return to room temperature before opening and weighting them to avoid moisture. Aliquot them to avoid repeated temperature changes.

2.2 Sample Preparation (See Note 1)

1. Stop solution: ice-cold 1× PBS with 2 mM Na_4VO_3, 2 mM NaF.
2. 1× Lysis buffer (for 7 ml): 0.5% IGEPAL, 50 mM HEPES (pH 7.4), 150 mM NaCl, 1 mM EDTA, 2 mM EGTA (*see* **Notes 2** and **3**), 2 mM Na_4VO_3, 2 mM NaF, cOmplete™, Mini Protease Inhibitor Tablet (1 tablet), 10 μg/ml leupeptine, 2 μg/ml pepstatin A. Avoid the use of irreversible PTP inhibitors such as Zn^{2+} which can irreversibly affect PTP activity at a nanomolar scale, probably by direct binding to amino-acid residues in the catalytic pocket [32, 33], or pervanadate which irreversibly oxidizes the catalytic cysteine [34].
3. 4× loading buffer: 40% glycerol, 258 mM Tris–HCl pH 6.8, 8% SDS, 0.008% bromophenol blue, 20% 2-mercaptoethanol (*see* **Note 4**).

2.3 In-Gel PTP Assay (See Note 5)

1. PBS (phosphate buffered saline) 1×: Dissolve in 500 ml water: 8 g NaCl, 0.2 g KCl, 2.68 g Na_2HPO_4, 0.24 g KH_2PO_4. Adjust volume to 1 l, filter through a 0.22 μm filter. The pH should be at 7.4.
2. GST-FER: Initially, take 2 μg and store the rest at −80 °C (*see* **Notes 6** and **7**).
3. Glutathione Sepharose 4B solution: When needed, resuspend beads in the stock solution, take 50 μl and wash three times with 0.5 ml PBS. Keep in 0.5 ml PBS until use.
4. poly-(Glu:Tyr) [4:1]: resuspend in water at 20 mg/ml, aliquot in 50 μl portions and freeze at −80 °C.
5. 1 M Imidazole pH 7.2 buffer: Weigh 3.044 g imidazole and dissolve in 25 ml water. Adjust the pH to 7.2, complete to 50 ml with H_2O. Filter through a 0.22 μm Corning filter and store at 4 °C until use.
6. G-25 Sephadex PD-10 Desalting column: Sephadex G-25 resin with an exclusion limit of 5000 M_r and an 8.3 ml bed volume.
7. Buffer A: 50 mM Imidazole pH 7.2, 10 mM DTT, 0.1% 2-mercaptoethanol.

8. Buffer B: 50 mM Imidazole pH 7.2, 30 mM $MgCl_2$, 10 mM $MnCl_2$, 10 mM DTT, 1 mM Na_3V0_4, 1% Triton X-100, 0.2 mM ATP, 0.1% 2-mercaptoethanol.
9. 0.4 mCi/14.8 MBq [γ-^{32}P] ATP.
10. TCA 100%: 5 g TCA, 2.27 ml H_2O.
11. 1 M Tris pH 11 buffer: Weigh 6.057 g Tris base and dissolve it in 35 ml water. Adjust the pH to 11 and complete to 50 ml.
12. G-25 Sephadex PD-10 column.
13. PROTEAN II Xi Cell Electrophoresis system (16 × 16 cm gels) with cooling system and 1.5 mm spacers (*see* **Note 8**).
14. Ammonium persulfate: 10% in H_2O.
15. Acrylamide (29:1) solution: Weigh 29 g of acrylamide monomers and 1 g *N,N'*-Methylenebis (acrylamide) (a cross-linker) and dissolve in 50 ml of water. Adjust volume to 100 ml and filter through a 0.45 μm filter. Store at 4 °C, in a bottle wrapped in aluminum foil (*see* **Note 9**).
16. 1× Running buffer: 25 mM Tris, 250 mM glycine, 0.1% SDS. Store at 4 °C overnight before use.
17. Buffer C: 50 mM Tris, pH 8, 20% 2-propanol.
18. Buffer D: 50 mM Tris, pH 8, 0.3% 2-mercaptoethanol.
19. Buffer E: 50 mM Tris, pH 8, 0.3% 2-mercaptoethanol, 6 M guanidine HCl (*see* **Notes 10** and **11**).
20. Buffer F: 50 mM Tris, pH 8, 0.04% Tween 40 (*see* **Note 12**), 1 mM EDTA, 0.3% 2-mercaptoethanol.
21. Buffer G: 50 mM Tris, pH 8, 0.04% Tween 40 (*see* **Note 12**), 1 mM EDTA, 0.3% 2-mercaptoethanol, 3 mM DTT.
22. Staining solution: 50% Methanol, 10% acetic acid, 40% H_2O, 0.2% Coomassie blue R250.
23. De-staining solution: 50% Methanol, 10% acetic acid, 40% H_2O.
24. Softener Buffer: 3% glycerol, 40% Methanol, 10% acetic acid, 47% H_2O.
25. Sheets of cellophane.
26. Gel dryer with vacuum and heat or gel drying frames.

2.4 Colorimetric PTP Assay (See Note 13)

1. Clear, flat-bottom 96-well plates.
2. 10× PTP assay buffer (stock solution): 250 mM HEPES pH 7.4, 1 mM EDTA, 1 μg/ml BSA, 0.1% Tween 20, 20 mM NaF. Ensure that the pH is at 7.4. Store at 4 °C up to 1 month (*see* **Note 14**).
3. 1× PTP assay buffer: 1 ml 10× PTP assay buffer, 20 μl 1 M DTT, freshly made, and 8.98 ml water.

4. pNPP: store powder at −20 °C. Make a fresh solution in 1× PTP assay buffer at 0.095 g/ml.
5. 100 mM Na_3VO_4.

3 Methods

3.1 Sample Preparation

1. Stimulate cells with your agonist and then stop the reaction on ice. For adherent cells, remove supernatants and add the appropriate volume of the stop solution. Scrape cells on ice and centrifuge at 4 °C.
2. Resuspend the pellets in ice cold 1× lysis buffer by gently pipetting up and down, using a volume allowing final protein concentration around 1 μg/μl. Avoid vortexing.
3. Incubate lysates 30 min at 4 °C under gentle agitation before freezing or, if they are to be used immediately, centrifuge to remove membranes (25 min, 13,000 × *g*) to obtain cleared lysates.
4. Determine protein amount. We use Bradford protein assays and we dilute cleared lysates in water (1 in 3) to avoid interference with IGEPAL and leupeptin, among others.
5. If needed, immunoprecipitate PTPs using specific antibodies and protein A or protein G-sepharose beads (*see* **Notes 15** and **16**), according to the manufacturer's instructions. A negative control must be included with the use of isotype control antibodies for the immunoprecipitation. Keep aliquots of cleared lysates and immunoprecipitated protein in order to perform Western blots for normalization of the PTP activity and store them at −20 °C until use, after addition of loading buffer. Also keep cleared lysate aliquots as positive controls for the PTP assays. See each PTP assay method for further details.

3.2 In-Gel PTP Assay

3.2.1 Poly [Glu:Tyr] Labeling

1. Incubate for 30 min at room temperature: 2 μg GST-FER (activity around 9.5 nmol/min, according to the lot) with the Glutathione Sepharose 4B bead solution and DTT (2 mM final).
2. After 30 min, centrifuge 2 min at 2 × *g* and wash two times with 1 ml of buffer A.
3. Resuspend GST-FER-conjugated beads in 800 μl in buffer B.
4. Add 0.4 mCi/14.8 MBq [γ-^{32}P] ATP and 50 μl of the 20 mg/ml poly (Glu:Tyr) [4:1] solution.
5. Incubate 6–16 h at room temperature, with gentle rotation, behind Plexiglas shield (*see* **Note 17**).
6. Stop the reaction and concentrate the radiolabeled proteins by the addition of 100 μl 100% TCA. Mix by inversion.

7. Incubate 30 min on ice.
8. Centrifuge 15 min at 12,000 × *g* at 4 °C (*see* **Note 18**).
9. Remove the supernatant and resuspend pellets in 200 μl 1 M Tris pH 11 by gentle up and down pipetting.
10. Centrifuge 15 min at 12,000 × *g* at 4 °C (*see* **Note 18**). This step allows the removal of GST-FER coupled to sepharose beads.
11. During the centrifugation period, equilibrate the G-25 Sephadex PD-10 Desalting column with 10 ml of the 50 mM imidazole pH 7.2 buffer prepared by diluting 1 ml of the 1 M imidazole pH 7.2 buffer for a final volume of 20 ml.
12. Load supernatants recovered from the last centrifugation onto the equilibrated column. Allow supernatant absorption.
13. Add 8 ml of 50 mM imidazole pH 7.2 and collect 1 ml aliquots in numbered tubes.
14. For each aliquot, put 2 μl scintillation liquid and count.
15. Pool fractions with the highest activity (usually around the fraction 4, a count of 2.5×10^5 cpm/μl is expected) and divide into 100 μl aliquots. Aliquots can be stored at 4 °C and should be used within 2 weeks.

3.2.2 Gel Casting

1. Mix all solutions well, except TEMED and APS, needed for a 10% acrylamide SDS-PAGE gels. For a 16 × 16 cm gel, prepare 30 ml per gel. As seen in Fig. 1, this gel size allows good resolution for proteins ranging between 180 and 10 kDa, covering molecular weights of almost all PTPs that can be detected with this method, from LWM-PTP (18 kDa) [35] to PTP-PEST (115 kDa) [36].
2. Add the radiolabeled poly [Glu-Tyr] at a ratio of 1.1×10^5 cpm per ml of resolving gel. Mix well.
3. Add TEMED and APS, mix well and pour gels.
4. Cover gel surface with hydrated isobutanol and allow gel polymerization for at least 120 min for maximum reproducibility in pore size [37] (*see* **Note 19**).
5. Prepare stacking gels according to standard protocols with a polymerization time of at least 1 h.
6. Load samples onto gels. If samples are immunoprecipitated proteins, then a negative control (immunoprecipitation with the isotype control antibodies) and positive controls (cleared lysates) should also be loaded. Stained protein ladder is also loaded to follow protein migration and help to determinate presence of different isoforms. It is also possible to use recombinant proteins for identification of the different isoforms, but first ensure of their purity. A fraction of immunoprecipitated

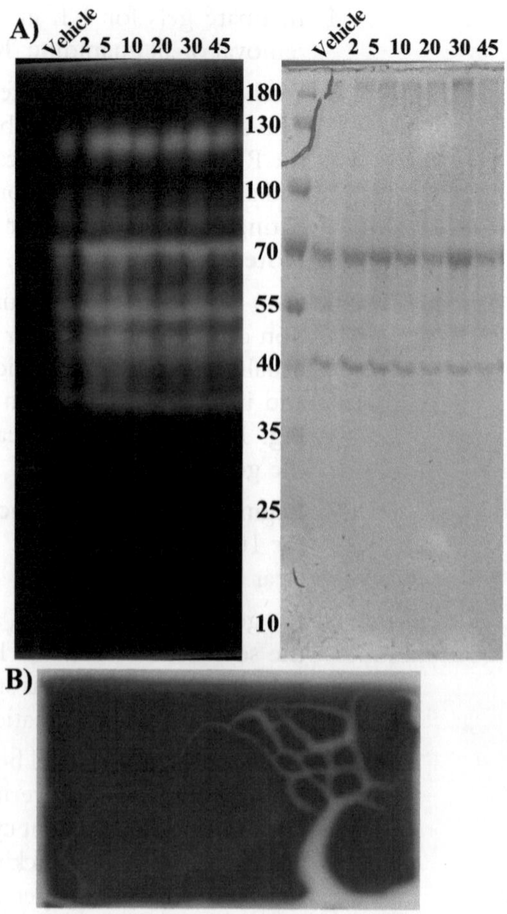

Fig. 1 In-gel PTP assay: (**a**) Typical result obtained with 10 μg of cleared lysate obtained from human monocyte-derived dendritic cells stimulated with vehicle (ethanol) or 10 nM PAF for indicated times. In-gel assay allows the detection of PTPs according to their molecular weight and gel staining with Coomassie blue (right panel) allows the detection of variations that could be due to unequal loading. (**b**) The importance of removing all bubbles: under the vacuum pressure, bubbles will fissure the gel

proteins should be kept for Western blots. This ensures that similar levels of the pertinent proteins have been used in each condition.

7. Add cold 1× running buffer and set cooling system if needed.

3.2.3 PTP Assay

1. After migration, incubate gels in buffer C under gentle agitation for 3–16 h. This will allow SDS removal.
2. Wash gels two times by incubating in buffer D under gentle agitation for 30 min.
3. Incubate gels 90 min in buffer E under agitation. This will allow a complete protein denaturation (*see* **Note 20**).

4. Incubate gels for 1 h in buffer F. This wash step allows the removal of all guanidine. Repeat.
5. Wash gels for 1 h in buffer G (renaturation/reaction buffer). This step allows the equilibration of the gel and the initiation of the PTP refold. Then, discard carefully (*see* **Note 21**) the buffer and allow dephosphosphorylation reaction to proceed by incubating gels in buffer G for a minimum of 1 h or up to 24 h, *see* **Note 21** for more details.
6. Stop PTP reaction by staining gels. Protein loading determination can be done by silver staining [35] (*see* **Note 22**) but we usually use the Coomassie staining solution described above and incubate gels 45 min under gentle agitation. As seen in Fig. 1a this step will reveal the protein quantity loaded onto the gel.
7. Begin the destaining by incubating gels in the destaining buffer for 10 min with sponges. Change solution until you obtain a clear background.
8. If a gel dryer will be used, after the first 10 min, soak the gel in the softener solution for 1–2 h. Change sponges regularly to help the destaining. This step allows the gel to keep its flexibility, avoiding crack formation during gel drying.
9. Drying gels: Put the gel between two cellophane sheets on the gel drying frames for overnight drying or use a gel dryer, using vacuum and the gradient cycle for 3 h at 70 °C, as final temperature, to avoid gel cracking. Alternatively, gels can be dried between Whatman paper and a cellophane sheet, in the same conditions as described for the gel dryer. Carefully remove all bubbles found between the gel and cellophane to avoid fissuring, as shown in Fig. 1b.
10. Expose the gels at room temperature or at −80 °C to a high-sensitivity X-Ray film for different times (4 h to overnight or more) in order to obtain an optimal signal. *See* **Note 23** for further considerations.

3.3 Colorimetric PTP Assay

1. Resuspend freshly immunoprecipitated PTP with 110 μl of 1× PTP assay buffer. For cleared lysates, dilute samples 1:2 in 2× PTP assay buffer for a final concentration around 0.3 μg/μl.
2. Incubate samples for 10 min at room temperature. Put 30 μl of sample in three wells (triplicates) of the 96-well plate. Negative controls such as isotype control, 1× PTP assay buffer alone and cleared lysate or immunoprecipitated protein with 2 mM Na_3V0_4 (final concentration) and positive controls such as cleared lysates or recombinant proteins should also be used (*see* **Note 24**).
3. Add 20 μl of pNPP per well.

4. A first reading of absorbance is done at 405 nm with reference at 800 nm. This is time 0.
5. Incubate plates at 30 °C with gentle stirring and measure absorbance every 5 or 10 min until saturation is reached (usually around 60 min) (*see* **Note 25**).
6. Add 15 μl of 4× loading buffer and perform Western blots to ensure equal protein quantities (*see* **Note 26**).
7. Determine PTP activity: Subtract time 0 values from optical density (O.D.) values found for each time point. Then, mean values of replicates are plotted against time. Slopes of the linear part of the curve are used to calculate pNPP hydrolysis rate:

$$\text{Enzymatic activity}: \frac{[\Delta O.\text{D.} \times \text{Volume}]}{[\varepsilon \times L \times \mu g \text{ lysate}]}.$$

where:

Δ O.D.: the slope of variation of O.D with time (Δ O.D. per min).

ε: the molar extinction coefficient of pNPP: 18,000 M^{-1} cm^{-1}.

L: the length of the light path in cm, usually 1 cm.

Volume: the reaction volume (μl).

μg lysate: μg of protein in lysate used for assays.

4 Notes

1. Sample preparation is similar for in-gel and colorimetric PTP assays. However, the storage of samples differs depending of the PTP assays that will be done. For in-gel PTP assays, whole cell lysates can be frozen before immunoprecipitation or the addition of loading buffer. Moreover, cleared lysates or immunoprecipitated proteins can be stored in the 1× loading buffer for 5–7 days at 4 °C before the assays. For colorimetric PTP assays, we obtained better results, at least for the 48 kDa isoform of PTPN2, when immunoprecipitation and PTP assays were made on the same day as the cell lysis.
2. A stock solution (7 ml) can be made by mixing: 0.5% NP-40, 50 mM HEPES (pH 7.4), 150 mM NaCl, 1 mM EDTA, 2 mM EGTA and cOmplete™, Mini Protease Inhibitor Tablet (1 tablet). Aliquots can be frozen at −80 °C and kept at least 2 months. Other protease and phosphatase inhibitors are added fresh, just before lysis.
3. For some PTPs, better results were obtained with 1% IGEPAL. Optimization may be needed.
4. A stock solution can be done without 2-mercaptoethanol and kept at 4 °C in small aliquots: 50% glycerol, 323 mM Tris–HCl

pH 6.8, 10% SDS, 0.01% bromophenol blue. When needed, warm an aliquot and add 200 μl 2-mercaptoethanol to 800 μl of this stock solution to obtain the 4× loading buffer solution.

5. We adapted our protocol from Meng et al. [30] and Burridge and Nelson [31]. Briefly, given the radioactive substrate, all manipulations must be done with the appropriate personal protection equipment (gloves, lab coats and safety glasses, when needed). Radioactive materials must be manipulated in a designed radioisotope laboratory with the appropriate protection against ß-radiation.
6. The manufacturer specifies that formats <21 μg do not need to be aliquoted since the kinase will support freeze/thaw until exhaustion of the stock whereas for formats >20 μg, 20 μl aliquot can be done (Invitrogen, data sheet).
7. Some protocols suggest that other kinases such as GST-Src [35] or GST-EGFR [31] can replace GST-FER.
8. Mini-protean III electrophoresis system can also be used to determine optimal conditions, such as protein quantity needed, renaturation times or pH of renaturation solution, but we obtained a better resolution with this electrophoresis system and 1.5 mm spacers, which facilitate handling.
9. Acrylamide solution is made fresh in order to avoid degradation products that could interfere with protein separation or PTP assays. We usually prepare a stock solution the day before the first gels are cast. We store this solution at 4 °C and use it for a maximum of 2 weeks after its preparation.
10. 8 M urea can be used instead of the 6 M guanidine HCl, but we obtained better results for cleaved PTP1B fragments with guanidine than urea, even though other authors report no difference between these two conditions for other PTPs [38, 39].
11. For 1 l, dissolve 573.18 g guanidine HCl in 475 ml H_2O, add 25 ml 2 M Tris pH 8.0, 3 ml 2-mercaptoethanol and adjust volume to 1 l.
12. Other non-ionic detergent such as 0.04% Tween 20 [40] can be used instead of Tween 40.
13. In this method, the p-nitrophenyl phosphate (pNPP) is dephosphorylated by PTPs, generating phosphates and p-nitrophenol, a yellow-colored product. Monitoring the change in the absorbance at 405 nM allows the determination of PTP activity. There are many different protocols where pNPP is used as substrate and many of these use an end-point method with the addition of NaOH [41]. NaOH gives a maximal signal since it quenches the protonated p-nitrophenol, thus generating the highly chromogenic de-protonated form which gives an intense yellow color [41, 42]. Here, we describe a continuous PTP assay. This is

useful when stimulation conditions result in different hydrolysis rates and saturation is reached at very different time points or when there are great differences in activity between experiments, such as when primary cells are used. Moreover, this method can be used to determinate the initial velocity and can be more sensitive than the quenched assay in detecting time-dependent inhibition, when inhibitors are added to the PTP reactions [41].

14. We also tested another 10× PTP assay buffer recipe (2 mM NaCl, 250 mM HEPES pH 7.4, 2 mM EDTA, 1 μg/ml BSA, 0.1% Tween 20, 20 mM NaF with pH adjusted to 7.4). We found that, even if a deeper yellow color was visually observed, slopes of PTP hydrolysis rates were similar. Tests should be done to find which recipe is more convenient.
15. We also tested the Dynabead system, however, the elution protocol with the acid glycine buffer disturbs the hydrolysis reaction, even after pH neutralization.
16. For colorimetric PTP assays, we found that endogenous PTP1B immunoprecipitated from 300 μg of total protein from cleared HEK293 cell lysates resulted in enough protein to obtain O.D. that differed sufficiently from the background signal (negative control); whereas 200 μg of total protein from cleared cell lysates were needed when the overexpressed 45 kDa PTPN2 was immunoprecipitated, given the low quantities recovered due to its nuclear localization. For in-gel PTP assays, we found that 10–15 μg of total protein, from HEK293 cells, monocyte or Mo-DC lysates, gave good results. We usually immunoprecipitated endogenous PTP1B from 150 μg total protein from HEK293 or Mo-DC cell lysates.
17. We obtained better results with a 16 h incubation.
18. Even if it is preferable to centrifuge at 4 °C to prevent protein degradation due to warming, centrifugation with a soft stop at room temperature can be done if installations do not allow access to a 4 °C centrifuge.
19. For overnight storage, remove isobutanol after 1 h polymerization, wash gel surfaces with 1.5 M Tris pH 8.8 and cover the gel with 1.5 M Tris pH 8.8. Wrap gels in a plastic bag and wet paper to avoid drying.
20. This step can be omitted, but for several PTPs it allows an optimal renaturation [31].
21. This step allows the PTP renaturation and the dephosphorylation of the radioactive substrate. Most of the radioactivity will be released from gels during these steps. Times and composition of the buffer G proposed here are the general conditions for PTP renaturation and dephosphorylation reactions. Some

PTPs could need longer or shorter reaction times to reach the optimal conditions which would allow the detection of changes of PTP activity between experimental conditions. Moreover, some authors report that the optimal pH for the renaturation buffer could change according to the PTP of interest, thus this should also be taken into consideration [35].

22. *See* [43] for an easy silver staining protocol, compatible with visualization of proteins in addition to being compatible with matrix-assisted laser desorption/ionization and electrospray ionization-mass spectrometry.
23. An intensifying screen can also be used to decrease exposure times and must be used at −70 °C, but resolution may be lower [37]. Condensation should be removed before opening the gel casing and we found that insertion of plastic sheets between the gel and the film minimizes the adhesion between them. A room temperature exposure without the intensifying screens is recommended if proteins of similar molecular weights need to be detected (our observation and [37]).
24. Negative control with Na_3VO_4 allows the determination of the reaction specificity given that Ser/Thr phosphatases can also hydrolyze pNPP. In our hands, the presence of 2 mM NaF in the PTP assay buffer was sufficient to block any pNPP hydrolysis in the negative control when Na_3VO_4 was added. This was comparable to the control containing only solutions, without any lysate. It was valid for both immunoprecipitated PTPs and cleared lysate samples, suggesting that 2 mM NaF is sufficient to inhibit the Ser/Thr phosphatases found in these samples. If needed, NaF concentration can be increased to 10 mM or other Ser/Thr phosphatase inhibitors such as calyculin A or okadaic acid can be added. Cys-based PTPs are metal-insensitive which explains why okadaic acid has no effect; this inhibitor binds the binuclear metal center of Ser/Thr phosphatases abrogating the nucleophilic attack [44, 45].
25. Saturation is reached when the readings reach a plateau and happens when pNPP is exhausted, the detection limit of the spectrophotometer is reached or when hydrolyzed products reach an inhibiting concentration [46]. Instead of reading samples at various time points, it is also possible to make serial dilutions to determine the linear range [47], however, a higher amount of protein could be needed. The absorbance would then be determined only once, and NaOH can be used to quench the reaction.
26. If, given the addition of the pNPP solution, protein concentrations are too low for an optimal detection by Western blotting: increase the amount of immunoprecipitated protein by 25% and resuspend freshly immunoprecipitated PTPs with 135 μl of

1× PTP assay buffer instead of 110 μl and keep 20 μl for Western blots. This should increase the amount of proteins loaded onto the SDS gel by two times without any significant change in the amount of protein used for the PTP assay.

Acknowledgments

This work was supported by Canadian Institutes of Health Research grant (#MT6822).

References

1. Prabakaran S, Lippens G, Steen H et al (2012) Post-translational modification: nature's escape from genetic imprisonment and the basis for dynamic information encoding. Wiley Interdiscip Rev Syst Biol Med 4:565–583
2. Attwood PV (2013) Histidine kinases from bacteria to humans. Biochem Soc Trans 41:1023–1028
3. den HJ (2003) Regulation of protein phosphatases in disease and behaviour. EMBO Rep 4:1027–1032
4. Allen KN, Dunaway-Mariano D (2004) Phosphoryl group transfer: evolution of a catalytic scaffold. Trends Biochem Sci 29:495–503
5. Patterson KI, Brummer T, O'brien PM et al (2009) Dual-specificity phosphatases: critical regulators with diverse cellular targets. Biochem J 418:475–489
6. Tonks NK (2006) Protein tyrosine phosphatases: from genes, to function, to disease. Nat Rev Mol Cell Biol 7:833–846
7. Pao LI, Badour K, Siminovitch KA et al (2007) Nonreceptor protein-tyrosine phosphatases in immune cell signaling. Annu Rev Immunol 25:473–523
8. Tonks NK (2013) Protein tyrosine phosphatases: from housekeeping enzymes to master-regulators of signal transduction. FEBS J 280:346–378
9. Kang N-I, Yoon H-Y, Kim H-A et al (2011) Protein kinase CK2/PTEN pathway plays a key role in platelet-activating factor-mediated murine anaphylactic shock. J Immunol 186:6625–6632
10. Kim H-A, Kim K-J, Seo KH et al (2012) PTEN/MAPK pathways play a key role in platelet-activating factor-induced experimental pulmonary tumor metastasis. FEBS Lett 586:4296–4302
11. Ni Y, Sinnett-Smith J, Young SH et al (2013) PKD1 mediates negative feedback of PI3K/Akt activation in response to G protein-coupled receptors. PLoS One 8:e73149
12. Lapointe F, Stiffel M, Auger J-P et al (2015) Platelet-activating factor induces du-al-specificity phosphatase 1 and 5 gene expression. Pharmacol Pharm 6:442–450
13. Suen JY, Gardiner B, Grimmond S et al (2010) Profiling gene expression induced by protease-activated receptor 2 (PAR2) activation in human kidney cells. PLoS One 5:e13809
14. Mancini AD, Battista JAD (2011) The cardinal role of the phospholipase A2/cyclooxygenase-2/prostaglandin E synthase/prostaglandin E2 (PCPP) axis in inflammostasis. Inflamm Res 60:1083–1092
15. Venema RC, Venema VJ, Eaton DC et al (1998) Angiotensin II-induced tyrosine phosphorylation of signal transducers and activators of transcription 1 is regulated by Janus-activated kinase 2 and Fyn kinases and mitogen-activated protein kinase phosphatase 1. J Biol Chem 273:30795–30800
16. Willoughby EA, Collins MK (2005) Dynamic interaction between the dual specificity phosphatase MKP7 and the JNK3 scaffold protein β-arrestin 2. J Biol Chem 280:25651–25658
17. Xu Y, Fisher GJ (2012) Receptor type protein tyrosine phosphatases (RPTPs) – roles in signal transduction and human disease. J Cell Commun Signal 6:125–138
18. Nunes-Xavier CE, Martín-Pérez J, Elson A et al (2013) Protein tyrosine phosphatases as novel targets in breast cancer therapy. Biochim Biophys Acta 1836:211–226
19. Zhu JW, Doan K, Park J et al (2011) Receptor-like tyrosine phosphatases CD45 and CD148 have distinct functions in chemoattractant-mediated neutrophil migration and response to S. aureus. Immunity 35:757–769
20. Lee J-H, Choi S-H, Lee B-H et al (2013) Activation of lysophosphatidic acid receptor

by gintonin inhibits Kv1.2 channel activity: involvement of tyrosine kinase and receptor protein tyrosine phosphatase α. Neurosci Lett 548:143–148

21. Imbrici P, Tucker SJ, D'Adamo MC et al (2000) Role of receptor protein tyrosine phosphatase alpha (RPTPalpha) and tyrosine phosphorylation in the serotonergic inhibition of voltage-dependent potassium channels. Pflugers Arch 441:257–262
22. Tsai W, Morielli AD, Cachero TG et al (1999) Receptor protein tyrosine phosphatase alpha participates in the m1 muscarinic acetylcholine receptor-dependent regulation of Kv1.2 channel activity. EMBO J 18:109–118
23. Chernock RD, Cherla RP, RK G (2001) SHP2 and cbl participate in alpha-chemokine receptor CXCR4-mediated signaling pathways. Blood 97:608–615
24. Vila-Coro AJ, Rodríguez-Frade JM, Martín De Ana A et al (1999) The chemokine SDF-1alpha triggers CXCR4 receptor dimerization and activates the JAK/STAT pathway. FASEB 13:1699–1710
25. Lodeiro M, Theodoropoulou M, Pardo M et al (2009) c-Src Regulates Akt Signaling in Response to Ghrelin via β-Arrestin Signaling-Independent and -Dependent Mechanisms. PLoS ONE 4:e4686
26. Lodeiro M, Alén BO, Mosteiro CS et al (2011) The SHP-1 protein tyrosine phosphatase negatively modulates Akt signaling in the ghrelin/GHSR1a system. Mol Biol Cell 22:4182–4191
27. Hamel-Côté G, Gendron D, Rola-Pleszczynski M et al (2017) Regulation of platelet-activating factor-mediated protein tyrosine phosphatase 1B activation by a Janus kinase 2/calpain pathway. PLoS One 12:e0180336
28. Shen Y, Schneider G, Cloutier JF et al (1998) Direct association of protein-tyrosine phosphatase PTP-PEST with paxillin. J Biol Chem 273:6474–6481
29. Kappert K, Peters KG, Böhmer FD et al (2005) Tyrosine phosphatases in vessel wall signaling. Cardiovasc Res 65:587–598
30. Meng T-C, Hsu S-F, NK T (2005) Development of a modified in-gel assay to identify protein tyrosine phosphatases that are oxidized and inactivated in vivo. Methods 35:28–36
31. Burridge K, Nelson A (1995) An in-gel assay for protein tyrosine phosphatase activity: detection of widespread distribution in cells and tissues. Anal Biochem 232:56–64
32. Haase H, Maret W (2003) Intracellular zinc fluctuations modulate protein tyrosine phosphatase activity in insulin/insulin-like growth factor-1 signaling. Exp Cell Res 291:289–298
33. Bellomo E, Birla KS, Massarotti A et al (2016) The metal face of protein tyrosine phosphatase 1B., The metal face of protein tyrosine phosphatase 1B. Coord Chem Rev 327–328:70–83
34. Huyer G, Liu S, Kelly J et al (1997) Mechanism of inhibition of protein-tyrosine phosphatases by vanadate and pervanadate. J Biol Chem 272:843–851
35. Markova B, Gulati P, Herrlich PA et al (2005) Investigation of protein-tyrosine phosphatases by in-gel assays. Methods 35:22–27
36. Hallé M, Liu Y-C, Hardy S et al (2007) Caspase-3 Regulates catalytic activity and scaffolding functions of the protein tyrosine phosphatase PEST, a novel modulator of the apoptotic response. Mol Cell Biol 27:1172–1190
37. Walker JM (1996) The protein protocols handbook. Springer Science & Business Media, Berlin
38. Kameshita I, Ishida A, Okuno S et al (1997) Detection of protein phosphatase activities in sodium dodecyl sulfate-polyacrylamide gel using peptide substrates. Anal Biochem 245:149–153
39. Kameshita I, Baba H, Umeda Y et al (2010) In-gel protein phosphatase assay using fluorogenic substrates. Anal Biochem 400:118–122
40. Heneberg P, Dráberová L, Bambousková M et al (2010) Down-regulation of protein-tyrosine phosphatases activates an immune receptor in the absence of its translocation into lipid rafts. J Biol Chem 285:12787–12802
41. McCain DF, Zhang Z-Y (2002) Assays for protein-tyrosine phosphatases. Methods Enzymol 345:507–518
42. Lorenz U (2011) Protein tyrosine phosphatase assays. Curr Protoc Immunol 11:Unit 11.7
43. Yan JX, Wait R, Berkelman T et al (2000) A modified silver staining protocol for visualization of proteins compatible with matrix-assisted laser desorption/ionization and electrospray ionization-mass spectrometry. Electrophoresis 21:3666–3672
44. Zhang M, Yogesha SD, Mayfield JE et al (2013) Viewing serine/threonine protein phosphatases through the eyes of drug designers. FEBS J 280:4739–4760
45. Lim W, Mayer B, Pawson T (2014) Cell signaling: principles and mechanisms. Taylor & Francis, Routledge
46. Montalibet J, Skorey KI, BP K (2005) Protein tyrosine phosphatase: enzymatic assays. Methods 35:2–8
47. Mercan F, Bennett AM (2010) Analysis of protein tyrosine phosphatases and substrates. Curr Protoc Mol Biol Chapter 18:Unit 18.16

Chapter 14

Measurement of β-Arrestin Recruitment at GPCRs Using the Tango Assay

Geneviève Laroche and Patrick M. Giguère

Abstract

Intracellular signal transduced by G protein-coupled receptors (GPCRs) is tightly controlled by a guanine nucleotide-binding complex made of G protein Gα, Gβ, and Gγ subunits, as well as a growing array of regulatory and accessory proteins such as arrestins. G protein-independent β-arrestin recruitment at GPCRs is universally accepted as the canonical interactor system and it has been found to be a powerful tracker of most GPCRs activation. Pharmacological concepts have evolved remarkably after the finding that different ligands, binding at the same receptor, can selectively activate specific subsets of signaling pathways among all pathways activated by balanced ligands. This new paradigm referred to as functional selectivity or biased signaling, has opened new avenues for the design of tailored drugs with enhanced therapeutic efficacies and reduced side effects. Here, we describe a unique platform for the interrogation of GPCR using a transcriptional-based assay to measure transient β-arrestin recruitment called Tango.

Key words G protein-coupled receptor, β-Arrestin, Cell-based assay, Tango, GPCRs signaling

1 Introduction

G protein-coupled receptors characteristically transduce the signal through activation of at least one member of the Gα guanine nucleotide-binding protein family, hence leading to their name. The intracellular signals transduced by these heptahelical proteins are themselves tightly controlled by the heterotrimer Gαβγ subunits, as well as regulatory and accessory proteins such as arrestins. Once activated by an extracellular trigger, the GPCR undergoes a conformational change that allosterically activated the bound heterotrimer leading to their dissociation. Signal termination occurs classically following the recruitment of a family of regulators of G protein signaling (RGS) but also through the recruitment of arrestin that terminates signal transduction by sterically inducing G protein dissociation followed by the recruitment of the GPCR–arrestin complex into the clathrin-coated internalization pathway [1]. Many biochemical and pharmacological data have

Mario Tiberi (ed.), *G Protein-Coupled Receptor Signaling: Methods and Protocols*, Methods in Molecular Biology, vol. 1947,
https://doi.org/10.1007/978-1-4939-9121-1_14,

provided evidence of ligand-dependent dynamicity between different functional states [2]. This novel concept in GPCR pharmacology has led to the renewing of our conception of the linear ligand–GPCR activation complex; where one ligand activates one receptor leading to activation of signaling pathways with all the same efficacy and potency (balanced signaling) [3]. This new paradigm, referred to as functional selectivity or biased signaling, has opened new avenues for the design of tailored drugs with enhanced therapeutic efficacies and reduced side effects [3]. Biased activity is generally quantified by comparing the activation of the canonical G protein-regulated pathway (cAMP or Ca^{2+}) with β-arrestin recruitment using specifically designed cell-based assays. Characterization of the signaling pathways enlisted by the targeted receptor and measurement of ligand biased activity toward those must be considered to understand drug's action, to establish the pharmacological profiling of a receptor and to identify the therapeutically relevant signaling pathway(s). To address the pharmacological selectivity of newly discovered compounds toward the entire GPCR-ome, as well as to decipher the polypharmacological profile of some effective drugs, a unique screening platform called Tango (or Presto-Tango) [4, 5] provides the capacity to simultaneously interrogate the entire druggable GPCR-ome (i.e. ~350 non-olfactory GPCRs) via a G protein-independent β-arrestin recruitment assay (Fig. 1). The interrogation of the GPCR-ome in a parallel and simultaneous fashion was technologically and economically challenging before the development of this assay. This β-arrestin recruitment assay is also a powerful tool for large-scale interrogation of potential interactors at orphan and non-orphan GPCRs. The Tango arrestin-recruitment assay provides some advantages over other orthologous assays. It allows the measurement of a signal specifically activated by the target receptor via the expression of the receptor fused to the bacterial transcription factor tTA preceded by a Tobacco Etch virus protease cutting site (TEVcs). Once activated, the receptor recruits the β-arrestin–TEV fusion protein that cleaves its TEVcs releasing the tTA. The artificial transcription factor translocated to the nucleus where it binds to the tetracyclin-response element (TRE) driving the expression of the reporter Firefly Luciferase (Luc). This reporter is thus amplifying the signal received from one receptor leading to a very sensitive assay. Another advantage of this assay is that it does not require any information on the coupling partners, most of the time not available for orphan GPCRs allowing parallel interrogation.

Herein, we comprehensively describe two different experimental approaches using the β-arrestin Tango assay, the β-Arrestin recruitment for one GPCR interrogation with multiple drugs and the parallel approach for β-Arrestin recruitment at multiple GPCRs with one drug.

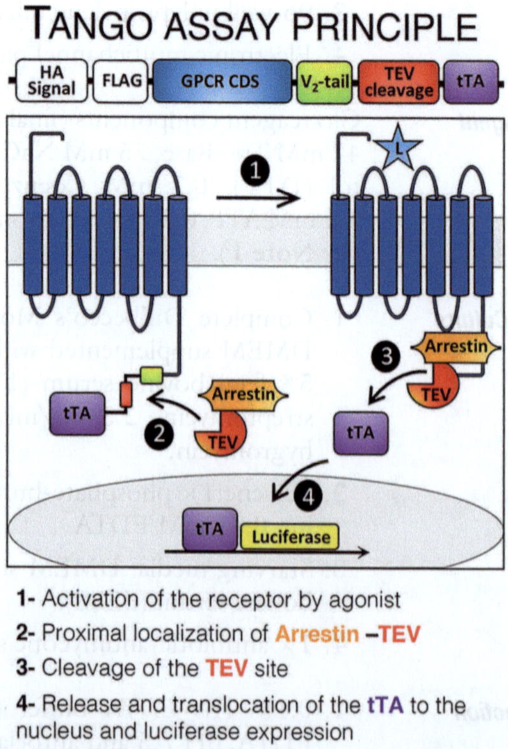

Fig. 1 Overview of the β-arrestin recruitment Tango assay. Top, Template of the codon-optimized cDNA encoding GPCR coding sequence fused to required Tango component. HA, hemagglutinin signal sequence; Flag, epitope DYKDDDDA; V2-Tail, vasopressin receptor 2 arrestin recruitment domain; TEV, Tobacco Etch virus protease; tTA, tetracycline transactivator protein. Bottom, Overview of the β-arrestin recruitment Tango assay

2 Materials

Prepare all solutions using autoclaved deionized water (18 MΩ cm at 25 °C); stock solutions are stored at room temperature, −20 °C or 4 °C as indicated.

2.1 Poly-L-Lysine-Coated Plate

1. 25 mg/ml (1000×) stock solution of Poly-L-lysine (PLL), average mol. wt 30,000–70,000. Store aliquots at −20 °C.
2. 100× antibiotic-antimycotic is diluted to 1× and use to rinse and store the PLL-coated plate.
3. White or black 384-well optical bottom microplates.

2.2 Drug Dilutions

1. 1× Hank's Balanced Salt Solution (HBSS) in 20 mM HEPES, pH 7.4.
2. 10 mM stock solutions of agonists of interest in DMSO. Store aliquots at −20 °C.

3. 96-well polypropylene microplates (drug plates).
4. Electronic multichannel pipets.

2.3 Glo Reagent

Glo reagent components (final concentration): 108 mM Tris–HCl, 42 mM Tris-Base, 75 mM NaCl, 3 mM $MgCl_2$, 5 mM Dithiothreitol (DTT), 0.2 mM Coenzyme A, 0.14 mg/ml D-Luciferin, 1.1 mM ATP, 0.25% v/v Triton X-100, 2 mM Sodium hydrosulfite (*see* **Note 1**).

2.4 Tissue Culture Reagents

1. Complete Dulbecco's Modified Eagle's Medium (DMEM): DMEM supplemented with 5% bovine calf serum (BCS) and 5% fetal bovine serum (FBS), 100 μg/ml of penicillin and streptomycin, 2.5 μg/ml of puromycin and 50 μg/ml of hygromycin.
2. Versene:1× phosphate-buffered saline (PBS), pH 7.4 containing 0.53 mM EDTA.
3. Starving media: DMEM supplemented with 1% dialyzed fetal bovine serum (dFBS).
4. 1× antibiotic-antimycotic solution.

2.5 Transfection Reagents

1. 0.1× Tris–EDTA buffer (TE): 1 mM Tris–HCl, 0.1 mM EDTA, pH 7.5 and autoclaved. Store at 4 °C.
2. 2.5 M $CaCl_2$, autoclaved and store at 4 °C.
3. 2× HEPES Buffer (HB): 50 mM HEPES, 280 mM NaCl, 1.5 mM Na_2HPO_4. Adjust pH exactly to 7.05, sterilize by filtration and keep at 4 °C (*see* **Note 2**).

3 Methods

3.1 β-Arrestin Recruitment for One GPCR Interrogation with Multiple Drugs

3.1.1 Cell Culture

Human Embryonic Kidney cell line (HEK293T) expressing TEV fused-β-Arrestin2 and a tetracycline transactivator (tTA)-driven luciferase (called HTLA cells, kindly provided by Dr. Richard Axel) are maintained in complete DMEM at 37 °C in a humidified atmosphere containing 5% CO_2. For HTLA cells stably expressing a receptor, zeocin is added at a final concentration of 200 μg/ml (*see* **Note 3**).

1. Culture HTLA cells in 15 or 10 cm dishes and maintain in complete DMEM.
2. Pass cells twice a week at 1/10 dilution factor.
3. Twenty-four hours before the transfection, subculture cells in 15 cm dishes at a density of 12×10^6 cells in 22 ml of complete media (*see* **Note 4**).

3.1.2 Transfection

Calcium phosphate precipitation method is used as described by Jordan et al. in 1996 [6].

1. Pre-warm the materials needed at room temperature (*see* **Note 2**).
2. Prepare the amount needed of 0.1 × TE with $CaCl_2$: (for one 15 cm dish use: 900 μl 0.1 × TE + 100 μl 2.5 M $CaCl_2$) and vortex.
3. Add 1 ml of the solution (0.1 × TE + $CaCl_2$) to 18 μg of cDNA and vortex (*see* **Note 5**).
4. Add 1 ml of 2× Hepes buffer solution.
5. Shake vigorously (do not vortex) and wait 1 min (*see* **Note 6**).
6. Add 2 ml of the solution dropwise to the cells.
7. Gently rock the plate back and forth, then side to side to evenly distribute precipitate (do not swirl around).
8. Incubate at 37 °C for 24 h.

3.1.3 Plate Coating

1. Coat white or black 384-well optical bottom plates with 20 μl/well of 25 μg/ml PLL using a reagent dispenser and incubate for a minimum of 30 min (up to 2 h) at room temperature.
2. Remove PLL by flicking the plate and gently taping it over a paper towel.
3. Add 40 μl of 1× antibiotic-antimycotic in each well.
4. Store PLL-coated plates at 4 °C. Use plates within 2 months (*see* **Note 7**).

3.1.4 Plate Seeding

1. Rinse gently transfected cells once with Versene.
2. Detach cells using 5 ml of 0.05% trypsin-0.53 mM EDTA per dish and transfer to a centrifuge tube containing at least 5 ml complete DMEM.
3. Spin and resuspend cells at a density of 0.4 × 10^6 cells/ml in starving media. Seed cells into white 384-well poly-L-Lysine-coated plates at a density of 20,000 cells/well (~45 μl) using an electronic multichannel pipet and put plates at 37 °C overnight (*see* **Note 8**). If stimulation is performed on the same day of plate seeding, seed 25,000 cells/well and incubate at least 4 h before stimulation.

3.1.5 Drug Plate Preparation

1. Pipet 270 μl of drug buffer (1× HBSS, 20 mM HEPES pH 7.4, 1× antibiotic-antimycotic) into a drug plate except for the first row which is 300 μl (*see* **Note 9**).
2. Calculate the amount of drug needed to be added for the "Low" and "High" well for each drug (Fig. 2a). Prepare 3× drug concentrations and add 20 μl of drug dilutions to 40 μl of

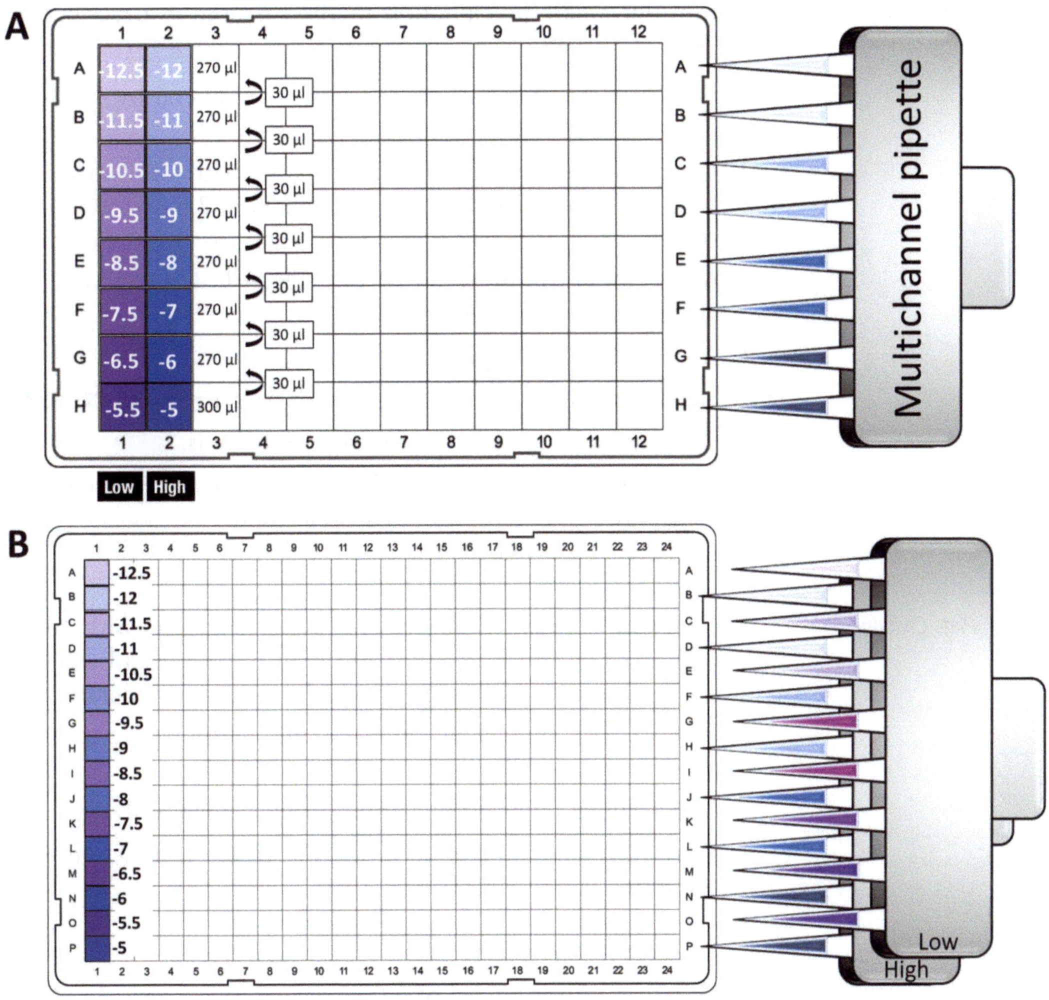

Fig. 2 Schematic design of (**A**) drug plate preparation and (**B**) cell stimulation. Visual representation of a drug plate preparation for a 16-point dose curve starting at 10 µM (final concentration) with half-log between each point

media/cells (*see* **Note 10**). For a dose–response curve starting with a final concentration of 10 µM, prepare the high concentration at 30 µM and the low concentration at 9.49 µM (High divided by half-log or 3.16, *see* **Note 11**).

3. Add the amount of drug calculated in **step 2** in the last row of the drug plate (row H).
4. Mix solution from the last row using a multichannel pipet and by pipetting up and down and discard tips.
5. Do a serial dilution by pipetting 30 µl solution from the last row (row H) to the previous row (row G) and mix, then repeat up to the first and most diluted row (row A) (*see* **Note 12**).

3.1.6 Cell Stimulation and Luminescence Reading

1. Pipet 20 μl of the column "Low" to well A-O and 20 μl of the column "High" to well B-P of a 384-well drug plate previously seeded with transfected cells (*see* **Note 13**). Repeat for each drug (Fig. 2b).
2. Return at 37 °C and incubated for at least 16-h (*see* **Note 14**).
3. After 16–24 h incubation at 37 °C and 5% CO_2, decant media containing drugs and add 20 μl of Glo reagent per well (*see* **Note 15**).
4. Incubate with Glo reagent for 5–20 min at room temperature and read plates using a microplate luminescence counter (*see* **Note 16**).

3.2 β-Arrestin Recruitment for Multiple GPCRs Interrogation with One Drug

3.2.1 Plate Seeding

1. Gently rinse cells once with 1× PBS.
2. Detach cells using 5 ml of 0.05% trypsin/0.53 mM EDTA spin and resuspend at a density of 0.2 × 10^6 cells/ml in complete DMEM.
3. Seed cells into ***black*** 384-well PLL-coated plates at a density of 10,000 cells/well (~45 μl) using an electronic multichannel pipet.
4. Incubate plates at 37 °C overnight.

3.2.2 DNA-Plate Preparation and Transfection

1. Pre-warm the transfection reagents needed at room temperature.
2. Distribute plasmids cDNA encoding GPCRs of interest through a 96-well plate at a concentration of 50 ng/ml in 0.1× TE.
3. Transfer 10 μl of the DNA solution from **step 2** to a 384-well plate (Fig. 3). To perform the experiment in quadruplicate, the 96-well cDNA plate is distributed in a four-quadrant pattern. Thus, half of 96-well plate with cDNAs is required for one 384-well plate where each receptor will be tested in quadruplicate for two conditions (± stimulation). This step can be performed using a multichannel pipet or using automated 96-channel pipettor with the ability to perform quadrant dispensing.
4. Transfer 40 μl of 0.313 M $CaCl_2$ (2.5 M stock dilute 1/8 in 0.1 × TE, vortex) and mix by pipetting up and down.
5. Add 50 μl of 2× HBS solution to the **step 3** using automated 384-channel pipettor and mix by pipetting up and down and let stand for 1 min.
6. Transfer 10 μl of **step 5** to seeded cells (*see* **Note 17**).
7. Incubate cells at 37 °C in 5% CO_2 overnight.

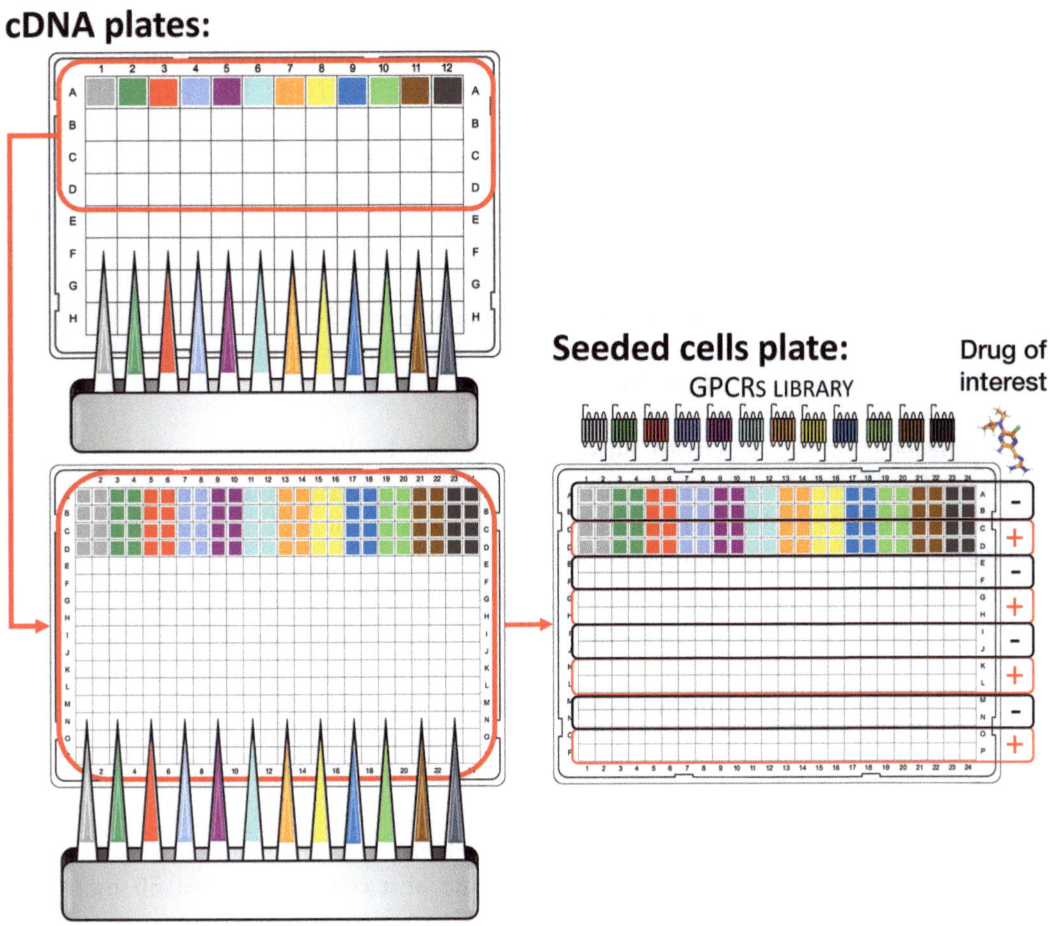

Fig. 3 Schematic design of cDNA plate preparation and cell stimulation. Visual representation of a cDNA plate preparation and transfection in 384-well plate for parallel interrogation using the Tango assay. For a final reading in quadruplicate, with and without the drug, cells are transfected in octuplicate

3.2.3 Cell Stimulation and Luminescence Reading

1. Remove transfected media by flicking the 384-well plate and taping it gently on a paper towel, and add slowly 40 μl of starving media without touching the cell directly.
2. Pipet 20 μl of the drug of interest at a 3× concentration into the even number column and add drug buffer into the odd number column. Each receptor will be assayed in quadruplicate with and without the drug.
3. Return at 37 °C in 5% CO_2 and incubated for at least 16 h (*see* **Note 14**) and proceed to the luminescence reading.

4 Notes

1. Modified from [7]. Stock solutions can be made for 2× Tris–HCl/Tris-Base, 2.5 M NaCl and 1 M $MgCl_2$, and

store at room temperature. Stock solution of 0.5 M DTT, 0.1 M ATP, and 0.01 M Coenzyme A are aliquoted and stored at −20 °C. Stock solution of 1 M sodium hydrosulfite is kept at 4 °C. D-Luciferin is always freshly added from powder. D-Luciferin can be aliquoted and kept at −20 °C but activity decrease rapidly.

2. All transfection buffers are stored at 4 °C. Warm up all buffers at room temperature few hours before transfection.
3. All cDNA encoding GPCR-Tango are available from Addgene.org and are cloned in pcDNA3.1-Neomycin plasmid. Those constructs work for transient transfection only. If a stably expressing receptor is desired in HTLA cells, we recommend sub-cloning the GPCR-Tango within pIRESbleo3 vector (Clontech) and select cells using zeocin. To facilitate the direct transfer from pcDNA3.1 to pIRESbleo3, we have introduced a NheI restriction site in the multi-cloning site of the pIRES vector by site-directed mutagenesis (replacing the BamHI). GPCR-Tango cDNA can be extracted from pcDNA3.1 by cutting at restriction sites NotI and XbaI and cloned into the modified pIRESbleo3 at NotI-NheI.
4. For high transfection efficiency 50–70% cell confluency is optimal.
5. Calcium phosphate precipitation works well with HEK293-derived cells. Highly pure plasmid cDNA has to be used. One to two microgram of a plasmid encoding a fluorescent protein can be used to track transfection efficiency without interfering with the assay.
6. Incubation time at **step 5** can vary from 1 to 10 min and has to be determined empirically.
7. 384-well optical bottom *white* plates are used for most experiment except for parallel interrogation whereas *black* plates are preferred as they reduced the bleed-through between wells. Increasing the amount of D-Luciferin in the Glo-reagent will compensate for the reduced luminescence.
8. PLL-coated plates are incubated at 37 °C and storage solution is removed by flicking over the sink and gently taping the plate on paper towel.
9. Drug buffer can be supplemented as needed depending on the drug interrogated. For peptides, colloidal molecules, low water solubility molecules, the addition of 0.1–1% BSA is suggested. To prevent drug oxidation, up to 0.01% ascorbic acid can be used.
10. 3× concentration is for 20 μl of drug addition to 40 μl of media (cells). The proportion has to be adjusted in function of the

amount of drug solution added to the cells or the volume of cells plated.

11. A 3.16-fold ($10^{0.5}$-fold) serial dilution also called a half-log dilution will give an accurate 16 points dose curve.
12. Automated multichannel can be used on the function "pipet and mix". Starting from high to low concentration allows using the same tips during the drug plate preparation. If working with "sticky" molecule, BSA is suggested as well as changing tips between each row.
13. If 45 μl of media is used for seeding the cells, we used 40 μl for calculation considering evaporation.
14. The 16 h represents the optimal incubation time found for most GPCRs but up to 24-h can be used.
15. A commercial reagent can also be used; Bright-Glo from Promega can be diluted 1/20 in 1× HBSS. In that case, the plate is incubated for 15–20 min at room temperature for completed cell lysis before reading. We found that dilution of Bright-Glo up to 1/20 works well.
16. If using Bright-Glo at 1×, 1–2 min incubation is sufficient for complete cell lysis.
17. Using this protocol, one cDNA plate can be used to transfect up to nine seeded-cells plates with a residual volume of 10 μl (dead volume). **Steps 4–6** have to be performed using automated pipettor and cDNA plate from **step 4** should not be incubated for more than 15 min allowing the time to perform the transfer to the nine seeded-cells plates.

Acknowledgments

This work was supported by the Canadian Institutes of Health Research MOP142219.

References

1. Luttrell LM, Gesty-Palmer D (2010) Beyond desensitization: physiological relevance of arrestin-dependent signaling. Pharmacol Rev 62(2):305–330. https://doi.org/10.1124/pr.109.002436
2. Katritch V, Cherezov V, Stevens RC (2013) Structure-function of the G protein-coupled receptor superfamily. Annu Rev Pharmacol Toxicol 53:531–556. https://doi.org/10.1146/annurev-pharmtox-032112-135923
3. Urban JD, Clarke WP, von Zastrow M, Nichols DE, Kobilka B, Weinstein H, Javitch JA, Roth BL, Christopoulos A, Sexton PM, Miller KJ, Spedding M, Mailman RB (2007) Functional selectivity and classical concepts of quantitative pharmacology. J Pharmacol Exp Ther 320(1):1–13. https://doi.org/10.1124/jpet.106.104463
4. Kroeze WK, Sassano MF, Huang XP, Lansu K, McCorvy JD, Giguere PM, Sciaky N, Roth BL (2015) PRESTO-Tango as an open-source resource for interrogation of the druggable human GPCRome. Nat Struct Mol Biol. https://doi.org/10.1038/nsmb.3014
5. Barnea G, Strapps W, Herrada G, Berman Y, Ong J, Kloss B, Axel R, Lee KJ (2008) The

genetic design of signaling cascades to record receptor activation. Proc Natl Acad Sci U S A 105(1):64–69. https://doi.org/10.1073/pnas.0710487105

6. Jordan M, Schallhorn A, Wurm FM (1996) Transfecting mammalian cells: optimization of critical parameters affecting calcium-phosphate precipitate formation. Nucleic Acids Res 24 (4):596–601

7. Baker JM, Boyce FM (2014) High-throughput functional screening using a homemade dual-glow luciferase assay. J Vis Exp (88). https://doi.org/10.3791/50282

Chapter 15

Probing the Interactome of Corticotropin-Releasing Factor Receptor Heteromers Using Mass Spectrometry

Burcu Hasdemir, Juan A. Oses-Prieto, Alma Burlingame, and Aditi Bhargava

Abstract

Mass spectrometry is a sensitive technique used in the field of proteomics that allows for simultaneous detection and characterization of several proteins in a sample. It is also a powerful methodology to elucidate protein–protein interactions in a sequence-dependent and unbiased manner. G protein-coupled receptors (GPCRs) seldom function in isolation and characterization of proteins present in the receptor complex (or its interactome) is critical for understanding the vast spectrum of functions these receptors perform in a context-dependent manner. Here, we describe a mass spectrometry-based method to sequence and characterize proteins present in heteromeric complexes formed by corticotropin-releasing factor (CRF) receptors that belong to class B GPCRs. CRF receptor heteromeric complexes were identified in HEK293 cells co-transfected with tagged CRF receptors 1 and 2. CRF receptors were immunoprecipitated using antibodies against the tags from transfected HEK293 cells and proteins in their interactome were identified using liquid chromatography mass spectrometry method (LC-MS/MS). Both CRF receptors were identified in this interactome. A few of the proteins identified in the CRF receptor interactome using MS were confirmed to be true interactions using traditional co-immunoprecipitation and Western blotting methods.

Key words Antibody, Computational analysis, Co-immunoprecipitation, GPCRs, Proteome, Tagged receptors, Western blot analysis

1 Introduction

Many GPCRs, including CRF receptors exist in heteromeric assemblies, which allow them to interact with one another, cross-regulate each other, and respond to different stimuli in an integrated and balanced way [1–3]. Antibody specificity for GPCRs remains a significant challenge in isolating receptor complexes from in vivo systems or tissue lysates. To overcome this caveat, tagged GPCRs are often used in heterologous in vitro cell culture systems to delineate their function. Here, we describe steps for analyzing the interactome of CRF_1R + $CRF_{2\beta}R$ heteromers obtained from HEK293 cells co-transfected with HA-tagged CRF_1R and Flag-

Mario Tiberi (ed.), *G Protein-Coupled Receptor Signaling: Methods and Protocols*, Methods in Molecular Biology, vol. 1947, https://doi.org/10.1007/978-1-4939-9121-1_15,

tagged $CRF_{2\beta}R$. While fluorescence resonance energy transfer (FRET) and other methods have been used to identify protein–-protein interactions, these techniques have their limitations. FRET can identify only proteins that directly interact with the receptor (one-on-one) and are present in proximity [4]. Other proteins that are present in the complex that do not directly interact with the receptor or protein of interest cannot be identified using FRET.

The first step in identifying proteins present in the CRF receptor interactome was immunoprecipitation of CRF_1R and $CRF_{2\beta}R$ and their protein partners using either anti-HA or anti-Flag antibody. The next step was to separate protein complexes on native or SDS-PAGE followed by mass spectrometry for the identification of proteins in the CRF receptor interactome. Identified proteins may either be interacting directly or indirectly with the CRF receptors or with only one or both receptor subtypes in the heteromeric complex. Western blots of resolved co-immunoprecipitated proteins were subsequently performed to validate a subset of candidate proteins using mass spectrometry.

To identify proteins using mass spectrometry (MS), they first need to be digested into their component peptides. Trypsin is the most frequently used protease for digesting proteins for MS for a number of reasons. Most importantly, it results in robust and efficient digestions and it generates peptides with C-terminal positively charged residues that produce strong C-terminal sequence ion series. The resultant tryptic peptides are then separated by reverse phase HPLC column using an acetonitrile gradient (solvent of choice). The eluate of the column is interfaced with the MS using an electrospray source and the peptides present analyzed using tandem mass spectrometry or MS/MS. In the first step (MS1), the mass to charge (m/z) ratios of co-eluted components is measured. In the second step (MS2), a number of the co-eluted peptide ions detected in MS1 are selected then sequentially isolated and used as precursors for fragmentation using different energy-deposition mechanisms and the mass of the resulting fragment ions measured and recorded. Then a new cycle of MS1 and MS2 acquisition is performed, and this is repeated during the full elution time of the peptides from the column. Data are analyzed using search engines such as Mascot (Matrix Science) or Protein Prospector.

The availability of a specific antibody against the target protein that works well for IPs is a pre-requisite for this entire process. It can be challenging to find reliable and specific antibodies for GPCR subtypes [5]. To overcome the issue of antibody specificity, we used the anti-HA antibody to immunoprecipitate CRF receptor complex from HEK293 co-transfected with HA-CRF_1R + Flag-$CRF_{2\beta}R$. This is a good and very commonly used starting point for the analysis of a desired heteromeric complexes in cells, while tools are being developed to specifically recognize the heteromeric complex in endogenously expressing tissues [6]. The

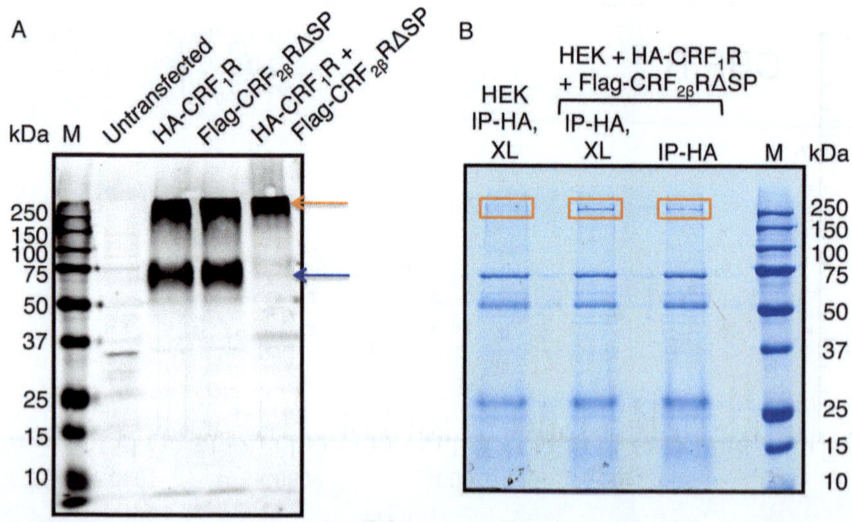

Fig. 1 (**a**) Western blot analysis of HEK293 using anti-CRFR$_{1/2}$ antibody showed bands at ~70 kDa (blue arrow, predicted monomer size) and ~250 kDa (orange arrow) in cells transfected with either HA-CRF$_1$R or Flag-CRF$_2$βRΔSP, whereas only a prominent band at ~250 kDa was seen in cells co-transfected with both HA-CRF$_1$R + Flag-CRF$_2$βRΔSP, suggesting heteromerization and formation of a supercomplex (orange arrow). Untransfected cells were used as negative control. (**b**) Coomassie Blue-stained gel of IPs showed a band at ~250 kDa in lysates from co-transfected, but not untransfected cells (orange boxes). IPs using anti-HA antibody was performed from co-transfected HA-CRF$_1$R + Flag-CRF$_2$βRΔSP HEK293. All visible bands were excised and processed as individual samples for MS analysis. *XL* cross-linked, *M* marker ($n = 3$–4 for IP and 2–3 for MS) (Reproduced from ref. 1 with permission from *Molecular Biology of the Cell*)

co-immunoprecipitation (co-IP) was resolved on SDS-PAGE and a complex at ~250 kDa was detected (Fig. 1). Subsequent excision, digest, and analysis of this band by MS revealed that it contained both CRF$_1$R and CRF$_{2\beta}$R (Fig. 2) and several other interacting proteins [1]. Some of these protein interactions were statistically significant between mock transfected and cell transfected with CRF receptors and included actin, tubulin, and heat shock protein 70 (Fig. 3). These interactions were confirmed to be specific using standard immunoblotting techniques (Fig. 4). Several investigators have also obtained results on other GPCR heteromers using similar techniques [7–9].

2 Materials

2.1 Co-transfection of HEK 293 Cells

1. cDNA constructs: HA-CRF$_1$R and Flag-CRF$_{2\beta}$RΔSP in pcDNA-FRT-5.0 vector as described by us previously [1, 10] or appropriately tagged constructs.
2. Lipofectamine 2000.
3. Opti-MEM I Reduced Serum Media.

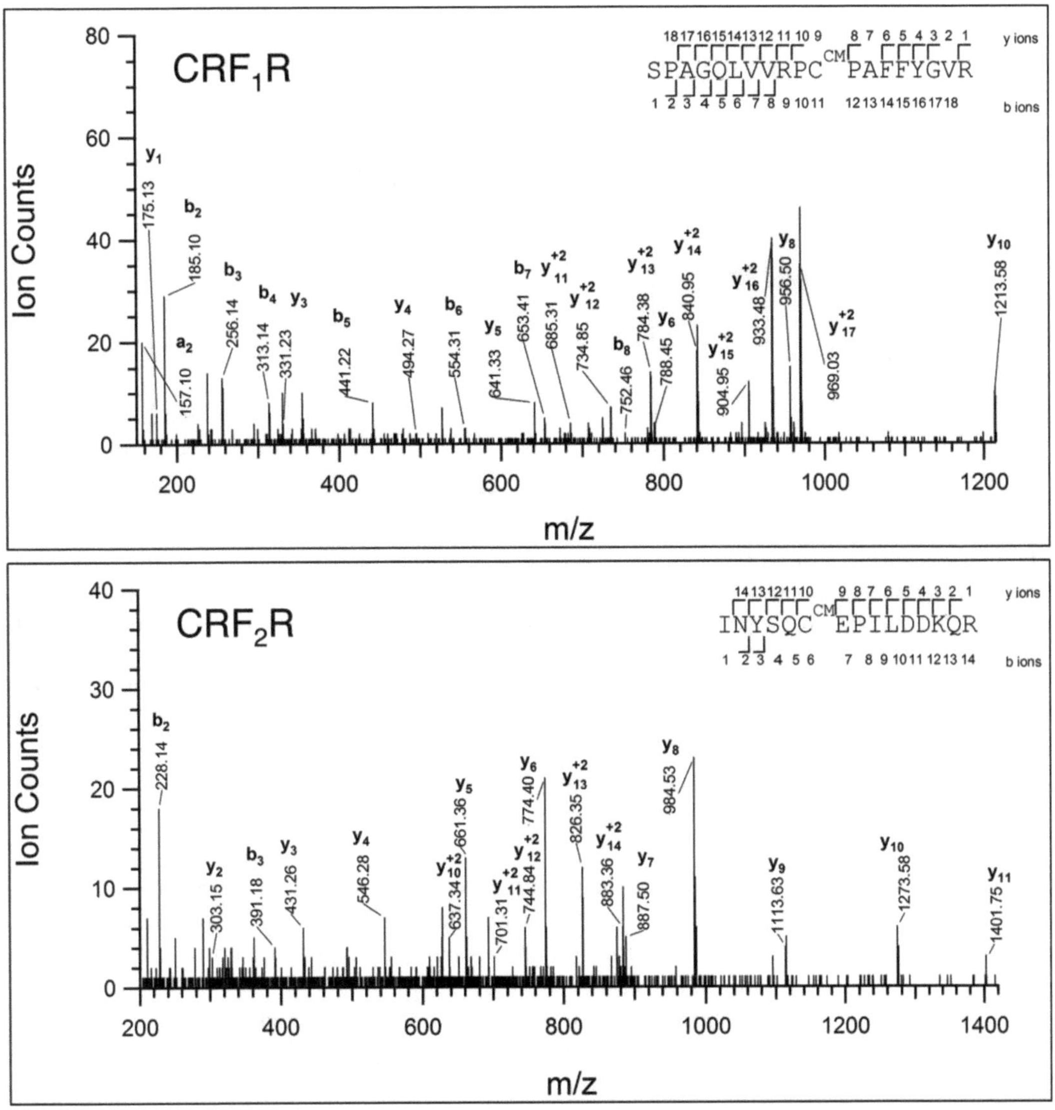

Fig. 2 Representative MS-MS demonstrating the identification-specific peptides for CRF_1R and CRF_2R. High-energy collision dissociation–tandem mass spectra obtained from precursor ions with mass 707.7146^{+3} (panel 1) and 626.9720^{+3} (panel 2) found in tryptic digests of immunoaffinity pulldowns of HA-CRF_1R corresponding to peptides spanning residues S58 to R76 of human CRF_1R and I93 to R107 of human CRF_2R. b- and y-type ion series are labeled in the figure. (Reproduced from ref. 1 with permission from *Molecular Biology of the Cell*)

4. Complete Dulbecco's Modified Eagle Medium (DMEM), high glucose (4.5 g/L): DMEM with 10% heat-inactivated fetal bovine serum. Store at 4 °C (*see* **Note 1**).

5. Sterile calcium- and magnesium-free phosphate-buffered saline (PBS): 137 mM NaCl, 10 mM phosphate, 2.7 mM KCl, pH 7.4. Store at 4 °C.

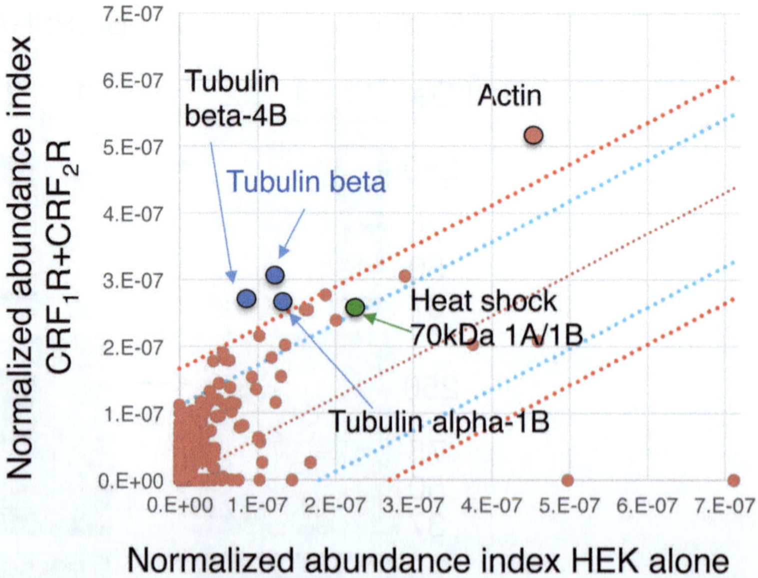

Fig. 3 Scatter plot showing relative enrichment of HA-CRF_1R-associated proteins in anti-HA vs. mock pull downs. Intervals of confidence for 95% (blue lines) and 99.7% (red lines) are indicated. Tubulin α/β-chain, actin, and heat shock protein 70 (Hsp70) were proteins that were specifically enriched in $CRF_1R + CRF_2\beta R$ heteromeric complexes (reproduced from ref. 1 with permission from *Molecular Biology of the Cell*)

6. Cellstripper solution. Pre-warm to 37 °C before use. Store at 4 °C.
7. Human embryonic kidney (HEK) 293 cells at low passage (passage <6) to allow for efficient transfection.
8. Cell culture incubator: 95% air, 5% CO_2, 37 °C.
9. Cell culture hood.
10. 1.5 mL sterile microcentrifuge tubes.
11. 15 mL sterile falcon tubes.
12. Cell scraper.
13. Centrifuges.

2.2 Immunoprecipitation Assay

1. Ice-cold PBS.
2. RIPA buffer: 50 mM Tris–HCl, pH 7.4, 150 mM NaCl, 5 mM $MgCl_2$, 1 mM EGTA, 10 mM NaF, 10 mM $Na_4P_2O_7$, 0.1 mM Na_3VO_4, 0.5% Nonidet P-40; supplemented with protease inhibitor cocktail (complete Mini, EDTA-free) and phosphatase inhibitor cocktail 3.
3. Immunoprecipitating antibody: rabbit anti-HA11 (Sigma Aldrich) or rat anti-HA11 (Roche).
4. Protein A/G PLUS agarose beads.

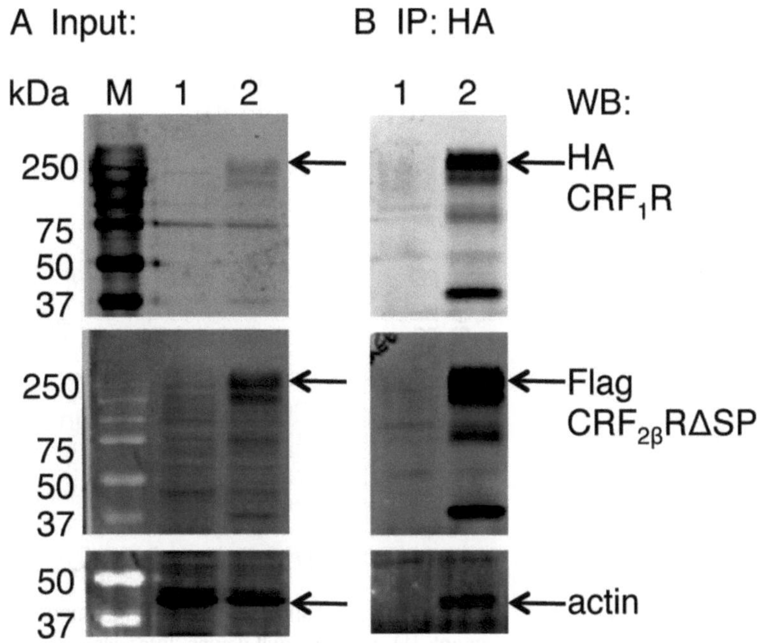

Fig. 4 Western blot analysis of CRF_1R immunoprecipitation and validating co-immunoprecipitation of CRF_2R and actin. (**a**) Input from untransfected cells do not show presence of CRF receptors (Lane 1, top 2 blots), whereas input from CRF receptor co-transfected cells showed presence of CRF receptors (Lane 2, top 2 blots) when anti-HA or anti-Flag antibodies were used (arrows). Inputs from both untransfected and co-transfected cells showed equivalent levels of actin confirming that similar levels of total protein were present in both lysates. (**b**) IP with anti-HA did not pull down any protein from untransfected cells (Lane 1), but confirmed pull down of CRF_1R from co-transfected cells (Lane 2, top blot). Anti-flag confirmed co-immunoprecipitation of CRF_2R (Lane 2, middle blot) and anti-actin confirmed presence of actin in the complex (Lane 2, bottom blot) (reproduced from ref. 1 with permission from *Molecular Biology of the Cell*)

5. Microcentrifuge (Eppendorf) tubes 1.5 mL, low retention.
6. Mixer rotator (preferred for efficient end-on-end mixing) or rocking platform at 4 °C.

2.3 Proteomics Sample Preparation (See Note 2)

1. 0.2 M glycine (pH 2.0).
2. 1 M Tris–HCl (pH 8.0).
3. 2× Laemmli sample buffer with 5% β-Mercaptoethanol.
4. Electrophoresis equipment.
5. 10% SDS-PAGE.
6. Coomassie Staining Solution: 0.1% Coomassie Blue R-250, 10% acetic acid, 50% methanol, 40% H_2O.
7. Destaining Solution: 10% acetic acid, 50% methanol, 40% H_2O.

8. Belly dancer platform shaker.
9. Trypsin solution for in-gel digestion: 1–10 ng/μL trypsin (Gold, Mass Spectrometry Grade) in 25 mM ammonium bicarbonate (NH_4HCO_3, 100 mg/50 mL).
10. 25 mM NH_4HCO_3 (100 mg/50 mL).
11. 25 mM NH_4HCO_3 in 50% acetonitrile (ACN).
12. 50% ACN/5% formic acid (*see* **Note 2**). Used in peptide extraction.
13. Solution a: 10 mM DTT in 25 mM NH_4HCO_3 (*see* **Note 3**).
14. Solution b: 55 mM iodoacetamide in 25 mM NH_4HCO_3 (*see* **Note 3**).
15. Solvent A for chromatography: 0.1% formic acid in water.
16. Solvent B for chromatography: 0.1% formic acid in ACN.
17. 0.65 mL siliconized tubes.
18. C18 ZipTips (Millipore).
19. 75-μm × 150-mm reverse phase BEH C18 nanoACQUITY Column (Waters).
20. NanoAcquity™ Ultra performance liquid chromatography (UPLC) system (Waters).

2.4 Proteomics Analysis (or LC-MS/MS Analysis)

1. Mass spectrometer: LTQ Orbitrap XL (Thermo Scientific) or QSTAR Elite (Applied Biosystems).
2. PAVA software [11] (if LTQ Orbitrap XL was used) or Mascot Distiller version 2.1.0.0 (Matrix Science, Boston, MA) (if QSTAR Elite was used).
3. Protein Prospector version 5.2.2 (a public version is available online).

2.5 Western Blot Analysis

1. Primary antibodies: rabbit anti-HA11, rabbit anti-Flag, mouse anti-Flag, rat anti-HA11 (Roche); goat anti-CRFR1/2 (Santa Cruz Biotechnology) (*see* **Note 4**).
2. Secondary antibodies: anti-mouse or anti-rat or anti-rabbit IgG coupled to Alexa Fluor 680 (Invitrogen), and coupled to IRDye 800 (Rockland Immunochemicals) (*see* **Note 5**).
3. 2× Laemmli sample buffer with 5% β-Mercaptoethanol.
4. 10% SDS-PAGE.
5. PBS.
6. PBT: PBS with 0.1% Tween 20.
7. Polyvinylidene difluoride (PVDF) membranes (Immobilon-FL; Millipore).
8. Odyssey Blocking Buffer (TBS) (Li-Cor Biosciences).

9. Odyssey Infrared Imaging System (Li-Cor Biosciences).
10. Belly dancer platform shaker.
11. NewBlot™ Stripping Buffer for PVDF membranes (LI-Cor Biosciences).

3 Methods

3.1 Co-transfection of HEK293 Cells

1. Grow HEK293 cells in DMEM culture medium in six-well plates in a cell culture incubator until ~80% confluent (*see* **Note 1**). This can be done by seeding cells at a density of ~1×10^6 cells per well and growing them for 12–24 h before transfecting (*see* **Note 6**).
2. Transient transfection of cells. Prepare plasmid DNA and transfection reagent mix (Lipofectamine 2000 or similar reagent) in a sterile tube as follows (amount described is for transfecting one well in a 6-well plate—scale up if multiple wells/condition are needed): in one tube (A), assemble 1 μg of HA-CRF_1R + 1 μg of Flag-$CRF_{2\beta}R\Delta SP$ plasmid DNA with 0.5 mL Opti-MEM, mix gently, spin briefly. In a second tube (B), add 5 μL Lipofectamine into 0.5 mL Opti-MEM, mix gently, spin briefly. Leave for 5 min at room temperature (RT) and then combine contents of the tubes A and B to obtain 1 mL transfection mix. Incubate for an additional 30 min at RT.
3. Aspirate the medium from the cells, wash twice with PBS and apply 1 mL transfection mix. Cover plate and gently mix by making an eight motion. Add 1 mL of pre-warmed complete DMEM (final volume 2 mL per well) and mix again by making an eight motion. Incubate cells in the cell culture incubator for 6 h.
4. In parallel, transiently transfect cells with empty vector (pcDNA5, 2 μg of DNA) as negative control using similar methodology as in **steps 2** and **3**.
5. After 6 h of transfection, aspirate medium and wash cells twice with sterile PBS. Apply 0.5 mL Cellstripper (*see* **Note** 7) per well and incubate for ~5 min in cell culture incubator (or until cells become loose). Add 2 mL of complete DMEM per well and transfer cells into a 15 mL Falcon tube (tilt the plate for easy removal of all medium; cells from identical transfection conditions/well can be pooled at this step in to a 50 mL Falcon tube). Spin down cells in a centrifuge at 500 × *g* for 5 min. Discard supernatant and resuspend the cells in 10 mL DMEM. Plate them on 10 cm dishes at ~2×10^6 cells per dish in 10 mL of complete DMEM (mix cells by pipetting up and down gently before plating as cells will settle to the bottom of the

tube). Cells can be counted in a hemocytometer before seeding on 10 cm dishes or an equal aliquot can be split between the dishes. Place in a cell culture incubator for 24–48 h before using them for immunoprecipitation (Subheading 3.2).

3.2 Receptor Co-IP

1. Place 10 cm dishes of HEK293 cells co-transfected with receptors (Subheading 3.1, **step 5**) on ice. Wash cells twice with ice-cold PBS, tilt the plate and aspirate all PBS. At this point, proteins can be cross-linked (*see* **Note 8**) before cells are lysed or cells can be lysed without cross-linking. Lyse cells in 500 μL RIPA buffer per plate. Ensure that the entire plate is covered with RIPA buffer by swirling and tilting the plate back and forth. Scrape cells using a cell scraper or pipette tip. Collect the cell lysate from a plate and transfer to a 1.5 mL microcentrifuge tube. Perform the same procedure for pcDNA5-transfected cells growing in 10 cm dishes that will serve as negative control for the IP (*see* **Note 9**).
2. Pipette lysate up and down (~10 times) using a 1 mL tip to reduce viscosity and to mix well. Centrifuge at 17,000 × *g* at 4 °C for 15 min and transfer supernatant to a fresh tube on ice.
3. Save ~20 μL of the supernatant from **step 2** to be used as "IP input" and store at −20 °C for short-term or −80 °C for long-term storage. Use the rest of the supernatant for IP (*see* **Notes 10** and **11**).
4. To immunoprecipitate, use primary antibody against protein of interest at a concentration of 2.5 μg antibody per 100 μL RIPA buffer. Add 100 μL antibody mix to each supernatant (total volume at this point will be ~600 μL). To IP HA-CRF_1R, we used rat anti-HA antibody at a concentration of 2.5 μg antibody per 100 μL RIPA buffer (*see* **Note 12**).
5. Rotate the tubes end-over-end on a rotator overnight at 4 °C.
6. Wash and equilibrate A/G PLUS agarose beads for pull down the following morning. You will need 30 μL beads slurry per IP tube. First mix A/G PLUS beads with 1 mL of RIPA buffer, invert the tube at least 6 times to mix, spin at 2000 × *g* for 30 s and discard the supernatant. Repeat three times. After the final wash, resuspend beads at a concentration of 100 μL RIPA buffer per 30 μL bead slurry. Add 100 μL of bead slurry mix to each IP tube from **step 5**.
7. Rotate the tubes end-over-end on a rotator for 2 h at 4 °C.
8. Spin down the beads at 2000 × *g* for 30 s and then wash three times with 1 mL RIPA buffer.
9. Resuspend the beads in elution buffer. Composition and volume of elution buffer will depend upon whether MS analysis or Western blot analysis will be performed next. Sample elution

for MS analysis, *see* Subheading 3.3, **step 1** and for Western blot analysis, *see* Subheading 3.5, **step 1**.

3.3 Sample Preparation for Mass Spectrometry

1. For MS analysis, resuspend the beads from Subheading 3.2 after **step 8** in 50 μL of 0.2 M glycine (pH 2.0).
2. Vortex and incubate with agitation by occasional flicking of the tube for 10 min at 4 °C. Centrifuge (9000 × *g*, 4 °C) for 2 min. Recover supernatant. Add 5 μL of 1 M Tris–HCl (pH 8.0) to each tube to neutralize the pH. This supernatant contains eluted IP proteins.
3. Resolve eluted proteins from IP on a 10% SDS-PAGE (*see* **Note 13**).
4. Coomassie stain the gel by immersing it in at least 5 volumes of Coomassie staining solution and place on a gently shaking belly dancer platform shaker for 2 h at RT. Destain the gel with destaining solution with gentle shaking on the belly dancer platform for at least 4 h or overnight. Change solution if needed until the background is nearly clear. After destaining, gels can be stored in water.

3.3.1 In-Gel Digestions

Perform in-gel digestion of proteins with 100 ng of trypsin solution made in 25 mM NH_4HCO_3 (*see* **Note 14**). To do this, cut out gel pieces containing protein bands of interests using a razor blade or scalpel (*see* **Notes 3** and **15**).

1. Dice each gel piece into small pieces (1 mm^3) and place into 0.65 mL siliconized tubes.
2. Add 100 μL of 25 mM NH_4HCO_3/50% ACN and vortex for 10 min.
3. Briefly spin down (a low speed pulse spin in a picofuge will work). Using a pipette tip, remove the supernatant and discard.
4. Repeat **steps 2** and **3**.
5. Speed Vac the gel pieces to complete dryness (~20–40 min).
6. Add trypsin solution to just barely cover the gel pieces to rehydrate and incubate on ice or at 4 °C for 10 min. Estimate the gel volume and add about 3× volume of trypsin solution. Gel weight roughly equals its volume and can be determined by subtracting the weight of an empty tube of same size. This volume will vary between samples (depending on the gel piece originally excised, but is typically between ~10–100 μL (final trypsin concentration is 100 ng/digestion).
7. Spin briefly. Add 25 mM NH_4HCO_3 as needed to cover the gel pieces. Mix gently by inverting, spin briefly, and incubate at 37 °C for 4 h to overnight. The in-gel digestion method described here was originally derived from Rosenfeld et al. [12] and is also outlined in [13].

3.3.2 Peptide Extraction

1. Spin the tubes containing the gel pieces. Carefully transfer the aqueous supernatant into a clean 0.65–1.5 mL siliconized microfuge tube.
2. To the gel pieces, add 100–400 μL (depends on the gel volume; add enough to cover) of 50% ACN/5% formic acid, vortex 20–30 min, spin and sonicate 5 min. Repeat and pool supernatants.
3. Vortex the extracted digests (pooled supernatants from **steps 1** and **2**, above), spin and Speed Vac to reduce volume to less than 10 μL. Take care not to completely evaporate or over dry the extracted proteins.
4. If the sample is completely dry, add 20 μL of 0.1% formic acid. If some volume is still in the tube, add 2–5 μL of 5% formic acid. Proceed with C18 ZipTip (Millipore) cleanup. Resuspend in a final volume of 3–10 μL of 0.1% formic acid. Proceed with analysis by LC-MS/MS.

3.4 LC-MS/MS Analysis

1. Use 5 μL of sample from **step 4** of Subheading 3.3.2 (remaining can be stored at −20 °C should MS analysis need to be re-run). Separate sample by nano-flow liquid chromatography using a 75-μm × 150-mm reverse phase BEH C18 nanoACQUITY Column (Waters) at a flow rate of 300 nL/min in a NanoAcquity™ Ultra performance UPLC system. To do this, first equilibrate the column in 5% solvent B and then inject an aliquot of each digest (5 μL). The organic content of the mobile phase gets linearly increased to 25% over 40 min (if samples are analyzed in LTQ Orbitrap XL) or over 80 min (if samples are analyzed in QStar Elite). In either case, it then gets further increased to 50% in 2 min.
2. Couple the liquid chromatography eluate to a nanoelectrospray ion source attached to the mass spectrometer. Analyzed peptides in positive ion mode and in information-dependent acquisition mode to automatically switch between MS and MS/MS acquisition.
3. Acquire MS spectra using a LTQ Orbitrap XL or a QStar Elite.

3.4.1 Acquisition of MS Spectra Using LTQ Orbitrap XL

1. In LTQ Orbitrap XL, MS spectra are acquired in profile mode using the Orbitrap analyzer in the m/z range between 325 and 1500.
2. For each MS spectrum, select the six most intense multiple charged ions over a threshold of 1000 counts to perform collision-induced dissociation experiments.
3. Analyze product ions on the linear ion trap in centroid mode. Set the collision-induced dissociation energy to 25%. Use a dynamic exclusion window of 0.5 Da to exclude the same m/z from being selected for 45 s after its acquisition.

3.4.2 Acquisition of MS Spectra Using QStar Elite

1. In QStar Elite, MS spectra are acquired for 0.75 s in the *m/z* range between 310 and 1400. MS acquisitions are followed by 1-s collision-induced dissociation (CID) experiments in information-dependent acquisition mode.
2. For each MS spectrum, select the two most intense multiple charged ions (charge 2–5) over a threshold of 20 counts for generation of CID mass spectra. The collision energy gets automatically set according to mass to charge (*m/z*) ratio and charge state of the precursor ion.
3. Apply a dynamic exclusion window to prevent the same *m/z* from being selected for 1 min after its acquisition.

3.4.3 Generation of MS Peak Lists

1. Generate peak lists using either PAVA software [11] (if data were acquired in the LTQ Orbitrap XL) or Mascot Distiller version 2.1.0.0 (if data were acquired in the QStar Elite).
2. Set the parameters for MS processing as follows: peak half-width, 0.02; data points per Da, 100.
3. Set the parameters for MS/MS data as follows: peak half-width, 0.02; data points per Da, 100.
4. Search the peak lists against the human subset of the SwissProt database using Protein Prospector version 5.2.2.
5. Concatenate a randomized version of all entries to the database to estimate false discovery rates in the searches. Peptide tolerance in searches is 20 ppm (LTQ Orbitrap XL) or 30 ppm (QStar Elite) for precursor ions, and 0.8 Da (LTQ Orbitrap XL) or 0.2 Da (QStar Elite) for product ions, respectively. Peptides containing two miscleavages should be allowed and included. Allow carbamidomethylation of cysteine as constant modification, allow acetylation of the N terminus of the protein, pyroglutamate formation from N terminal glutamine, and oxidation of methionine as variable modifications. Limit the number of modifications to two per peptide.
6. Use Protein Prospector thresholds for identification criteria as follows: minimal protein score of 15, minimal peptide score of 15, maximum expectation value of 0.1, and minimal discriminant score threshold of 0.0. Limit the false discovery rate to 1%.
7. Consider protein hits as significant when two or more peptide sequences match a protein entry (*see* **Note 16**).

3.5 Western Blot Verification

1. The co-IP is performed as described in Subheading 3.2, **steps 1–8**. To elute proteins from the beads, resuspend beads in 30 μL of 2× Laemmli sample buffer + β-Mercaptoethanol, boil for 5 min (*see* **Note 17**), chill briefly on ice, spin briefly, load samples on a 10% gel and separate using SDS-PAGE.

2. Transfer proteins onto a polyvinylidene difluoride membrane using standard Western transfer protocol [14, 15].
3. Wash the membrane in PBS and block in LI-COR Odyssey Blocking Buffer with gentle shaking on a belly dancer platform shaker for 1 h at RT [15].
4. Incubate the membrane with primary antibodies. Use rabbit anti-HA and mouse anti-Flag at 1:1000 dilution in blocking buffer, incubate for 2 h at RT. Wash with PBT for 30 min (3× 10 min).
5. Incubate the membrane with secondary antibodies. Use anti-rabbit IgG coupled to Alexa Fluor 680 and anti-mouse IgG coupled to IRDye 800 at 1:10,000 dilution in blocking buffer, incubate for 1 h at RT. Wash twice with PBT for 10 min at RT and once with PBS for 10 min (*see* **Note 18**).
6. To detect other proteins of interest that are predicted to co-IP, the membrane can be stripped (*see* **Note 19**) and re-probed with the antibody of interest (repeat **steps 4** and **5**). In case of CRF receptors, anti-actin antibody (1:10,000 dilution) was used to verify pull-down and confirm MS findings (Fig. 4).
7. Scan membrane using the Odyssey Infrared Imaging System (*see* **Note 18**).

4 Notes

1. The culture medium should be antibiotic-free. While antibiotics are included in medium for cell maintenance and passage, the cells should be switched to antibiotic-free medium when they are seeded for transfection. This is done because antibiotics can be toxic to cells that are going through the process of transfection. For preparing the transfection mix (DNA and Lipofectamine 2000) Opti-MEM I Reduced Serum Media is used and it should be noted that Opti-MEM is light-sensitive and should be kept dark (wrap foil around the bottle). Pre-warm medium and PBS solutions to 37 °C before adding to cells to prevent temperature shock, but do not leave medium at RT or at 37 °C for prolonged periods as FBS and antibiotics in the medium will lose potency.
2. Water, acetonitrile, and formic acid used in preparing the various solutions listed in Subheading 2.3 Proteomics Sample Preparation, points 8–13, need to be high-performance liquid chromatography (HPLC) grade. Formic acid should be used with the solvents since trifluoroacetic acid usually interferes with the ionization process.

3. For low-level proteins (<1 pmol) reduction and alkylation is recommended by performing the following procedures before Speed Vac drying the gel pieces.
   (a) Add 25 μL of 10 mM DTT in 25 mM NH_4HCO_3 (Solution a, Subheading 2.3, **item 10**) to dried gels. Vortex and spin briefly. Allow reaction to proceed at 56 °C for 1 h.
   (b) Remove supernatant, add 25 μL 55 mM iodoacetamide (Solution b, Subheading 2.3, **item 11**) to the gel pieces. Vortex and spin at 2000 × *g* for 30 s. Allow reaction to proceed in the dark for 45 min at room temperature.
   (c) Remove supernatant (discard). Wash gels with ~100 μL NH_4HCO_3, vortex 10 min and spin at 2000 × *g* for 30 s.
   (d) Remove supernatant (discard). Dehydrate gels with ~100 μL of 25 mM NH_4HCO_3 in 50% ACN, vortex 5 min, spin at 2000 × *g* for 30 s. Repeat one time.
   (e) Speed Vac the gel pieces to complete dryness (~20 min). Proceed with trypsin digest.
4. When available, primary antibody from one species should be used for IP and from another species for Western blot analysis to avoid cross-reactivity and detection of IgG. When antibodies from two species are not available for a given protein, IgG bands on Western blots will be unavoidable, but data can still be interpreted.
5. Secondary antibodies from different species are used for the same reasons as described above for primary antibodies.
6. Typically, six-well plates are used for transfection as a high transfection efficiency requires cells to be ~80% confluent at the time of transfection. Transfected cells from one well of a 6-well plate (~80% confluent) can be split onto two 10 cm dishes thereby providing identical conditions for replicates and avoiding overgrowth in the 10 cm dishes.
7. Cellstripper is a non-enzymatic cell dissociation solution. It is a frequently used alternative to trypsin and either one can be used for splitting cells or dislodging adherent cells in culture. It needs to be considered however that some proteins are cleaved by trypsin and some GPCRs are cleaved and activated by trypsin, such as protease-activated receptors (PARs) [16]. To rule out these potential effects of trypsin, we used Cellstripper in our protocol.
8. In vivo most protein–protein interactions occur transiently and crosslinking reagents allow capturing these brief interactions by making them stable for isolation and characterization. To do this, 4% formaldehyde is applied to cells (1 mL/dish) [17]. We, however, found that formaldehyde crosslinking resulted in less numbers of proteins identified in the interactome by MS. For

other GPCRs interactomes crosslinking may be beneficial though.

9. Another unrelated GPCR construct that contains both tags such as Flag-PAR2-HA was used as an additional control as described previously [1, 16]. This served as a positive control for the pull down of a HA or Flag tagged protein, but potentially as a negative control for the co-IP of interacting proteins.
10. Make sure not to use up all the lysate for the IP but save some for "IP input". This allows for checking for co-transfection efficiency by Western blot analysis, using anti-HA and anti-Flag antibodies (Fig. 4).
11. Once the cell lysates are obtained (Subheading 3.2, **step 2**), determination of protein concentration using the Protein Assay reagent (Bio-Rad laboratories) is recommended before proceeding with IP that is done typically with ~500 μg of total protein as starting material. This is useful for reproducibility and minimizing variability between technical replicates (when the same sample is tested in duplicate or triplicate) and biological replicates (repeating the entire experiment using identical condition on another day to capture biological variation).
12. IPs should be performed with antibodies against tags on both receptors. For CRF receptors, IPs were performed using anti-HA and anti-Flag antibodies in parallel.
13. Gel handling in preparation for MS needs to be done in the cleanest possible way. Wipe down all surfaces with methanol/water moistened lint-free cloth. Wear clean gloves and sleeve protectors. Common contaminants are keratin proteins from handler's skin.
14. If larger volume of trypsin solution is needed for gel hydration, adjust stock solution concentration such that the final trypsin concentration added to each digest is 100 ng. Trypsin is often the first protease used for several reasons. First, it cleaves proteins into peptides with an average size of 700–1500 Da (the ideal size for mass spec). Second, it cleaves at the carboxyl side of arginine and lysine residues, thereby providing a charged peptide that is easily detectable by mass spectrometry. Third, trypsin is a hardy protease that can tolerate and function under a variety of conditions. And lastly, trypsin can be modified by the methylation of lysines to prevent self-digestion at these sites.
15. Any material that will come in contact with the gel for MS analysis needs to be cleaned and wiped down with methanol/water moistened lint-free cloth, including all surfaces and containers, glass plates used for PAGE, the outside of tubes and the inside of centrifuges, vortex, tube racks, etc.

16. Further details about identification of fragments and criteria to be used can be found in our previous publication [18].
17. Boiling ensures complete elution and denaturation of proteins, but with GPCRs it can also promote protein aggregation. If this is a concern, an alternative to boiling the samples can be to let the samples sit in 2× Laemmli sample buffer + -β-Mercaptoethanol at room temperature for 30 min before loading.
18. A full description of the Odyssey Infrared Imaging System can be found in the website listed in ref. 19. If this system is not available to you, the more traditional chemiluminescence method [20] can serve as an alternative method. If using chemiluminescent methodology, HRP-conjugated secondary antibodies can be used.
19. Membranes can be stripped up to three times. To strip a membrane, use NewBlot™ Stripping Buffer. Complete usage instructions are given in the pack insert. Make sure to keep the membrane wet at all times.

Acknowledgments

This work was supported by NIH grants DK080787 to A. Bhargava and GM8P41GM103481 to A. Burlingame. B.H. was supported by T32 AT003997.

References

1. Hasdemir B, Mahajan S, Oses-Prieto J, Chand S, Woolley M, Burlingame A, Grammatopoulos DK, Bhargava A (2017) Actin cytoskeleton-dependent regulation of corticotropin-releasing factor receptor heteromers. Mol Biol Cell 28(18):2386–2399. https://doi.org/10.1091/mbc.E16-11-0778
2. Milligan G (2009) G protein-coupled receptor hetero-dimerization: contribution to pharmacology and function. Br J Pharmacol 158 (1):5–14. https://doi.org/10.1111/j.1476-5381.2009.00169.x
3. Vischer HF, Watts AO, Nijmeijer S, Leurs R (2011) G protein-coupled receptors: walking hand-in-hand, talking hand-in-hand? Br J Pharmacol 163(2):246–260. https://doi.org/10.1111/j.1476-5381.2011.01229.x
4. Shrestha D, Jenei A, Nagy P, Vereb G, Szollosi J (2015) Understanding FRET as a research tool for cellular studies. Int J Mol Sci 16 (4):6718–6756. https://doi.org/10.3390/ijms16046718
5. Michel MC, Wieland T, Tsujimoto G (2009) How reliable are G-protein-coupled receptor antibodies? Naunyn Schmiedeberg's Arch Pharmacol 379(4):385–388. https://doi.org/10.1007/s00210-009-0395-y
6. Gomes I, Gupta A, Bushlin I, Devi LA (2014) Antibodies to probe endogenous G protein-coupled receptor heteromer expression, regulation, and function. Front Pharmacol 5:268. https://doi.org/10.3389/fphar.2014.00268
7. Chung KY, Day PW, Velez-Ruiz G, Sunahara RK, Kobilka BK (2013) Identification of GPCR-interacting cytosolic proteins using HDL particles and mass spectrometry-based proteomic approach. PLoS One 8(1):e54942. https://doi.org/10.1371/journal.pone.0054942
8. Ciruela F, Burgueno J, Casado V, Canals M, Marcellino D, Goldberg SR, Bader M, Fuxe K, Agnati LF, Lluis C, Franco R, Ferre S, Woods AS (2004) Combining mass spectrometry and pull-down techniques for the study of receptor heteromerization. Direct epitope-epitope

electrostatic interactions between adenosine A2A and dopamine D2 receptors. Anal Chem 76(18):5354–5363. https://doi.org/10.1021/ac049295f

9. Matsubara S, Shiraishi A, Sakai T, Okuda T, Satake H (2017) Heterodimerization of the prostaglandin E2 receptor EP2 and the calcitonin receptor CTR. PLoS One 12(11): e0187711. https://doi.org/10.1371/journal.pone.0187711
10. Hasdemir B, Mahajan S, Bunnett NW, Liao M, Bhargava A (2012) Endothelin-converting enzyme-1 actions determine differential trafficking and signaling of corticotropin-releasing factor receptor 1 at high agonist concentrations. Mol Endocrinol 26(4):681–695. https://doi.org/10.1210/me.2011-1361
11. Guan S, Price JC, Prusiner SB, Ghaemmaghami S, Burlingame AL (2011) A data processing pipeline for mammalian proteome dynamics studies using stable isotope metabolic labeling. Mol Cell Proteomics 10(12): M111 010728. https://doi.org/10.1074/mcp.M111.010728
12. Rosenfeld J, Capdevielle J, Guillemot JC, Ferrara P (1992) In-gel digestion of proteins for internal sequence analysis after one- or two-dimensional gel electrophoresis. Anal Biochem 203(1):173–179
13. UCSF in-gel digestion protocol. http://msf.ucsf.edu/protocols.html. Accessed 24 May 2018
14. LI-COR (Biosciences) protein electrotransfer methods. https://www.licor.com/documents/t932473b94z25454q3kqazh3t830xfoy. Accessed 24 May 2018
15. Sambrook J (2000) Molecular cloning, a laboratory manual, 3rd edn. Cold Spring Harbor Laboratory Press, Cold Spring Harbor, NY
16. Hasdemir B, Bunnett NW, Cottrell GS (2007) Hepatocyte growth factor-regulated tyrosine kinase substrate (HRS) mediates post-endocytic trafficking of protease-activated receptor 2 and calcitonin receptor-like receptor. J Biol Chem 282(40):29646–29657. https://doi.org/10.1074/jbc.M702974200
17. Klockenbusch C, Kast J (2010) Optimization of formaldehyde cross-linking for protein interaction analysis of non-tagged integrin beta1. J Biomed Biotechnol 2010:927585. https://doi.org/10.1155/2010/927585
18. Clauser KR, Baker P, Burlingame AL (1999) Role of accurate mass measurement (+/− 10 ppm) in protein identification strategies employing MS or MS/MS and database searching. Anal Chem 71(14):2871–2882
19. LI-COR (BIOSCIENCES) Odyssey classic application protocols. https://www.licor.com/documents/o2lbjk5lhkb0ehh5e89x. Accessed 24 May 2018
20. LI-COR (BIOSCIENCES) Introduction to chemiluminescence western detection. https://www.licor.com/bio/applications/chemiluminescent_western_blot/index.html. Accessed 24 May 2018

Part V

Visualizing Subcellular Compartmentalization of GPCRs

Chapter 16

Monitoring the Aggregation of GPCRs by Fluorescence Microscopy

Samuel Génier, Jade Degrandmaison, Christine L. Lavoie, Louis Gendron, and Jean-Luc Parent

Abstract

G protein-coupled receptors (GPCRs) contain highly hydrophobic domains that are subject to aggregation when exposed to the crowded environment of the cytoplasm. Many events can lead to protein aggregation such as mutations, endoplasmic reticulum (ER) stress, and misfolding. These processes have been widely known to impact GPCR folding, maturation, and localization. Protein aggregates are transported toward the microtubule-organizing center via dynein to form a large juxta-nuclear structure called the aggresome, and in due course, are then targeted for degradation. Here, we describe a method to study aggregation of GPCRs by fluorescence microscopy.

Key words GPCR, Aggregation, Aggresome, Misfolding, PROTEOSTAT®

1 Introduction

With over 800 members, GPCRs represent one of the largest family of membrane proteins and are the therapeutic targets of approximately 34% of the drugs approved by the US Food and Drug Administration [1]. These receptors are subjected to several post-translational modifications (PTMs) and active transport mechanisms in order to be properly expressed, folded and localized [2]. The folding of GPCRs is intricate due to their seven-transmembrane topology. Minor alterations such as point mutations can have a significant impact on the receptor maturation and functionality [3]. For example, the naturally occurring Pro23His mutation in human rhodopsin leads to its aggregation in a structure called the aggresome causing retinitis pigmentosa, a degenerative condition ultimately resulting in blindness [4]. Furthermore, many wild-type GPCRs undergo sub-optimal folding, are retained intracellularly and eventually degraded [3]. We recently reported that other GPCRs, such as the β isoform of the thromboxane A_2

Mario Tiberi (ed.), *G Protein-Coupled Receptor Signaling: Methods and Protocols*, Methods in Molecular Biology, vol. 1947, https://doi.org/10.1007/978-1-4939-9121-1_16, © Springer Science+Business Media, LLC, part of Springer Nature 2019

receptor and the β_2-adrenergic receptor, accumulate in the aggresome when the CCT7 subunit of the cytosolic chaperonin complex (CCT/TRiC) is depleted [5], indicating that this is a process relevant to multiple GPCRs undergoing improper folding and maturation. The aggresome is formed in the cytoplasm by protein aggregates that have been transported along the microtubules to the centrosome via the recognition of ubiquitinated proteins by HDAC6 and its association with the dynein motor protein [6]. The formation of this juxta-nuclear structure is a natural cellular response to prevent cytotoxicity of protein aggregates by sequestration and, ultimately, autophagy-dependent degradation [7].

Here, we use the human delta opioid receptor (DOPr) as a model to visualize aggregation of a GPCR using the PROTEOSTAT® Aggresome Detection Kit in a confocal microscopy assay. To induce aggresome formation, HEK293 cells stably expressing HA-tagged DOPr were treated with the proteasome inhibitor MG-132. Since a large population (over 60%) of newly synthesized DOPrs is primarily retained in the ER, ubiquitinated and retro-translocated in the cytosol to, ultimately, be degraded by the proteasome [8, 9], inhibition of the proteasome instigates DOPr aggregation. After fixation, the cells are treated with the PROTEOSTAT® Aggresome Detection Reagent, a fluorescent dye designed to recognize aggregated proteins. Subsequently, standard labeling of the receptor with fluorophore-conjugated antibodies is carried out for visualization by confocal microscopy. We also provide a comprehensive procedure to analyze and quantify co-localization between the PROTEOSTAT® dye and the receptor. Overall, this method can be used to detect the aggregation of GPCRs, or of other polytopic proteins, caused by mutations, ER stress, chaperone depletion, or any condition that potentially results in improper folding and intracellular accumulation of the proteins under study.

2 Materials

Prepare all solutions using ultrapure deionized water (18 MΩ cm at 25 °C). All solutions need to be freshly prepared if no conservation instructions are specified.

2.1 Cell Culture and Generation of the HA-Tagged Delta Opioid Receptor (HA-DOPr) Stable Cell Line

1. Six-Well Clear Flat Bottom TC-Treated Multiwell Cell Culture Plate (Falcon).
2. Cell culture dish: TC Dish 100 × 20 mm, Standard (Sarstedt).
3. 50 mL Polypropylene conical Tube.
4. Human embryonic kidney 293 (HEK293) cells (ATCC).
5. Cell culture media: Dulbecco's modified Eagle's medium (DMEM) supplemented with 10% fetal bovine serum (FBS) (Gibco). Conserve at 4 °C up to 1 month.

6. Opti-MEM®I Medium.
7. *Trans*IT®-LT1 Transfection Reagent (Mirus).
8. Trypsin/EDTA (0.05% Trypsin/0.53 mM EDTA). Conserve at −20 °C in sterile aliquots.
9. Phosphate-buffered saline (PBS): 8 g NaCl, 0.2 g KCl, 1.44 g Na_2HPO_4, 0.2 g KH_2PO_4. Adjust to 1 L with H_2O. Conserve at 4 °C up to 1 month.
10. pcDNA3-HA-DOPr plasmid (*see* **Note 1**).
11. Geneticin (G418) solution: Dissolve 200 mg of active G418 sulfate (Wisent) in H_2O according to the potency of the lot indicated on the bottle and adjust to a final volume of 1 mL. Conserve at −20 °C in sterile aliquots up to 6 months.

2.2 Cell Seeding and Treatment

1. Six-Well Clear Flat Bottom TC-Treated Multiwell Cell Culture Plate (Falcon).
2. Micro Cover Glass (25 mm Circle No. 1).
3. Poly-L-lysine (PLL) solution: Dissolve 5 mg of Poly-L-lysine hydrobromide in sterile H_2O and adjust to a final volume of 50 mL. Conserve at −20 °C up to 3 months.
4. 5 mM MG-132: Dissolve 120 nmol of MG-132 in 24 μL of dimethyl sulfoxide (DMSO) in sterile conditions. The MG-132 is provided in the PROTEOSTAT® Aggresome Detection Kit (Enzo). Conserve at −20 °C up to 3 months.
5. HEK293 cells stably expressing human HA-DOPr (*see* Subheading 3.1).

2.3 Cell Labeling for Microscopy

1. PBS (*see* Subheading 2.1, **item 9**).
2. Fixation buffer (PFA 4%): Dilute 2.66 mL of 12% (v/v) paraformaldehyde (PFA) solution in ~5 mL of PBS. Adjust to 8 mL with PBS.
3. Permeabilization buffer (0.1% PBS-T: Add 100 μL of Triton X-100 in 100 mL of PBS.
4. Blocking buffer (PBS-T + 2% BSA): Solubilize 0.5 g of bovine serum albumin (BSA) in 25 mL of 0.1% PBS-T.
5. Parafilm.
6. Microscope Slides (25 × 75 × 1.0 mm).
7. ProLong® Gold antifade reagent (Life technologies).
8. Primary antibody: Purified α-HA Epitope Tag Antibody (1 mg/mL, mouse IgG, BioLegend).
9. Alexa Fluor® conjugated secondary antibody: α-Mouse IgG conjugated to Alexa Fluor® 633 (goat, ThermoFisher).
10. PROTEOSTAT® Aggresome Detection Kit (Enzo).

2.4 Fluorescence Microscopy and Image Analysis

1. Leica TCS SP8 STED 3X/WLL/FLIM/FCS microscope and Leica LASX. The microscope is equipped with a 8 kHz tandem scanner, a motorized stage, five channels spectral detector (400–800 nm), HC PLAPO 63× immersion-oil objective and a white light laser (470–670 nm) (*see* **Note 2**).
2. Type F immersion liquid.
3. LAS X software (Leica) for image processing, co-localization analysis and quantification.
4. Prism 7 software (GraphPad) for statistical analysis.

3 Methods

3.1 Cell Culture and Generation of the HA-DOPr Stable Cell Line

Perform all cell culture manipulations under a biological safety cabinet to maintain a sterile environment. The stable HEK293 cell line is cultured in DMEM medium supplemented with 10% FBS at 37 °C in a humidified atmosphere containing 5% CO_2.

3.1.1 Cell Culture

1. Place the Trypsin/EDTA and the DMEM medium supplemented with 10% FBS in a water bath at 37 °C for prewarming.
2. Aspirate the culture medium of HEK293 cells from a confluent cell culture dish (100 × 20 mm).
3. Wash carefully the cells with 2 mL of sterile PBS.
4. Gently rock the cell culture dish to eliminate any traces of culture medium and dead cells.
5. Remove the PBS and add 2 mL of Trypsin/EDTA.
6. Incubate at 37 °C with 5% CO_2 for 2 min (*see* **Note 3**).
7. Rock the cell culture dish until cells are detached from the surface and dispersed.
8. Add 8 mL of DMEM containing 10% FBS and mix gently by pipetting up and down 2–3 times.
9. Transfer the solution in a sterile 50 mL Falcon Tube and centrifuge at 450 × *g* for 5 min at 4 °C (*see* **Note 4**).
10. Discard the supernatant and resuspend the pellet with 15 mL of DMEM containing 10% FBS by gently pipetting up and down.
11. Count the cells using a hemacytometer or any other cell counting device.
12. Seed the required amount of cells in fresh DMEM containing 10% FBS in the appropriate cell culture dish or plate (*see* below).

3.1.2 Generation of the HA-DOPr Stable Cell Line

1. Seed 3×10^5 HEK293 cells/well of a six-well cell culture plate in a final volume of 2 mL of DMEM containing 10% FBS (*see* **Note 5**).
2. Incubate at 37 °C with 5% CO_2 for 24 h.
3. Place the DMEM containing 10% FBS and the Opti-MEM Medium in a water bath at 37 °C for prewarming.
4. Warm the transfection reagent (*Trans*IT®-LT1) to room temperature and vortex gently.
5. In a microtube, add 4 μL of *Trans*IT®-LT1 reagent to 200 μL of Opti-MEM. Agitate gently and incubate at room temperature for 8 min (*see* **Note 6**).
6. In parallel, prepare a microtube containing 1 μg of the pcDNA3-HA-DOPr plasmid (*see* **Note 7**).
7. Add 102 μL of the Opti-MEM/*Trans*IT®-LT1 mix previously prepared (**step 5**) in the microtube containing the DNA (**step 6**). Mix by gently pipetting up and down and incubate for 20 min at room temperature.
8. Aspirate carefully the culture medium from the six-well culture plate and replace it with 2 mL of prewarmed DMEM containing 10% FBS.
9. Add the mix from **step 7** drop by drop in a well of the six-well culture plate and agitate gently. For the "kill control" condition, add the remaining 102 μL of the Opti-MEM/*Trans*IT®-LT1 mix from **step 5** in another well.
10. Incubate at 37 °C with 5% CO_2 for 48 h.
11. Aspirate carefully the culture medium and replace it with 2 mL of prewarmed DMEM containing 10% FBS.
12. Add 8 μL of the G418 solution (200 mg/mL) in each well to obtain a final concentration of 0.8 mg/mL (*see* **Notes 8** and **9**).
13. Mix gently and incubate at 37 °C with 5% CO_2.
14. Change the culture medium every 48–72 h with prewarmed DMEM containing 10% FBS and 0.8 mg/mL of G418 for 2 weeks (*see* **Note 10**).
15. When the selection is completed (*see* **Note 11**), replace the medium with prewarmed DMEM containing 10% FBS and 0.25 mg/mL of G418. Maintain the cells for an additional week prior to use.
16. Confirm HA-DOPr expression (*see* **Notes 12** and **13**).
17. To maintain HA-DOPr stable expression, culture the cells in DMEM containing 10% FBS and a final concentration of 0.25 mg/mL of G418.

3.2 Cell Culture and Treatment

1. Thaw the PLL solution in a water bath at 37 °C.
2. Deposit a glass coverslip in each well of a six-well cell culture plate.
3. Carefully add 850 μL of the PLL solution on the coverslip using a micropipette. Try to cover the entire coverslip surface without breaking the drop (*see* **Note 14**). Incubate at room temperature for 10 min.
4. Remove the PLL solution and let the coverslip air-dry in the cell culture cabinet for 15 min. The PLL solution can be reused up to ten times.
5. Seed 5×10^4 HEK293 cells stably expressing HA-DOPr in each well of the culture plate in 2 mL of DMEM containing 10% FBS (*see* Subheading 3.1).
6. Incubate at 37 °C with 5% CO_2 for 48 h.
7. Treat the cells with 2 μL of 5 mM MG-132 to inhibit the proteasome or 2 μL of DMSO as control directly in the culture medium. Incubate at 37 °C with 5% CO_2 for 12 h (*see* **Note 15**).

3.3 Cell Labeling for Microscopy

The following procedures can be done at the bench but some steps will require a dark environment to minimize exposition of fluorophores to light.

1. Aspirate the culture medium from the wells of the culture plate containing the coverslips.
2. Wash each well by adding 1 mL of PBS and gently rocking the plate.
3. Fix the cells by adding 1 mL of fixation buffer. Incubate 30 min at 4 °C.
4. Wash with 1 mL of PBS. Repeat once.
5. Permeabilize the cells by adding 1 mL of permeabilization buffer. Incubate 15 min at room temperature.
6. Discard the permeabilization buffer.

   The following steps should be done in a dark environment to protect fluorophores from light.
7. Prepare the aggresome dye solution by adding 0.2 μL of PROTEOSTAT® Aggresome Detection Reagent to 600 μL of permeabilization buffer.
8. Cut a piece of parafilm (~4″ × 8″) and stick it on the bench with scotch tape. The parafilm piece needs to be as flat as possible. Select and identify an area on the parafilm for each experimental condition.
9. Deposit 70 μL of the aggresome dye solution per experimental condition on the parafilm.

10. Carefully remove the coverslips from the wells and lay them *face down* to have cells bathing in the aggresome dye solution. Incubate 30 min at room temperature.
11. Place the coverslips back in the culture plate *face up* in their respective well.
12. Wash with 1 mL of permeabilization buffer by gently rocking the plate. Repeat once.
13. Add 1 mL of blocking buffer. Incubate 30 min at room temperature.
14. Prepare the primary antibody solution by adding 1 μL of α-HA antibody to 499 μL of blocking buffer (*see* **Note 16**).
15. Prepare a parafilm piece as mentioned in **step 8**.
16. Deposit 70 μL of the primary antibody solution per condition on the parafilm.
17. Carefully remove the coverslips from the wells and lay them *face down* on the solution. Incubate 1 h at room temperature (*see* **Note 17**).
18. Place the coverslips back in the culture plate *face up* in their respective well.
19. Wash with 1 mL of permeabilization buffer and gently rock the plate. Repeat once.
20. Add 1 mL of blocking buffer. Incubate for 10 min at room temperature.
21. Prepare the secondary antibody solution by mixing 3 μL of α-Mouse IgG Alexa Fluor® 633 in 597 μL of blocking buffer.
22. Prepare a parafilm piece as mentioned in **step 8**.
23. Add 70 μL of the secondary antibody solution per condition on the parafilm.
24. Carefully remove the coverslips from the wells and lay them *face down* on the solution. Incubate 1 h at room temperature.
25. Place the coverslips back in the culture plate *face up* in their respective well.
26. Wash with 1 mL of permeabilization buffer and gently rock the plate. Repeat twice.
27. Wash with 1 mL of PBS and gently rock the plate. Repeat once.
28. Label a microscope slide for each experimental condition.
29. Place a small drop (~10 μL) of ProLong® Gold antifade reagent on the microscope slide and make sure that there is no bubble.
30. Lay the coverslips *face down* on the ProLong® Gold drop slowly to avoid the formation of bubbles.
31. Protect the microscope slides from light and let them dry flat for the night at room temperature.

32. Apply a thin layer of nail polish on the edges of the coverslips. Let it dry 30 min to 1 h at room temperature in the dark.
33. Keep the microscope slides in a light tight box at 4 °C. For better results, examine slides under the microscope within a week (*see* **Note 18**).

3.4 Fluorescence Microscopy

1. Turn on the confocal system and the lasers (White light laser or 488 and 633 nm lasers).
2. Put a small drop of type F immersion oil on the microscope slide coverslip.
3. Insert the microscope slide on the stage tray and select the 60× or 63× objective depending of your confocal system. In order to observe the aggresome formation, the imaging parameters for the aggresome dye should be first optimized on a slide with a MG-132- treated condition (or any aggregation-inducing conditions).
4. Set the focus using the normal light.
5. Verify the quality of the cell labeling using the epifluorescence imaging (*see* **Note 19**).
6. For the receptor (Alexa Fluor® 633) visualization, set the excitation laser to 633 nm and the detection of the emission should be performed with an Alexa Fluor 633 filter (643–765 nm). Adjust the *Z* position and the laser intensity using live imaging to obtain the optimal signal with a few saturated pixels.
7. For the aggresome dye visualization, set the excitation laser to 488 nm and the detection of the emission should be performed using a Texas-Red filter (572–623 nm) because of its long Stokes shift. Adjust laser intensity using live imaging to obtain the optimal signal with a few saturated pixels.
8. Once you find the optimal excitation and detection parameters, acquire single slice images with a resolution of ~100 nm per pixels.
9. Acquire images of at least 50 representative cells for each condition without changing the acquisition parameters between slides (*see* **Note 20**).

3.5 Image Processing

To process images, any imaging software allowing co-localization analysis provided with your confocal system can be used. The following procedure describes a step-by-step walkthrough for processing acquired images using LAS X software provided by Leica.

1. Open LAS X software and select the "Process" tab.
2. In the "Open projects" section, import the project containing your acquired images.

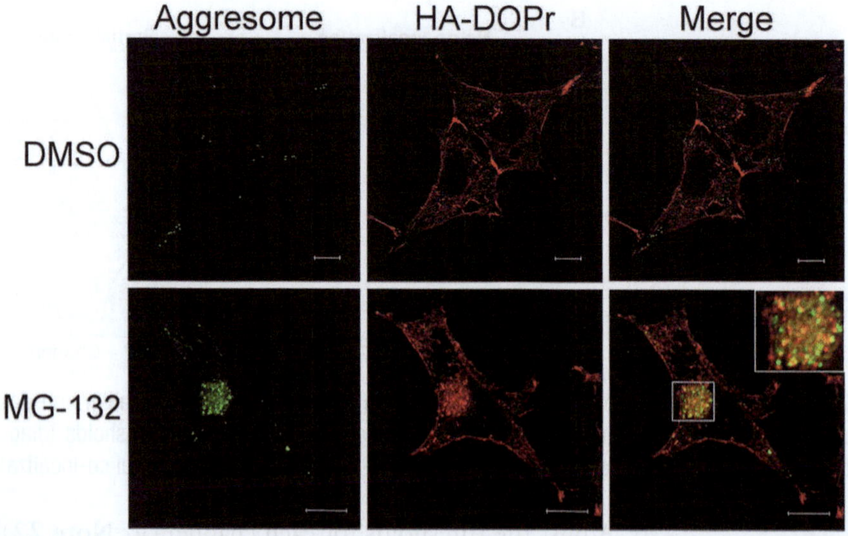

Fig. 1 Detection of HA-DOPr aggregation by confocal microscopy using the PROTEOSTAT® aggresome dye. HEK293 cells stably expressing HA-DOPr were treated for 12 h with 5 μM of MG-132 or DMSO. The cells were fixed, permeabilized, treated with the PROTEOSTAT® Aggresome Detection Reagent and labeled with HA-specific mouse IgG and Alexa Fluor® 633 anti-mouse IgG. The third panel on the right represents merged images of the green and red signals and the areas with high degree of co-localization appear in yellow. These images are single slices confocal microscopy representative of over 200 observed cells. Bars, 10 μm

3. In the project folder, select a representative image of each condition, in this case DMSO and MG-132 treated, and switch to the "Process Tools" section.
4. Adjust the background for each channel and apply the changes to the image. The background values should be kept constant during the image processing of each condition.
5. Select the modified images that should now appeared in the project folder.
6. Export the images as Tiff files with the overlay channels and a micron scale.
7. Final figures can be generated using any image processing software (e.g., Photoshop, GIMP, Illustrator) (Fig. 1).

3.6 Co-localization Analysis and Quantification

1. Open LAS X software and select the "Quantify" tab.
2. In the "Open projects" section, import the project containing your acquired images.
3. In the project folder, select a representative image of each condition and switch to the "Tools" section (*see* **Note 21**).
4. Select the co-localization tool. A window with a fluorogram and images with co-localization pixels should appear.
5. In the "Parameters" section, check the box to link the background for channel 1 and 2 and adjust the background level.

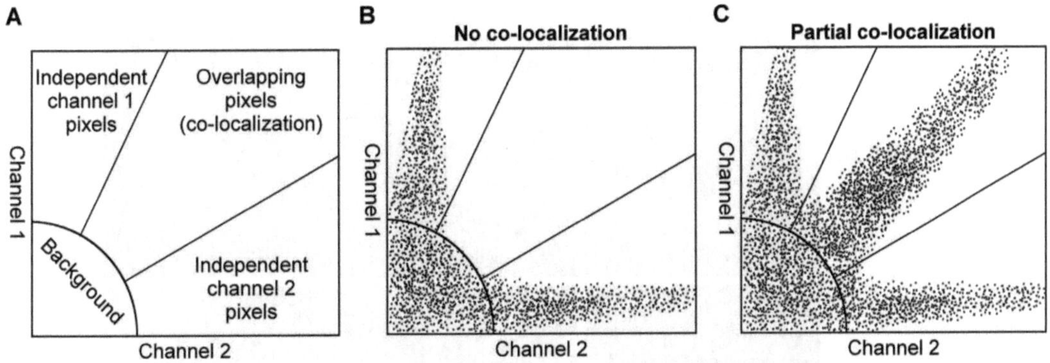

Fig. 2 Guide to adjust the background and thresholds on a fluorogram. (**a**) Identification and description of each area composing a fluorogram delimited by the background and the two thresholds (diagonal lines). Representative fluorograms of situations showing (**b**) no co-localization and (**c**) partial co-localization

6. Adjust the thresholds for each channel (*see* **Note 22**). See the examples in Fig. 2 as a guide to adjust the background and thresholds (Fig. 2).
7. Once the background and thresholds are optimized, the software will automatically output the co-localization rate in percentage in the middle window. This percentage represents the number of overlapping pixels delimited by the thresholds reported on the total number of pixels above background from both channels (*see* **Note 23**).
8. Save the co-localization pixels images, the fluorogram, and all the co-localization values by clicking the "Report" button (Fig. 3a).
9. Repeat **step 8** for each of the acquired images by selecting them in the "Open projects" section. The background and thresholds values should remain the same for all the co-localization analysis giving the fact that all parameters were the same during the image acquisitions.
10. For the statistical analysis, all co-localization rate percentages were imported in the Prism 7 software and separated in a two-column data table for the DMSO and MG-132 conditions (*see* **Note 24**).
11. Click the "analyze" button, select "t tests" in column analyses, select "unpaired" in experimental design and click on the "OK" button (*see* **Note 25**).
12. In the results section, you will find all the important values such as mean ± standard error of mean (SEM) for each column and the *P* value of your statistical analysis.
13. In the graphs section, you will find a histogram representative of your data. You can edit the graphic by right clicking on it and selecting "Format graph".

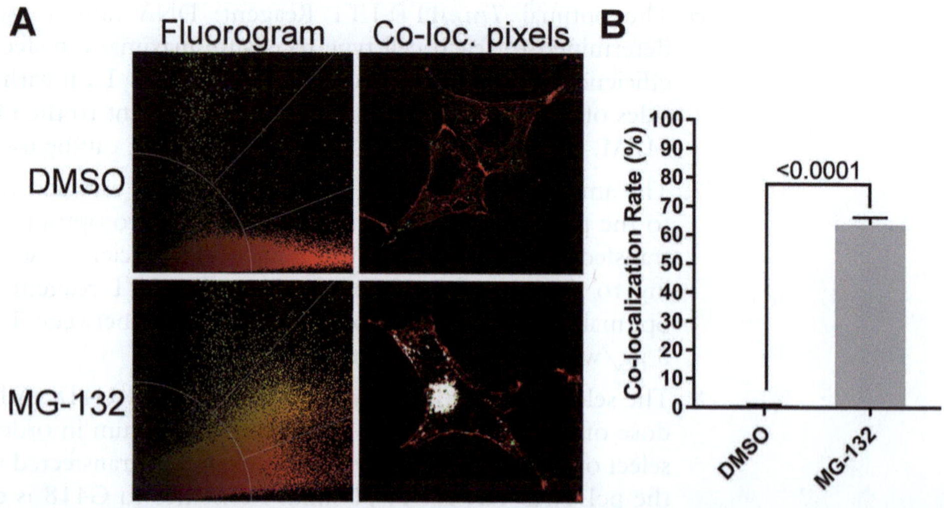

Fig. 3 Co-localization analysis between HA-DOPr and the PROTEOSTAT® aggresome dye. (**a**) Resulting fluorograms of the co-localization analysis done by the LAS X software and extraction of the co-localization pixels (white) layered on merge images of the green and red signals. (**b**) Statistical analysis (unpaired Student's *t*-test) of the co-localization rates between the red and green signals of over 50 cells per condition. The results are presented in the histogram as the mean ± SEM. *P* value ≤ 0.0001

14. Export the final histogram as a Tiff file (Fig. 3b), or as the file type required for publication.

4 Notes

1. The accession number EU883570.1 sequence of the human DOPr was used in our study. The sequence encoding the CDS of the human DOPr was amplified using primers containing the HA epitope in frame with the N-terminal open reading frame. The fragment was then inserted into the pcDNA3 vector (Invitrogen) between BamHI and EcoRI.
2. Any confocal system equipped with the necessary lasers and filters will do.
3. Depending on the cell type, trypsin concentration and trypsinization time may vary.
4. This step is performed in order to remove trypsin from the cell culture medium.
5. Each well will ultimately result in a stably transfected cell population. One well treated only with the transfection reagent must be added to serve as a "kill control", that is, cells will not acquire resistance to G418 resulting in death of all cells after a few days.

6. The optimal *Trans*IT®-LT1 Reagent: DNA ratio must be determined for each cell type to ensure maximal transfection efficiency. Avoid any contact of the *Trans*IT®-LT1 with the sides of the plastic tube while adding the reagent to the Opti-MEM. Any other transfection method/reagent can be used.
7. The amount of DNA used in transfection may vary according to the plasmid construct. It is recommended to optimize the transfection to obtain maximal transfection efficiency. According to the manufacturer of the *Trans*IT®-LT1 reagent, the optimal DNA concentration generally ranges between 1 and 3 μg/well for a six-well plate.
8. The selection is initiated by the addition of the G418. A high dose of antibiotic is added to the culture medium in order to select only the cells that have been successfully transfected with the pcDNA3-HA-DOPr plasmid. Resistance to G418 is conferred by the Neomycin resistance gene contained in the pcDNA3 vector. It is therefore normal to observe cellular mortality in all conditions since not all transfected cells will properly integrate the pcDNA3 construct in their genome. Depending on how recombination occurs during integration of plasmid DNA in the genome, it is possible that cells acquire G418 resistance without expressing the gene of interest, in this case, HA-DOPr.
9. The optimal G418 concentration required to generate a stable cell line may vary according to the cell types and the culture conditions. It is recommended to perform a concentration–response curve using increasing amounts of G418 (usually between 0.1 and 2.0 mg/mL) to determine the optimal working concentration prior to generation of the stable cell line.
10. Cells should be checked daily to ensure proper growth.
11. Transfer the cells when they reach over 80% of confluence in a larger cell culture dish (60 × 15 mm or 100 × 20 mm) with culture medium containing 0.8 mg/mL of G418. The selection can be considered completed when all the cells in the "kill control" condition are dead and when cells from your stable cell line are confluent.
12. Receptor expression can be assessed in several ways. Here, expression of the HA-DOPr receptor was analyzed by Western blotting, cell surface ELISA, and confocal microscopy to assess proper intracellular distribution of the receptor protein.
13. At this step, the generation of a stably transfected cell population is complete. However, it is possible to further process the cells in order to generate a clonal stable cell line by several methods (Fluorescence Activated Cell Sorting (FACS), serial dilution, cloning rings, etc.).
14. If the drop breaks in the well, just add more PLL solution to cover the coverslip surface.

15. The MG-132 concentration and incubation time may vary depending on the receptor turn-over and the cell type due to its cytotoxicity.

16. The antibody concentration may need optimization to obtain maximal staining and minimal non-specific background.

17. The incubation time with the primary antibody solution can vary depending on the antibody specificity, the targeted protein abundance, and the epitope accessibility.

18. The microscope slides can be kept for a longer period but to the detriment of the signal intensity due to fluorophore fading.

19. The acquisition of the Alexa Fluor 633 and the aggresome dye should be done sequentially. The imaging of the Alexa Fluor 633 should be done first to avoid photobleaching and Förster resonance energy transfer (FRET). If your system allows for time-gating detection, we recommend a time gate between 0.3 ns (start) and 6 ns (end) to prevent autofluorescence detection.

20. The excitation and detection parameters should not be changed during the whole experiment to allow the quantitative comparison between each condition.

21. Giving the fact that the MG-132 induces both the aggregation of the receptor and the formation of the aggresome, we used the DMSO-treated condition to determine the background level and thresholds for each channel.

22. The background and thresholds adjustments depend on the quality of the labeling, the antibodies specificity, and the autofluorescence. Here, the objective is to sequestrate the independent pixel populations using the background level and thresholds for each channel. The pixels contained in the area between the two thresholds represent the overlapping or co-localization pixels.

23. The LAS X software will also show other co-localization values such as the Pearson's correlation coefficient and the Overlapping coefficient. We choose the co-localization rate, also known as the Mander's coefficient, as the main outcome because it directly indicates the overlapping of the two signals and represents the true degree of co-localization [10]. For example, a co-localization rate of 50% means that 50% of all the pixels above the background are overlapping.

24. Any software able to do statistical analyses and generate graphs could be used.

25. In this case, an unpaired Student's *t*-test was used because two conditions were compared. If you compare more than two conditions (e.g., multiple mutant receptors, many small interfering RNAs) you should use an unpaired one-way ANOVA with post hoc analyses.

Acknowledgments

This work was supported by a grant from the Natural Sciences and Engineering Research Council of Canada and by the André-Lussier Research chair to J.-L.P, and by Ph.D. scholarships from NSERC (S.G.) and from the Fonds de Recherche du Québec-Santé (S.G. and J.D.). The authors are grateful to Leonid Volkov for his expertise in confocal microscopy.

References

1. Hauser AS, Attwood MM, Rask-Andersen M et al (2017) Trends in GPCR drug discovery: new agents, targets and indications. Nat Rev Drug Discov 16:829–842
2. Dupré DJ, Hébert TE (2006) Biosynthesis and trafficking of seven transmembrane receptor signalling complexes. Cell Signal 18:1549–1559
3. Tao YX, Conn PM (2014) Chaperoning G protein-coupled receptors: from cell biology to therapeutics. Endocr Rev 35:602–647
4. Saliba RS, Munro PMG, Luthert PJ et al (2002) The cellular fate of mutant rhodopsin: quality control, degradation and aggresome formation. J Cell Sci 115:2907–2918
5. Génier S, Degrandmaison J, Moreau P et al (2016) Regulation of GPCR expression through an interaction with CCT7, a subunit of the CCT/TRiC complex. Mol Biol Cell 27:3800–3812
6. Zaarur N, Meriin AB, Gabai VL et al (2008) Triggering aggresome formation: dissecting aggresome-targeting and aggregation signals in synphilin 1. J Biol Chem 283:27575–27584
7. Hao R, Nanduri P, Rao Y et al (2013) Proteasomes activate aggresome disassembly and clearance by producing unanchored ubiquitin chains. Mol Cell 51:819–828
8. Gendron L, Cahill CM, von Zastrow M et al (2016) Molecular pharmacology of δ-opioid receptors. Pharmacol Rev 68:631–700
9. Petäjä-Repo UE, Hogue M, Laperrière A et al (2001) Newly synthesized human δ opioid receptors retained in the endoplasmic reticulum are retrotranslocated to the cytosol, deglycosylated, ubiquitinated, and degraded by the proteasome. J Biol Chem 276:4416–4423
10. Zinchuk V, Zinchuk O, Okada T (2007) Quantitative colocalization analysis of multicolor confocal immunofluorescence microscopy images: pushing pixels to explore biological phenomena. Acta Histochem Cytochem 40:101–111

Chapter 17

Combining RNAi and Immunofluorescence Approaches to Investigate Post-endocytic Sorting of GPCRs into Multivesicular Bodies

Xuezhi Li, Stéphanie Rosciglione, Andréanne Laniel, and Christine Lavoie

Abstract

Following stimulation, G protein-coupled receptors (GPCRs) are internalized and transported to early endosomes where they are either recycled back to the plasma membrane for another round of activation or targeted to the lysosomes for degradation and long-term signal termination. This latter requires internalization of receptors into intraluminal vesicles (ILVs) of multivesicular bodies (MVBs) for complete degradation following fusion with lysosomes. This endosomal sorting step is highly regulated and has profound functional consequences. This chapter describes how RNAi and confocal microscopy methods can be combined to evaluate whether a protein of interest (herein Gαs) is involved in GPCR sorting into ILVs of MVBs.

Key words GPCR, Immunofluorescence, Gαs, Confocal microscopy, Endosome, Intraluminal vesicles, Multivesicular bodies, siRNA

1 Introduction

G protein-coupled receptors (GPCRs) are the most abundant class of signaling receptors expressed in the mammalian genome and are involved in cellular responses to numerous physiological stimuli. The trafficking of GPCRs through the endocytic pathway is critical for controlling the temporal and spatial fidelity of GPCR signaling responses. Once activated at the plasma membrane, GPCRs are rapidly desensitized and internalized in early endosomes. In this sorting compartment, GPCRs are either recycled back to the cell surface promoting resensitization of cellular responsiveness or targeted to lysosomes for degradation, producing a prolonged attenuation of cellular signaling [1]. Lysosomal sorting of receptors requires their transfer from the early endosome limiting membranes to intraluminal vesicles (ILVs) of multivesicular bodies

Mario Tiberi (ed.), *G Protein-Coupled Receptor Signaling: Methods and Protocols*, Methods in Molecular Biology, vol. 1947, https://doi.org/10.1007/978-1-4939-9121-1_17, © Springer Science+Business Media, LLC, part of Springer Nature 2019

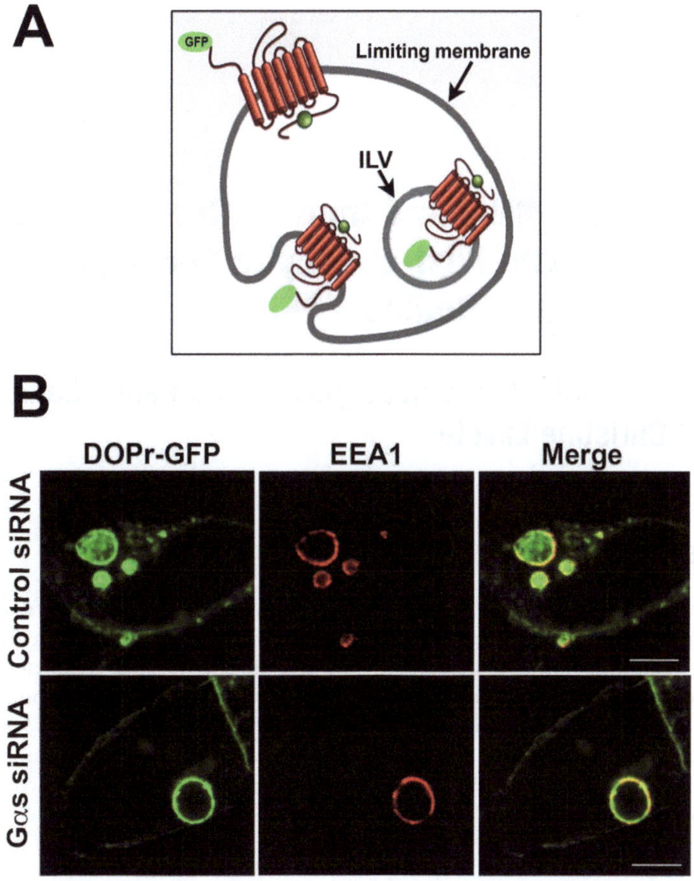

Fig. 1 Detection of DOPr-GFP in multivesicular bodies by confocal microscopy. (**a**) Diagram illustrating the distribution of a GPCR with a C-terminal GFP tag on the outer membrane of an endosome and in ILVs. (**b**) Effect of Gαs knockdown on DOPr-GFP distribution on the endosome-limiting membrane and ILVs. Representative confocal images of control and Gαs-depleted HEK293 cells co-transfected with DOPr-GFP along with Rab5-Q79L (to create enlarged endosomes) and treated with agonist DPDPE for 90 min. The early endosome marker EEA1 was used to label the limiting membranes of early endosomes. Cells were fixed, permeabilized, incubated with antibodies and processed for confocal microscopy. Scale bars, 5 μm

(MVBs) [1, 2] (Fig. 1a). The MVBs eventually fuse with lysosomes to degrade the ILVs and their protein cargo [2]. GPCR sorting into ILVs is highly regulated and involves recognition of the receptors, usually via an ubiquitin tag, by a conserved machinery called the endosomal sorting complex required for transport (ESCRT) [3–6].

GPCRs are classically known to signal via the activation of heterotrimeric G proteins at the plasma membrane. However, emerging findings indicate that GPCRs can also signal from the surface of endosomes [7, 8]. Indeed, GPCRs interaction with endosomal β-arrestin was reported to mediate scaffolding of

downstream kinase cascades (such as MAPK) and thereby producing G protein-independent signaling events [9, 10]. Furthermore, recent evidence has revealed the presence of Gαs on endosomes and its capacity to elicit a "second wave" of signaling by GPCR internalized in endosome following stimulation [8, 11]. Therefore, GPCR internalization into ILVs and degradation in lysosomes is a critical step in the regulation of receptor signaling and an alteration of this sorting step can lead to aberrant signaling and contribute to various pathological conditions, including cancer progression [12, 13]. However, the precise mechanisms that regulate GPCRs transfer within the ILVs are still poorly defined.

Our group recently reported that, in addition to an endosomal signaling function, $G\alpha_s$ is involved in endosomal sorting of GPCRs [14]. Using siRNA and confocal microscopy, we showed that cellular depletion of Gαs specifically delays the lysosomal degradation of GPCRs by disrupting the transfer of GPCRs into the ILVs of MVBs [14] (Fig. 1b). Here, we describe the conditions and procedures used to evaluate whether Gαs alters the sorting of human delta opoid receptor (DOPr) into ILVs using a confocal microscopy assay. DOPr tagged at its C terminus with green fluorescent protein (DOPr-GFP) is expressed in HEK293 cells transfected with control or Gαs siRNA. A constitutively active form of the small GTPase Rab5 (Rab5-Q79L) is co-expressed in these cells to create enlarged endosomes and facilitate the detection of DOPr–GFP on the limiting and intraluminal membranes of MVBs. Following stimulation with agonist, confocal microscopy is used to visualize the internalization of DOPr-GFP in the lumen of enlarged endosomes. The extent of intraluminal distribution of this receptor is then quantified using line scan analysis of confocal cross-sections of endosomes. This method is equally applicable to evaluate the role of Gαs or any other protein of interest on the sorting of various GPCRs, as well as single transmembrane receptors (such as EGFR), into ILVs of MVBs.

2 Materials

Prepare all solutions using ultrapure water (prepared by purifying deionized water, to attain a sensitivity of 18 MΩ cm at 25 °C) and analytical grade reagents. Prepare and store all reagents at room temperature (unless indicated otherwise). Diligently follow all waste disposal regulations when disposing waste materials.

2.1 Cell Culture

1. Human embryonic kidney 293 (HEK293) cells from American Type Culture Collection (ATCC, #CRL-1573).
2. Complete medium: Dulbecco's modified Eagle's medium (DMEM) high glucose supplemented with 10% (v/v) fetal

bovine serum (FBS) and 1% (v/v) PSG (penicillin (10,000 U), streptomycin (10,000 μg) and L-glutamine (29.2 mg/mL)) solution. Conserve at 4 °C (*see* **Note 1**).

3. Trypsin/EDTA solution: 0.05% trypsin with 0.48 mM EDTA 4Na and phenol red. Store at 4 °C.
4. 10× Phosphate-buffered saline (PBS) stock: Dissolve 25.62 g Na_2HPO_4 in 200 mL H_2O; dissolve 225 g $Na_2HPO_4 \cdot 7H_2O$ in 800 mL H_2O and dissolve 878.6 g NaCl in 8 L H_2O. Mix the three solutions and bring final volume to 10 L with H_2O. Filter sterilize and store at room temperature. Dilute 1:10 for use. pH will be at 7.4.
5. 100-, 60-, or 35-mm tissue culture dishes.
6. Sterile hood for conducting all preparation steps.
7. Sterile incubator set at 37 °C containing 5% CO_2 in the atmosphere.

2.2 Cell Transfection with cDNA

1. Opti-MEM®I Medium.
2. X-Treme GENE™ HP DNA transfection reagent.
3. Plasmids at 1 μg/μL DNA concentration. Plasmids DNA used here are pEGFP-C1-DOPr and pcDNA3-Rab5QL (*see* **Note 2**).
4. Sterile 1.5 mL microfuge tubes.

2.3 Cell Transfection with Small Interfering RNA (siRNA)

1. Sterile and RNAse-free microtubes of 1.5 mL.
2. Opti-MEM®I Medium.
3. Lipofectamine™ 2000 transfection reagent (Invitrogen).
4. siRNA oligonucleotides: siRNA duplex targeted against our protein of interest (Gαs) and negative control (scrambled) siRNA (Dharmacon), 20 μM stock concentration in RNAse-free solution (*see* **Note 3**).

2.4 Cell Stimulation with Agonists

1. Dulbecco's modified Eagle's medium (DMEM) high glucose.
2. Stimulation medium: DMEM containing 25 mM Hepes pH 7.4 and 0.2% bovine serum albumin (BSA). Make fresh and filter sterilize.
3. GPCR's agonists. We used (D-Pen^{2,5})-Enkephalin (DPDPE) as an agonist for DOPr. Dilute at 5 μM in stimulation medium.

2.5 Indirect Immunofluorescence Reagents

1. 12 mm round glass coverslips, #1 poly-L-lysine coated (Corning BioCoat™) (*see* **Notes 4** and **5**).
2. Glass microscope slides such as VWR Superfrost Plus micros slide.
3. Sodium phosphate buffer 0.1 M pH 7.4: $Na_2HPO_4 \cdot 7H_2O$ and $NaH_2PO_4 \cdot H_2O$. Filter sterilize and store at room temperature.

4. PBS (*see* Subheading 2.1).
5. Fixation buffer (PFA 3%): Dilute paraformaldehyde (PFA) 16% (aqueous solution, methanol free) in cold sodium phosphate buffer pH 7.4. Prepare fresh.
6. Quenching solution: 50 mM NH_4Cl in PBS. Prepare fresh (*see* **Note 6**).
7. Permeabilization buffer: 0.1% Triton X-100 in PBS. Prepare fresh.
8. Blocking solution: 10% goat serum in PBS. Prepare fresh (*see* **Note 7**).
9. Washing buffer: 1% goat serum in PBS. Prepare fresh (*see* **Note 8**).
10. Primary antibodies: Anti-GFP monoclonal antibody (1:300, Clontech) and anti-EEA1 polyclonal antibody (1:100, Thermo Scientific). Dilute in blocking buffer. Prepare fresh.
11. Secondary antibodies: Goat anti-mouse IgG (H + L) highly cross-adsorbed secondary antibody Alexa Fluor® 488 conjugate (2 mg/mL) and goat anti-rabbit IgG (H + L) highly cross-adsorbed secondary antibody Alexa Fluor® 594 conjugate (2 mg/mL) diluted in blocking buffer. Prepare fresh (*see* **Notes 9** and **10**).
12. Mounting medium: 1% N-Propyl gallate in glycerol:PBS (1:1). Sonicate in waterbath sonicator to dissolve the solution. Keep aliquots at 4 °C (short term) or −20 °C (long term) (*see* **Note 11**).

2.6 Confocal Microscopy Analysis

1. Inverted confocal laser scanning microscope (FV1000; Olympus, Tokyo, Japan) equipped with a PlanApo 60×/1.42 oil immersion objective, Argon 458/488/515 nm, HeNe 543 nm, HeNe 633 nm and diode 405 nm lasers (*see* **Note 12**).
2. Type F immersion oil (IMMOIL-F30CC Olympus).
3. Olympus FluoView version 1.6a to acquire and analyze the images.
4. Excel software.
5. Prism 7 software (GraphPad) for statistical analysis and creation of graphs and histograms.

3 Methods

Carry out all procedures at room temperature unless otherwise specified.

3.1 Cell Culture and Transfection

1. *Day 1*: seed HEK293 cells (~80,000 cells) onto 35 mm Petri dishes in 2 mL complete medium and incubate at 37 °C and 5% CO_2. Cells should be well spread and at ~30–40% confluency the next day (*see* **Note 13**).
2. *Day 2*: Transfection with siRNA. In a sterile and RNAse-free microtube, add 2 μL of Lipofectamine 2000 in 100 μL of prewarmed OptiMEM. In another sterile RNAse-free microtube, add 5 μL of 20 μM siRNA (100 nM final concentration) in 100 μL of prewarmed OptiMEM. Incubate 5 min at room temperature. Prepare a set of tubes for each siRNA to transfect (control and Gαs siRNA) (*see* **Note 14**).
3. Combine diluted siRNA to diluted Lipofectamine reagent and mix by gently pipetting up and down two to three times. Incubate 20 min at room temperature.
4. Aspirate carefully the cell culture medium of the 35-mm Petri dish and replace it with 800 μL of prewarmed DMEM containing 10% FBS without antibiotics (*see* **Note 15**).
5. Add the transfection mixture (200 μL) dropwise over the entire surface of the culture cells in each Petri dish. Mix by gently rocking the plates back and forth and incubate at 37 °C with 5% CO_2 (*see* **Note 16**).
6. Five hours later, aspirate carefully the culture medium and replace it with 2 mL of prewarmed complete medium.
7. *Day 3*: Transfection with cDNA. Allow the transfection reagent (*X-TremeGENE HP*) to reach room temperature and briefly vortex. In a sterile microtube, dilute 2 μg of DNA (1 μg of pEGFP-DOPr and 1 μg of pcDNA3-Rab5QL) in 200 μL of prewarmed OptiMEM and mix gently. Constitutively active Rab5 (Rab5-Q79L) is used to create enlarged endosomes in order to facilitate optical resolution of the intraluminal space of endosomes (*see* **Notes 17** and **18**).
8. Add 2 μL of X-tremeGENE HP directly into the diluted DNA solution without coming into contact with the walls of the plastic tube. Mix gently and incubate 15 min at room temperature (*see* **Note 19**).
9. Add the transfection mixture (200 μL) dropwise over the entire surface of the culture cells in each Petri dish. Mix by gently rocking the plates back and forth to ensure even distribution over the entire plate surface and incubate at 37 °C with 5% CO_2 (*see* **Note 20**).

3.2 Trypsinization and Reseeding

1. *Day 4*: Seed the cells on coverslips. Wash cells twice with 2 mL warm PBS, add 0.5 mL of prewarmed Trypsin/EDTA and return to the incubator for 2–3 min. Rock the cell culture

dish until cells are detached from the surface and dispersed (observe under an inverted microscope) (*see* **Note 21**).

2. Add 2 mL of complete medium and mix gently by pipetting up and down to achieve a homogenous cell suspension.
3. Seed the transfected cells (the entire 2.5 mL) into a 35 mm petri dish containing up to 5 poly-L-lysine-coated 12 mm coverslips. Mix by gently rocking the plates and incubate at 37 °C (*see* **Note 22**).

3.3 Receptor Stimulation

1. *Day 5*: Seventy-two hours after the siRNA transfection, cells are serum starved. Rinse cells with warm DMEM and incubate with 1 mL of warm stimulation medium (DMEM +25 mM Hepes pH 7.4 + 0.2% BSA) for 1 h at 37 °C with 5% CO_2 (*see* **Note 23**).
2. Aspirate the medium and replace with stimulation medium supplemented with 5 μM DPDPE. Incubate 90 min at 37 °C with 5% CO_2 to allow receptor internalization into endosomes and MVBs (*see* **Notes 24** and **25**).

3.4 Immunofluorescence

1. Aspirate carefully the medium and slowly add 1.5 mL of ice-cold fixation buffer. Be careful not to disturb the cells. Incubate 30 min at room temperature (*see* **Note 26**).
2. Remove the fixing solution and place in biohazard waste container. Wash the cells with 2 mL of PBS.
3. Use fine tweezers to carefully transfer each coverslip (cells face up) in a pre-identified 12 wells plate containing 1 mL PBS (*see* **Note 27**).
4. Wash another time with 1 mL PBS. Incubate 10 min at room temperature (*see* **Note 28**).
5. After the third PBS wash, add 1 mL of quenching solution and incubate 10 min at room temperature.
6. Cells are next permeabilized by adding 1 mL of permeabilization buffer. Incubate 10 min at room temperature.
7. Remove the permeabilization buffer and add 0.5 mL of blocking solution. Incubate 30 min at room temperature.
8. Stick a piece parafilm of the size of 12 well dish on the bench. Using a pen, draw and identify 12 squares on the parafilm corresponding to the 12 wells (*see* **Note 29**).
9. Prepare the primary antibodies solution by diluting anti-mGFP (to detect the receptors) and anti-pEEA1 (to detect the early endosomes) in a microtube containing blocking buffer (50 μL is needed per coverslips) (*see* **Note 30**).
10. Deposit 50 μL of primary antibodies solution on the corresponding squares of the parafilm.

11. Use tweezers to pick-up the coverslip and carefully touch the edge of a Kimwipe to remove the liquid.
12. Rapidly and carefully invert the coverslip (cell side down) on the drop of antibodies solution. Repeat for each coverslip. Put the 12-well dish cover on the top of the parafilm to protect the coverslips. Incubate 1 h at room temperature.
13. Put back the coverslips (cell side up) into the corresponding 12 wells containing 1 mL of washing buffer. Incubate 10 min.
14. Wash two more times 10 min with 1 mL of washing buffer.
15. Prepare the secondary antibodies solution by diluting goat anti-mouse Alexa Fluor® 488 (1:300) and goat anti-rabbit Alexa Fluor® 594 (1:300) in a microtube containing blocking buffer (50 μL is needed per coverslips). Protect tube from light.
16. Prepare a new piece of parafilm (as described in **step 8**) and deposit 50 μL of secondary antibodies solution on the corresponding parafilm squares.
17. As described in **steps 11–13**, transfer the coverslips (cells side down) on the drop of secondaries antibodies solution and incubate for 45–60 min in a dark environment (*see* **Note 31**).
18. Put back the coverslips (cell side up) on the corresponding 12 wells and wash three times 10 min with 1 mL of washing buffer. Protect from light.
19. Prelabel microscope slides and add one drop of 5 μL mounting medium (avoid bubbles) on each slide.
20. Use tweezers to pick-up the coverslip and carefully touch the edge of a Kimwipe to remove the liquid.
21. Carefully invert the coverslip (cells side down) on the drop of mounting medium, starting with one edge of the coverslip and slowly deposit on the mounting medium, in order to avoid the formation of bubbles (*see* **Note 32**).
22. Apply small drops of transparent nail polish on the edges of the coverslips. Transfer the microscope slides into a dark cardboard slide holder or tray. Let dry the nail polish 30–60 min before analysis at the microscope.
23. For long-term storage, transfer the microscope slides in a dark slide tray or box at 4 °C. Ideally, analyze the cells by microscopy soon after the staining procedure (*see* **Note 33**).

3.5 Fluorescence Microscopy

1. Visualize cells on a confocal microscope. Herein, images were acquired with a Fluoview FV1000 scanning confocal microscope using the 60× Plan Apochromat 1.42 oil objective and Argon 458/488/515 nm and HeNe 543 nm lasers (*see* **Note 34**).

2. Cells expressing DOPr-GFP and enlarged endosomes (labeled with EEA1) are first located using epifluorescence imaging.
3. In the Image Acquisition Control panel, select the dyes from the dye database.
4. In order to minimize bleaching of the sample, low-resolution and high-speed scan is first used to set focus and zoom on the region of interest.
5. Select endosomes labeled with both GFP (DOPr) and EEA1. The size of endosomes should be large enough to clearly see the lumen of endosome. EEA1 labeling is used to confirm the nature of the endosome. Furthermore, given that EEA1 should only label the cytoplasmic surface of the outer membrane of the endosome, and not the ILVs, it is also used to find the best confocal plan that crosses the center of the lumen of the endosome (*see* **Note 35** and Fig. 1b).
6. Adjust the laser power and detector sensitivity (voltage (HV) and offset) for individual wavelength channels to get good quality image and avoid saturation of the signal. The goal is to have a dark background (by changing the offset) and no saturated pixels (highlighted in red, by adjusting the HV). Given that the DOPr (Alexa 488) signal on endosomal membranes or ILVs will be quantified (in Subheadings 3.6 and 3.7), it is crucial to adjust the imaging settings to avoid saturated pixel. Therefore, depending on the intensity of the signal, the acquisition setting for each endosome can be different (*see* **Note 36**).
7. Sequential scanning mode is selected to minimize spectral overlap (false-positive signal) in multi-color imaging.
8. Z-stack acquisition can be performed to select the best optical plan that crosses an endosome or to obtain optical plans for many endosomes present in the same cell (*see* **Note 37**).
9. Adjust the confocal aperture, and scan size and speed to achieve the desired resolution. These settings should be kept for the acquisition of all the images (*see* **Notes 38** and **39**).
10. When all imaging parameters are set up, start image acquisition of one optical slice or of z-stack of optical slices.
11. Save image in ".oib" format. This Olympus file format contains all acquisition metadata.
12. Acquire images of at least 100 endosomes per condition (control siRNA and Gαs siRNA in our study). Ideally, pictures of 30–50 endosomes should be obtained from multiples cells in three independent experiments (for each condition).

3.6 Image Processing

1. Open image file ".oib" format with Fluoview Olympus confocal software (*see* **Note 40**).

2. Using line tool, draw a straight-line across the diameter of the DOPr-GFP labeled endosome, precisely passing through its center (Fig. 2a, b) (*see* **Note 41**).
3. Click on the analysis button (upper bar panel) and select intensity profile.
4. A window showing a graph of the intensity of pixel detected along the line drawn across the endosome will appear.
5. Multiple lines can be drawn on the same image. Clicking on a specific line will show the corresponding intensity graph.
6. Save the data as an excel file. This file will contain a column with the pixel number, the length of the line, and another column with the corresponding pixel intensity along this line (Table 1).
7. Repeat these steps for each endosome.
8. Representative images for each condition can also be selected and save as ".Tiff" format using the Fluoview software.
9. Final figures (using these .Tiff images) can be generated using various imaging softwares (e.g., Photoshop, Illustrator) (Fig. 1b).

3.7 Quantification of GPCR Localization in the Endosomal Lumen

1. Using the excel software, open the excel file containing the data collected in Subheading 3.6.
2. To define the limiting membrane of the endosome, identify the two maximum pixel picks intensities in the "Intensity Average" column (highlighted in yellow in column C, Table 1).
3. Create a new column (Normalized intensity (Column E, Table 1)) in which the intensities are expressed in percentage relative to the highest intensity pixel of the limiting membranes (which is set to 100; *see* **Note 42**).
4. The endosomal diameter is next normalized to account for endosomes varying in size. To do so, the pixel numbers (Column A, Table 1) corresponding to the limiting membranes of the endosome defined in **step 2** will be normalized to 0 and 100.
5. Create a new column (Normalized diameter (column D, Table 1)) in which the location across the line of pixel 0 is subtracted from each position situated on the line. This value is then divided by the total diameter (amount of pixel between the two limiting membranes) of the endosome and expressed as a percentage (*see* **Note 43**).
6. This generate normalized pixel distances corresponding to the distance across the line occupied by each pixel and was expressed as a percentage.
7. Finally, to calculate the middle-fluorescence value in the endosomal lumen, the pixel intensity values corresponding to pixels

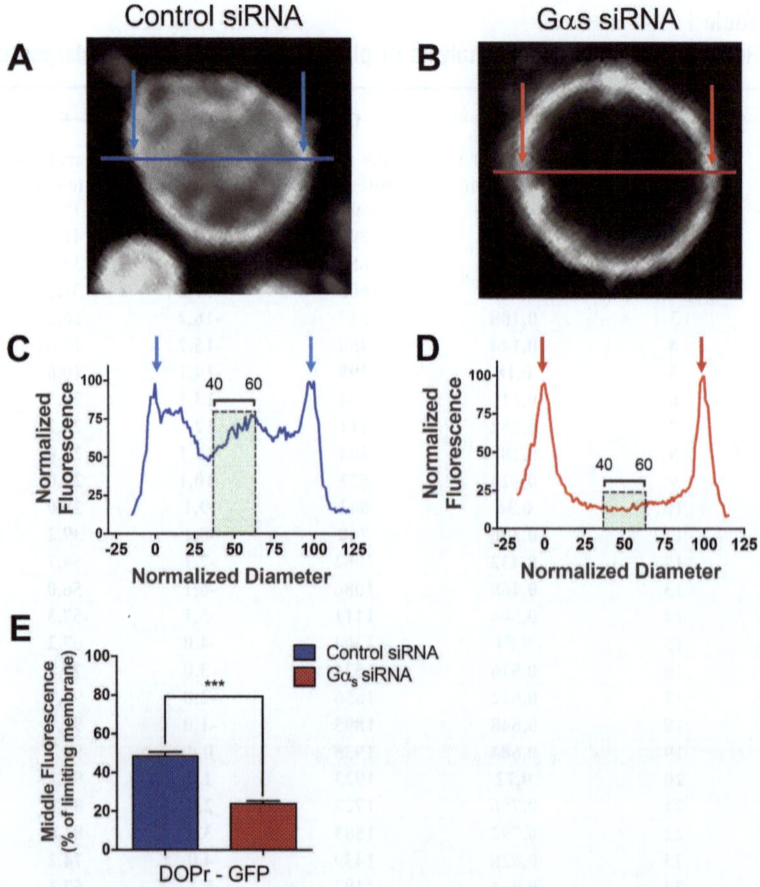

Fig. 2 Line scan analysis to quantify DOPr-GFP localization into the lumen of enlarged endosomes. (**a**, **b**) Representative optical sections of DOPr-GFP labeled endosomes imaged in Fig. 1b (control siRNA (**a**) and Gαs siRNA (**b**) treated cells). Blue and red lines overlaid on each image indicate the position of line scan used for quantification. Arrows indicate the regions where the lines cross the endosome limiting membranes. (**c**, **d**) Resulting graphs obtained from line scan analysis of the endosomes showed in panels **a** and **b**. Normalized fluorescence intensity represents the pixel intensity measured across the blue and red lines (shown in **a** and **b**), where maximum intensity is set to 100. Normalized distance represents the diameter measured across the blue and red lines, where maxima representing the limiting membrane (arrows in panels **a** and **b**) are labeled as 0 and 100 (arrows on the graphs). The green hatched box highlights the normalized fluorescence values of pixels from 40% to 60% of the normalized diameter that were used to determine the mean intraluminal fluorescence (or middle fluorescence) for each endosome. (**e**) Compiled results of the middle fluorescence obtained by line scan analysis of endosomes. The data are presented in the histogram as the mean ± s.e. of $n \geq 100$ endosomes from three independent experiments. Statistical analysis was performed using the Student's *t*-test. ***$P < 0.001$

Table 1
Representative line scan analysis of pixel intensity across an enlarged endosome

A	B	C	D	E	
RAW DATA Pixel Position No	RAW DATA Length (um)	RAW DATA Intensity Average	Normalized diameter	Normalized Intensity (%)	
0	0	803	-19,2	41,4	
1	0,036	687	-18,2	35,4	
2	0,072	608	-17,2	31,3	
3	0,108	517	-16,2	26,6	
4	0,144	458	-15,2	23,6	
5	0,18	399	-14,1	20,6	
6	0,216	394	-13,1	20,3	
7	0,252	411	-12,1	21,2	
8	0,288	504	-11,1	26,0	
9	0,324	523	-10,1	27,0	
10	0,36	543	-9,1	28,0	
11	0,396	760	-8,1	39,2	
12	0,432	1062	-7,1	54,7	
13	0,468	1086	-6,1	56,0	
14	0,504	1111	-5,1	57,3	
15	0,54	1304	-4,0	67,2	
16	0,576	1534	-3,0	79,1	
17	0,612	1856	-2,0	95,7	
18	0,648	1895	-1,0	97,7	
19	0,684	1935	0,0	99,7	← LIMITING MEMBRANE
20	0,72	1923	1,0	99,1	
21	0,756	1727	2,0	89,0	
22	0,792	1583	3,0	81,6	
23	0,828	1439	4,0	74,2	
24	0,864	1107	5,1	57,1	
25	0,9	941	6,1	48,5	
26	0,936	976	7,1	50,3	
27	0,972	897	8,1	46,2	
28	1,008	818	9,1	42,2	
29	1,044	795	10,1	41,0	
30	1,08	808	11,1	41,6	
31	1,116	763	12,1	39,3	
32	1,152	718	13,1	37,0	
33	1,188	724	14,1	37,3	
34	1,224	633	15,2	32,6	
35	1,26	661	16,2	34,1	
36	1,296	689	17,2	35,5	
37	1,332	717	18,2	37,0	
38	1,368	625	19,2	32,2	
39	1,404	618	20,2	31,9	
40	1,44	637	21,2	32,8	
41	1,476	657	22,2	33,9	
42	1,512	659	23,2	34,0	
43	1,548	708	24,2	36,5	
44	1,584	867	25,3	44,7	
45	1,62	877	26,3	45,2	
46	1,656	887	27,3	45,7	
47	1,692	1069	28,3	55,1	
48	1,728	1314	29,3	67,7	
49	1,764	1361	30,3	70,2	
50	1,8	1408	31,3	72,6	

(continued)

Table 1
(continued)

51	1,836	1542	32,3	79,5	
52	1,872	1440	33,3	74,2	
53	1,908	1346	34,3	69,4	
54	1,944	1308	35,4	67,4	
55	1,98	1270	36,4	65,5	
56	2,016	1115	37,4	57,5	
57	2,052	1012	38,4	52,2	
58	2,088	993	39,4	51,2	
59	2,124	975	40,4		
60	2,16	983	41,4	50,7	
61	2,196	890	42,4	45,9	
62	2,232	832	43,4	42,9	
63	2,268	807	44,4	41,6	
,0	2,304	783	45,5	40,4	
65	2,34	834	46,5	43,0	
66	2,376	871	47,5	44,9	
67	2,412	844	48,5	43,5	
68	2,448	818	49,5	42,2	MEAN INTENSITY (%)
69	2,484	849	50,5	43,8	46,0
70	2,52	879	51,5	45,3	
71	2,556	1015	52,5	52,3	
72	2,592	978	53,5	50,4	
73	2,628	942	54,5	48,6	
74	2,664	1100	55,6	56,7	
75	2,7	947	56,6	48,8	
76	2,736	914	57,6	47,1	
77	2,772	881	58,6	45,4	
78	2,808	814	59,6	42,0	
79	2,844	788	60,6	40,6	
80	2,88	694	61,6	35,8	
81	2,916	690	62,6	35,6	
82	2,952	686	63,6	35,4	
83	2,988	783	64,6	40,4	
84	3,024	858	65,7	44,2	
85	3,06	871	66,7	44,9	
86	3,096	884	67,7	45,6	
87	3,132	863	68,7	44,5	
88	3,168	794	69,7	40,9	
89	3,204	724	70,7	37,3	
90	3,24	718	71,7	37,0	
91	3,276	713	72,7	36,8	
92	3,312	675	73,7	34,8	
93	3,348	621	74,7	32,0	
94	3,384	625	75,8	32,2	
95	3,42	629	76,8	32,4	
96	3,456	592	77,8	30,5	
97	3,492	654	78,8	33,7	
98	3,528	669	79,8	34,5	
99	3,564	665	80,8	34,3	
100	3,6	661	81,8	34,1	

(continued)

Table 1
(continued)

101	3,636	672	82,8	34,6	
102	3,672	740	83,8	38,1	
103	3,708	718	84,8	37,0	
104	3,744	696	85,9	35,9	
105	3,78	660	86,9	34,0	
106	3,816	616	87,9	31,8	
107	3,852	649	88,9	33,5	
108	3,888	642	89,9	33,1	
109	3,924	636	90,9	32,8	
110	3,96	796	91,9	41,0	
111	3,996	963	92,9	49,6	
112	4,032	997	93,9	51,4	
113	4,068	1031	94,9	53,1	
114	4,104	1468	96,0	75,7	
115	4,14	1693	97,0	87,3	
116	4,176	1915	98,0	98,7	
117	4,212	1927	99,0	99,3	
118	4,248	1940	100,0	100,0	← LIMITING MEMBRANE
119	4,284	1789	101,0	92,2	
120	4,32	1562	102,0	80,5	
121	4,356	1251	103,0	64,5	
122	4,392	1111	104,0	57,3	
123	4,428	971	105,1	50,1	
124	4,464	782	106,1	40,3	
125	4,5	601	107,1	31,0	
126	4,536	544	108,1	28,0	
127	4,572	487	109,1	25,1	
128	4,608	427	110,1	22,0	
129	4,644	311	111,1	16,0	
130	4,68	235	112,1	12,1	
131	4,716	235	113,1	12,1	

position that lay 40–60% across the endosomal diameter were averaged (highlighted in green in column E, Table 1).

8. Figure 2c, d show graphs of representative line scan analysis with normalized distance and pixel intensity measured along lines crossing endosomes from control and Gαs siRNA-treated cells, respectively. The hatched box highlights the normalized fluorescence values of pixels from 40% to 60% of the normalized diameter that are used to determine the mean intraluminal fluorescence for each endosome.
9. The intraluminal (or middle) fluorescence values for multiple endosomes ($n \geq 100$) quantified in multiple cells and various experiments ($n = 3$ independent experiments) for each condition (control siRNA versus Gαs siRNA in our study) are compiled and saved in an excel file.
10. For the statistical analysis, the endosomal middle-fluorescence values quantified for our two conditions (control siRNA versus Gαs siRNA) are imported in a two-column data table in the Prism 7 software (*see* **Note 44**).

11. Select unpaired *t*-tests in the "column analysis" panel.
12. *P* value of the statistical analysis as well as mean ± standard error of mean (s.e.m.) for each column are obtained in the "Results" panel (*see* **Note 45**).
13. A histogram of the data (represented as the mean ± s.e. and annotated with the statistical analysis) is created with the Prism software (in the "Graphs" panel) (Fig. 2e).
14. Export the final histogram as .Tiff.

4 Notes

1. Use of penicillin–streptomycin helps to minimize bacterial contamination. Unlike most other amino acids, L-glutamine is not stable in liquid media. At temperature above 4 °C, the L-glutamine within cell culture media non-enzymatically degrades. This short half-life is why stock solutions of L-glutamine must be stored frozen at −20 °C, and why L-glutamine should be added to freshly prepared media.
2. Supercoiled plasmids DNA with an *A260/A280* ratio between 1.8 and 2.0 is recommended for transfection. Good quality DNA can be obtained using commercial plasmid preparation kits from companies such as QIAGEN.
3. Target sequence of human Gαs: Gαs-sense, 5′-GGC GCA GCG UGA GGC CAA CdTdT; Gαs-antisense, 5-GUU GGC CUC ACG CUG CGC CdTdT [15]. Scrambled RNA oligos (scramble II duplex; Dharmacon Research) were used as non-targeting siRNA control. Many pre-designed siRNA for genes of interest are commercially available. Furthermore, siRNA design services or algorithms are available for custom siRNA sequence synthesis. Negative control siRNAs should be included in every experiment to help monitor siRNA efficiency and non-specific (off-target) effects. To validate the specificity of the siRNA targets, it is also recommended to perform a rescue assay in which siRNA-resistant forms of the gene of interest is reintroduced in siRNA-treated cells.
4. Thickness of coverslips #1 is 0.13–0.17 mm. Use the coverslip thickness required for your objective. The correct cover glass thickness is important to achieve the best resolution with a given objective.
5. Poly-L-Lysine (PLL) are synthetic molecules used to enhance cell attachment to plastic or glass surfaces. Indeed, most cells do not adhere to glass coverslips as well as to plastic culture dishes. Coverslips can also be coated with PLL by the users before plating the cells [16]. In particular case such as double transfections, even cells known for efficiently attaching on glass

should be considered as sensible, and PLL use is highly recommended.

6. Ammonium chloride is used to quench the free aldehyde groups of paraformaldehyde and to reduce autofluorescence/ background staining (aldehyde fixatives react with amines and proteins to generate fluorescent products). As an alternative, 0.1 M glycine in PBS can be used.
7. 1% BSA or 10% FCS (fetal calf serum) in PBS can also be used as blocking solution. As a general rule, it is recommended to block in serum from the same host species as the secondary antibodies have been raised.
8. Washing buffer is a ten times dilution of the blocking buffer. If BSA is used as the blocking agent, the washing buffer will be 0.1% BSA in PBS.
9. Secondary antibodies should be selected based on the species in which the primary antibodies are developed. The fluorescent dyes must have minimal spectral overlap and be selected based on the wavelength of the laser lines.
10. A minimum of controls is recommended to do prior or in parallel to these experiments: (a) an autofluorescent control (no antibodies); (b) non-specific binding of the secondary antibodies (no primary antibodies), and (c) non-transfected cells to assess nonspecific primary and secondary antibodies.
11. Many commercial mounting mediums are also available.
12. Any confocal system equipped with the necessary lasers and filters can be used.
13. For an efficient transfection, it is crucial that cells are not too confluent, are well spread, and do not form aggregates. The amount of cells seeded needs to be adapted according to the cell type and mortality induced by the double transfections protocol.
14. Use RNAse-free tubes and tips to manipulate and mix siRNA. To obtain the highest transfection efficiency and low non-specific (off-target) effects, optimize transfection conditions by varying siRNA and Lipofectamine concentrations. As an alternative to transfection with Lipofectamine, cells can be electroporated using Nucleofector™ (Amaxa) [17].
15. Do not add antibiotics to media during transfection as this causes cell death.
16. A softer alternative for sensitive cells is to add 800 μL of prewarmed DMEM containing 10% FBS without antibiotics to the transfection mix (prepared in **step 2**). Slowly and carefully add this mixture to the cells (on the side of the dish) after aspirating the culture medium.

17. According to the plasmids and cells used, the amount of DNA used in transfection will need to be optimized to obtain maximal transfection efficiency.
18. As an alternative, Rab5QL tagged with a fluorescent protein (such as RFP-Rab5QL) could be transfected and directly used to visualize enlarged endosomes (instead of using anti-EEA1 antibody when cells are transfected with untagged Rab5QL).
19. To avoid adversely affecting transfection efficiency, the manufacturer recommends to avoid undiluted X-tremeGENE HP reagent to come into contact with plastic surface and avoid the use of siliconized pipette tips or tubes.
20. Removal of growth medium or antibiotic is not necessary. Once the transfection reagent:DNA complex has been added to the cells, there is no need to replace with fresh medium (as necessary with Lipofectamine). Furthermore, as an alternative to the dropwise addition, dishes can be lean on the side and the transfection mixture added to the media accumulated on this side and mix by gently pipetting up and down two to three times.
21. Trypsin concentration and time of incubation will depend on the cell type used.
22. Make sure the cells are well detached and do not form aggregates. This step is crucial to obtain cell nicely sparse and spread on the coverslips and to acquire nice confocal images.
23. Serum starvation is required to remove serum components that could affect the outcome of experiments. It can be done in DMEM without serum. Some cells are very sensitive to serum deprivation (they will start to die, round up, and detach). BSA and Hepes are added to help stabilize the cells and maintain the pH.
24. When working with other GPCRs or receptors, the stimulation time required for the receptors to reach the MVBs should be determined in order to optimize the assay (usually between 30 min and 2 h).
25. Fluorescent ligands can also be used instead of fluorescent receptors. In our study, GFP-tagged GPCR was used to visualize the internalization of receptors into MVBs. However, receptors labeled at the cell surface with fluorescent ligands could also be used to follow internalization in MVBs (such as performed for EGFR in [17]).
26. As an alternative, each coverslip can be carefully removed from the 35-mm petri dish and transfer to 12-well plates containing 0.5 mL fixation buffer. The cells remaining in the 35-mm dishes can be washed and lysed in order to test the knockdown efficiency by Western blotting.

27. In order to avoid damaging the cells, grip the coverslip with the tweezers on the outer edge of the coverslip.
28. To avoid cell damaging and detachment, carefully add the buffer (and other solutions) on the side of the well and not directly on the coverslips.
29. A few drops of water under the parafilm help it stick to the bench. Make sure the parafilm is as flat as possible by flattening it with a kimwipe.
30. Even if the use of anti-GFP antibody is not necessary to visualize the receptor, we found that it increases and stabilizes the signal (less bleaching and fading) during image acquisition and when the coverslips are stored for a few days before analysis at the microscope. A third antibody against the knockdown protein (in this case Gαs) can also be added at this step to confirm protein knockdown in the cells (as we showed in [17]). This third antibody has to be raised in another species. Given that we previously showed the efficient knockdown of Gαs by immunofluorescence and Western blot, we did not add it in this protocol.
31. The secondary antibodies are coupled to fluorophores and are thus sensitive to light. Protect them by covering the piece of parafilm or 12 wells with a dark box or aluminum foil.
32. Two coverslips (12 mm) can be mounted on the center of the same microscope slide. Do not place them too close to the ends of the slide because they will be difficult to observe at the microscope.
33. The samples can be kept at 4 °C in a light protecting box for many weeks. However, some deterioration of staining will occur (fading of the signal intensity, higher background). Images should thus be captured as soon as possible.
34. Since each instrument setup is different, the reader is advised to seek the guidance of the imaging facility manager for assistance in optimizing instrument settings.
35. EEA1 labeling should only be on the outer membrane of the endosome. EEA1 labeling in the lumen indicates an altered endosome or that the confocal plan is not on the center of the endosome or that the endosome is too small and is included in the optical section.
36. If the HV needs to be set much above 700, the images will be noisy. To collect more signals, increase the laser power or use another numerical aperture objective if possible. Noise can also be decreased by lowering scan speed using kalman filtering (averaging). Confocal aperture can also be open to collect more light but the z resolution will be decreased.

37. For maximum resolution, the z-step should be 2–3× smaller than the optical z-resolution.
38. In our study, a confocal aperture size of 110 μm, a 1024 × 1024 frame size and scan speed of 10 μs/pixel were selected to collect all the final images.
39. Usually, the confocal aperture size is set automatically by the confocal software to obtain the optimal size to reject out-of-focus light (best resolution) without losing too much signal. This size can be changed manually in the Image Acquisition Control panel.
40. Other software such as Image J can also be used for similar kind of image analysis.
41. Ideally, this step should be performed by the same person in a single blind way. Only mid-focal plane images of endosomes should be analyzed (as described in **step 5** of the Subheading 3.5). The line should be drawn to pass in the center of the endosomal lumen and cross two regions of the limiting membrane with similar fluorescence intensity.
42. In the example (Table 1), the highest intensity is 1940. Therefore, to calculate the normalized intensity, divide the raw intensity (column C) by 1940 and multiply by 100 to express as percentage.
43. In the example (Table 1), the first limiting endosomal membrane is at position #19 and the second limiting endosomal membrane is at the position #118. The diameter of the endosome is thus 99. Therefore, to calculate the normalized diameter (shown in column D), subtract 19 from the position # in column A and divide by 99 and then multiply by 100 to express as percentage.
44. Any other software able to do statistical analyses can be used.
45. A value of $P < 0.05$ is usually considered as significant. Indicate with stars (*) the level of significance. Herein, ***$P < 0.001$, ****$P \leq 0.0001$.

Acknowledgments

This work was supported by a grant from the Natural Sciences and Engineering Research Council of Canada to C.L.

References

1. Hanyaloglu AC, von Zastrow M (2008) Regulation of GPCRs by membrane trafficking and its potential implications. Annu Rev Pharmacol Toxicol 48:537–568
2. Piper RC, Katzmann DJ (2007) Biogenesis and function of multivesicular bodies. Annu Rev Cell Dev Biol 23:519–547. https://doi.org/10.1146/annurev.cellbio.23.090506.123319

3. Hasdemir B, Bunnett NW, Cottrell GS (2007) Hepatocyte growth factor-regulated tyrosine kinase substrate (HRS) mediates post-endocytic trafficking of protease-activated receptor 2 and calcitonin receptor-like receptor. J Biol Chem 282(40):29646–29657
4. Marchese A, Raiborg C, Santini F, Keen JH, Stenmark H, Benovic JL (2003) The E3 ubiquitin ligase AIP4 mediates ubiquitination and sorting of the G protein-coupled receptor CXCR4. Dev Cell 5(5):709–722
5. Henry AG, White IJ, Marsh M, von Zastrow M, Hislop JN (2011) The role of ubiquitination in lysosomal trafficking of delta-opioid receptors. Traffic 12(2):170–184. https://doi.org/10.1111/j.1600-0854.2010.01145.x
6. Hislop JN, Henry AG, von Zastrow M (2011) Ubiquitination in the first cytoplasmic loop of mu-opioid receptors reveals a hierarchical mechanism of lysosomal down-regulation. J Biol Chem 286(46):40193–40204. https://doi.org/10.1074/jbc.M111.288555
7. Eichel K, von Zastrow M (2018) Subcellular organization of GPCR signaling. Trends Pharmacol Sci 39(2):200–208. https://doi.org/10.1016/j.tips.2017.11.009
8. Irannejad R, Tomshine JC, Tomshine JR, Chevalier M, Mahoney JP, Steyaert J, Rasmussen SG, Sunahara RK, El-Samad H, Huang B, von Zastrow M (2013) Conformational biosensors reveal GPCR signalling from endosomes. Nature 495(7442):534–538. https://doi.org/10.1038/nature12000
9. Shenoy SK, Lefkowitz RJ (2011) Beta-Arrestin-mediated receptor trafficking and signal transduction. Trends Pharmacol Sci 32 (9):521–533. https://doi.org/10.1016/j.tips.2011.05.002
10. Irannejad R, von Zastrow M (2014) GPCR signaling along the endocytic pathway. Curr Opin Cell Biol 27:109–116. https://doi.org/10.1016/j.ceb.2013.10.003
11. Tsvetanova NG, Irannejad R, von Zastrow M (2015) G Protein-coupled receptor (GPCR) signaling via heterotrimeric G proteins from endosomes. J Biol Chem 290 (11):6689–6696. https://doi.org/10.1074/jbc.R114.617951
12. Booden MA, Eckert LB, Der CJ, Trejo J (2004) Persistent signaling by dysregulated thrombin receptor trafficking promotes breast carcinoma cell invasion. Mol Cell Biol 24 (5):1990–1999
13. Li YM, Pan Y, Wei Y, Cheng X, Zhou BP, Tan M, Zhou X, Xia W, Hortobagyi GN, Yu D, Hung MC (2004) Upregulation of CXCR4 is essential for HER2-mediated tumor metastasis. Cancer Cell 6(5):459–469. https://doi.org/10.1016/j.ccr.2004.09.027
14. Rosciglione S, Theriault C, Boily MO, Paquette M, Lavoie C (2014) Galphas regulates the post-endocytic sorting of G protein-coupled receptors. Nat Commun 5:4556. https://doi.org/10.1038/ncomms5556
15. Zheng B, Lavoie C, Tang TD, Ma P, Meerloo T, Beas A, Farquhar MG (2004) Regulation of epidermal growth factor receptor degradation by heterotrimeric Galphas protein. Mol Biol Cell 15(12):5538–5550
16. Génier S, Degrandmaison J, Lavoie C, Gendron L, Parent J-L (2018) Monitoring the aggregation of GPCRs by fluorescence microscopy. Methods Mol Biol. this issue
17. Li X, Letourneau D, Holleran B, Leduc R, Lavigne P, Lavoie C (2017) Galphas protein binds ubiquitin to regulate epidermal growth factor receptor endosomal sorting. Proc Natl Acad Sci U S A 114(51):13477–13482. https://doi.org/10.1073/pnas.1708215114

Chapter 18

Super-Resolution Imaging of G Protein-Coupled Receptors Using Ground State Depletion Microscopy

Fabiana A. Caetano Crowley, Bryan Heit, and Stephen S. G. Ferguson

Abstract

G protein-coupled receptors (GPCRs) comprise the largest family of integral membrane proteins, which are coupled to heterotrimeric G proteins to influence cell signaling. Subsequent to G protein activation, agonist-stimulated G protein-coupled receptor kinase (GRK) phosphorylation results in the recruitment of β-arrestin proteins, which form both stable and unstable complexes with GPCRs. β-Arrestins when bound to GPCRs not only contribute to the uncoupling of G protein signaling but also to the redistribution of GPCRs to clathrin-coated pits via their association with both clathrin and β2-adaptin facilitating GPCR endocytosis. This allows β-arrestins to couple GPCRs to additional cell signaling proteins allowing a second wave of receptor signaling. Importantly, the β-arrestin-regulated subcellular localization of these complexes also plays a critical role in regulating how these signals are transduced and which proteins are recruited. Here, we describe a methodology for assessing the GPCR subcellular localization by super-resolution microscopy and suggest that this methodology can be extended to the study of GPCR/protein complexes.

Key words β-Arrestin, RalGDS, Green fluorescent protein, G protein-coupled receptor, Ground state depletion microscopy, Translocation, Endocytosis, Agonist

1 Introduction

G protein-coupled receptors (GPCRs) are guanine nucleotide exchange factors (GEFs) for heterotrimeric G proteins that stimulate the activation of a diverse array of signal transduction pathways [1]. Activation of a GPCR not only results in the exchange of GDP for GTP on the α-subunit of heterotrimeric G proteins to allow for the activation of a variety of G protein-regulated effectors of intracellular signaling, but it promotes the attenuation of GPCR signaling as the consequence of G protein-coupled receptor kinase (GRK)-mediated phosphorylation and β-arrestin protein binding to the GPCR [1]. β-Arrestin when bound to a GPCR promotes the clathrin-mediated endocytosis of the GPCR due to its ability to recruit both clathrin and β2-adaptin subunit of the AP2 clathrin

Mario Tiberi (ed.), *G Protein-Coupled Receptor Signaling: Methods and Protocols*, Methods in Molecular Biology, vol. 1947, https://doi.org/10.1007/978-1-4939-9121-1_18,

adaptor complex [2, 3]. In addition to function of β-arrestins as proteins that contribute to both GPCR desensitization and endocytosis, β-arrestins are now recognized as proteins linking GPCRs to signal transduction pathways that are activated following the desensitization of G protein signaling. The cell biological investigation of GPCR-dependent regulated alterations in β-arrestin subcellular localization in response to agonist activation by microscopy has represented an intense area of scientific investigation over the last 25 years [4]. Multiple studies have provided evidence using confocal microscopy to suggest that GPCRs colocalize both at the cell surface and in endosomes with β-arrestin [5–8]. However, these studies only infer a potential interaction in these subcellular compartments by low-resolution confocal microscopic imaging. Several fluorescence-based assays have been developed to confirm the inference that GPCRs and β-arrestins can colocalize in different subcellular compartments [9, 10]. However, until recently, the methodology to confirm these interactions using visual imaging techniques has not been possible. We have recently reported a technique to evaluate the subcellular localization of GPCRs by super-resolution confocal microscopy using ground state depletion microscopy [11]. The methodology for this investigation is reported here and is applicable to the investigation of GPCR/protein interactions.

2 Materials

2.1 Coverslip Preparation

1. #1.5 thickness, 18 mM diameter circular glass coverslips.
2. 0.22 μm syringe filter.
3. 30 or 60 mL syringe.
4. 70% and 100% ethanol.
5. Deionized water.
6. Concentrated hydrochloric acid.
7. Hot plate.
8. Thermometer.
9. 250 mL glass beaker.
10. Kimwipes or other lint-free absorbent paper.

2.2 Cell Culture and Transfection

1. 100 mM circular plastic dishes.
2. 6-well cell culture plates.
3. Human embryonic kidney (HEK 293) cells from ATCC®.
4. Eagle's minimal essential medium: MEM (Gibco®) supplemented with 10% Fetal bovine serum (FBS) in absence of antibiotic.
5. Trypsin-EDTA (0.25%), phenol red (Gibco®).

6. Sterilized water.
7. 2.5 M $CaCl_2$.
8. 2× HEPES buffered saline (HBS): 0.05 M HEPES, 0.28 M NaCl and 1.5 mM Na_3PO_4, pH 7.1 (this pH is essential for transfection efficiency). Sterilize and store aliquots at −20 °C.
9. Phosphate-buffered saline (PBS): 2.7 mM KCl, 1.5 mM KH_2PO_4, 138 mM NaCl, 8 mM Na_2HPO_4. Add 500 mL of distilled water to a glass beaker. Weigh 0.2 g KCl, 0.24 g KH_2PO_4, 8 g NaCl, 1.15 g Na_2HPO_4 and transfer to the beaker. Add water to a volume of 900 mL. Mix and adjust pH to 7.4 make up to 1 L with water. Sterilize by filtration using a disposable vacuum filter system with a 0.45 μm membrane. Store at 4 °C until use.
10. DNA of interest.
11. Acid-washed coverslips.
12. Epitope-tagged or GFP-conjugated plasmid cDNAs of choice.

2.3 Cell Preparation and Staining

1. PBS.
2. Hank's balanced salt solution (HBSS): 5.4 mM KCl, 0.3 mM Na_2HPO_4, 0.4 mM KH_2PO_4, 4.2 mM $NaHCO_3$, 1.3 mM $CaCl_2$, 0.5 mM $MgCl_2$, 0.6 mM $MgSO_4$, 138 mM NaCl, 5.6 mM D-glucose. Add 500 mL of water to a glass beaker. Weigh 0.4 g KCl, 0.042 g Na_2HPO_4, 0.055 g KH_2PO_4, 0.35 g $NaHCO_3$, 0.191 g $CaCl_2$, 0.048 g $MgCl_2$, 0.072 g $MgSO_4$, 8 g NaCl, 1 g D-glucose and transfer to the beaker. Add water to a volume of 900 mL. Mix and adjust pH to 7.4. Make up to 1 L with water (*see* **Note 1**). Sterilize by filtration using a disposable vacuum filter system with a 0.45 μm membrane.
3. Fixing solution: 4% paraformaldehyde (PFA) in PBS (*see* **Note 2**).
4. Permeabilizing solution: 0.2% Triton X-100 in PBS (*see* **Note 3**).
5. Blocking solutions: 5% bovine serum albumin (BSA) or 5% serum in PBS. Antibodies should be tested in both blocking solutions to determine the most appropriate one for your conditions.
6. Diluent solution: 0.1% BSA and 0.1% Triton X-100 in PBS (*see* **Note 4**).
7. Primary antibody of interest.
8. Secondary antibody of interest labeled with appropriate fluorophore.

2.4 Sample Mounting

1. Embedding media: 100 mM cysteamine in PBS. Add 40 mL of 1× PBS to a glass beaker. For 100 mM concentration, weigh 0.39 g cysteamine and transfer to the beaker. Mix using a stirring bar and adjust pH to 7.4. Make up to 50 mL with

PBS. Filter the solution using a 0.45 μm syringe filter. Make 1 mL aliquots and store at −80 °C (*see* **Note 5**).

2. Water-compatible rapid-curing silicone sealant (Twinsil®, Picodent, Wipperfürth, Germany).
3. 70% ethanol.
4. Kimwipes or other lint-free absorbent paper.

2.5 Ground State Depletion Microscopy (GSDM) Imaging

1. Objective oil.
2. Leica SR GSD microscope.

2.6 Fluorophore Characterization

1. Secondary antibody or recombinant protein conjugated to fluorophore of interest.
2. Cell line expressing fluorescent protein of interest.
3. PBS.
4. Triton X-100.

3 Methods

3.1 Acid Washing Coverslips

The detection of single molecules in GSDM imaging makes this form of microscopy highly sensitive to contamination by fluorescent particulates. As such, it is critical that all potentially fluorescent organic materials be removed from coverslips prior to culturing cells or otherwise mounting the target cell/tissue. Using a concentrated acid wash, at high temperatures, will hydrolyze most organic materials, thus removing a significant portion of the background fluorescence from the slides (*see* **Note 6**).

1. Using syringes and 0.22 μm syringe filters, filter 100 mL of 70% ethanol into a clean beaker.
2. Individually place coverslips into the filtered ethanol solution, ensuring both sides of the coverslip are whetted.
3. Soak coverslips in ethanol for 20 min with occasional agitation.
4. During the ethanol soak, prepare 200 mL of a 2 M hydrochloric acid solution, using filtered deionized water to dilute the acid.
5. Decant the ethanol solution and rinse the coverslips three times with filtered deionized water.
6. Add the 2 M hydrochloric acid solution to the coverslips, and place on a hot plate. Heat to 60 °C and hold at this temperature overnight.
7. Decant the acid solution and rinse coverslips at least three times with filtered deionized water. Be sure to wash out the acid from between the coverslips.

8. Rinse once in 100% ethanol, and dry between Kimwipe sheets.
9. Place in a clean container and autoclave to sterilize.

3.2 Cell Culture and Transfection

GSDM imaging can be performed on any cultured immortalized cell line, cultured primary cells, and on thin tissue sections. As an example procedure, commonly used HEK293 cells can be used as follows:

1. Seed 3–4 × 10^6 cells in 100 mM dish 1 day prior to transfection (*see* **Note 7**).
2. Add growth media (MEM + 10% FBS) to a total of 8–10 mL final volume.
3. The day of transfection, dilute 1 μg of flag-tagged ß2-adrenergic receptor (β2AR-flag) expression plasmid or expression plasmid of interest in 450 μL of sterilized water in a sterile tube.
4. Add 50 μL of 2.5 M $CaCl_2$ to the diluted DNA solution.
5. Slowly drip 500 μL of 2× HBS over the DNA/$CaCl_2$ solution.
6. Mix this solution immediately by slowly dripping the solution into the tube two times.
7. Slowly drip this solution over the cells.
8. Incubate the cells for 18 h in the incubator.
9. Aspirate the media and wash the cells two to three times with sterile PBS buffer.
10. Add 10 mL of growth media and allow cells to recover from transfection for 24 h.
11. The next day, rinse the plate with 2 mL of trypsin-EDTA.
12. Quickly aspirate the trypsin-EDTA and add fresh 2 mL of this solution.
13. Gently tap the dish until the cell start to detach.
14. Add 8 mL of growth media and gently homogenize the cell suspension.
15. Add 8 × 10^5 cells in six-well plates containing pre-washed coverslips (*see* **Note 8**).
16. Add growth media to each well to make up for a 2 mL final volume.
17. Allow cells to recover for 24 h before proceeding with the staining protocol.

3.3 Cell Preparation and Staining

Samples must be plated on #1.5 high quality glass coverslips. Square or round coverslips can be used as long as they are at least 22 mM in width (to cover the cavity of the depression slide).

Acid-washed coverslips need to be used for all samples (*see* Subheading 3.1). Alternatively, pre-cleaned coverslips from a

manufacturer can be used. It is important to ensure the cells present normal morphology and they are not contaminated before proceeding with the staining protocol. This can be done through visualization of cultures using a bench microscope. Carry out all procedures at room temperature unless otherwise specified. If samples were transfected with DNA for fluorescent proteins, cover the samples with foil throughout the procedure. If not, cover the samples with foil during and after incubation with secondary antibody.

1. Rinse samples with HBSS buffer. For six-well plates, use 2 mL per well.
2. Carry out any required treatment in the HBSS buffer. After treatment, rinse samples once with 2 mL of PBS buffer.
3. Aspirate PBS buffer and add 1.5 mL of fixing solution. Incubate samples for 15 min.
4. Rinse samples 3× with PBS buffer.
5. Wash samples 3× with PBS buffer, 5 min each.
6. Permeabilize the samples using 0.2% Triton X-100 for 10 min.
7. Rinse the samples three times with PBS buffer.
8. Incubate the samples for 1 h with the most appropriate blocking solution (*see* **Note 9**).
9. Following the blocking, add appropriate primary antibody to the samples (diluted in diluent solution) and incubate for either 1 h, 4 h or overnight at 4 °C (*see* **Note 10**).
10. Wash as in **steps 4** and **5**.
11. Add appropriate secondary antibody at appropriate dilution (1:1000–10,000 diluted in diluent solution) and incubate for 30 min (*see* **Note 11**).
12. Wash samples 5× with PBS buffer, 5 min each.
13. Leave sample on PBS buffer until imaging.

3.4 Sample Mounting

It is recommended to mount coverslips on glass depression slides filled with an embedding media containing an oxygen scavenger such as glucose-oxidase, polyvinyl alcohol and cysteamine. To avoid unnecessary oxidation of the embedding media, samples should be mounted just prior to imaging.

1. Add 50–75 μL of embedding media into the well of a depression slide (*see* **Note 12**).
2. Carefully place the stained coverslip in the depression slide with cells facing the cavity.
3. Centre the coverslip in the depression slide using tweezers. The cavity should be well covered by the coverslip to avoid leakage of embedding media.

4. Remove air bubbles (if any) by gently moving the coverslip around the cavity using tweezers (*see* **Note 13**).
5. Remove excess of embedding media using tissue paper.
6. Seal the coverslip using a water-compatible rapid-curing silicone sealant (*see* **Note 14**).
7. Gently clean mounted coverslip with 70% ethanol or an objective cleanser.
8. Image the samples within 5 h (*see* **Note 15**).

3.5 GSDM Imaging

It is recommended to turn the system on at least 2–3 h before imaging to ensure stable temperature of all components, reducing noise. The proper imaging configuration for each fluorophore depends on the equipment used. However, some key components are (1) a high-intensity excitation light source, (2) high numerical aperture lens (>1.4 NA) and a total magnification of 100× to 200×, (3) high sensitivity electron multiplied charged-coupled device (EM-CCD) camera, and (4) anti-vibration table. Regardless of the equipment used, GSDM imaging consists of a depletion phase followed by photon collection/imaging phase. The following are general steps but must be adapted to the system used.

1. Add a drop of high quality immersion oil to objective (*see* **Note 16**).
2. Place mounted coverslip on the microscope stage. Ensure the slide is secure on stage to avoid the effects of drift on the image.
3. Slowly bring objective closer to the coverglass.
4. Select the appropriate filter set and switch to microscope view mode (*see* **Note 17**).
5. Move the stage or objective while focusing.
6. Move the stage in X, Y to define the area to be imaged.
7. Switch to computer view.
8. Select the appropriate filter set for each fluorophore.
9. Adjust the acquisition parameters according to the fluorescence intensity on the sample.
10. Acquire a preview image for all the fluorophores (*see* **Note 18**).
11. Switch to GSDM mode.
12. Adjust the acquisition parameters for the GSDM image (*see* **Note 19**).
13. Start the depletion phase by illuminating the sample with a high-intensity light source. The wavelength used should be the closest to the maximum excitation peak for each particular fluorophore. The intensity of the light source will depend on the concentration and stability of fluorophores in the sample.

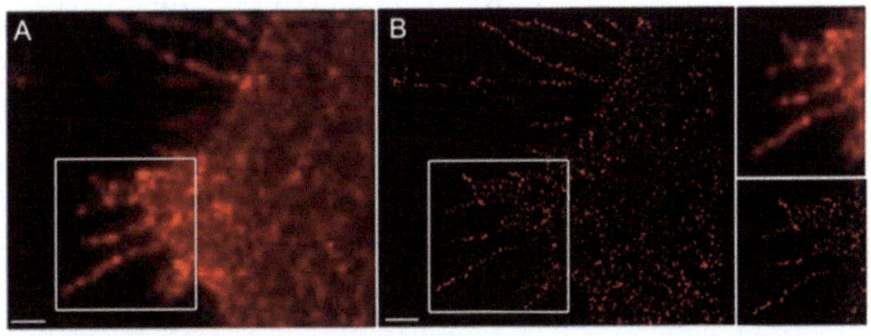

Fig. 1 Comparison of the cellular distribution of β_2-adrenergic receptor under epifluorescence versus super-resolution microscopy. HEK 293 cells were transfected with β2AR-Flag and labeled with anti-Flag primary antibody and Alexa Fluor 647-conjugated secondary antibody. A GSD microscope was used to acquire the preview image in epifluorescence mode (**a**) and the Super-resolution image (**b**). The right panels present a magnified view of the regions indicated in the images. The image is representative of at least 20 cells from three different experiments. Scale bar = 2 μm

Typically, 50–80% power is used to push the majority of the fluorophores into the triplet and dark states (*see* **Note 20**).

14. Switch to acquisition mode (*see* **Note 21**). Adjust the acquisition parameters to optimize the return and fluorescence detection of fluorophores.
15. Acquire 10,000 to 50,000 frames (Fig. 1).

3.6 Fluorophore Characterization

Selection of appropriate fluorophores is critical for successful GSDM imaging. While many fluorophores have been identified in the literature as appropriate for GSDM imaging, it may be necessary to test unreported fluorophores, or to characterize individual fluorophores on your GSD microscope system. The suitability of a fluorophore for GSDM imaging is largely determined by three characteristics: (1) fluorophore brightness, which determines the precision of the image, (2) fluorophore recovery following depletion, which determines the portion of labeled proteins detected in the GSDM image, and (3) number of times a fluorophore will "blink" before bleaching, which impacts analyses affected by oversampling. This characterization is performed by coating coverslips with the fluorophore at high dilution, such that individually resolvable fluorophores are deposited on the coverslip. GSDM acquisition of these fluorophores is then used to determine their characteristics.

3.6.1 Preparation of Chemical Fluorophores for Characterization

1. Using PBS, dilute an antibody or protein labeled with the fluorophore of interest to a concentration where, when coated on a coverslip, produces a dispersed field of fluorophores, with each fluorophore separated from its neighbors by ~300 nm. For most secondary antibodies, this is a dilution of 1:10,000 to 1:100,000 (10 ng/mL to 100 ng/mL).

2. Absorb the labeled protein onto a coverslip by placing a drop of the diluted protein (~100 μL) onto the center of an acid cleaned coverslip, and incubate at room temperature for 30 min.
3. Gently rinse the coverslip three times with PBS and mount for GSDM imaging as per Subheading 3.4.

3.6.2 Preparation of a Fluorescent Protein for Characterization

1. Into a 1.5 mL microcentrifuge tube, collect ~1×10^6 mammalian cells expressing the fluorescent protein of interest (*see* **Note 22**).
2. Pellet the cells using a 500 × *g*, 5 min centrifugation and resuspend in 250 μL PBS + 0.1% Triton X-100.
3. Vortex the cells for 30 s at maximum speed.
4. Clarify the lysate by centrifuging at >10,000 × *g* for 5 min. Transfer the supernatant into a clean microcentrifuge tube.
5. Absorb the fluorescent protein onto a coverslip by placing a drop of the cell lysate (~100 μL) onto the center of an acid cleaned coverslip, and incubate at room temperature for 30 min.
6. Gently rinse the coverslip three times with PBS and mount for GSDM imaging as per Subheading 3.4.
7. Reserve the remaining cell lysate, as a dilution may be necessary to dilute the sample. All dilutions should be made using PBS + 0.1% Triton X-100.

3.6.3 Acquiring Fluorophore Images

1. Focus on the dispersed fluorophores and adjust the microscope acquisition settings to produce clear images of the fluorophores.
2. Move to a new field of view and acquire a single image. This image will be used to identify all fluorophores present prior to GSDM acquisition. Using image analysis software or manual quantification, count the number of discrete fluorophores in the image.
3. Deplete and acquire a GSDM image of the same region as per Subheading 3.5. Count the number of discrete fluorophores in the image as above, and export both the GSDM image acquisition video and raw detection data (e.g. X/Y/Z coordinates, brightness, precision, etc.) from your GSDM software.
4. Repeat **steps 2** and **3** for an additional 5 to 10 fields.

3.6.4 Fluorophore Analyses

1. Fluorophore recovery is defined as the ratio between the number of discrete fluorophores detected in the GSDM acquisition and the number of discrete fluorophores present in the pre-GSDM image (Fig. 2). Fluorophores with higher recovery (>0.8) will provide the most representative images. Recoveries

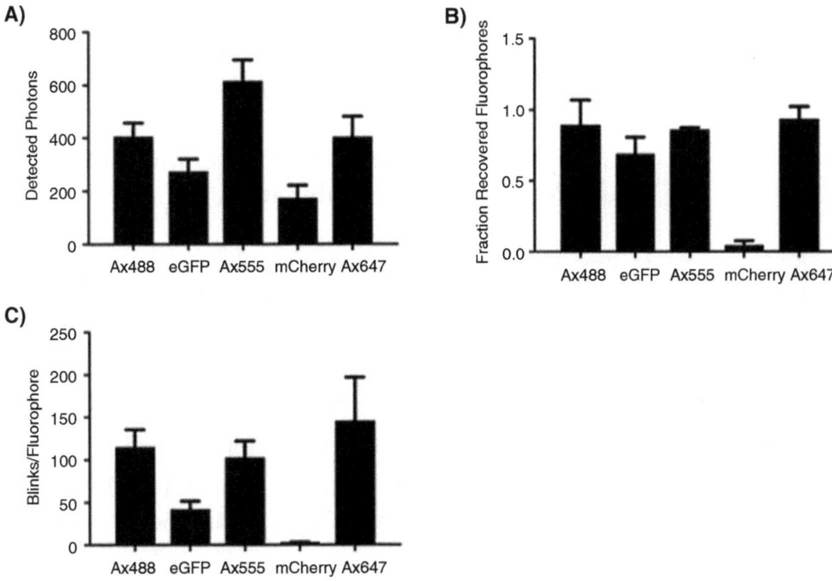

Fig. 2 Fluorophore GSDM acquisition characteristics. Monodispersed fields of the indicated fluorophores were imaged using a GSD microscope. (**a**) Average brightness of each fluorophore, measured as photons collected/ 10 ms image frame at maximum laser intensity, prior to fluorophore depletion. (**b**) The fraction of fluorophores present prior to depletion which were subsequently detected in the GSDM image collected post-depletion. (**c**) Number of times each fluorophore was repeatedly detected as a discrete "blink" during GSDM image acquisition. All fluorophores but mCherry display characteristics amenable to GSDM imaging. Data are presented as mean ± SEM, $n = 6$ independent experiments with 10 images/condition collected in each experiment

below this level may not provide an accurate representation of the labeled sample, especially if the labeled target is not abundant or is diffusely localized within the cell.

2. Most GSDM software will provide a measure of fluorophore brightness in their raw detection data—e.g. Leica GSD software reports the number of photons detected from each fluorophore and the resulting precision of detection (Fig. 2), whereas quickPALM provides pixel intensity. These data can be used to determine the brightness of a fluorophore. To ensure that comparisons between fluorophores are equivalent, images must be acquired of different fluorophores using identical acquisition parameters—e.g. identical laser power, camera/detector gain, and exposure time settings. Brighter fluorophores produce images with higher precision.
3. Blinking frequency and blinking rate can be determined by manually counting the number of blinks performed by individual fluorophores throughout an acquisition video. At least 100 fluorophores should be quantified to ensure an accurate measure of fluorophore blinking dynamics. Fluorophores which undergo significant blinking tend to produce better

images and have better fluorophore recovery. However, the repeat detection of the same fluorophore can result in over-sampling of the image, which can complicate some forms of analysis.

4 Notes

1. It is recommended that this solution be prepared fresh each time. However, it can be stored at 4 °C for up to a week. In this case, add the D-glucose reagent right before use.
2. This solution should be prepared fresh each time. Add 70 mL of PBS buffer to a glass beaker. Weigh 4 g of paraformaldehyde and transfer to the beaker. Mix solution using a stirring bar and heat it to 70 °C. Slowly add a few drops of 5 N NaOH until paraformaldehyde is dissolved. Let the solution cool. Adjust pH to 7.4 using varied concentrations of HCl. Add PBS to a volume of 100 mL in a graduated cylinder. Sterilize this solution using a 0.45 μm syringe filter. Alternatively, a 16% paraformaldehyde solution can be purchased from Electron Microscopy Sciences and diluted to a 4% working solution.
3. Allow enough mixing time to ensure the detergent Triton X-100 is properly dissolved. Use a stirring bar for mixing.
4. If staining intracellular proteins, or if intracellular fluorescent proteins have been expressed, samples must be permeabilized during the blocking step through the addition of 0.1% Triton X-100 to the blocking solution. Do not use Triton X-100 if you are not permeabilizing the samples.
5. When adjusting the pH of the embedding solution, be careful not to overshoot it. If the pH value is passed the desired point of7.4, discard the solution and start over. The concentration of cysteamine impacts the blinking dynamics of the fluorophores. This reagent quenches molecular oxygen, aiding fluorophores to transition to the dark states. In this protocol we recommend using a final concentration of 100 mM. However, the titration of cysteamine is recommended to optimize the concentration to the dyes used.
6. It is easy to contaminate cleaned slides with dust and other organic materials that may re-introduce unwanted background fluorescence. All cleaning solutions should be filtered through 0.22 μm syringe filters prior to use, and once cleaned, coverslips should be handled in a biosafety cabinet to prevent particulate contamination.
7. The ability of cells to uptake DNA is related to the surface area of the cells exposed to the medium. Thus, at the time of transfection, aim for 70–80% cell confluency.

8. From our experience, pre-coating coverslips with poly-L-lysine produced high background signal, particularly when imaging in Total Internal Reflection Fluorescence (TIRF) mode. For immortalized cell lines such as HEK293, no coating was used. However, for primary cell cultures, the glass coverslips were coated with this polymer to facilitate cell adhesion.
9. From our experience, 5% BSA blocking solution is ideal for most secondary antibodies. However, some may perform better in 5% serum host-specific solution. Background noise tends to be higher when using serum as the blocking reagent.
10. The dilution factor depends on the type of antibody. For epitope tag antibodies such as anti-flag and anti-HA, we recommend a 1:10,000 dilution and 1 h incubation time. For protein-specific primary antibodies, we recommend a 1:500 dilution and incubation time of 4 h as a starting point. For protein with low expression, we recommend a 1:300 dilution and overnight incubation at 4 °C as a starting point.
11. Several fluorophores are suitable for GSD imaging. We have been successful using Alexa Fluor 488, 532, 647, yellow and green fluorescent proteins. Cy3 and mCherry fluorophores did not work well in our experience. An excitation light in the 405 nm wavelength is commonly used for back pumping fluorophores from the dark to the ground state. Thus, it is important not to use dyes in this range (including nuclei staining such as DAPI) to avoid nonspecific signal bleeding through the channels.
12. Frozen work aliquots of cysteamine must be thawed before mounting the samples. Once thawed, replace the aliquot after 5 h to avoid reduced efficiency of the cysteamine due to its oxidation.
13. In case of major air bubbles, carefully remove the coverslip, clean the depression slide, add fresh embedding media, and mount the coverslip again.
14. One example of sealant that can be used is Twinsil®. Mix the yellow and blue components in a ratio of 1:1 and apply to the edges of the coverslip using a disposable pipetting tip. Allow 5–10 min for the glue to harden.
15. If necessary, the sealant can be removed and coverslip can be stored or mounted again using fresh embedding media. Before storing, rinse samples three times with abundant PBS buffer.
16. Ensure objective is clean before imaging to avoid optical aberrations. Objectives can be cleaned using isopropanol or a manufactured cleanser.
17. For highly fluorescent samples, fluorescent light can be used to focus. Please note that far-red dyes are hardly visible through

the eyepieces. Alternatively, transmitted light can be used to focus the sample.

18. A preview image can be acquired using the TIRF or Epifluorescence mode. The TIRF mode presents the advantage of minimizing the fluorescence background from labeled structures located deeper in the cells. However, this mode is prone to more background signal coming from the coverslips. This can be minimized by using high quality, clean clover glass, by decreasing the incubation times with primary and secondary antibody and by increasing the number of washes after antibody incubation. An alignment of the system is necessary to image in TIRF mode. This can be done following the manufacturer's instructions or by following the built-in TIRF Auto Alignment wizard. The Epifluorescence mode does not require the auto alignment step. For thin samples, both modes will offer comparable results. The preview image is to be used for comparison with the GSDM image. Thus, it is important to use the same mode to acquire both images.
19. We recommend to use the maximum value for the EM gain and intensity parameters but use a low exposure time, as a starting point. A minimum detection threshold of 10 photons/pixel is recommended. Lower detection threshold values favor the collection of background signal. Most systems allow for changes on this threshold while acquiring the image.
20. When imaging dual or triple labeled samples, start by depleting and imaging the longer wavelength fluorophores. The time needed to bring the majority of the fluorophores into the triplet state varies according to the chemical properties of each fluorophores. This time can vary from a few seconds to minutes and can be optimized by adjusting the intensity of the light source and the dilution ratio of the primary and secondary antibody. Fluorophores that are not sent to the dark state are not well resolved in the detection phase, tending to form large size clusters in the image. From our experience, Cy3 dyes possess great blinking properties but are the most resistant to switching states.
21. Depending on the system used, the image acquisition settings must be defined prior to the depletion phase. Thus, it is important to select the illumination mode (TIRF or Epifluorescence), the direction of the evanescent wave if using TIRF mode, the proper excitation and emission filters, and the intensity of the light source for excitation. To avoid having too many fluorophores returning simultaneously to the ground state is recommended to start the image acquisition using a low light power. Adjust the intensity of the light source such as that about 400 photons per fluorophore are captured.

The backpumping light source accelerates the returning time of fluorophores from the dark to the ground state. If used, the intensity of this source should also be adjusted to achieve the 400 photons per fluorophore ratio. Increasing the intensity of the excitation light source can result in photobleaching of dyes and loss of signal and resolution. A balance between the power of the backpumping and the excitation light sources can reduce this problem.

22. Soluble (cytosolic) proteins are preferred over membrane-bound proteins as they are less likely to aggregate. Bacterial cells can also be used to express fluorescent proteins. However, cell numbers and lysis protocols will have to be optimized for the bacterial cells.

References

1. Ferguson SSG (2001) Evolving concepts in G protein-coupled receptor endocytosis: the role in receptor desensitization and signaling. Pharmacol Rev 53:1–24
2. Laporte SA, Oakley RH, Zhang J, Ferguson SSG, Caron MG, Barak LS (1999) Recruitment of adaptin AP-2 to β2-adrenergic receptor/βarrestin complexes during endocytosis. Proc Natl Acad Sci U SA 96:3712–3717
3. Goodman OB Jr, Krupnick JG, Santini F, Gurevich VV, Penn RB, Gagnon AW, Keen JH, Benovic JL (1996) β-Arrestin acts as a clathrin adaptor in endocytosis of the β_2-adrenergic receptor. Nature 383:447–450
4. Peterson YK, Luttrell LM (2017) The diverse roles of Arrestin scaffolds in G protein-coupled receptor signaling. Pharmacol Rev 69:256–297
5. Oakley RH, Laporte SA, Holt JA, Barak LS, Caron MG (1999) Association of β-arrestin with G protein-coupled receptors during clathrin-mediated endocytosis dictates the profile of receptor resensitization. J Biol Chem 274:32248–32257
6. Anborgh PH, Seachrist J, Dale L, Ferguson SSG (2000) Receptor/β-arrestin complex formation and the differential trafficking and resensitization of β_2-adrenergic and angiotensin II type 1A receptors. Mol Endocrinol 14:2040–2053
7. Seachrist JL, Laporte SA, Dale LB, Babwah AV, Caron MG, Anborgh PH, Ferguson SSG (2002) Rab5 association with the angiotensin II type 1A receptor promotes Rab5 GTP-binding and vesicular fusion. J Biol Chem 277:679–685
8. Rajagopal S, Shenoy SK (2018) GPCR desensitization: acute and prolonged phases. Cell Signal 41:9–16
9. Barak LS, Ferguson SSG, Zhang J, Caron MG (1997) A β-arrestin/green fluorescent protein biosensor for identifying G protein-coupled receptor activation. J Biol Chem 272:27497–27500
10. Marullo S, Bouvier M (2007) Resonance energy transfer approaches in molecular pharmacology and beyond. Trends Pharmacol Sci 28:362–365
11. Caetano FA, Cavanagh PC, Tam JHK, Dirk BS, Ferguson SSG, Pasternak SH, Dikeakos JD, DeBruyn J, Heit B (2016) Quantitative analysis of protein interactions, dynamics and formation of multi-protein structures by super-resolution imaging. PLoS Comp Biol 11(12): e1004634

Chapter 19

Analysis of Spatial Assembly of GPCRs Using Photoactivatable Dyes and Localization Microscopy

Kim C. Jonas and Aylin C. Hanyaloglu

Abstract

Super-resolution imaging has provided unprecedented insight in the molecular complexities of fundamental cell biological questions. For G protein-coupled receptors (GPCRs), its application to the study of receptor homomers and heteromers have unveiled the diversity of complexes these GPCRs can form at the plasma membrane at a structural and functional level. Here, we describe our methodological approach of photoactivated localization microscopy with photoactivatable dyes (PD-PALM) to visualize and quantify the spatial assembly of GPCR heteromers at the plasma membrane.

Key words G protein-coupled receptor, Dimer, Oligomer, Heteromer, Homomer, PD-PALM, Super-resolution microscopy

1 Introduction

1.1 G Protein-Coupled Receptor Di/Oligomerization

Developments in imaging technologies have contributed to our current understanding of the molecular regulation of G protein-coupled receptor (GPCR) activity. This is particularly pertinent for the study of receptor complexes, from GPCR engagement with its cognate G protein and key adaptor proteins such as arrestin, to GPCR-GPCR associations either as homomers or heteromeric complexes. Thus, techniques such as cryo-EM, single molecule tracking, fluorescence correlation spectroscopy, and super-resolution imaging have unveiled interaction interfaces, stoichiometry, and single-molecule kinetics of these associations, particularly in the study of GPCR homo/heteromers [1–6].

That GPCRs not only exist as monomers at the plasma membrane, but also as dimers and higher order oligomers has been reported for more than 20 years. Initially these studies employed classical biochemical approaches and then rapidly evolved with the advent of resonance energy transfer methods, particularly bioluminescence resonance energy transfer (BRET); enabling real time receptor–receptor interactions to be measured in intact cells

Mario Tiberi (ed.), *G Protein-Coupled Receptor Signaling: Methods and Protocols*, Methods in Molecular Biology, vol. 1947, https://doi.org/10.1007/978-1-4939-9121-1_19, © Springer Science+Business Media, LLC, part of Springer Nature 2019

[7, 8]. However, the controversy in this area of GPCR biology has primarily stemmed from the cell systems used (heterologous cells overexpressing GPCRs) to the methodology themselves [9–17]. This fruitful debate has facilitated the development of standardized methods and criteria in the study of GPCR homomers and particularly heteromers [18]. There have been numerous recent reports demonstrating physiologically relevant GPCR homomeric and heteromeric interactions, the transient and flexible nature of these associations and equally as important, that the propensity of certain GPCRs to associate as dimers or oligomers are likely to be receptor-specific [19–21].

We and others have previously descr ibed distinct single-molecule imaging modalities used to study GPCRs, and the advantages and limitations of each of these methods. Thus, we refer the reader to the following previous reviews [6, 22]. Here, we will discuss our methodology of the super-resolution imaging technique of photoactivatable localization microscopy (PALM) using photoactivatable dyes (PD-PALM), within the context of applying this technology to study GPCR heteromers and the reorganization of these complexes following ligand activation.

1.2 PD-PALM Unveils Spatial Organization and Stoichiometry of GPCR Homo/Heteromers

The ability of PALM and PD-PALM to visualize single molecules with <10 nm resolution, compared to diffraction limited approaches where the maximal resolution is 200 nm, is based on the following properties of employing photoswitchable or photoactivatable fluorophores. These probes, which in the case of PD-PALM are photoactivatable dyes, remain in the dark state until they undergo an uncaging process following exposure to UV light. This enables these probes to then be excited by the appropriate fluorophore-defined wavelength but then are subsequently, and irreversibly, photobleached into the dark state. In order to resolve individual molecules at higher densities than diffraction-limited single-molecule approaches, these uncaging/activation/photobleaching steps are carried out in a stochastic manner and repeated until all fluorophores have been activated and bleached in to the dark state.

We have employed PD-PALM to study the organization of complexes between the gonadotrophin hormone Class A GPCRs, specifically luteinizing hormone receptor (LHR) di/oligomers and LHR heteromers with follicle-stimulating hormone receptor (FSHR) [23]. Our prior studies, and reviews, in this area have discussed the physiological importance of these interactions to reproduction [22–24]. In brief, we have shown that by using distinct LHR ligand binding-deficient and signal-deficient mutants, which alone are non-functional, but when co-expressed, can activate Gαs-adenylate cyclase-cAMP signaling (the primary G protein pathway for this receptor) via functional complementation. This activity via receptor di/oligomerization was able to rescue the

infertile phenotype in LHR knockout male mice [20]. Subsequently, PD-PALM was employed to identify the organizational cell surface landscape of the wild-type and mutant LHRs undergoing functional complementation, as it enabled us to use these mutants as tools to directly visualize the protomers responsible for binding ligand, and those that mediate signaling. Furthermore, PD-PALM enabled us to resolve single receptor protomers at expression levels equivalent to those previously reported in vivo, in contrast to single-molecule tracking approaches that require very low expression or labeling in order to image single molecules. Using this approach, we observed that multiple forms of these receptors were observed (monomer, dimer, and various oligomeric forms) and that 'heteromers' of the two mutant receptors were organized in functionally asymmetric oligomers that favored more active receptor complexes, i.e. trimers and tetramers with an excess of binding-deficient (signal-able) receptor over the signaling-deficient (ligand-binding) mutant receptor. An additional application of the PALM imaging studies was the integration with molecular modeling approaches to identify possible LHR–LHR interfaces. Surprisingly, this suggested that multiple interfaces for distinct di/oligomers were predicted, and is consistent with the concept that these receptors form diverse physical and functional complexes in our study and more recently in work with distinct Class A GPCRs [25–27].

Our more recently published studies employing PD-PALM to visualize heteromers of LHR and FSHR also provided further support that the complexes formed are highly diverse in arrangement and potentially that asymmetric complexes favor certain cellular signaling profiles [23]. Unlike our studies with LHR homomers, ligand activation of LHR induced a highly specific increase in the number of LHR/FSHR heterotetramers, and favoring an increase in the proportion of FSHR protomers within a single tetramer [23]. This demonstrates the intricate molecular detail that can be obtained with this method for protein interactions with this super-resolution method. The next sections will cover the materials, methods, and troubleshooting steps to consider when undertaking PD-PALM experiments.

2 Materials

2.1 Plasmids

N-terminally FLAG-tagged LHR and N-terminally HA-tagged FSHR, both in pcDNA3.1 plasmids, as previously described [23, 28].

2.2 Cell Lines and Transfection

HEK293 cells stably expressing FLAG-LHR, generated as previously described [23, 28], with transient transfection of HA-FSHR (*see* **Note 1**).

2.3 Antibody Labeling

1. CAGE 500 and CAGE 552 NH_3 esters (Abberior GMBH).
2. HA.11 mouse clone 16B12 (Biolegend), and FLAG rabbit (Sigma Aldrich). Buffers utilized during labeling of HA.11 and FLAG antibodies with CAGE dyes were purchased from Sigma Aldrich and Sephadex purification columns from GE Healthcare (UK). The labeling protocol (*see* **Note 2**) and required materials are as prescribed by Abberior, in Subheading 2.1 of their 'Labeling protocols' information section (Abberior GMBH, Germany, http://www.abberior.com/site/assets/files/1856/20120316-labeling_protocol.pdf).

2.4 Transient Transfections and Cell Culture

1. OptiMEM and Lipofectamine 2000, both from Thermo Fisher.
2. 1.5 ml microfuge tubes, 6-well cell TC plates, 8-well glass-bottom chamberslides (#1.5, LabTech).
3. DMEM, 10% fetal bovine serum, 100 iU/0.1 mg penicillin/streptomycin, D-PBS and Trypsin.
4. 4% paraformaldehyde with 0.2% glutaraldehyde (4 g in 100 ml D-PBS). Weigh paraformaldehyde using a mask to minimize exposure, and heat to 50–60 °C in a fume hood until dissolved. Leave to cool, check pH to ensure it is between 7.2 and 7.4, and add glutaraldehyde solution.
5. D-PBS with 1% sodium azide.

3 Methods

3.1 Transfection of Receptors for PD-PALM Studies

Our transfection protocols were as follows:

1. Plate HEK293 cells stably expressing LHR into 1 × 35-mm well to ensure a confluency of 80–90% the following day. We tend to plate using confluency rather than cell counts, but typically ~400,000 cells would achieve this.
2. The following day, check cells for confluency. If >90% confluent or <70%, it is not advisable to proceed with the transfection, as transfection efficiency will be compromised.
3. Using 2 separate 1.5 ml microfuge tubes, prepare HA-FSHR plasmid DNA in OptiMEM and lipofectamine 2000 in OptiMEM (*see* Table 1), and incubate at RT for 5 min.
4. Combine the two OptiMEM DNA and lipofectamine 2000 containing solutions, and gently mix. Incubate at room temperature for a further 5 min.
5. Carefully add the OptiMEM-Lipofectamine 2000-DNA mix to the cells in a dropwise motion, and incubate at 37 °C.

Table 1
Lipofectamine 2000-mediated transfection conditions employed for transient transfection of FSHR

Vol. of Lipofectamine 2000 (μl)	Amount of FSHR DNA (μg)	Vol. of OptiMEM (μl)
10		125
	2.5	125

6. The following day, harvest the cells from the well, and re-plate into an eight-well glass-bottomed chamber slide and incubate overnight at 37 °C (*see* **Note 3**).
7. Following overnight incubation, dilute CAGE 552-HA and CAGE-500-FLAG antibodies (we use 1:500) in 10% serum containing culture media (*see* **Note 4**).
8. From this point onwards, it is essential to switch the light off of the Class II cabinet and carry out the following steps in a darkened environment, to minimize any photoactivation of the CAGE dyes and preserve as much signal as possible.
9. Aspirate the media from cells and replace with CAGE-labeled antibody containing media and incubate at 37 °C for 30 min.
10. Use ~300 μl of antibody-media per well, but if the antibody were particularly precious, ~200 μl would be sufficient.
11. During this time, prepare working concentrations of ligands in cell culture media. For our studies, we use LH and FSH at a concentration of 10–100 nM.
12. At the end of the 30-min incubation, aspirate media, carefully wash cells using D-PBS (containing calcium) and stimulate with ligand for desired time. For our studies, we tend to utilize a combination of time points, from very acute (30 s), to longer time points of 2 and 5 min.
13. Following stimulation, aspirate the media, wash cells twice in D-PBS, and fix for 30 min in 4% paraformaldehyde/0.02% glutaraldehyde (*see* **Note 5**). Again, use ~300 μl of fixative, to ensure efficient fixation and a sufficient ratio of fixative to cell surface area.
14. Wash cells three times in D-PBS, and add 300 μl of D-PBS containing 1% sodium azide, to prevent growth of bacteria, etc. Store at 2–8 °C, in darkness (using a light-controlled box and foil) until imaged. We typically image within 24–72 h of labeling.

3.2 PD-PALM Image Acquisition

For our previous work, we have utilized a custom-made super resolution adapted total-internal reflection fluorescence (TIRF) microscope, the specification of which is previously described [22, 24]. For our more recent studies, we have moved to a

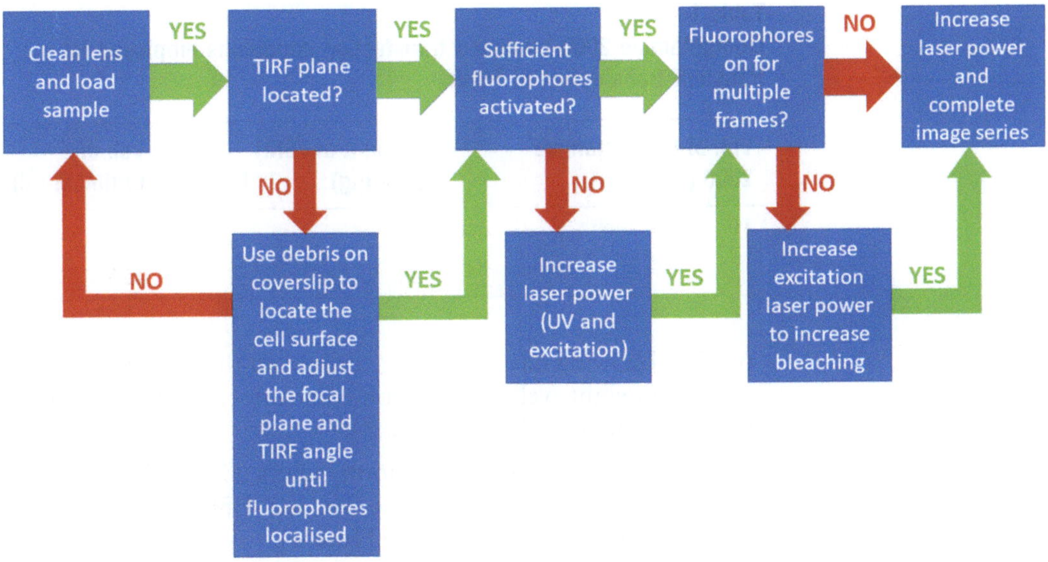

Fig. 1 Trouble shooting guide for PD-PALM imaging. Blue arrows depict the standard workflow of image acquisition for PD-PALM, detailing common issues experienced at each step of imaging (red arrows), and solutions to these problems (green arrows)

commercially available microscope, the Zeiss Elyra PS1 (Zeiss, Germany), a somewhat more user-friendly and automated microscopy platform. Both have their merits and downsides, however, for the purpose of this Chapter, we will describe our imaging methods using the commercially available Elyra microscope. The principles of acquiring and resolving single GPCR molecules are the same, whichever platform used. The Elyra microscope we used was custom designed to enable simultaneous dual-color imaging, via the provision of an additional EM-CCD camera (two cameras in total, one per fluorophore). Should only one camera be available, sequential imaging of differential fluorophores with each frame will be required. In terms of other technical set-up points, the microscope is surrounded by a temperature controlled, dark-box, for maximizing stabilization of the imaging environment and preventing light exposure (both to sample, and from exposure to high laser power to the user). The lasers that we utilized have the following power capabilities; 405 nm–50 mW, 488 nm–200 mW, and 561 nm–200 mW. The imaging of single GPCR molecules is somewhat difficult to describe a set protocol for. However, below is an approximate 'how-to' with tips and pointers that we have found useful through our imaging endeavors (*see* Fig. 1 for additional trouble shooting guide):

1. Open the Zen software and ensure two camera imaging is selected.

2. Turn on the UV (405 nm), 488 nm and 561 nm laser lines and allow to warm up and stabilize for 30 min.
3. Set up the following image parameters: TIRF mode, X100 objective, a time-lapse series (we typically use 10,000 frame imaging at 10 frames/s) with simultaneous imaging using 405 and detection within green and red excitation/emission spectra. The size of your imaging area is also something to consider—the larger this is, the larger your file size. We typically image a 256×256 pixel area.
4. Clean the X100 1.4NA objective using microscope tissue and 70% ethanol and add a drop of emersion oil to the lens. Be sure to allow the 70% ethanol to fully evaporate from the lens, and also to use emersion oil with the correct refractive index for D-PBS.
5. Place the chamber slide onto the microscope and using the locate function tab in the Zen software, scroll through the Z plane to locate the cells.
6. Once cells are located, using minimal UV and low power of 488 nm and 561 nm laser lines, use a combination of adjusting the focal plane, and TIRF angle until distinct activated fluorophores are detected.
7. At this point, switch on the definite focus function to minimize drift, and increase UV, 488 and 561 laser powers to achieve sufficient activation of spatially separate fluorophores. Having spatially separate fluorophores is imperative to being able to localize individual receptor molecules, so is very important. The selection of laser power is a delicate balance between achieving sufficient photoactivation and efficient photobleaching, and thus will vary dependent on fluorophores used, system set, and sensitivity of the camera, so it is difficult to be prescriptive with exact laser settings.
8. Whilst imaging is occurring, you can monitor the localization events within the analysis part of the Zen software. This acts as a good guide for quality of the imaging and for laser settings in terms of effectiveness of photobleaching.

3.3 Extracting xy Coordinates from Acquired PD-PALM Movies

For obtaining *xy* coordinates download the freely available, open source Fiji imaging software and QuickPALM plug-in [29]:

1. Open .zsi image file obtained from Zeiss Elyra in Fiji.
2. Adjust brightness of image to visualize your activated fluorophores and delete any frames from the beginning of the movie where initial adjustments were made.
3. Separate the movie into two channels; within the 'Image' tab, go to 'Color' and 'Split channels'. This will allow the two receptors to be separately analyzed.

4. Create 7 × 7 μM squares within the cell boundaries.
5. Create a separate sub-stack under Image tab and Duplicate function, making sure to select the box for sub-stack.
6. Select quickPALM function within Plugins tab.
7. Within the dropdown menu enter the pixel size, which is obtainable from each individual image on the Zeiss Zen software, the full width half maximum of the individual localized fluorophores, and the signal to noise ratio (*see* **Note 6**).
8. Run the analysis and save the generated .csv *xy* coordinates file. You can also save the image generated alongside this.
9. Repeat the process for the second channel and for other cells within the image (*see* **Note 7**).

3.4 Quantitation of GPCR Complexes Via PD-Interpreter

For quantitation of GPCR complexes we designed a custom Java-based software package designed to analyze localization microscopy super-resolution data called PD-Interpreter. PD-interpreter utilizes the localization coordinates generated (as described in Subheading 3.3), to analyze the number of monomers/dimers/oligomers for 2-channel acquired PALM data. The software analysis is based on a model proposed by Getis and Franklin, which uses a modified second order neighborhood analysis approach (*see* **Note 8**).

1. Download PD-interpreter from www.superimaging.org.
2. Open program and select the parameters tab.
3. Select the desired radius for analysis denoted as spatial scale (default =10 nm yet ranges to 1000 nm). For LHR and FSHR we have used 50 nm (*see* **Note 9**).
4. Select the file tab and load the first channel of the data set into 'load right data set', remembering which wavelength this channel data represents.
5. Select the file tab and load the second channel of the data set into 'load left data set'.
6. Select the 'analyze' tab and 'calculate clusters' command. The progress of the analysis can be monitored using the live updating bar at the bottom of the page. The analysis is complete when this bar states 'got left/right clusters'.
7. Once completed, select the 'analyze' tab and 'output cluster data'. Selecting this will save the folder containing all data (Excel spreadsheet with localization data and accompanying images) into the folder from which the original data sets were obtained. The PD-interpreter folder will be saved with the date and time of the analysis (*see* **Note 10**).
8. To continue analyzing more data sets, select 'file', 'reset'. Further localization data sets can then be loaded and analyzed. The radius set in **step 2** will remain at your chosen setting until the

program is closed. The Excel spreadsheets contain a summary page with total self-associating and co-associating molecules and from which the monomer population can be obtained. A breakdown of the number of each complex type (dimer, trimer, etc., and labeled as 2, 3, etc. in the Summary tab) is also provided. For heteromers, the number of protomers within each individual complex is also detailed within this summary page. Additional tabs contain the individual *xy* coordinates of each assigned self-associating and co-associating complexes, and non-associating molecules. Data are also outputted as several different individual image files, initially depicting all molecules in the localization data sets from the two channels. From the analysis, co-localization plots are generated using blue and yellow colors to differentiate the two channel populations, and a multi-color plot depicting complex sizes and representing clusters with two or more molecules in represented by different colors (both omitting non-associating molecules).

4 Notes

1. Before beginning our PD-PALM experiments, the amount of HA-FSHR plasmid transfected was firstly optimized to ensure we could detect FSHR expression at the cell surface, and for effects on FSH and LH-dependent cAMP production. Our transfection reagent of preference is Lipofectamine 2000 (Invitrogen). We find that this gives a good transfection efficiency—typically 60–70% of cells and has very low toxicity to the cells.
2. To quantitate individual GPCR molecules, it is essential to have a 1:1 labeling of CAGE dye:antibody. Both the FSHR and LHR were differentially N-terminally tagged with small epitopes, namely HA and FLAG, respectively, therefore we utilized HA and FLAG antibodies for our studies. The choice of CAGE dyes is an important factor to consider, when wanting to label and visualize two different receptor subtypes, as the dyes will need to be spectrally separate to minimize bleed through between channels. The CAGE dyes we utilized are CAGE 500 and CAGE 552, detected by 488 nm and 561 nm laser lines, respectively, with spectral properties that are sufficiently separate to ensure specificity of detection. The CAGE dyes are available in a number of different modified forms from the manufacturer, e.g., malamide or NH3 ester. For our studies, we utilized the NH3 ester label and differentially labeled the HA.11 and FLAG antibodies follow the step-by-step guide provided by the manufacturer's, thus we will not describe this in detail here. An important factor for achieving a 1:1 ratio of CAGE dye:antibody is the molar ratio of label:antibody used.

For this, we used a ratio of 1:10, which in our hands with our antibodies was sufficient to achieve a labeling density of 1.3 ± 0.1 for HA-CAGE 552, and 1.0 ± 0.2 for FLAG-CAGE 500. Labeling efficiency was determined, again following manufacturer's instructions, by employing a derivate of the beer–Lambert law.

3. We routinely use 60% of the harvested cells to re-plate into the chamber slides, to achieve a confluency of approximately 50–60%. This ensures cells are a distinct monolayer, and not too confluent to image. Incubate overnight at 37 °C.
4. When optimizing our labeling conditions, we trialed many blocking agents. A comprehensive discussion of this can be found elsewhere [22].
5. The addition of glutaraldehyde is important as it has been shown to minimize lateral diffusion of transmembrane proteins, that cross-linking reagents such as paraformaldehyde have been previously shown to fail to prevent [30, 31].
6. For a description of how to calculate the full width half maximum and signal to noise ratio values, please refer to our previous publications [22, 24].
7. Filter any fluorophores that have remained on for multiple frames due to inefficient bleaching via the Zen software (during image acquisition) or via freely available ImageJ plugin ThunderSTORM.
8. This model determines if two (or more) molecules are located within a defined search radius, if so, they are identified as interacting, ultimately assessing the association and localization patterns of molecules of interest [32].
9. We utilize a 50 nm radius based on the size of the large extracellular domain of the gonadotrophin receptors, the size of the antibody, and localization precision.
10. It is essential to save the output data after each run, or data will not be saved.

References

1. Liang YL, Khoshouei M, Radjainia M, Zhang Y, Glukhova A, Tarrasch J, Thal DM, Furness SGB, Christopoulos G, Coudrat T, Danev R, Baumeister W, Miller LJ, Christopoulos A, Kobilka BK, Wootten D, Skiniotis G, Sexton PM (2017) Phase-plate cryo-EM structure of a class B GPCR-G-protein complex. Nature 546(7656):118–123
2. Baidya M, Dwivedi H, Shukla AK (2017) Frozen in action: cryo-EM structure of a GPCR-G-protein complex. Nat Struct Mol Biol 24 (6):500–502
3. Mayer JP, Tschöp MH, DiMarchi RD (2017) Once blind, now we see GLP-1 molecular action. Cell Metab 26(2):289–291
4. Martínez-Muñoz L, Rodríguez-Frade JM, Barroso R, Sorzano CS, Torreño-Pina JA, Santiago CA, Manzo C, Lucas P, García-Cuesta EM, Gutierrez E, Barrio L, Vargas J, Cascio G, Carrasco YR, Sánchez-Madrid F, García-Parajo MF, Mellado M (2018)

Separating actin-dependent chemokine receptor nanoclustering from dimerization indicates a role for clustering in CXCR4 signaling and function. Mol Cell 70(1):106–119. e110

5. Briddon SJ, Kilpatrick LE, Hill SJ (2018) Studying GPCR pharmacology in membrane microdomains: fluorescence correlation spectroscopy comes of age. Trends Pharmacol Sci 39(2):158–174
6. Scarselli M, Annibale P, McCormick PJ, Kolachalam S, Aringhieri S, Radenovic A, Corsini GU, Maggio R (2016) Revealing G-protein-coupled receptor oligomerization at the single-molecule level through a nanoscopic lens: methods, dynamics and biological function. FEBS J 283(7):1197–1217
7. Pfleger KD, Eidne KA (2005) Monitoring the formation of dynamic G-protein-coupled receptor-protein complexes in living cells. Biochem J 385(Pt 3):625–637
8. Milligan G, Ramsay D, Pascal G, Carrillo JJ (2003) GPCR dimerisation. Life Sci 74 (2-3):181–188
9. Bouvier M, Heveker N, Jockers R, Marullo S, Milligan G (2007) BRET analysis of GPCR oligomerization: newer does not mean better. Nat Methods 4(1):3–4. author reply 4
10. Bouvier M, Hébert TE (2014) CrossTalk proposal: weighing the evidence for class A GPCR dimers, the evidence favours dimers. J Physiol 592(12):2439–2441
11. Bouvier M, Hébert TE (2014) Rebuttal from Michel Bouvier and Terence E. Hébert. J Physiol 592(12):2447
12. Lambert NA, Javitch JA (2014) CrossTalk opposing view: weighing the evidence for class a GPCR dimers, the jury is still out. J Physiol 592(12):2443–2445
13. Lambert NA, Javitch JA (2014) Rebuttal from Nevin A. Lambert and Jonathan A Javitch. J Physiol 592(12):2449
14. Chabre M, le Maire M (2005) Monomeric G-protein-coupled receptor as a functional unit. Biochemistry 44(27):9395–9403
15. James JR, Oliveira MI, Carmo AM, Iaboni A, Davis SJ (2006) A rigorous experimental framework for detecting protein oligomerization using bioluminescence resonance energy transfer. Nat Methods 3(12):1001–1006
16. Fotiadis D, Jastrzebska B, Philippsen A, Müller DJ, Palczewski K, Engel A (2006) Structure of the rhodopsin dimer: a working model for G-protein-coupled receptors. Curr Opin Struct Biol 16(2):252–259
17. Salahpour A, Masri B (2007) Experimental challenge to a 'rigorous' BRET analysis of GPCR oligomerization. Nat Methods 4 (8):599–600. author reply 601
18. Kenakin T, Agnati LF, Caron M, Fredholm B, Guidoli D, Kobilka B, Lefkowitz RW, Lohse M, Woods A, Fuxe K (2010) International workshop at the Nobel forum, Karolinska Institutet on G protein-coupled receptors: finding the words to describe monomers, oligomers, and their molecular mechanisms and defining their meaning. Can a consensus be reached? J Recept Signal Transduct Res 30(5):284–286
19. Albizu L, Cottet M, Kralikova M, Stoev S, Seyer R, Brabet I, Roux T, Bazin H, Bourrier E, Lamarque L, Breton C, Rives ML, Newman A, Javitch J, Trinquet E, Manning M, Pin JP, Mouillac B, Durroux T (2010) Time-resolved FRET between GPCR ligands reveals oligomers in native tissues. Nat Chem Biol 6 (8):587–594
20. Rivero-Müller A, Chou YY, Ji I, Lajic S, Hanyaloglu AC, Jonas K, Rahman N, Ji TH, Huhtaniemi I (2010) Rescue of defective G protein-coupled receptor function in vivo by intermolecular cooperation. Proc Natl Acad Sci U S A 107(5):2319–2324
21. Viñals X, Moreno E, Lanfumey L, Cordomí A, Pastor A, de La Torre R, Gasperini P, Navarro G, Howell LA, Pardo L, Lluís C, Canela EI, McCormick PJ, Maldonado R, Robledo P (2015) Cognitive impairment induced by Delta9-tetrahydrocannabinol occurs through Heteromers between cannabinoid CB1 and serotonin 5-HT2A receptors. PLoS Biol 13(7):e1002194
22. Jonas KC, Huhtaniemi I, Hanyaloglu AC (2016) Single-molecule resolution of G protein-coupled receptor (GPCR) complexes. Methods Cell Biol 132:55–72
23. Jonas KC, Chen S, Virta M, Mora J, Franks S, Huhtaniemi I, Hanyaloglu AC (2018) Temporal reprogramming of calcium signalling via crosstalk of gonadotrophin receptors that associate as functionally asymmetric heteromers. Sci Rep 8(1):2239
24. Jonas KC, Fanelli F, Huhtaniemi IT, Hanyaloglu AC (2015) Single molecule analysis of functionally asymmetric G protein-coupled receptor (GPCR) oligomers reveals diverse spatial and structural assemblies. J Biol Chem 290 (7):3875–3892
25. Martínez-Muñoz L, Barroso R, Dyrhaug SY, Navarro G, Lucas P, Soriano SF, Vega B, Costas C, Muñoz-Fernández M, Santiago C, Rodríguez Frade JM, Franco R, Mellado M (2014) CCR5/CD4/CXCR4 oligomerization prevents HIV-1 gp120IIIB binding to the cell

surface. Proc Natl Acad Sci U S A 111(19): E1960–E1969

26. Jin J, Momboisse F, Boncompain G, Koensgen F, Zhou Z, Cordeiro N, Arenzana-Seisdedos F, Perez F, Lagane B, Kellenberger E, Brelot A (2018) CCR5 adopts three homodimeric conformations that control cell surface delivery. Sci Signal 11(529)
27. Dijkman PM, Castell OK, Goddard AD, Munoz-Garcia JC, de Graaf C, Wallace MI, Watts A (2018) Dynamic tuneable G protein-coupled receptor monomer-dimer populations. Nat Commun 9(1):1710
28. Jean-Alphonse F, Bowersox S, Chen S, Beard G, Puthenveedu MA, Hanyaloglu AC (2014) Spatially restricted G protein-coupled receptor activity via divergent endocytic compartments. J Biol Chem 289(7):3960–3977
29. Henriques R, Lelek M, Fornasiero EF, Valtorta F, Zimmer C, Mhlanga MM (2010) QuickPALM: 3D real-time photoactivation nanoscopy image processing in ImageJ. Nat Methods 7(5):339–340
30. Tanaka KA, Suzuki KG, Shirai YM, Shibutani ST, Miyahara MS, Tsuboi H, Yahara M, Yoshimura A, Mayor S, Fujiwara TK, Kusumi A (2010) Membrane molecules mobile even after chemical fixation. Nat Methods 7 (11):865–866
31. Annibale P, Vanni S, Scarselli M, Rothlisberger U, Radenovic A (2011) Identification of clustering artifacts in photoactivated localization microscopy. Nat Methods 8 (7):527–528
32. Rossy J, Cohen E, Gaus K, Owen DM (2014) Method for co-cluster analysis in multichannel single-molecule localisation data. Histochem Cell Biol 141(6):605–612

Part VI

Investigating In Vivo Functions of GPCRs

Chapter 20

Optical Modulation of Metabotropic Glutamate Receptor Type 5 In Vivo Using a Photoactive Drug

Marc López-Cano, Joan Font, Amadeu Llebaria, Víctor Fernández-Dueñas, and Francisco Ciruela

Abstract

Optopharmacology is a very promising approach based on the use of light-deliverable drugs, which allows manipulating physiological processes with high spatiotemporal resolution. Light-dependent drugs (i.e. caged-compounds) targeting G protein-coupled receptors (GPCRs) have been developed to provide great pharmacological precision on the control of pain. Metabotropic glutamate type 5 ($mGlu_5$) receptors are widely expressed through the pain neuraxis and play a key role in pain transmission. In line with this, selective $mGlu_5$ receptor negative allosteric modulators (NAMs) have consistently shown analgesic activity in experimental animal models of inflammatory pain. Accordingly, we synthesized a light-deliverable drug (i.e. caged compound) using the chemical structure of raseglurant, a $mGlu_5$ receptor NAM, as a molecular scaffold. And thereafter, we evaluated the analgesic activity of the caged compound in formalin-injected (hind paw) mice upon light irradiation (405 nm). Of note, light was both delivered at the peripheral (i.e. hind paw) and central level (i.e. thalamus), by means of brain-implanted fiber-optics. The novel light-deliverable drug, JF-NP-26, showed antinociceptive activity upon violet light irradiation either of the hind paw or the thalamus, demonstrating the ability of precisely activating, in time and space, the caged compound. Here, we describe in detail the protocol used to perform a reliable and reproducible formalin nociception test in mice using an optopharmacology approach (i.e. light-deliverable compounds).

Key words $mGlu_5$ receptor, Optopharmacology, Pain, Analgesia, Mice

1 Introduction

Optopharmacology is a novel pharmacological approach that allows a high spatiotemporal resolution of drug actions. In brief, it consists of using photoactivable compounds that can bind and modulate their target in a light-dependent manner [1]. Interestingly, G protein-coupled receptors (GPCRs) constitute the largest family of membrane receptors and represent the main drug targets in pharmacology. Accordingly, the design and synthesis of new light-deliverable drugs (i.e. optodrugs) targeting GPCRs may enrich the pharmacological arsenal and tackle GPCR-related

Mario Tiberi (ed.), *G Protein-Coupled Receptor Signaling: Methods and Protocols*, Methods in Molecular Biology, vol. 1947, https://doi.org/10.1007/978-1-4939-9121-1_20,

pathological conditions. Metabotropic glutamate (mGlu) receptors are GPCRs involved in a myriad of physiological processes, including pain transmission [2]. And, importantly, it has been shown that negative allosteric modulators (NAM) of $mGlu_5$ receptors display analgesic effects in animal models of inflammatory pain [3–6].

Pain research is mostly based on experimental assays using animal models [7]. These assays are developed both to decipher the physiopathology of the pain condition and to propel drug discovery. The formalin nociception test is considered one of the most predictive assays to study acute pain in rodents [8]. This test permits monitoring pain-related responses (i.e. itch) caused by a subcutaneous injection of an inflammatory agent, namely 2.5% formalin solution, in the hind paw. After the injection, two distinct time periods or phases of licking/biting behavior occur, which are separated by a quiescent period. Importantly, these phases differ in duration and underlying mechanisms. Hence, the initial acute phase (phase I), commonly recorded for 5 min just after formalin administration, reflects acute peripheral pain, probably due to direct activation of nociceptors through the Transient Receptor Potential subtype A1 (TRPA1) channel. On the other hand, the phase II, which starts after the quiescent period (5–15 min) and is commonly recorded for 15–30 min, is due to the ongoing inflammatory input and central nociceptive sensitization [9].

Here, we provide a complete description of the optopharmacological approach used in combination with the formalin nociception test to facilitate its implementation by other scientists.

2 Materials

2.1 Animals

1. Animals. Adult male CD-1 mice weighing 20–25 g were used (*see* **Note 1**).
2. 0.93% formaldehyde solution.
3. Syringes. BD Micro-Fine™ Demi, U-100 Insulin, 30G × 1/2″ −0.33 mm × 8 mm (Fig. 1d).
4. Analgesic drug to be tested. In the example provided here we used JF-NP-26, a light-activable (i.e. caged-derivative) NAM of the $mGlu_5$ receptor [3].
5. Saline solution: 0.9% NaCl in H_2O.
6. Eppendorff tubes.
7. Cleaning solution: 70% ethanol in H_2O.

2.2 Instrumentation and Equipment

1. Observational chamber. Circular (25 cm ø × 30 cm height) glass container (i.e. glass beaker) (Fig. 1a).
2. Weighing scale (Fig. 1c).

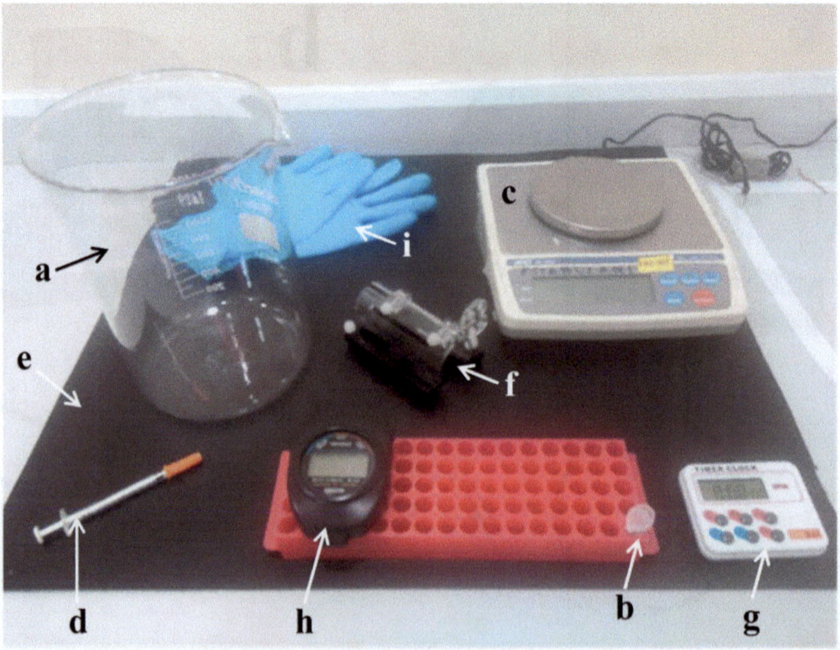

Fig. 1 Material needed for the formalin test. Glass container (**a**), freshly prepared 2.5% formalin solution (**b**), weighing machine (**c**), insulin syringe (**d**), black plastic surface (**e**), restrain apparatus for mice (**f**), timer (**g**), chronometer (**h**), and gloves (**i**). Adapted from [12]

3. Colored surface (Fig. 1e).
4. Restrain apparatus for mice (Fig. 1f). Alternatively, a falcon tube of 50 mL could be used.
5. Micropipette.
6. Timer (Fig. 1g) and chronometer (Fig. 1h).
7. Personal protective equipment (Fig. 1i). Laboratory coat, gloves, masks, etc.
8. Digital lab standard stereotaxic apparatus (Stoelting Co., Wood Dale, IL, USA) and Franklin and Paxinos's mouse brain atlas [10].
9. LED-based fiber-optic system (Doric Lenses Inc., Québec, Canada) composed by: (1) programmable LED driver (Fig. 2a); (2) dual LED + fiber-optic rotatory joint (Fig. 2b); (3) dual fiber-optic patch cords with slim magnetic connection (Fig. 2c); and (4) dual fiber-optic cannula with magnetic connection (Fig. 2d).
10. Doric Lenses software (OTPG4 Controller) and a computer/laptop to run it.

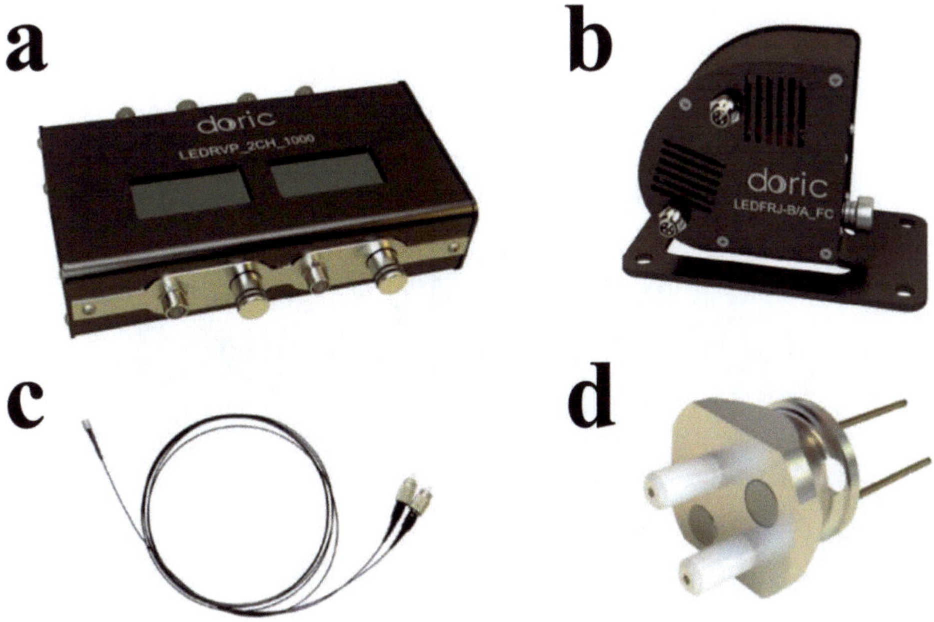

Fig. 2 Instrumentation and equipment needed for the light irradiation (Doric). Programmable LED driver (**a**), dual LED + fiber-optic rotatory joint (**b**), dual fiber-optic patch cords with slim magnetic connection (**c**), and dual fiber-optic cannula with magnetic connection (**d**)

3 Methods

3.1 Formalin Test

1. Habituation of animals. The day before performing the experiment, animals to be tested are individually exposed to the observational chamber (Fig. 1a) for 1 h. The glass container (i.e. observational chamber) should be cleaned with 70% ethanol first and then with sterile water to remove any animal/ ethanol odor. Air dry for 5 min before testing another animal. No access to food or water should be allowed (*see* **Note 2**).
2. Reagents preparation. The day of the experiment prepare the 2.5% formalin solution (which corresponds to a 0.93% formaldehyde solution) by mixing: 25 μL formaldehyde solution (36.5–38% in H_2O) and 975 μL sterile physiological saline solution (Fig. 1b) and the drug to be tested (not shown). Clean all the material needed.
3. Weighing animals. Remove the mouse from the house cage and measure its body weight using a weighing scale (Fig. 1c).
4. Administration of the analgesic drug. Administer (i.e. intraperitoneally, i.p.) the analgesic drug to be tested (i.e. JF-NP-26; Fig. 3) or the same volume of vehicle (saline) using an insulin syringe (Fig. 1d) (*see* **Note 3**).

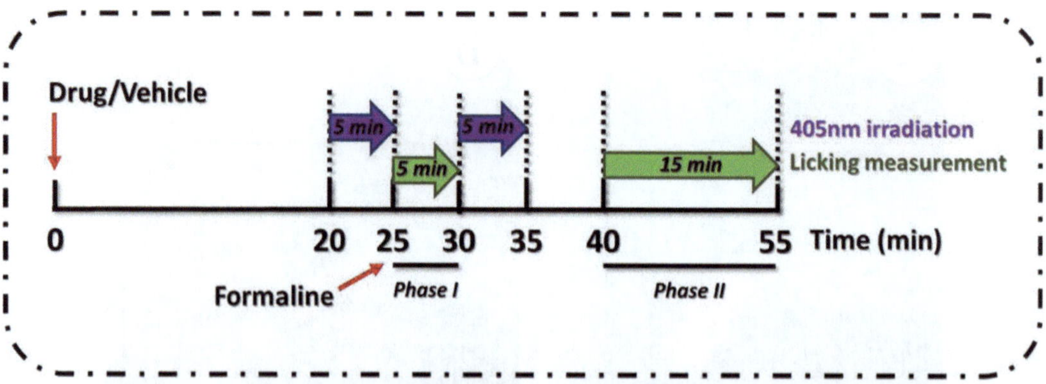

Fig. 3 Treatment schedule depicting the administration and irradiation regimen of drug, formalin, and the behavioral testing (phase I and II)

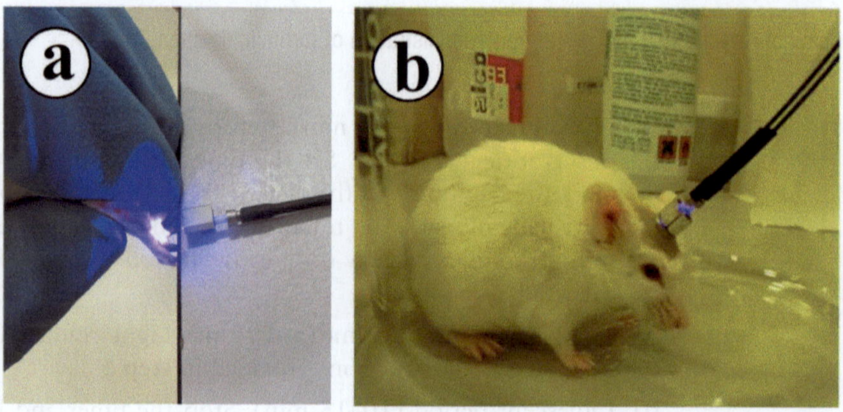

Fig. 4 Light irradiation in (**a**) peripheral tissue (i.e. hind paw) and (**b**) central brain area (i.e. thalamus)

5. Acclimation of the animals. The animal is placed in the observational chamber for 20 min before the irradiation regime (Fig. 3) (*see* **Note 4**).
6. First irradiation regime. Light irradiate (i.e. 405 nm wavelength, 2000 mA intensity and 1 Hz frequency) for 5 min the tissue: (1) hind paw (Fig. 4a) using the dual fiber-optic patch cord; (2) thalamus (Fig. 4b) using the dual fiber-optic cannula with magnetic connection (*see* **Note 5**).
7. Formalin administration. Load 20 μL of 2.5% formalin solution directly in the 0.3 mL insulin syringe equipped with a 30G needle (*see* **Note 6**). Take the animal from the observational cage and place it carefully inside the restrain module (Fig. 1f) with the selected hind paw (i.e. right) exposed outside (Fig. 5a). Inject slowly the 20 μL of 2.5% formalin solution using the insulin syringe in the mid-plantar surface of the hind paw (Fig. 5b) (*see* Note 7).

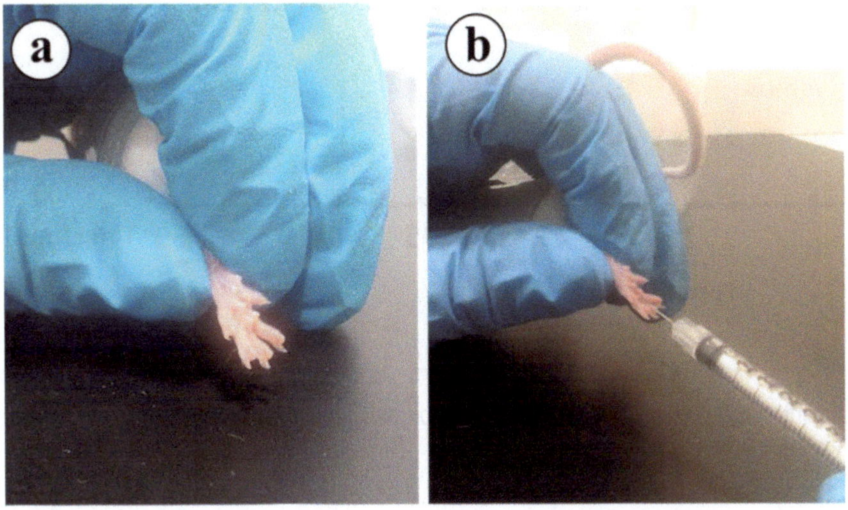

Fig. 5 Hind paw of a restrained mouse (**a**) and administration of formalin solution (**b**)

8. Phase I recording (0–5 min). Return the injected animal to the observational chamber. Set the timer in 5 min and start it to mark the beginning of the phase I (Fig. 3). Use a chronometer to visually measure the total time the animal spends licking or biting the injected paw during this period (5 min). Annotate the result (*see* **Note 8**).
9. Second irradiation regime (5–10 min). Light irradiate for 5 min following the instructions provided in **step 6**.
10. Quiescent period (10–15 min). Stop the timer and leave the animal without disturbing it for 5 min (Fig. 3).
11. Phase II recording (15–30 min). Set the timer in 15 min and start it to mark the beginning of the phase II (Fig. 3). Use a chronometer to visually measure the total time the animal spends licking or biting the injected paw during this period (15 min). Annotate the result.
12. Finishing. Return the animal to its home cage (*see* **Note 9**).
13. Cleaning. Clean all the material used (*see* **Note 10**).

3.2 Data Analysis

1. Antinociception induced by drug treatment in the formalin test can be calculated with the following formula:

$$\text{Antinoceptive effect (\%)} = [(\text{LTV} - \text{LTD})/\text{LTV}] \times 100$$

where LTV and LTD represent the licking/biting time in the vehicle- and drug-treated animals, respectively. The antinociceptive effect is expressed as percentage (mean ± S.E.M.) of the

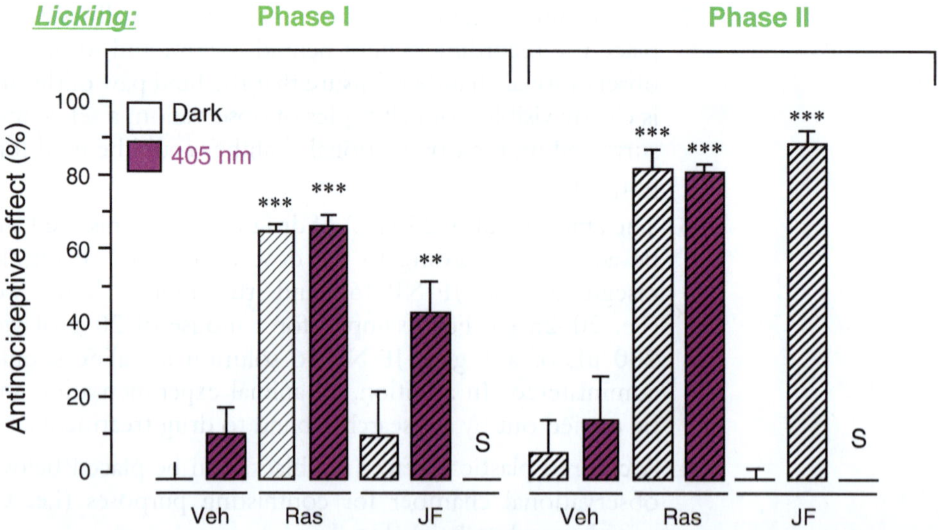

Fig. 6 Representative results. Central light-dependent JF-NP-26-mediated antinociception in mice. Animals were intraperitoneally injected with vehicle (Veh, 20% DMSO +20% tween 80 in saline), raseglurant (Ras, 10 mg/kg) or JF-NP-026 (JF, 10 mg/kg) 20 min before the thalamus was irradiated at 405 nm (or dark) for 5 min. As a control, the striatum (S) was also irradiated. Subsequently, the total hind paw licking was measured for 0–5 min (Phase I) and 15–30 min (Phase II) after intraplantar injection of 20 μL of formalin solution (2.5% paraformaldehyde). The antinociceptive effect was calculated as the percentage of the maximum possible effect (mean ± S.E.M., $n = 5$–6 mice per group). $^{**}p < 0.01$ and $^{***}p < 0.001$, one-way ANOVA with Dunnett's multiple comparison test using Veh + dark as a control. Reproduced from Font et al. [3]

maximum effect observed (Fig. 6). The results are analyzed either by Student's *t* test or by one-way analysis of variance (ANOVA) followed by Dunnett's post-hoc test using vehicle-treated animals as a control (Fig. 6). Statistical significance is set as $p < 0.05$.

4 Notes

1. In the example, animals were housed in groups of five in standard cages with ad-libitum access to food and water and maintained under 12 h dark/light cycle (starting at 7:30 AM), 22 °C temperature, and 66% humidity (standard conditions). All manipulations were carried out between 9:00 h and 16:00 h. the University of Barcelona Committee on animal use and care approved the protocol. Animals were housed and tested in compliance with the guidelines described in the guide for the care and use of laboratory animals [11] and following the European Union directives (2010/63/EU), FELASA and ARRIVE guidelines. Efforts were made to minimize suffering and reduce the number of animals used in the experiments.

2. A recording system (i.e. camera) can be optionally used. Thus, place the recording system behind, below, and/or above the observational chamber. Ensure that the hind paw of the animal is clearly visible from all angles of observation; a set of mirrors surrounding the observational chamber could be used for this purpose.
3. The amount (200–250 μL) of drug to be administered should be adjusted according to the dose tested (i.e. 10 mg/kg of raseglurant or JF-NP-26) and the mouse body weight (i.e. 20–25 g). For example, for a mouse of 20 g of weight 200 μL of a 1 g/L JF-NP-26 solution in saline should be administered. In addition, all animal experimentation should be carried out by a researcher blind to drug treatments.
4. A colored plastic surface (i.e. black) can be placed below the observational chamber for contrasting purposes (i.e. CD-1 mice have white hair) (Fig. 1e).
5. For central nervous irradiation a stereotaxic surgery must be done to implant the fiber-optic cannula in the brain. Briefly, mice were anesthetized with a combination of ketamine/xylacine (dose of 100 mg/kg body weight for ketamine and 10 mg/kg for xylacine). Subsequently, the dual fiber-optic cannula with magnetic connection (Fig. 2d) was implanted in the target brain areas using a stereotaxic apparatus. The site of implantation was the left and right thalamus (coordinates: −1.8 mm posterior to the bregma, ±1.5 mm lateral to the midline, 3.5 mm ventral from the surface of skull), following the Franklin's and Paxinos mouse brain atlas [10]. Finally, fiber-optic cannulas were fixed in the skulls of the animals using dental cement and chirurgic screws of 0.8 mm diameter (Agnthos, Lidingö, Sweden).
6. Optionally, the 20 μL of 2.5% formalin solution can be taken with a 2–20 μL micropipette, placed in a 100 μL Eppendorf tube and then loaded with the insulin syringe.
7. The needle should be placed between toes and ankle and inserted beneath the surface of the skin. Any tissue damage should be avoided. The injection should be done firmly and time-optimized. Otherwise, animals may perceive the restrain as a stressor stimulus, which can increase the variability of the results.
8. If a recording system is used, switch on and off the system accordingly. We strongly recommend the investigator to perform some training at this step to familiarize themselves with the licking/biting behavior.
9. If a recording system is used switch off the camera. Check whether animals need any special attention or health care (i.e. bleeding, extra tissue damage, etc.).

10. Repeat the experiment with another animal if required. If a camera is used, visualize the videos and measure the total time the animal spends licking or biting the injected paw during each period (phase I and II).

Acknowledgments

This work was supported by MINECO/ISCIII (SAF2017-87349-R and PIE14/00034), the Catalan government (2017 SGR 1604), Fundació la Marató de TV3 (Grant 20,152,031), FWO (SBO-140028) to FC. MINECO (PCIN-2013-017-C03-01 and CTQ2014-57020-R), the Catalan government (2014SGR109 and 2014CTP0002) to AL. We thank Esther Castaño and Benjamín Torrejón, from the CCiT-Bellvitge Campus of the University of Barcelona for the technical assistance.

References

1. Fajardo O, Friedrich RW (2013) A light switch for pain. Nat Chem Biol 9:219–220. https://doi.org/10.1038/nchembio.1203
2. Kolber BJ (2015) mGluRs head to toe in pain. Prog Mol Biol Transl Sci 131:281–324. https://doi.org/10.1016/bs.pmbts.2014.12.003
3. Font J, López-Cano M, Notartomaso S et al (2017) Optical control of pain in vivo with a photoactive mGlu 5 receptor negative allosteric modulator. elife 6:e23545. https://doi.org/10.7554/eLife.23545
4. Bhave G, Karim F, Carlton SM, Gereau RW (2001) Peripheral group I metabotropic glutamate receptors modulate nociception in mice. Nat Neurosci 4:417–423. https://doi.org/10.1038/86075
5. Palazzo E, Marabese I, de Novellis V et al (2014) Supraspinal metabotropic glutamate receptors: a target for pain relief and beyond. Eur J Neurosci 39:444–454. https://doi.org/10.1111/ejn.12398
6. Lea PM 4th, Faden AI (2006) Metabotropic glutamate receptor subtype 5 antagonists MPEP and MTEP. CNS Drug Rev 12:149–166. https://doi.org/10.1111/j.1527-3458.2006.00149.x
7. Mogil JS (2009) Animal models of pain: progress and challenges. Nat Rev Neurosci 10:283–294. https://doi.org/10.1038/nrn2606
8. Tjolsen A, Berge OG, Hunskaar S et al (1992) The formalin test: an evaluation of the method. Pain 51:5–17
9. Hunskaar S, Hole K (1987) The formalin test in mice: dissociation between inflammatory and non-inflammatory pain. Pain 30:103–114
10. Franklin KBJ, Paxinos G (2008) The mouse brain in stereotaxic coordinates, 3rd edn. Academic Press
11. Clark JD, Gebhart GF, Gonder JC et al (1997) Special report: the 1996 guide for the care and use of laboratory animals. ILAR J 38:41–48
12. López-Cano M, Fernández-Dueñas V, Llebaria A, Ciruela F (2017) Formalin murine model of pain. Bio-Protocols 7:e2628

Chapter 21

Imaging of Tissue-Specific and Temporal Activation of GPCR Signaling Using DREADD Knock-In Mice

Dmitry Akhmedov, Nicholas S. Kirkby, Jane A. Mitchell, and Rebecca Berdeaux

Abstract

Engineered G protein-coupled receptors (DREADDs, designer receptors exclusively activated by designer drugs) are convenient tools for specific activation of GPCR signaling in many cell types. DREADDs have been utilized as research tools to study numerous cellular and physiologic processes, including regulation of neuronal activity, behavior, and metabolism. Mice with random insertion transgenes and adeno-associated viruses have been widely used to express DREADDs in individual cell types. We recently created and characterized *ROSA26-GsDREADD* knock-in mice to allow Cre recombinase-dependent expression of a Gαs-coupled DREADD (GsD) fused to GFP in distinct cell populations in vivo. These animals also harbor a CREB-activated luciferase reporter gene for analysis of CREB activity by in vivo imaging, ex vivo imaging, or biochemical reporter assays. In this chapter, we provide detailed methods for breeding GsD animals, inducing GsD expression, stimulating GsD activity, and measuring basal and stimulated CREB reporter bioluminescence in tissues in vivo, ex vivo, and in vitro. These animals are available from our laboratory for non-profit research.

Key words Gs-DREADD, GPCR, cAMP, CREB luciferase reporter, In vivo bioluminescence imaging, Clozapine N-oxide

1 Introduction

G protein-coupled receptors (GPCRs) are the largest class of receptors in mammals and regulate multiple physiologic processes [1, 2]. Dysregulated GPCR signaling underlies many diseases including cancer, diabetes, and fertility disorders [3–5]. Endogenous GPCRs are commonly expressed in numerous cell types, so results of pharmacologic experiments in vivo using ligands for endogenous GPCRs can be difficult to interpret due to activation of signaling in numerous cell types. In addition, many GPCRs activate more than one intracellular signaling pathway [6–8]. To enable mechanistic and physiologic studies of GPCR-activated G-protein signaling, DREADD (designer receptors exclusively

Mario Tiberi (ed.), *G Protein-Coupled Receptor Signaling: Methods and Protocols*, Methods in Molecular Biology, vol. 1947, https://doi.org/10.1007/978-1-4939-9121-1_21,

activated by designer drugs) technology was developed as a chemical-genetic tool [9–13].

DREADDs (also originally termed RASSLs, receptors activated solely by synthetic ligands [9]) are derived from M3- or M4-muscarinic receptors and have point mutations in the transmembrane domains rendering the receptors unresponsive to the endogenous ligand acetylcholine and yet highly responsive to synthetic drug clozapine *N*-oxide (CNO) [10–13]. DREADD receptors coupled to each class of G proteins Gαs, Gαq, and Gαi have been developed [14]. For in vivo applications, viral DREADD expression vectors or transgenic animals expressing DREADDs are commonly used [15–18]. In DREADD-expressing mice, activation of DREADD receptor signaling is achieved by administering CNO by injections or in drinking water.

Two *ROSA26*-based knock-in mice have been generated to accomplish Cre recombinase-dependent expression of Gαq or Gαi [19, 20]. We recently reported generation of a third *ROSA26*-based knock-in mouse line expressing Gαs-coupled DREADD fused to GFP in a Cre-dependent manner for tissue-specific activation of cAMP signaling (Fig. 1a). Without Cre recombinase expression, the GsD gene is silenced due to an intervening lox-stop-lox (LSL) sequence between the promoter and the GsD cDNA [21]. The GsD receptor can be expressed in any tissue or cell type for which a tissue-specific Cre driver or suitable viral vector is available.

The knock-in allele in the *ROSA26-GsD* mice also encodes luciferase under control of a minimal promoter harboring four full cAMP-response elements (CRE) [21], which are directly bound by CREB [22], to monitor CREB activity by in vivo or ex vivo bioluminescence assays. The reporter is encoded on the opposite strand, does not require Cre recombinase activity to be expressed, and is sensitive to increased CREB activity due to GsD activation by CNO as well as by endogenous cAMP signaling. The vast diversity of transgenic Cre drivers and viral vectors encoding Cre and widely available instrumentation for bioluminescence reporter imaging should enable investigators to easily utilize these mice to manipulate and monitor cAMP signaling in select cell types in vivo.

Our laboratory has utilized this mouse line to drive cAMP pathway activity selectively in mouse hepatocytes and skeletal muscle using AAV-Cre vectors as well as a genetic skeletal muscle-specific Cre driver ([21] and unpublished). In this chapter, we detail the methods to breed and utilize *ROSA26-GsD* animals for stimulation of Gαs signaling in vivo and bioluminescence reporter assays to assess resultant CREB activity. We also discuss consideration of control groups and highlight methods to confirm appropriate GsD expression and activity.

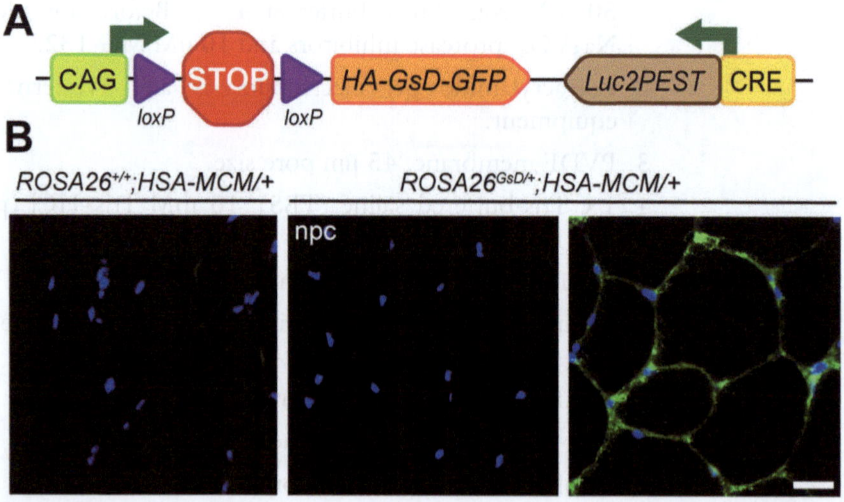

Fig. 1 GsD expression is induced by Cre recombinase in a tissue-specific manner. (**a**) *ROSA26-LSL-GsDREADD-CRE-luc* allele. Arrows, transcriptional start sites. (**b**) GsD expression in skeletal muscle fibers of *ROSA26*$^{GsD/+}$;*HSA-MerCreMer/+* mice following tamoxifen treatment. HSA-MerCreMer is a tamoxifen inducible Cre recombinase under control of the human α-skeletal muscle promoter [28]. Green = GFP immunolabeling, blue = DAPI. npc = no primary antibody control. Scale bar = 20 μm

2 Materials

2.1 Induction of GsD Receptor Expression In Vivo

1. Male or female heterozygous *ROSA26*$^{GsD/+}$ mice (*see* **Note 1**), aged 6–14 weeks, housed at 22 °C in individually ventilated cages with free access to food and water with a 12 h light/dark cycle (7 AM/7 PM) in a facility approved by the local Institutional Animal Care and Use Committee (IACUC). We recommend maintaining the line on the albino B6 (*B6(Cg)-Tyr*$^{c-2J}$*/J*) background (Jackson Laboratories, stock #000058) for optimum imaging sensitivity (*see* **Note 2**).
2. Mouse chow diet.
3. Adeno-associated viruses (AAV), adenoviruses (AdV) or transgenic mice expressing Cre recombinase.
4. Tamoxifen if using a vector or mouse line encoding Cre-ER. Make a 100 mg/ml solution in pure ethanol. To facilitate TMX solubilization heat the tube at 37 °C for 2–3 min. Store this solution at −20 °C and use within 2 weeks.
5. Syringes and needles. Small gauge needles (25–30G) are suitable for luciferin injection. Larger gauge needles (20–22G) are more appropriate for tamoxifen (*see* **Note 3**).

2.2 Detection of GsD Receptor Expression

1. Modified RIPA-T buffer: 20 mM HEPES (pH 7.5), 137 mM NaCl, 1% Triton X-100, 2% SDS, 0.5% Na-DOC, 0.5 mM EDTA, 5 mM Na_4P2O_7, 20 mM β-glycerophosphate,

50 mM NaF. Store buffer at 4 °C. Before use add 1 mM Na_3VO_4, protease inhibitors and 10 μM MG-132.

2. Polyacrylamide gel electrophoresis and Western blotting equipment.
3. PVDF membrane, 45 μm pore size.
4. 1× Tris-buffered saline (TBS): 10 mM Tris–HCl (pH 7.6), 100 mM NaCl.
5. 1× Tris-buffered saline containing 0.1% Tween 20 (TBS-T).
6. Bovine serum albumin (Fraction V): make 3% (w/v) solution in TBS-T.
7. Nonfat dry milk: make 5% (w/v) solution in TBS-T.
8. Anti-GFP antibody (Rockland Immunochemical).
9. Donkey anti-goat Alexa Fluor 488.
10. Donkey anti-goat HRP.

2.3 Stimulation and Monitoring of GsD Activity In Vivo and Ex Vivo

1. Dimethyl sulfoxide (DMSO).
2. 10 mM clozapine *N*-oxide (CNO) stock solution: approximately 0.2 ml of dimethyl sulfoxide (DMSO) per 25 mg CNO. Bring to the required final volume with 0.9% saline (final concentration of DMSO is 2.5%). Store stock solution at −20 °C in 0.5 ml aliquots for up to 4 months.
3. 1× phosphate-buffered saline (PBS).
4. 120 mg/ml D-luciferin potassium salt stock solution: reconstitute D-luciferin powder in sterile saline or 1× PBS to make a colloid. Mix well and make 0.2 ml aliquots, protect from light and store at −20 °C.
5. IVIS Lumina XR instrument (PerkinElmer) or similar (*see* **Note 4**) connected to isoflurane vaporizer with anesthesia chamber and equipped with 3 or 5 position nose cone manifold to maintain anesthesia during imaging.
6. Living Image® software (PerkinElmer).
7. pSer133 CREB antibody (87G3) (Cell Signaling clone 87G3).
8. Total CREB antibody (48H2) (Cell Signaling clone 48H2).
9. Direct cAMP ELISA kit (Enzo Life Sciences).

2.4 Ex Vivo Bioluminescence CREB Reporter Imaging of Mouse Tissues

1. NKH477 adenylyl cyclase activator.
2. 3-isobutylmethylxanthine (IBMX).
3. 120 mg/ml D-luciferin stock solution.
4. IVIS Lumina XR instrument.
5. Living Image® software.

3 Methods

3.1 Induction of GsD Receptor Expression In Vivo

The GsD receptor can be expressed in any tissue or cell type for which a tissue-specific Cre driver or suitable viral vector is available.

3.1.1 AAV-Cre Vectors

1. For liver-specific GsD expression inject intravenously recombinant AAV2/8-TBG-Cre (AAV8.TBG.PI.Cre.rBGH) or AAV2/8-TBG-GFP (AAV8.TBG.PI.eGFP.WPRE.bGH) using 2×10^{11} viral genomes (v.g.)/mouse.
2. Detect receptor expression and activity (*see* Subheadings 3.2 and 3.3). Typically, this can be detected as few as 7 days after AAV injection.

3.1.2 Activation of Cre-ER Recombinase by Tamoxifen

1. Dilute tamoxifen stock solution (*see* Subheading 2.1, **item 4**) in corn oil to 15 mg/ml.
2. Inject in mice subcutaneously or intraperitoneally at 75 mg/kg animal weight (*see* **Note 3**).

3.2 Detection of GsD Receptor Expression

We recommend testing GsD receptor expression by native GFP fluorescence, immunofluorescence using anti-GFP antibody (*see* Subheading 2.2, **items 8** and **9** and Fig. 1b) or Western blotting with an anti-GFP antibody (*see* **Note 5**).

1. For Western blotting, lyse tissues in modified RIPA-T buffer containing 2% SDS (*see* Subheading 2.2, **item 1**).
2. Resolve proteins on 7.5% polyacrylamide gels and transfer to PVDF membranes.
3. Block membranes by incubating in 5% milk in TBS-T for 30 min.
4. Incubate membranes with primary GFP antibody (diluted 1:2000 in 3% BSA in TBS-T) overnight at 4 °C.
5. Remove primary antibody and wash blots in TBS-T, 3×15 min.
6. Incubate with secondary antibody (donkey anti-goat HRP, diluted 1:4000 in 5% milk in TBS-T) for 45 min.
7. Remove secondary antibody and wash blots in TBS-T, 3×15 min.
8. Expose membranes to film and develop film.

3.3 Stimulation and Monitoring of GsD Activity In Vivo and Ex Vivo

The DREADD system is a chemical-genetic approach that theoretically requires both receptor expression and ligand delivery to initiate signaling to downstream effector pathways (*see* **Note 6**).

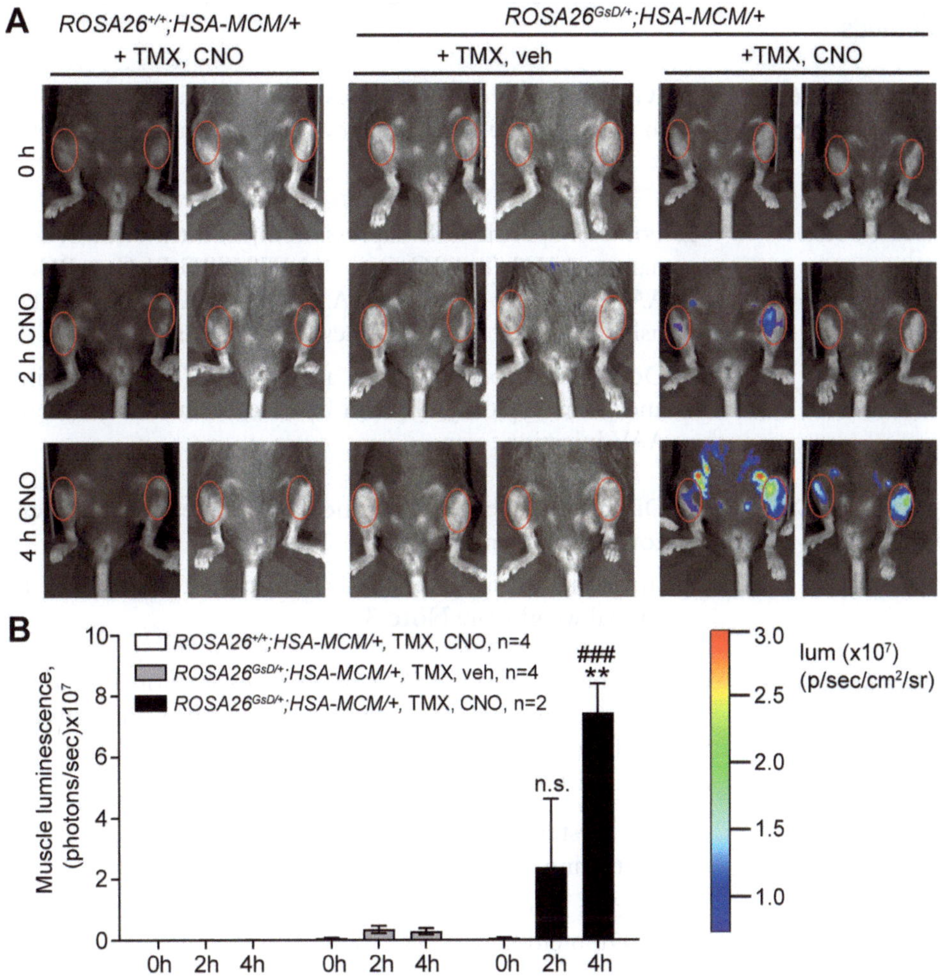

Fig. 2 (**a**) Bioluminescent imaging of CRE-luciferase reporter in *ROSA26*$^{GsD/+}$*;MCM/+* mice before (0 h) and after CNO injection (2 h, 4 h). (**b**) Quantification of bioluminescence in regions outlined in (**a**). Significance determined by *t*-test; *significance vs. 0 h within the same genotype, # significance vs. 0 h of TMX, vehicle control

1. Administer CNO 2–6 h before imaging to allow for luciferase cDNA transcription and translation (*see* **Note 7**). Strong bioluminescence signal is typically detected 6 h after CNO injection in mice expressing GsD in liver [21] and 4 h after CNO injection in mice expressing GsD in skeletal muscle fibers (Fig. 2).
2. Proceed with imaging (*see* Subheading 3.3.1) or biochemical analysis of GsD signaling (*see* Subheading 3.3, **steps 3** and **4**).
3. Lyse tissues and proceed with Western blot (*see* Subheading 3.2) using the following antibody dilutions: phospho-CREB (Ser133) 1:3000 in 3% BSA in TBS-T, total CREB 1:4000 in 3% BSA in TBS-T.

4. Determine cAMP concentration using cAMP kit (*see* Subheading 2.3, **item 9**). Follow the acetylation protocol provided by the manufacturer.

3.3.1 In Vivo Bioluminescence CREB Reporter Imaging

1. Turn on the IVIS Lumina XR instrument. Check oxygen supply and isoflurane level in the vaporizer.
2. Open Living Image® software and click 'Initialize' in the 'IVIS Acquisition Control Panel'. Wait until the temperature indicator turns from red to green, which indicates that the heated stage reached 37 °C.
3. Adjust image acquisition settings: select 'Luminescent' and 'Photograph' imaging modes, which will instruct the software to first take a photograph and then a luminescent image, and to overlay pseudocolored luminescence image onto the photograph.

   'Binning' has low, medium, or high settings. Binning increases signal-to-noise ratio and high binning helps in detection of low intensity signals but does so at the expense of reduced resolution. We use 'medium' binning (binning factor 4).

   'F/stop' adjusts the amount of light accessible to camera. Small aperture F/16 is usually used in 'Photograph' mode, while large aperture F/1 is used in 'Luminescent' mode to maximize sensitivity.

   'Field of View' adjusts the area captured by the camera and has four positions, A through D. Use position D to image whole mouse or multiple mice. Use other positions to capture smaller areas, such as part of mouse body.

   'Exposure time' should be adjusted according to the intensity of the signal. In our experience, liver bioluminescence induced by GsD-CNO requires only a 5–10 s exposure. When imaging for the first time, select auto exposure. This mode automatically sets optimal binning and F/stop. For analysis of the signal intensity the image should not be oversaturated (*see* **Note 8**).

   Once acquisition settings are selected for a group of animals, they should not be changed throughout the experiment. If acquisition settings are altered, signals may not be directly quantitatively compared among different experimental time points.
4. Anesthetize the mice in an isoflurane chamber, weigh the mouse and calculate volume of 24 mg/ml D-luciferin to be injected to achieve dose of 100 mg/kg animal weight (*see* **Note 9**).
5. For injections, mix colloid D-luciferin stock solution well and dilute to 24 mg/ml in sterile saline or 1× PBS. Inject D-luciferin intraperitoneally at 100 mg/kg animal weight. For

liver imaging, waiting 2 min before acquiring the images is sufficient.

6. Place the mice in the imaging chamber with the nose inserted in the nose cone for isoflurane inhalation. A 3-position manifold allows imaging up to 3 mice simultaneously. If several mice are imaged at the same time, separate the animals by black dividers (such as black construction paper folded like a fan) to prevent signal interference between mice.
7. Acquire image. It is helpful to acquire several images using different exposure times.

3.3.2 Image Analysis

1. Open images in Living Image® software. To compare signal intensity in separate images, make sure that acquisition parameters, especially exposure time, are identical for all images to be compared. Image settings and appearance can be controlled from the 'Tool Palette' menu in Living Image® software.
2. Switch units to photons. In this mode, luminescence is displayed as photons per second per centimeter squared per steradian (photons/second/cm^2/sr).
3. In 'Tool Palette' > 'Image Adjust' set minimum and maximum for visualization color. Changing color scale only affects signal visualization and does not alter actual photon data.
4. Using 'Tool Palette' > 'ROI Tools' outline the area producing signal, known as region of interest (ROI). To compare similar ROIs in multiple images, such as in images of one mouse after different treatments, use the same ROI area. The best way to select the same ROI is to use 'Copy ROI' and 'Paste ROI' functions. Multiple ROIs can be drawn in one image, which is useful when similar areas need to be quantified.
5. After outlining ROIs, click the 'Measure ROIs' icon. This measures all ROIs in all images that are currently open and generates table with ROI data: total flux (photons/s), minimal, maximal, and average radiance (photons/s/cm^2/sr). These results can be copied and analyzed in a different software such as Microsoft Excel. We analyze and report total bioluminescence flux expressed in photons/sec.

3.3.3 Testing the Responsiveness of the CRE-Luciferase Reporter

The responsiveness of the CRE-luciferase reporter gene in liver should be verified in all weanlings following exposure to CREB-activating stimuli such as fasting and/or glucagon (Fig. 3 and ref. [23]) (*see* **Note 10**).

1. Take ventral bioluminescence images under ad libitum fed conditions at approximately ZT10 (*see* Subheading 3.3.1).
2. After animals recover from anesthesia, fast them by moving the animals to fresh cages with synthetic bedding and no food, but

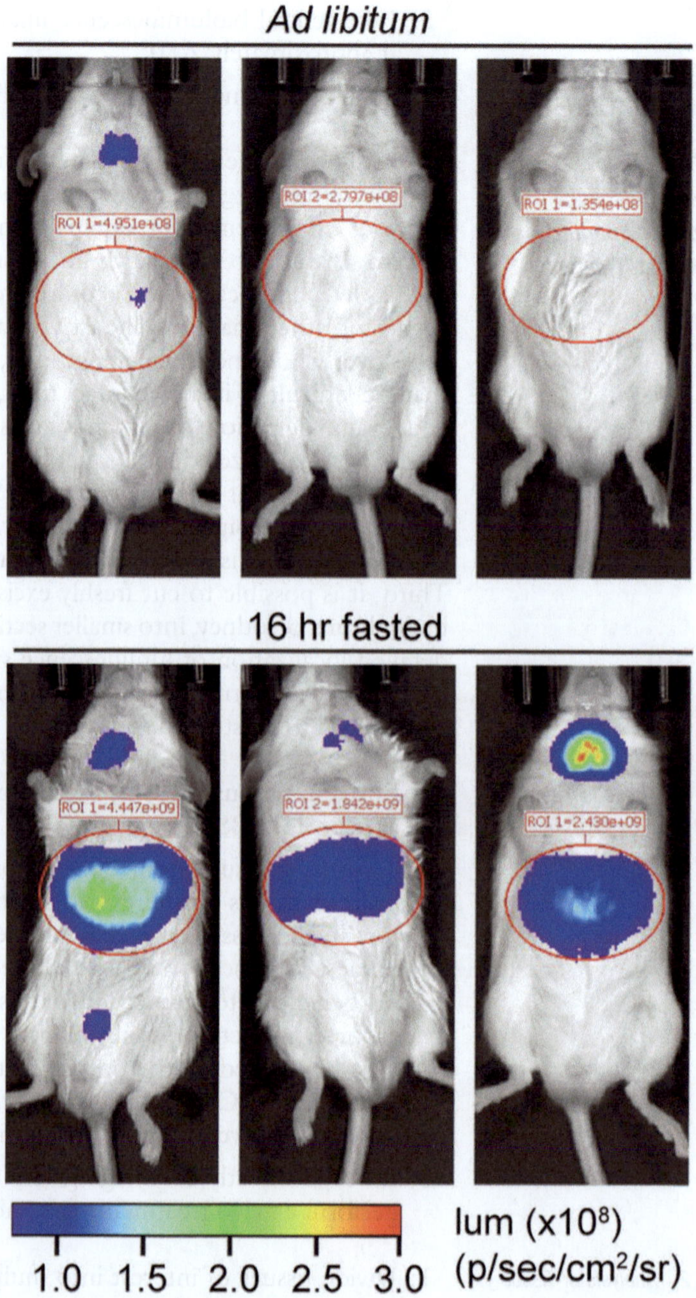

Fig. 3 CREB luciferase reporter imaging in *ROSA26*$^{GsD/+}$ mice. Bioluminescence signal in *ROSA26*$^{GsD/+}$ mice ad libitum fed (ZT10) and after 16 h fasting (ZT2)

with free access to water for 16 h (overnight) to activate endogenous CREB. An important consideration is the time of day images are collected, as CREB activity is under circadian control in many tissues (*see* **Note 11**).

3. Take ventral bioluminescence images under fasted conditions at approximately ZT0.
4. Analyze imaging data (*see* Subheading 3.3.2).

3.4 Ex Vivo Bioluminescence CREB Reporter Imaging of Mouse Tissues

Especially convenient for imaging of small organs, such as brown adipose, pituitary gland, and other specific regions of brain, this method entails removing organs from mice and imaging them ex vivo. It can also be used to image small pieces of larger organs such as liver and heart if it is required to keep the remainder of the organ for other analyses. The ex vivo approach has several advantages over whole mouse imaging. First, it allows imaging of organs that are difficult to image in intact mice, such as heart, kidney, lung, and brain, detection of which signals in vivo is limited by low intensity, small size, depth, shielding by bone, or other factors that attenuate light transmission [24]. Second, it allows the source of luminescence signals in vivo, which may arise from multiple overlapping organs, to be unambiguously identified (Fig. 4a). Third, it is possible to cut freshly excised complex organs such as brain, heart, or kidney, into smaller sections in order to obtain more detailed localization of luminescence signal in the organ. Fourth, using a 96-well format allows the effects of in vitro pharmacologic stimulation to be studied (Fig. 4b).

3.4.1 Tissue Isolation and Analysis of Tissue Immediately Post-Mortem

1. Euthanize mouse by CO_2 inhalation. Quickly excise tissues and rinse in 1× PBS.
2. To image tissues immediately post-mortem, lay out freshly isolated tissues of interest on a petri dish or multi-well plates. For smaller tissues this may be the entire tissue, or for larger tissues, a sub-section may be studied. In some cases, it may also be beneficial to sub-divide or dissect tissues so that CREB reporter induction within internal structures that may otherwise be masked by the opacity of the tissue can be visualized. For example, CREB reporter activity in the renal medulla can only be observed if the kidney is bisected.
3. Proceed directly to imaging (*see* Subheading 3.4.3) and/or perform ex vivo treatments (*see* Subheading 3.4.2).

3.4.2 Analysis of Tissue Responses to Ex Vivo Drug Treatment

1. Divide tissues of interest into multiple even pieces, ~1–2 mm across using a scalpel blade, being careful to consider that internal structure within organs may result in different signal intensities from different pieces.
2. Transfer tissue pieces into individual wells of a black 96-well microtiter plate containing DMEM media (100–200 μl). Add ex vivo treatments to activate GsD or native cAMP formation pathways. The water-soluble forskolin analogue, NKH477 (1 μM) may be used to provide a positive control of maximal

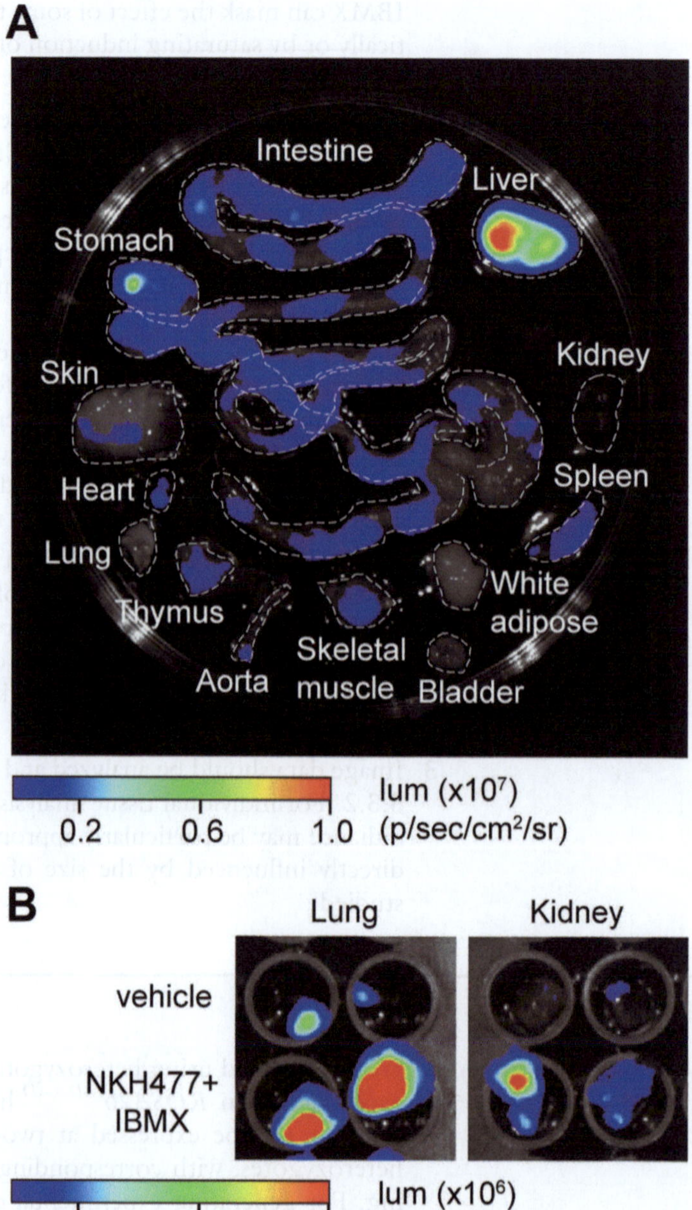

Fig. 4 Ex vivo imaging of CREB luciferase reporter in *ROSA26*$^{GsD/+}$ mice. Representative images showing (**a**) basal CREB-luciferase activity in a panel of tissues freshly excised from *ROSA26*$^{GsD/+}$ mice and (**b**) treatment of tissue ex vivo to induce CREB reporter activity using NKH477 (1 μM) and IBMX (100 μM)

CREB reporter induction (Fig. 4b). Signal intensity may be improved by addition of the non-specific phosphodiesterase inhibitor IBMX (100 μM), however, this requires careful consideration within the design of each experiment as addition of

IBMX can mask the effect of some treatments either mechanistically or by saturating induction of the CREB reporter.

3. Incubate plates in a 37 °C tissue culture incubator for the required amount of time. In our experience, 4 h provides an optimal time for induction of the CREB reporter in response to GsD activation by CNO and agents that acutely induce endogenous cAMP formation. However, the stimulation time may require further optimization depending on experimental design and the nature of the treatments.

3.4.3 Imaging

1. For both approaches, either immediately post-mortem or at the desired time point, add D-luciferin to the tissues to a final concentration of 15 mg/ml—either by addition of 120 mg/ml (8×) concentrated stock to the media, or by applying pre-diluted 15 mg/ml D-luciferin directly to the tissue surface. Transfer plate to IVIS Lumina XR machine.
2. Start acquiring images as described in Subheading 3.3.1. Typical image settings for this type of experiment are: Binning: 'Medium', Field of View: 'D', Exposure time: '1 min'. These settings may require optimization depending on experimental design. It may also be helpful to initially acquire several images using multiple exposure times.
3. Image data should be analyzed and described as in Subheading 3.3.2. For individual tissue analysis, quantification as maximal radiance may be particularly appropriate, as this measure is not directly influenced by the size of each tissue or tissue piece studied.

4 Notes

1. We recommend using heterozygote $ROSA26^{GsD/+}$ mice for all experiments. In $ROSA26^{GsD/GsD}$ homozygote mice, the GsD receptor will be expressed at two-fold higher levels than in heterozygotes, with corresponding increases in cAMP signaling. For generating experimental cohorts, we routinely cross homozygous $ROSA26^{GsD/GsD}$ with WT or transgenic Cre mice.
2. Hair attenuates bioluminescent signals. This is especially problematic in black mice. We recommend shaving mice with hair clippers to expose bare skin before imaging. This is not critical for very strong luminescent signals in albino mice, such as in liver-expressed GsD mice, but can be critical for imaging black mice especially for tissues that produce low intensity signals.
3. TMX solutions are not stable, so it is preferable to prepare only small amounts of solution sufficient for injections in the current animal cohort and keeping the remaining TMX powder at

−20 °C. For injections we recommend using large gauge needles, 20–22G, because of the high viscosity of oil.

4. In vivo imaging systems are also available from other manufacturers including Li-Cor (Pearl® Trilogy Imager) and Berthold Technologies (NightOWL II In Vivo Imaging System). Some instruments such as IVIS Lumina XR also allows taking X-ray images. For ex vivo imaging of small tissue slices or cells, more simple and less expensive systems, such as the Olympus LV200, can be used.
5. In most cases, native GFP fluorescence is difficult to observe in cells expressing GsD-GFP (unpublished observation). We recommend performing immunofluorescence staining using a GFP antibody. This is especially convenient for detection of GsD expressed in specific cell types, especially small cell populations, for which detection by Western blotting of whole tissue lysate is not appropriate. We successfully used immunolabeling to detect GsD expressed in skeletal muscle fibers (Fig. 1b).
6. The Gs-DREADD construct we employed has detectable basal signaling activity [21], though less than that of previous versions [25]. We observed elevated basal CREB activity in mice expressing GsD in liver, without CNO stimulation, as evidenced by increased CREB reporter bioluminescence and increased expression of CREB target genes (Fig. 2a, 2b, and 3c in ref. [21]). Therefore, we strongly recommend incorporation of a GsD + vehicle (no CNO) control group in the experimental design.
7. CNO dose and delivery method require optimization for the tissue targeted and physiology under study. We have successfully used 1 mg/kg CNO injected subcutaneously for activation of GsD expressed in liver and 10 mg/kg for activation of GsD in skeletal muscle satellite cells (unpublished). Other laboratories report DREADD activation by administering CNO in drinking water [18]. CNO is widely used to activate DREADD receptors in animals and has been considered to be pharmacologically inert. A recent report, however, demonstrates that CNO is converted to clozapine in mice when delivered at 10 mg/kg, a concentration often used to activate DREADD signaling in vivo [26]. In this study, mice were trained to associate clozapine with a food reward. When CNO, but not other drugs, was administered, the animals responded in the same way as to clozapine. Therefore, the authors concluded that CNO is not inert. To control for possible off-target effects of CNO, it is important to include control animals that do not express DREADD but are treated with CNO in parallel with the experimental animals.

8. For accurate analysis of signal intensity, the luminescent image should contain between 1000 and 60,000 counts. If 'image is oversaturated' message is displayed in Living Image®, decrease the exposure time.
9. We recommend weighing mice and calculating the volume of 24 mg/ml D-luciferin to be injected in each mouse before the experiment. This saves time during the imaging session, especially if a large cohort of animals is analyzed.
10. During analysis of CREB reporter signals in fasted animals [21], we observed animal-to-animal variability in bioluminescence. Moreover, although most *GsD/+* mice respond to fasting by a several-fold increase in hepatic luminescence, some mice do not have any increase in CREB reporter activity after fasting. We believe that this results from epigenetic silencing of the CRE-containing promoter or luciferase cDNA. We do not observe silencing of the GsD portion of the transgene. In experiments that incorporate the CREB reporter portion of the allele, it is therefore crucial to pre-screen all prospective experimental mice using the simple overnight fasting procedure to eliminate mice lacking bioluminescence response prior to assignment to cohorts. When planning mouse breeding for cohorts, we recommend breeding to obtain 20–25% extra mice to obtain a sufficient number of GsD animals after excluding non-responders. It is particularly important to use only well-responding mice as breeders, as reporter silencing appears to be heritable.
11. Hepatic cAMP/PKA/CREB signaling is regulated in a circadian manner [27]. We have not observed strong circadian regulation of CREB activity in liver in CREB luciferase reporter mice which express the same CREB reporter as GsD mice [23]. However, other tissues may demonstrate pronounced circadian regulation. We therefore recommend testing CREB reporter bioluminescence at different times during the light cycle.

Acknowledgments

The authors gratefully acknowledge financial support from the National Institutes of Health National Institute of Diabetes and Digestive and Kidney Diseases (R01-DK092590 to RB) and the British Heart Foundation (FS/16/1/31699 to NSK, PG/15/47/31591 to JAM and NSK, and RE/13/4/30184 to JAM and NSK). The funders had no role in the study design or preparation of this chapter.

References

1. Hanlon CD, Andrew DJ (2015) Outside-in signaling--a brief review of GPCR signaling with a focus on the Drosophila GPCR family. J Cell Sci 128(19):3533–3542. https://doi.org/10.1242/jcs.175158
2. Luttrell LM (2008) Reviews in molecular biology and biotechnology: transmembrane signaling by G protein-coupled receptors. Mol Biotechnol 39(3):239–264. https://doi.org/10.1007/s12033-008-9031-1
3. Heng BC, Aubel D, Fussenegger M (2013) An overview of the diverse roles of G-protein coupled receptors (GPCRs) in the pathophysiology of various human diseases. Biotechnol Adv 31(8):1676–1694. https://doi.org/10.1016/j.biotechadv.2013.08.017
4. Vassart G, Costagliola S (2011) G protein-coupled receptors: mutations and endocrine diseases. Nat Rev Endocrinol 7(6):362–372. https://doi.org/10.1038/nrendo.2011.20
5. Lappano R, Maggiolini M (2011) G protein-coupled receptors: novel targets for drug discovery in cancer. Nat Rev Drug Discov 10 (1):47–60. https://doi.org/10.1038/nrd3320
6. Kandola MK, Sykes L, Lee YS, Johnson MR, Hanyaloglu AC, Bennett PR (2014) EP2 receptor activates dual G protein signaling pathways that mediate contrasting proinflammatory and relaxatory responses in term pregnant human myometrium. Endocrinology 155 (2):605–617. https://doi.org/10.1210/en.2013-1761
7. Masuho I, Ostrovskaya O, Kramer GM, Jones CD, Xie K, Martemyanov KA (2015) Distinct profiles of functional discrimination among G proteins determine the actions of G protein-coupled receptors. Sci Signal 8(405):ra123. https://doi.org/10.1126/scisignal.aab4068
8. Michal P, El-Fakahany EE, Dolezal V (2007) Muscarinic M2 receptors directly activate Gq/11 and Gs G-proteins. J Pharmacol Exp Ther 320(2):607–614. https://doi.org/10.1124/jpet.106.114314
9. Coward P, Wada HG, Falk MS, Chan SD, Meng F, Akil H, Conklin BR (1998) Controlling signaling with a specifically designed Gi-coupled receptor. Proc Natl Acad Sci U S A 95(1):352–357
10. Armbruster BN, Li X, Pausch MH, Herlitze S, Roth BL (2007) Evolving the lock to fit the key to create a family of G protein-coupled receptors potently activated by an inert ligand. Proc Natl Acad Sci U S A 104(12):5163–5168. https://doi.org/10.1073/pnas.0700293104
11. Pei Y, Rogan SC, Yan F, Roth BL (2008) Engineered GPCRs as tools to modulate signal transduction. Physiology (Bethesda) 23:313–321. https://doi.org/10.1152/physiol.00025.2008
12. Conklin BR, Hsiao EC, Claeysen S, Dumuis A, Srinivasan S, Forsayeth JR, Guettier JM, Chang WC, Pei Y, McCarthy KD, Nissenson RA, Wess J, Bockaert J, Roth BL (2008) Engineering GPCR signaling pathways with RASSLs. Nat Methods 5(8):673–678. https://doi.org/10.1038/nmeth.1232
13. Dong S, Rogan SC, Roth BL (2010) Directed molecular evolution of DREADDs: a generic approach to creating next-generation RASSLs. Nat Protoc 5(3):561–573. https://doi.org/10.1038/nprot.2009.239
14. Wess J, Nakajima K, Jain S (2013) Novel designer receptors to probe GPCR signaling and physiology. Trends Pharmacol Sci 34 (7):385–392. https://doi.org/10.1016/j.tips.2013.04.006
15. Rossi M, Cui Z, Nakajima K, Hu J, Zhu L, Wess J (2015) Virus-mediated expression of DREADDs for in vivo metabolic studies. Methods Mol Biol 1335:205–221. https://doi.org/10.1007/978-1-4939-2914-6_14
16. Nakajima K, Cui Z, Li C, Meister J, Cui Y, Fu O, Smith AS, Jain S, Lowell BB, Krashes MJ, Wess J (2016) Gs-coupled GPCR signalling in AgRP neurons triggers sustained increase in food intake. Nat Commun 7:10268. https://doi.org/10.1038/ncomms10268
17. Alexander GM, Rogan SC, Abbas AI, Armbruster BN, Pei Y, Allen JA, Nonneman RJ, Hartmann J, Moy SS, Nicolelis MA, McNamara JO, Roth BL (2009) Remote control of neuronal activity in transgenic mice expressing evolved G protein-coupled receptors. Neuron 63(1):27–39. https://doi.org/10.1016/j.neuron.2009.06.014
18. Jain S, Ruiz de Azua I, Lu H, White MF, Guettier JM, Wess J (2013) Chronic activation of a designer G(q)-coupled receptor improves beta cell function. J Clin Invest 123 (4):1750–1762. https://doi.org/10.1172/JCI66432
19. Koehler S, Brahler S, Kuczkowski A, Binz J, Hackl MJ, Hagmann H, Hohne M, Vogt MC, Wunderlich CM, Wunderlich FT, Schweda F, Schermer B, Benzing T, Brinkkoetter PT (2016) Single and transient Ca(2+) peaks in Podocytes do not induce changes in glomerular filtration and perfusion. Sci Rep 6:35400. https://doi.org/10.1038/srep35400

20. Zhu H, Aryal DK, Olsen RH, Urban DJ, Swearingen A, Forbes S, Roth BL, Hochgeschwender U (2016) Cre-dependent DREADD (designer receptors exclusively activated by designer drugs) mice. Genesis 54 (8):439–446. https://doi.org/10.1002/dvg.22949
21. Akhmedov D, Mendoza-Rodriguez MG, Rajendran K, Rossi M, Wess J, Berdeaux R (2017) Gs-DREADD knock-in mice for tissue-specific, temporal stimulation of cyclic AMP signaling. Mol Cell Biol 37(9). https://doi.org/10.1128/MCB.00584-16
22. Altarejos JY, Montminy M (2011) CREB and the CRTC co-activators: sensors for hormonal and metabolic signals. Nat Rev Mol Cell Biol 12(3):141–151. https://doi.org/10.1038/nrm3072
23. Akhmedov D, Rajendran K, Mendoza-Rodriguez MG, Berdeaux R (2016) Knock-in luciferase reporter mice for in vivo monitoring of CREB activity. PLoS One 11(6):e0158274. https://doi.org/10.1371/journal.pone.0158274
24. Kirkby NS, Zaiss AK, Urquhart P, Jiao J, Austin PJ, Al-Yamani M, Lundberg MH, MacKenzie LS, Warner TD, Nicolaou A, Herschman HR, Mitchell JA (2013) LC-MS/MS confirms that COX-1 drives vascular prostacyclin whilst gene expression pattern reveals non-vascular sites of COX-2 expression. PLoS One 8(7): e69524. https://doi.org/10.1371/journal.pone.0069524
25. Guettier JM, Gautam D, Scarselli M, Ruiz de Azua I, Li JH, Rosemond E, Ma X, Gonzalez FJ, Armbruster BN, Lu H, Roth BL, Wess J (2009) A chemical-genetic approach to study G protein regulation of beta cell function in vivo. Proc Natl Acad Sci U S A 106 (45):19197–19202. https://doi.org/10.1073/pnas.0906593106
26. Manvich DF, Webster KA, Foster SL, Farrell MS, Ritchie JC, Porter JH, Weinshenker D (2018) The DREADD agonist clozapine N-oxide (CNO) is reverse-metabolized to clozapine and produces clozapine-like interoceptive stimulus effects in rats and mice. Sci Rep 8 (1):3840. https://doi.org/10.1038/s41598-018-22116-z
27. Zhang EE, Liu Y, Dentin R, Pongsawakul PY, Liu AC, Hirota T, Nusinow DA, Sun X, Landais S, Kodama Y, Brenner DA, Montminy M, Kay SA (2010) Cryptochrome mediates circadian regulation of cAMP signaling and hepatic gluconeogenesis. Nat Med 16 (10):1152–1156. https://doi.org/10.1038/nm.2214
28. McCarthy JJ, Srikuea R, Kirby TJ, Peterson CA, Esser KA (2012) Inducible Cre transgenic mouse strain for skeletal muscle-specific gene targeting. Skelet Muscle 2(1):8. https://doi.org/10.1186/2044-5040-2-8

Chapter 22

Laser Doppler Flowmetry to Study the Regulation of Cerebral Blood Flow by G Protein-Coupled Receptors in Rodents

Xavier Toussay, Mario Tiberi, and Baptiste Lacoste

Abstract

A large body of evidence suggests that G protein-coupled receptors (GPCRs) play an important role in the regulation of peripheral vascular reactivity. Meanwhile, the extent of GPCR influence on the regulation of brain vascular reactivity, or cerebral blood flow (CBF), has yet to be fully appreciated. This is of physiological importance as the modulation of CBF depends on an intricate interplay between neurons, astrocytes, pericytes, and endothelial cells, all of which partaking in the formation of a functional entity referred to as the neurovascular unit (NVU). The NVU is the anatomical substrate of neurovascular coupling (NVC) mechanisms, whereby increased neuronal activity leads to increased blood flow to accommodate energy, oxygen, and nutrients demands. In light of growing evidence showing impaired NVC in several neurological disorders, and the fact that GPCRs represent the most important targets of FDA-approved drugs, it is of utmost importance to use experimental approaches to study GPCR-induced regulation of NVC for the future development of pharmaceutical compounds that could normalize CBF function. Herein, we describe a minimally invasive approach called laser Doppler flowmetry (LDF) that, when used in combination with a whisker stimulation paradigm in rodents, allows gauging blood perfusion in activated cerebral cortex. We comprehensively explain the surgical procedure and data acquisition in mice, and discussed about important experimental considerations for the study of CBF regulation by GPCRs using pharmacological agents.

Key words Laser doppler flowmetry, Cerebral blood flow, GPCRs, Rodents, Physiology

1 Introduction

The family of G protein-coupled receptors (GPCR) form the largest group of transmembrane proteins in most living organisms. GPCRs are topologically organized as seven transmembrane domains linked by extracellular and intracellular loops, and harbor an extracellular N-terminus and an intracellular C-terminus of varying length. These structural characteristics have poised GPCRs to discriminate a large spectrum of environmental cues including photons, ions, neurotransmitters, odorants, and hormones [1–5]. Upon interaction with a specific ligand, these

Mario Tiberi (ed.), *G Protein-Coupled Receptor Signaling: Methods and Protocols*, Methods in Molecular Biology, vol. 1947,
https://doi.org/10.1007/978-1-4939-9121-1_22,

versatile molecular devices transduce extracellular cues into intracellular signals via the recruitment of heterotrimeric GTP-binding (G) proteins or non-G protein partners. The wide range of signaling modalities enables GPCRs to regulate a plethora of physiological functions in the central nervous system and peripheral tissues. Not surprisingly, GPCRs have taken center stage in the single largest family of druggable targets of the human genome, as close to ~35% approved FDA-approved pharmaceutical drugs are aimed at GPCRs [6–8].

The homeostatic regulation of cerebral blood flow (CBF) relies on the neurovascular unit (NVU), which is composed of neurons, astrocytes, pericytes, and endothelial cells [9]. A complex interplay within the NVU enables spatio-temporal control of blood supply to activated brain areas in order to fulfill needs in energy, oxygen, and nutrients. This physiological phenomenon is known as neurovascular coupling (NVC), or *functional hyperemia* [10, 11]. While numerous GPCRs have been implicated in the fine-tuning of peripheral vascular reactivity, research investigating the contribution of GPCR biology in the modulation of NVC in the brain is still in its infancy. As compromised neurovascular function has recently become an important hallmark of neurological disorders such Alzheimer's disease, vascular dementia, or even depression, future studies focusing on the mechanistic regulation of NVC and CBF by GPCRs may prove useful in the development of novel therapeutic approaches.

Laser Doppler Flowmetry (LDF) is a minimally invasive method to monitor blood perfusion of various organs, particularly their microcirculation [12–14]. A low intensity beam of monochromatic light, emitted from the laser diode of a flowmeter (or 'tissue perfusion monitor,' TPM), travels via a probe's fiber optic light guide and illuminate the tissue under study. The laser is scattered by reflective components within the tissue, and a portion of light is reflected back into the probe's light guide, and onto the photodetector of the TPM (Fig. 1). The received light has been reflected both by stationary tissue structures and by moving particles (i.e. red blood cells, RBC). Through a Doppler effect, RBCs modify the signal's frequency (wave length). The spectrum of the received signal is processed in the TPM to calculate overall blood flow (as well as RBC velocity and abundance depending on the TPM model used) within the tissue volume sampled by laser Doppler Probe. While the actual volume of tissue sampled by LDF varies with the optical properties of the tissue and the TPM used, it is approximately one cubic millimeter (for details about LDF theory, *see* [15]).

Herein, we will describe a method relying on the use of LDF to investigate the regulation of CBF by GPCRs in mice.

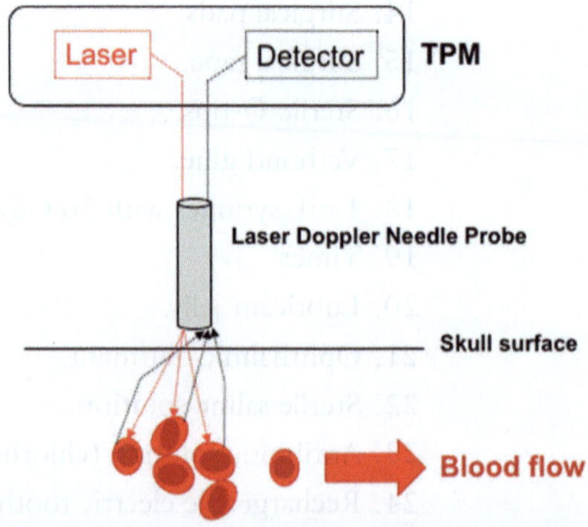

Fig. 1 Laser Doppler principles. A single-frequency light, emitted from the laser diode of a tissue perfusion monitor (TPM), travels via a probe's fiber optic light guide and illuminates a tissue sample. As laser is scattered by moving red blood cells within the tissue, a portion of light is reflected back into the probe's light guide and onto the photodetector of the TPM. By processing the frequency distribution of the backscattered light, an estimate of the blood flow can be achieved

2 Materials

2.1 Items for Surgery

1. Stereotaxic apparatus: standard stereotaxic Instrument for mice (or rats). Our team uses a Harvard Apparatus U-frame, which comes with a base plate, a 3-axis manipulator arm, a mouse adapter, ear bars, and a holder with corner clamp.
2. Mouse heating pad: homeothermic blanket with temp controller and flexible probe (Harvard Apparatus). This feedback loop warming system that allows for the animal's temperature to be maintained.
3. Small animal clipper: autoclaved.
4. Size 3 Scalpel: autoclaved.
5. Size 15 Blades: autoclaved.
6. Tweezers/Forceps: autoclaved.
7. Drill (e.g. Microtorque, Ram Products Inc.): autoclaved.
8. Surgical microscope: adjustable, on wheels, with illumination.
9. Ketamine/xylazine cocktail (*see* Subheading 3.1 for dosing).
10. Xylocaine jelly (lidocaine hydrochloride).
11. Bupivacaine.
12. Buprenorphine hydrochloride.
13. Weighing scale.

14. Surgical pads.
15. Surgical tape.
16. Sterile Q-tips.
17. Vetbond glue.
18. 1 mL syringes with 26 G5/8 needles.
19. Timer.
20. Lubricant jelly.
21. Ophthalmic ointment.
22. Sterile saline solution.
23. Antibiotic solution (chlorhexidine).
24. Rechargeable electric tooth brush.

2.2 Hardware and Software (See Note 1)

1. Flowmeter: BLF22, single-channel tissue perfusion monitor (Transonic Scisence Inc.). This model displays blood flow, mass, and velocity.
2. LDF probe: optic fiber laser needle probe ABLPHN18 (Transonic Scisence Inc.)
3. Data acquisition hardware: Powerlab 8/35, PL3508/P, 8 channels (ADInstruments).
4. BNC-BNC cable: 6 ft FC-BNC-6 (Transonic Scisence Inc.).
5. Heart rate/breathing rate monitor: pulse transducer TN1012/ST (ADInstruments). It uses a piezo-electric element to convert force applied to the active surface of the transducer into an electrical analog signal.
6. Blood pressure monitor: physiological pressure transducer (MLT844), pressure gauge (MLA1052), bridge amp (FE221), all by ADInstruments.
7. Blood gas machine: ABL80 FLEX (Radiometer Copenhagen).
8. Computer: laptop, Mac or PC, equipped with analysis software.
9. Data acquisition and analysis software: LabChart8 or LabChartPro (ADInstruments). Previous versions (e.g. LabChart7) can be used, but the latest version is presented here (*see* **Note 2**).

3 Methods

This procedure relies on stimulation of the whisker-to-barrel cortex pathway and it can be performed on any rodent. However, for the purpose of this protocol, we will present steps to be followed for work in mice. All procedures should be carried out at room temperature unless otherwise specified. Turn all systems ON before

Table 1
Preparation of the ketamine/xylazine cocktail

	Concentration	Dose	Volume
Ketamine	100 mg/mL	100 mg/kg	1.0 mL
Xylazine	20 mg/mL	10 mg/kg	0.5 mL
Saline	0.9%		8.5 mL

starting (i.e. computer, TPM, homeothermic blanket, etc.), as well as the LabChart software.

3.1 Surgery

1. Prepare injectables (ketamine/xylazine cocktail) as described in Table 1.
2. Weigh the animal.
3. Anesthetize the animal with an intraperitoneal (i.p.) injection of the ketamine/xylazine cocktail (0.1 mL/100 g) (*see* **Note 3**).
4. Place the animal in an incubator chamber settled at 37 °C to facilitate the drowsiness. Start timer. The experiment should be performed within 45 min, as the first anesthesia will last a maximum of 60 min (*see* **Note 4**).
5. When full anesthetized, shave off fur on animal's head with the clipper. Make sure you protect the eyes and the whiskers (*see* **Note 5**). Clean the scalp with gauze soaked with antibiotic solution, hydrate eyes with ophthalmic ointment (completely cover eyes, so that they do not dehydrate).
6. Apply a significant amount of xylocaine jelly (lidocaine hydrochloride) to each ear using Q-tips to prevent pain from stereotaxic ear bars.
7. Move the animal to the surgical area (i.e. stereotaxic frame) and place it on a surgical pad that covers the homeothermic blanket. Apply lubricant jelly to temperature probe and insert it in the animal's rectum. Cover or wrap animal to maintain body temperature.
8. Fix the mouse head using ear bars. By pinching the side of the head, open the animal's mouth and slide upper teeth onto tooth bar (*see* **Note 6**). Place pulse transducer under the mouse's abdomen.
9. Make straight incision on the scalp with sterile scalpel (*see* **Note 7**). Incision should be at the midline, and run anterior/posterior, from between the animal's eyes to lambda. Cut through skin and periosteum. Clean the incised area with sterile saline using Q-Tips. Use new Q-tips to dry the region and push

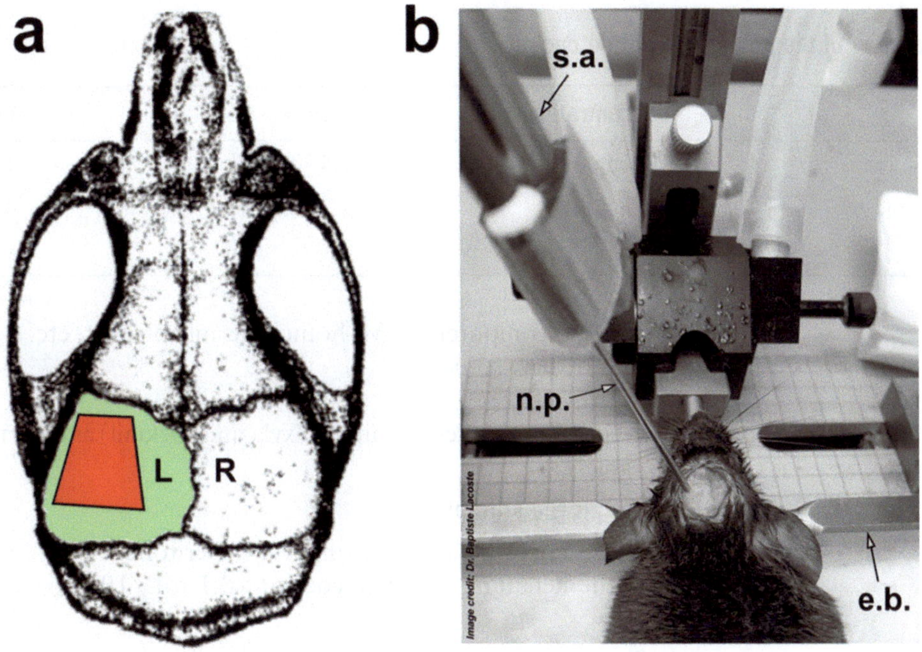

Fig. 2 Positioning of the laser Doppler needle probe over the mouse somatosensory cortex. (**a**) The portion of the skull located above the parietal cortex is indicated in green. The precise location where the primary somatosensory cortex will be found is depicted by the red trapezoid. R: right hemisphere, L: left hemisphere. (**b**) Photograph showing an anesthetized mouse, with open scalp and thinned skulled, fixed in a stereotaxic frame between ear bars (e.b.). The laser Doppler needle probe (n.p.) can be seen secured on the stereotaxic arm (s.a.) and positioned right above the left primary somatosensory cortex

scalp away to expose the skull surface. Make sure to expose the somatosensory cortex (*see* Fig. 2).

10. Under the surgical microscope, gently thin the skull bone to reach translucidity, clean well, and hydrate while thinning (*see* **Note 8**).

3.2 Cerebral Blood Flow Recording

1. Make sure the LDF system, computer, and software are turned ON.
2. Install the LDF needle probe on the stereotaxic arm and fix it using surgical tape.
3. Place the LDF needle probe right above the skull (3–500 μm in theory), perpendicular to the exposed cortex. Target the left somatosensory region (*see* Fig. 2), and avoid placing the probe right above big pial arteries to prevent biased (noisy) signal.
4. Turn laser ON (*see* **Note 9**) and press "start" in LabChart recording window. Read the various recorded channels and make sure the breathing trace is normal. Note that description of LabChart software's manual goes beyond the scope of this chapter (please refer to the manufacturer's instructions).

5. Make sure that the baseline flow trace is stable and in normal ranges. For a young healthy adult mouse (~2-month old), values measured above the somatosensory region should be around 15–20 Tissue Perfusion Units (TPU), and ideally around 15 TPU (*see* **Note 10**). The baseline flow trace should appear stabilized before starting any recording with whisker stimulation.

3.3 Whisker Simulation Paradigm to Monitor Neurovascular Coupling

1. Use electric tooth brush to stimulate whiskers, as described elsewhere [16, 17]. The LDF probe is positioned over the left somatosensory (S1) cortex during stimulations of the right whiskers.
2. By slowly moving the needle probe Look for the S1 cortex by stimulating whiskers: a ~15–20% blood flow increase from baseline should be observed in a healthy adult mouse.
3. Perform 20 s stimulations with 60 s intervals (*see* **Note 11**). Mark the start and end of each stimulation in the software (function keys can be designated for this task). Six recordings are acquired and averaged for each experimental animal (*see* Fig. 3).
4. After the last stimulation, record baseline TPU for 3 min (*see* **Note 12**).

3.4 Recovery of the Animal

1. Dampen sterile Q-tip with sterile saline and use it to push skin back and close incision site.
2. Apply Vetbond glue to incision site and use forceps to hold while drying (try to be quick, as glue dries fast). Avoid dropping glue on the mouse's eyes or ears.
3. Apply Bupivacaine to incision site and spread around with sterile Q-tip.
4. Inject a dose (0.05 mg/kg) of buprenorphine hydrochloride subcutaneously.
5. Allow the mouse to recover in a ventilated incubator at 37 °C.

3.5 Pharmacological Treatments

The high temporal resolution of LDF makes this method particularly suitable to study the regulation of CBF by GPCRs. Indeed, these receptors are tightly regulating vasomotor responses, both in brain and periphery [18]. Moreover, an important characteristic of GPCRs is their dynamic regulation by (1) ligand concentration, (2) desensitization/recycling mechanisms, and (3) subcellular compartmentalization [19, 20]. It is thus possible to investigate the role of several GPCRs in neurovascular regulation by assessing changes of LDF recordings (before, during, and after whisker stimulation) in response to various ligands (agonists or antagonists) and doses. Furthermore, the BLF22 perfusion monitor will allow to precisely assessing the timing of ligand effects using continuous

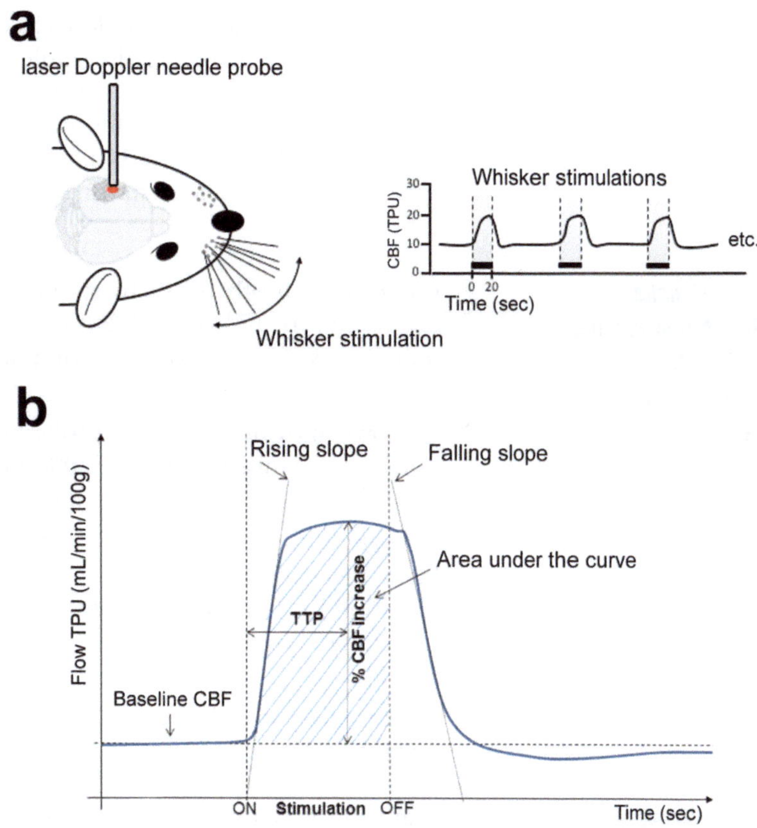

Fig. 3 LDF procedure to measure CBF responses to neuronal activation over the mouse primary somatosensory cortex. (**a**) Schematic representation of the experimental paradigm. The needle probe is positioned over the left primary somatosensory cortex, and the signal is recorded before, during, and after stimulations (20 s) of the right whiskers. Six stimulations, as well as baseline values, will be acquired and averaged for each mouse. (**b**) Parameters used to quantify and characterize CBF responses to sensory stimulation (i.e. to neuronal activation). Arrows indicate hemodynamic features that will be considered for in-depth analysis

LDF recordings with time constants of 0.1, 1.0, or 3.0 s. The following section does not aim at giving a precise protocol, as this highly depends on the researcher's scientific question. There are three major considerations for pharmacological studies involving LDF, which are as follows:

1. Choose the injection route, depending on the target and biological question. An i.p. injection may be efficient for a ligand that crosses the blood–brain barrier (BBB). However, it will be difficult to dissect the peripheral versus central consequences of the treatment. For (1) drugs that do not cross the BBB and (2) exclusively central targets it is essential to inject

the drug directly into the cerebrospinal fluid either in the cerebral ventricles (i.c.v.) or in the *cisterna magna*.

2. Design the experiment properly, depending on the ligand's bioavailability and half-life. It will be crucial to record both baseline and CBF responses before and after the injection. Moreover, a control experiment using the vehicle (i.e. solution in which the drug was dissolved) must be conducted in order to evaluate the effect(s) of the drug itself. Finally, important questions to consider will be, for instance: How long does the effect last? When is the maximal effect reached? Does the effect lose potency over time? Is the effect reversible? Etc.
3. Monitor the peripheral effects, which may strongly affect the conclusions of the study. It is considered mandatory to record mean arterial blood pressure (BP), heart rate, blood gazes (pCO_2 & pO_2), as well as blood pH before, during, and after the drug injection. These parameters are of fundamental importance in physiology and are often asked by reviewers. Note that an ideal way to both monitor BP and blood contents is through cannulation of the femoral artery coupled to a pressure transducer.

4 Notes

1. Performing this experiment does not require the use of a specific LDF flowmeter system. However, for the purpose of this protocol we describe usage of the TPM from Transonic Scisence Inc., with additional hardware from ADInstruments, material routinely used by our team.
2. The system requirements will vary depending on whether the experimenter uses PC or Mac. For PC users, LabChart 8.1.5 or later requires Microsoft Windows 10, Microsoft Windows 8.1 or Microsoft Windows 7. LabChart 8.0 to 8.1.4 versions are not compatible with Windows 10 and require Microsoft Windows 8.1 or Microsoft Windows 7. Furthermore, PC users will need Microsoft. NET Framework 4 or later, Internet Explorer 8.0 or later, a USB interface for operation with PowerLab, and a minimum screen resolution of 1280 × 768 to support common projector settings. For Mac users, OS × v10.8 Mountain Lion or later is needed. LabChart 8.1.1 or later is required for Mac OS × 10.11 El Capitan.
3. While no drug is perfect when it comes to impact on the brain vasculature, it is strongly recommended to avoid using isoflurane anesthesia during LDF recordings. Indeed, isoflurane induces potent vasodilation of pial arteries [21–24], which

can affect both baseline values and CBF responses to sensory stimulation.

4. Otherwise warm the animal under red light for 10–15 min. During the course of the experiment, always control animal's body temperature (maintained at 37 ± 1 °C). For longer procedures, to keep the animal well hydrated, inject 1 mL of sterile saline subcutaneously. Importantly, a second injection of 1/3 dose of anesthetic cocktail will be needed ~45 min after the start of the first anesthesia.
5. If animals are missing whiskers the experiment cannot be performed as the LDF procedure involves whisker stimulation-induced blood flow responses.
6. Ensure that head is straight medial-laterally and head is secure. Precise stereotaxic coordinates would not be needed; yet, the head has to be very well fixed.
7. It is preferable that this step is completed once animal is fixed in the stereotaxic frame.
8. Experimenter should be careful here as too much thinning warms the bone and underlying tissues. It is also critical not to perforate the bone, as the LDF measures are made through the bone.
9. Eliminate direct light sources on the tissue under study (e.g. turn OFF the surgical microscope light).
10. In older animals or in disease mouse models, the baseline may be either higher of lower. In addition, if the baseline is low (around or below 10 TPU) in a healthy mouse, it might mean that the skull bone is too thick.
11. To prevent recording artifacts, make sure to only contact whiskers with the tooth brush without touching the nose fur or the eye. Moreover, any slight touch to the stereotaxic frame will create noise. Do not perform more than # stimulations in each mouse as whisker hyperstimulation will saturate neurovascular coupling responses.
12. Monitoring baseline at the end of the stimulation paradigm will help assess the animal's stability. Note that these parameters can be asked by reviewers.

References

1. Wootten D et al (2018) Mechanisms of signalling and biased agonism in G protein-coupled receptors. Nat Rev Mol Cell Biol 19:638–653
2. Rosenbaum DM, Rasmussen SG, Kobilka BK (2009) The structure and function of G-protein-coupled receptors. Nature 459 (7245):356–363
3. Venkatakrishnan AJ et al (2013) Molecular signatures of G-protein-coupled receptors. Nature 494(7436):185–194
4. Husted AS et al (2017) GPCR-mediated signaling of metabolites. Cell Metab 25 (4):777–796

5. Vass M et al (2018) Chemical diversity in the G protein-coupled receptor superfamily. Trends Pharmacol Sci 39(5):494–512
6. Hauser AS et al (2017) Trends in GPCR drug discovery: new agents, targets and indications. Nat Rev Drug Discov 16(12):829–842
7. Hauser AS et al (2018) Pharmacogenomics of GPCR drug targets. Cell 172(1–2):41–54.e19
8. Sriram K, Insel PA (2018) G protein-coupled receptors as targets for approved drugs: how many targets and how many drugs? Mol Pharmacol 93(4):251–258
9. Andreone BJ, Lacoste B, Gu C (2015) Neuronal and vascular interactions. Annu Rev Neurosci 38:25–46
10. Cauli B, Hamel E (2010) Revisiting the role of neurons in neurovascular coupling. Front Neuroenerg 2:9
11. Lecrux C, Hamel E (2011) The neurovascular unit in brain function and disease. Acta Physiol (Oxf) 203(1):47–59
12. Micheels J, Alsbjorn B, Sorensen B (1984) Laser doppler flowmetry. A new non-invasive measurement of microcirculation in intensive care? Resuscitation 12(1):31–39
13. Rajan V et al (2009) Review of methodological developments in laser Doppler flowmetry. Lasers Med Sci 24(2):269–283
14. Bonner R, Nossal R (1981) Model for laser Doppler measurements of blood flow in tissue. Appl Opt 20(12):2097–2107
15. Fredriksson I, Fors C, Johansson J (2007) Laser doppler flowmetry—a theoretical framework. Department of Biomedical Engineering, Linköping University, Linköping. http://www.imt.liu.se/bit/ldf/ldfmain.html
16. Lacoste B et al (2013) Cognitive and cerebrovascular improvements following kinin B1 receptor blockade in Alzheimer'' disease mice. J Neuroinflammation 10:57
17. Toussay X et al (2013) Locus coeruleus stimulation recruits a broad cortical neuronal network and increases cortical perfusion. J Neurosci 33(8):3390–3401
18. Maguire JJ, Davenport AP (2005) Regulation of vascular reactivity by established and emerging GPCRs. Trends Pharmacol Sci 26 (9):448–454
19. Magalhaes AC, Dunn H, Ferguson SS (2012) Regulation of GPCR activity, trafficking and localization by GPCR-interacting proteins. Br J Pharmacol 165(6):1717–1736
20. Eichel K, von Zastrow M (2018) Subcellular organization of GPCR Signaling. Trends Pharmacol Sci 39(2):200–208
21. Larach DR, Schuler HG (1991) Direct vasodilation by sevoflurane, isoflurane, and halothane alters coronary flow reserve in the isolated rat heart. Anesthesiology 75(2):268–278
22. Schwinn DA, McIntyre RW, Reves JG (1990) Isoflurane-induced vasodilation: role of the alpha-adrenergic nervous system. Anesth Analg 71(5):451–459
23. Endoh H et al (2001) Cerebral autoregulation during sevoflurane or isoflurane anesthesia: evaluation with transient hyperemic response. Masui 50(12):1316–1321
24. Crystal GJ, Salem MR (2003) Isoflurane causes vasodilation in the coronary circulation. Anesthesiology 98(4):1030

Chapter 23

Preparation of Pulmonary Artery Myocytes and Rings to Study Vasoactive GPCRs

Martha Hinton, Anurag Singh Sikarwar, and Shyamala Dakshinamurti

Abstract

G protein-coupled receptors (GPCR) are crucial transducers of extracellular signals into changes in vascular tone. Vasoactive GPCR stimulation in the pulmonary circuit may be elicited by agonists released in acute tissue hypoxia or inflammation, as well as chronic disease. Acute responses involve activation of smooth muscle contraction or relaxation machinery causing changes in actomyosin interaction, thereby altering lumen diameter. Chronic responses may typically include activation of proliferation or fibrosis. Using pulmonary artery myocytes and pulmonary artery rings, we describe a general strategy for quantification of vasoconstrictor or vasodilator GPCR responses, and for comparison of signaling pathways in cultured cells and in contracted vessels using immunohistochemistry of contracting vessels.

Key words GPCR, Pulmonary artery, Smooth muscle cells, Vasoconstriction, Myography

1 Introduction

G protein-coupled receptors (GPCR) are transmembrane proteins important as drug targets and effectors of systemic signaling. Smooth muscle GPCRs transduce extracellular signals across the cell membrane via guanine-binding proteins (G-proteins) to generate second messengers specifically causing contraction or relaxation; or to trigger physiological processes such as phenotypic change, protein synthesis, proliferative activation or apoptosis.

Individual GPCR subtypes have characteristic functional coupling with effector G proteins; some receptors are capable of coupling to several Gα subunits. Among vasoactive GPCRs, those coupling to Gαq/11 subunits typically elicit calcium mobilization from the endoplasmic reticulum, via IP_3 or ryanodine receptor activation. Receptors coupling to Gαq/11 and Gα12/13 may also activate calcium sensitization pathways. Receptors activating Gαs trigger adenylyl cyclase activity elaborating cAMP; and those coupling to Gαi, attenuation of adenylyl cyclase activity [1].

Mario Tiberi (ed.), *G Protein-Coupled Receptor Signaling: Methods and Protocols*, Methods in Molecular Biology, vol. 1947, https://doi.org/10.1007/978-1-4939-9121-1_23,

Within the pulmonary circulation, GPCR expression varies from proximal to distal vessels, and is influenced by developmental stage and disease state [2]. Pulmonary artery myocytes are known to express over 135 GPCRs, many of which are orphan receptors of unknown function [3]. Pulmonary arterial myocytes respond to circulating vasoconstrictor and vasodilator factors, as well as those secreted from the lung parenchyma and endothelium [4]. In pulmonary hypertensive disorders, increased vascular resistance derives from sustained contraction of pulmonary arterial smooth muscle cells, and enhanced proliferation of cells in the vascular adventitia and media. Investigation of the functional activity of known and novel GPCRs in the pulmonary circuit is critical, both for greater insight into the pathophysiology of pulmonary hypertension, and for the search for new and exploitable therapeutic targets.

Clinical management of elevated pulmonary vascular resistance in pulmonary hypertensive disorders has led to a search for alternative GPCRs expressed in the pulmonary circuit, including orphan receptors, and candidate vasoactive agents that may be utilized to target them [5]. New GPCRs are identified by protein similarity to known receptor structures, using computational prediction; by whole genome screening; or by biochemical analyses of responses to known or predicted receptor ligands [6]. Proteins identified through these screening processes may include orphan receptors for which agonists or antagonists are yet unknown; this area holds greatest promise for the development of novel therapeutics [7]. In order to characterize known and novel GPCRs, easily standardized methods must be employed to compare constrictor versus dilator activities; and to characterize second messenger systems operative downstream of these receptors, in cell as well as tissue models. Comparisons of data derived from cell culture systems with tissue responses require standardization for cell phenotype, as the majority population of myocytes arterial wall exist in a contractile phenotype [8]. This phenotype typically is induced in cell culture by serum deprivation [9]. Proliferative or synthetic phenotype myocytes may express fewer vasoactive GPCRs at the cell surface, have attenuated responses to vasoactive agents or altered receptor coupling compared to contractile myocytes [10, 11].

The binary actions of vasodilation and vasoconstriction are primary functions of smooth muscle cells in the arterial media, are developmentally and spatially sensitive, but may be masked by structural changes due to arterial remodeling and rarefaction, which separately contribute toward circuit resistance. These actions may be endothelium-dependent, arising from GPCR or other receptor activation in endothelial cells and transduced by endothelial second messengers, or endothelium-independent, requiring activation of GPCRs in smooth muscle. In this chapter, while we acknowledge the importance of endothelial signaling, for simplicity

of purpose the focus is placed on characterization of smooth muscle GPCR responses—and here, specifically responses resulting in immediate mechanical contraction or dilation of cells or vessels, rather than the whole panoply of smooth muscle GPCR responses. Importantly, it is difficult to capture GPCR hyperreactivity to contractile agonists in tissue myography preparations, without controlling for smooth muscle bulk; as an increase in total smooth muscle content of the arterial media may also give rise to heightened force of contraction to agonist challenge, without a true change in contractile response per muscle, or per receptor. Non-standardized data should not be misinterpreted as a change in GPCR-specific activity. Standardization of agonist-induced force measurements to the maximal KCl-induced contraction of that vessel is appropriate where the calcium response to direct smooth muscle depolarization is unaffected by disease state. However, we propose that the standardization of force responses to smooth muscle volume within the same vessel may be a more unshakable approach to distinguish true hyperreactivity from hyperresponsiveness to agonist, in tissues from all disease conditions.

In this chapter, we outline methods for synchronization of cultured pulmonary arterial smooth muscle cells (PASMC) in a contractile phenotype, and a basic approach to determination of smooth muscle GPCR vasoactivity. Given how much is written elsewhere on GPCRs in PASMC with respect to cell proliferation and fibrosis signaling, we will focus on approaches to quantify the end-game for contractile or relaxant signaling (as intracellular calcium mobilization or cAMP generation respectively, in cultured cells; or using isometric myography). To ensure that PASMC data and pulmonary arterial tissue data are truly capturing the same phenomena, we also outline a novel method for capturing GPCR-mediated signaling in small conduit arteries, following isometric force quantification of vasoconstriction or vasodilation to GPCR agonists, by timed fixation of contracted or dilated arteries in situ, for interrogation of signaling pathway second messengers or phospho-proteins by immunohistochemistry. This approach may elucidate GPCR signaling cascades linked to vasoactive agonists in the pulmonary circuit.

2 Materials

2.1 PASMC Primary Culture

1. 10 cm^2 Tissue culture plates.
2. Tissue culture chamber (4-well slides; LabTek).
3. 500 ml Glass bottles.
4. 50 ml Falcon tubes.
5. 15 ml Corning tubes.

6. Vacuum sterilization filter Tops (Millipore).
7. Gentamycin (Amresco).
8. Antibiotic/Antimycotic (Ab/Am).
9. Fetal bovine serum (FBS).
10. Penicillin and Streptomycin (PS) Solution.
11. Insulin-Transferrin Selenium-X (ITS) (Invitrogen).
12. Trypsin.
13. Ham's F-12 nutrient mixture +L-glutamine.
14. Ca^{2+}-free Krebs-Henseleit (KH) buffer: 112.6 mM NaCl, 25 mM $NaHCO_3$, 1.38 mM, NaH_2PO_4, 4.7 mM KCl, 2.46 mM $MgSO_4$, 5.56 mM Dextrose, pH 7.4.
15. HEPES-buffered saline (HBS) solution: 130 mM NaCl, 5 mM KCl, 1.2 mM $MgCl_2$, 10 mM HEPES, 10 mM glucose; pH 7.4.
16. Digestion solution: 20 μM $CaCl_2$, 1750 U/ml type I collagenase, 1 mM dithiothreitol, 2 mg/ml bovine serum albumin, and 9.5 U/ml papain in HBS.
17. Class II biological safety cabinet (Thermo Fisher Scientific 1300 series A2).
18. Centrifuge.
19. Dissection tools (scalpel handle #4, scalpel blades #20, scissors, forceps; Fine Science Tools).
20. Fine dissection tools (microscissors, microforceps; Fine Science Tools).
21. Stereo microscope.
22. Inverted light microscope.
23. Shaking Water bath.
24. CO_2 Incubator.
25. Cell dissociation sieve—Tissue Grinder Kit.
26. Cell counter.
27. Trypan blue dye.

2.2 Quantification of Contractile Signal In Vitro by Live-Cell Calcium Mobilization (Fura-2AM)

1. Tissue culture chamber (4-well slides; LabTek).
2. Fura-2AM (Molecular Probes).
3. Pluronic acid.
4. Dimethyl sulfoxide.
5. HBSS: 1.26 mM $CaCl_2$, 0.493 mM $MgCl_2$-$6H_2O$, 0.407 mM $MgSO_4$-$7H_2O$, 5.33 mM KCl, 0.441 mM KH_2PO_4, 4.17 mM $NaHCO_3$, 137.93 mM NaCl, 0.338 mM $NaHPO_2$, with 0.1% BSA.

6. Desired GPCR agonists or antagonists for experimentation, in appropriate dilution.
7. Inverted digital microscope (Olympus).
8. Shutter Filters.
9. UltraPix FSI digital camera and Ultraview4 software (PerkinElmer).
10. CO_2 Incubator (Thermo Scientific).

2.3 Quantification of Relaxant Signal In Vitro by Adenylyl Cyclase Activity Assay

1. BD Falcon 96-well Black/Clear Tissue Culture Treated Plate, Flat Bottom (Reaction Plate)
2. Greiner Bio-One 96-well V-bottom Plate (source plate).
3. FlexStation 200 μl Pipette Tips
4. Terbium (III) chloride.
5. Norfloxacin.
6. Adenosine 5′-triphosphate magnesium salt.
7. Tris base.
8. Desired GPCR agonists or antagonists for experimentation, in appropriate dilution.
9. Forskolin.
10. Flex Station-3 fluorescence plate reader (Molecular Devices 1300 series A2).
11. SoftMax Pro 5.3 software (Molecular Devices).

2.4 Quantification of Contraction and Relaxation in Pulmonary Artery Rings by Isometric Wire Myography and Ring Fixation for Histology

1. Low Ca^{2+} KH buffer: 112.6 mM NaCl, 25 mM $NaHCO_3$, 1.38 mM NaH_2PO_4, 4.7 mM KCl, 2.46 mM $MgSO_4$, 1.91 mM $CaCl_2$, 5.56 mM dextrose).
2. High KCl Ca^{2+} KH buffer: 50 mM KCl, 73.1 mM NaCl, 25 mM $NaHCO_3$, 1.38 mM NaH_2PO_4, 2.46 mM $MgSO_4$, 5.56 mM dextrose, 2.5 mM $CaCl_2$).
3. Acetylcholine chloride.
4. 10% neutral buffered formalin.
5. Desired GPCR agonists or antagonists for experimentation, in appropriate dilution.
6. Multiwire myograph system, Danish Myo Technology (DMT) connected to 95% O_2 and 5% CO_2 source.
7. PowerLab 8/35, LabChart Pro Software (with DMT Normalization Module).
8. Mayo scissors (Fine Science Tools).
9. Wescott spring scissors (Fine Science Tools).
10. Vannus spring scissors (Fine Science Tools).

11. Dumont #5 and #5/45 forceps (Fine Science Tools).
12. Black Sylgard-coated petri dish.
13. Monoject™ standard hypodermic needles #200.
14. Dissecting microscope with calibrated eyepiece.
15. Mounting wires (40 μm or 25 μm) or mounting pins.
16. Steel wires or horsetail hair for endothelium denudation.

3 Methods

3.1 PASMC Primary Culture

1. Sacrifice animals by an intraperitoneal overdose of pentobarbital, and access chest by parasternotomy (*see* **Note 1**). Remove heart and lungs en bloc and immediately place in ice-cold, oxygenated calcium-free KH buffer. On black Sylgard-coated petri dish filled with ice-cold calcium-free KH buffer, remove heart and other tissues. Separate the lungs into right and left lobes. Under a dissecting microscope, dissect third-sixth generation PA tissue out of the lungs by removal of overlying bronchi and following the branching of the second generation PA (*see* **Note 2**). Allow PA to recover for 30 min at 4 °C in HBS + 1.5 mM $CaCl_2$ + 1% Ab/Am +0.1% gentamycin, and wash twice during that time period.
2. Incubate PA tissue in HBS + 20 μM $CaCl_2$ + 1% Ab/Am +0.1% gentamycin for 20 min at room temperature with three washes (*see* **Note 3**).
3. Transfer the pulmonary arteries to a sterile surface and remove as much HBS as possible; finely mince the PA tissue with a scalpel.
4. Transfer the minced PA tissue to 5 ml of digestion solution in a 15 ml Corning tube and incubate in a 37 °C shaking water bath for 30 min.
5. Centrifuge the Corning tube for 5 min at 200 × *g* at room temperature. Aspirate the Digestion Solution and replace it with HBS + 1% Ab/Am +0.1% gentamycin.
6. To wash away residual digestion solution, suspend the cells by gentle trituration and centrifuge once more for 5 min at 200 × *g*.
7. Resuspend cell pellet with Ham's F-12 with 10% FBS, 1% PS and then filter through a sterile metal mesh screen to get a single cell suspension.
8. Plate cells at 1×10^5 cells/cm^2 in Ham's F-12 + 10%FBS +1% PS in tissue culture plates or chambers.

9. Grow PASMC to 80% confluence with media changed every 72 h and then passage cells once using 0.05% Trypsin with 0.02% EDTA.

10. To synchronize the PASMCs in a contractile phenotype, at 80% confluence, serum deprive cells for 48 h in Ham's F-12 + 1% PS + 1% ITS (*see* **Note 4**). Thereafter, split cells into treatment conditions following a media change.

3.2 Quantification of Contractile Signal In Vitro by Live-Cell Calcium Mobilization (Fura-2AM)

The method is adapted from Grynkiewicz et al. [12]; all steps carried out in the dark or light-shielded.

1. After thawing the vial of fura-2 AM, add 50 μl DMSO directly to the vial and vortex. Add 10 μl of the solubilized dye and 1 μl of pluronic acid to 1 ml of HBSS to make the fura2AM loading media (*see* **Note 5**).

2. Aspirate the serum deprivation media. Following two washes of HBSS with 0.1% BSA (HBSA) (*see* **Note 6**), load PASMC with fura-2AM loading media for 1 h in a CO_2 incubator. After 1 h remove the loading media by suction and wash the cells with HBSA three times with gentle agitation to remove residual dye. Then allow cells to recover in a final volume of precisely 400 μl HBSA in the dark for 30 min at room temperature, allowing for complete intracellular cleavage of AM esters to release fura-2 dye. Any channel or receptor blockers to be tested can be added before this 30 min incubation.

3. Fix the plate onto the microscope stage and focus on a confluent monolayer of cells.

4. To obtain a recording, begin with a stable baseline. Add 100 μl of your GPCR agonist (100 μl and 5 times concentrated to give 1× when mixed with 400 μl of HBSA in the chamber), and allow peak and fall of Ca^{2+} response. Remove agonist by repeated washes with HBSS to return to baseline.

5. Analyze after subtracting background captured from a cell-free area. Calculate the peak Ca^{2+} mobilization in terms of 340/380 wavelength emission ratio by subtracting the stable baseline from the maximum 340/380 value, thus normalizing for loading conditions (*see* **Note 7**).

3.3 Quantification of Relaxant Signal In Vitro by Adenylyl Cyclase Activity Assay

The method is adapted from Spangler et al. [13].

3.3.1 Reagent Preparation

1. Make 5 ml of 1.0 mM Terbium III chloride in 20 mM Tris buffer (pH 7.4) supplemented with 20 mM $MgCl_2$, 40 μM $CaCl_2$, and 0.1 g BSA.

2. Prepare 0.22 mM Norfloxacin by first dissolving the powder in concentrated HCl, and then diluting in 5 ml of 20 mM Tris buffer (pH 7.4)0.3. Add the Terbium solution to the Norfloxacin Solution in a 1:1 ratio.
3. Challenge live PASMCs with GPCR agonists or antagonists, or 1 μM forskolin (receptor-independent adenylyl cyclase activator), or diluent for 30 min, then lyse cells (*see* **Note 8**). Determine lysate protein content by Bradford method, and adjust lysates to 5 μg protein/μl, in 20 mM Tris buffer without protease inhibitors (*see* **Note 9**).
4. Add 50 μl of 3× the desired concentration of ATP and 50 μl of Terbium-Norfloxacin solution to the reaction plate.
5. Add 100 μl of lysate/well to the source plate. Allow the source plate and reaction plate to equilibrate at 37 °C in the flex station. Set the FlexStation to take measurements immediately following transfer of 50 μl of the lysate to the reaction plate, and every 20 s thereafter for 3 min with excitation/emission wavelengths of 337/545 nm (*see* **Note 10**).

3.4 Quantification of Contraction and Relaxation in Pulmonary Artery Rings by Isometric Wire Myography and Ring Fixation for Histology

1. Collect lungs freshly in ice-cold calcium free KH buffer. Separate the right and left lung lobes with mayo scissors. Place the right or left median lobe into the black sylgard-coated petri dish filled with cold calcium-free KH buffer, gently secure the lobe using Monoject™ standard hypodermic needles, and using a binocular dissection microscope, expose pulmonary arteries proximally to distally (*see* **Note 11**).
2. Fill the myograph chamber with cold calcium-free KH buffer and cut the vessels carefully under the microscope into 1.5 to 2 mM length rings by using Vannus spring scissors. Note down the length and internal diameter of the rings using a calibrated eyepiece reticle.
3. To study pulmonary artery rings in the absence of the endothelium, remove endothelium by passing a steel wire or a horsetail hair through the lumen followed by a gentle rubbing around the circumference.
4. Pulmonary artery rings can be mounted on the myograph by one of two methods depending on their internal diameter. For rings having an internal diameter >500 μm, use mounting pins. For ring preparations having an internal diameter >60 μm but <500 μm, use mounting jaws with stainless steel or tungsten wires.
5. Put the chamber with mounted vessel on the myograph machine, fill the chamber with Low Ca^{2+}-KH buffer and adjust the temperature to 37 °C. Start oxygenation of the Low Ca^{2+} KH buffer in the chamber.

6. Let the rings rest for 20 min without handling; after which, follow the normalization procedure using Lab chart's DMT normalization module at transmural pressure of 20 mM Hg using a predetermined normalization factor (*see* **Note 12**).
7. After normalization, let the rings equilibrate untouched for 45 min. Test ring viability by stimulating the ring with High KCl Ca^{2+} KH buffer for 2–3 min followed by 3–4 washes with Low Ca^{2+} KH buffer. Let the vessel return to baseline, after which repeat this step one more time (*see* **Note 13**).
8. At this time endothelium denudation/function can be tested by stimulation of the pulmonary artery ring with a contractile agent such as thromboxane receptor agonist U46619 at a sub-maximal response concentration (30 nM) followed by relaxation to the serial concentrations of acetylcholine (10^{-9} M to 10^{-6} M). Pulmonary artery rings with intact endothelium show complete relaxation to acetylcholine whereas relaxation to acetylcholine is absent in endothelium-denuded pulmonary artery rings.
9. Proceed with the designed contraction-relaxation experiments, in the presence or absence of inhibitors. For studying dilator GPCR responses, arteries may be pre-contracted with 30 nM U46619 (thromboxane mimetic).
10. Fix rings in situ at maximum contraction or relaxation response (one ring for each agonist concentration, in steady state post agonist challenge) by quickly filling the myograph chamber with 10% formalin for 15 min (*see* **Note 14**). Retrieve fixed rings from the myograph and immediately embed in optimal cutting temperature (OCT) compound in an end-on orientation (i.e. perpendicular to the sides of the OCT block).
11. Cut tissues as 10 μm cryosections, and study rings if properly oriented in cross-section. Visualize tissues by routine hematoxylin–eosin staining, stain to highlight target proteins (e.g. polymerized actin, Fig. 1) or immunostain with antibodies to specific phosphoproteins (e.g. phospho-myosin light chain, phospho-Rho, or other transient signaling intermediates). Counterstain nuclei with Hoechst 33342. Capture fluorescence staining at 40× magnification and quantify by image intensity analysis, or by counting per cell. Fluorescence per cell can also be quantified by laser scanning microscopy as a peripheral integral intensity contour around each nucleus (Fig. 1) (*see* **Note 15**).

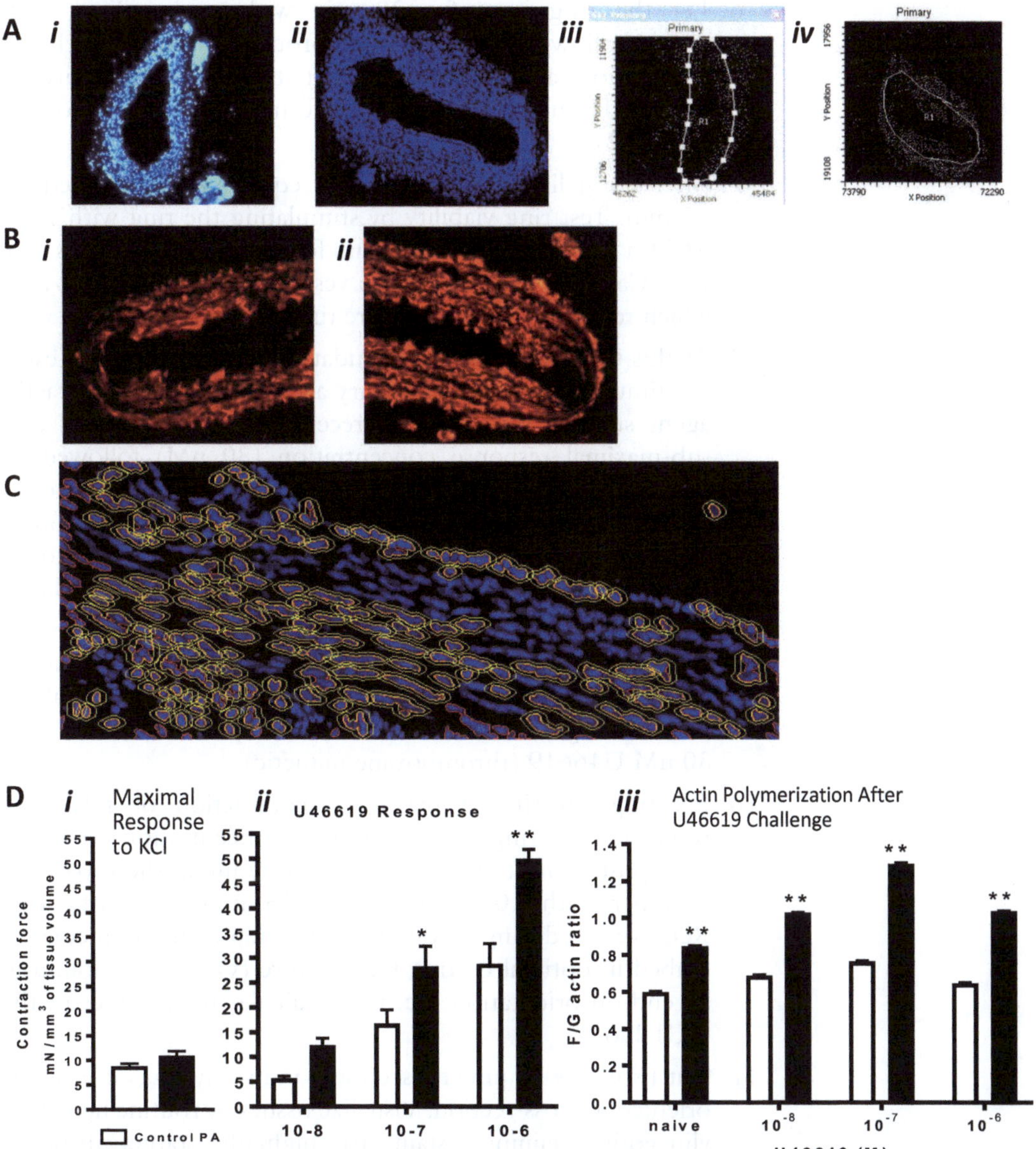

Fig. 1 Force myography and microscopy of contracted rings fixed in situ. Visualization of contracted arterial tissues by laser scanning microscopy. Nuclei stained with Hoechst 33342 to define primary contours. Slides are scanned at 40× magnification using blue channels to visualize nuclei. (**a**) Viewing arterial cross-sections using only the blue channel (**i** and **ii**), a region of interest encompassing the muscular layer is highlighted (**iii**) and selected (**iv**). An inner contour may be added to exclude the endothelium. For individual myocyte quantifications, scanning can be limited to nuclei within this region of interest. (**b**) Immunostaining with antibody to smooth muscle myosin heavy chain (red) delineates the full extent of the muscular layer in (**i**) control and (**ii**) PPHN pulmonary arteries. This cross-sectional area, multiplied by the thickness of the ring to give a muscle volume, can be used as a denominator to standardize isometric force measurements. (**c**) Quantification of immunostaining in individual myocytes in arterial media, by laser scanning microscopy. Single field image from an arterial ring, illustrating primary and peripheral contours. Slides are scanned to visualize blue channel (nuclei). Individual nuclei (20–200 μm^2) are selected for the primary contour (red line),

4 Notes

1. This cell culture method and all subsequently described procedures have been standardized using freshly obtained neonatal (day 0) porcine pulmonary artery, though methods are generalizable to all medium or large animal models. Up to sixth generation PA tissue may be obtained from larger lungs by careful removal of overlying tissues without applying traction to vessels. For small animal models, up to third generation pulmonary arteries may be obtained by microdissection.
2. Keep tissues fully submerged in cold Ca^{2+}-free KH buffer to maintain tissue integrity. Debris can be cleared from the dissecting field by drawing up a small amount of cold buffer in a syringe and squirting it gently over the vessel of interest.
3. Resting the isolated cells in a low calcium (but not calcium free) buffer improves viability and adherence of myocytes, yielding nearly a smooth muscle monoculture.
4. Contractile smooth muscle phenotype may be readily confirmed by immunocytochemistry staining for differentiated smooth muscle protein markers including alpha smooth muscle actin, smooth muscle type myosin heavy chain, desmin or SM-22 [9, 14].
5. This step must be carried out in the dark or with the vial wrapped in aluminum foil. Pluronic acid is a non-ionic surfactant that improves solubilization of fura-2AM ester. Omitting this addition results in in homogenous dye uptake and patchy appearance of extracellular fluorescent dye.
6. Prewarm HBSS to 37 °C for all washes.

Fig. 1 (continued) and used to derive a peripheral contours (yellow lines) comprising the estimated cytoplasmic area of each visualized cell, to calculate the integral pixel intensity of immunostaining per cell. (**d**) Isometric force of contraction of Control and Persistent Pulmonary Hypertension of the Newborn (PPHN) pulmonary arteries to (**i**) 100 mM KCl and to (**ii**) serial concentrations of U46619 (a thromboxane mimetic activating the TP GPCR). Force measurements normalized to arterial muscular volume, as determined by subsequent microscopy of the fixed vessel. (**iii**) Rings fixed at maximum contraction (15 min post-U46619 challenge) are cross-sectioned and stained with Phalloidin 633 (filamentous actin) and DNase 488 (globular actin) for laser-scanning cytometry. Nuclei are counterstained with Hoechst 33342 to define primary contours, and thence derive peripheral contours (as described above), for each visualized cell in the pulmonary artery muscular layer. F- and G-actin integral pixel intensity is quantified within each peripheral contour, and normalized to mean intensity of F-actin in Control (agonist-naïve) pulmonary artery. A composite F/G actin ratio is calculated from absolute integral pixel intensities, to quantify polymerization of actin in response to thromboxane receptor stimulation ($N = 3$, 3-day-old piglets; $^{*}p < 0.01$ and $^{**}p < 0.001$ compared with Control pulmonary arteries in same treatment group) (**a**) (ii), (**a**) (iv), (**c**) and (**d**) originally published in *Am J Physiol Lung Cell Mol Physiol* 302:L13–L26, 2012 [11]; reproduced by author with permission

7. For optimal data capture, create up to eight identically-sized regions of interest per microscope field, each containing portions of 3–5 cells, ensuring that regions of interest selected do not contain precipitate, and that the image is stable through the recording.
8. Lysates should be prepared immediately before experimentation, and maintained on ice.
9. A detailed protocol for determining concentration of solubilized protein in a lysate, following the Bradford method, is given on the Biorad website http://www.bio-rad.com/webroot/web/pdf/lsr/literature/LIT33.pdf.
10. The decrease in fluorescence is proportional to the increase in AC activity; AC activity data are thus represented as Δfluorescence (inversely)/mg lysate protein/min of measurement.
11. Pulmonary arteries run just beneath the airways; to expose them, carefully cut the airway using Wescott spring scissors and lift away from the field. Use Dumont #5/45 forceps to loosen the parenchyma around the vessel of interest. Use Vannus spring scissors to cut the desired length of the vessel, remove parenchyma around the vessel and lift vessel free without stretching. Store the vessel in a glass beaker filled with Ca^{2+}-free KH buffer, on ice.
12. Normalization factor, or IC_1/IC_{20} ratio, is a ratio between the vessel internal circumference at which active force generation is maximal, and the internal circumference obtained when that vessel is subject to an optimal physiologic transmural wall pressure. It should be determined for each vessel type under study. Step-by-step details of the normalization procedure are given on the DMT website www.dmt.dk/multi_myograph_620m.html, along with the method for determination of an artery-specific normalization factor. The objective of normalization is to set the vascular ring at a consistent physiological resting internal circumference which gives rise to maximal active force and agonist response. Vessels can also be normalized by the traditional method of length tension analysis, stretching the ring to incremental increases in circumference then subjecting the vessel to a KCl challenge at each set point, and selecting as the optimal internal circumference whichever one generates the maximum active force, without accrual of passive force. This method has gone out of current use in favor of the normalization procedure.
13. A ring that has been excessively handled or stretched longitudinally during dissection or mounting, and is therefore exhibiting lower than expected force responses, can be recovered by serial KCl challenge with intervening periods of rest. Repeat

cycles of KCl stimulation until the maximal force response has plateaued, to ensure a stable preparation.

14. The myograph chamber will need to be flushed repeatedly to clear formalin residue before loading the next vessel.
15. Force measurements may now also be normalized to the cross-sectional area of the tunica media (or 3-dimensionally, muscle tissue volume, calculated by multiplying tunica media cross-sectional area by length of ring) for each fixed vessel.

Acknowledgments

Supported by a grant-in-aid from Heart and Stroke Foundation and a mid-career operating grant from Research Manitoba (SD).

References

1. Iyinikkel J, Murray F (2018) Gpcrs in pulmonary arterial hypertension: tipping the balance. Br J Pharmacol 175(15):3063–3079
2. Jeffery TK, Morrell NW (2002) Molecular and cellular basis of pulmonary vascular remodeling in pulmonary hypertension. Prog Cardiovasc Dis 45(3):173–202
3. Insel PA, Wilderman A, Zambon AC, Snead AN, Murray F, Aroonsakool N, McDonald DS, Zhou S, McCann T, Zhang L et al (2015) G protein-coupled receptor (gpcr) expression in native cells: "novel" endogpcrs as physiologic regulators and therapeutic targets. Mol Pharmacol 88(1):181–187
4. Shimoda LA, Sham JS, Sylvester JT (2000) Altered pulmonary vasoreactivity in the chronically hypoxic lung. Physiol Res 49 (5):549–560
5. Insel PA, Snead A, Murray F, Zhang L, Yokouchi H, Katakia T, Kwon O, Dimucci D, Wilderman A (2012) Gpcr expression in tissues and cells: are the optimal receptors being used as drug targets? Br J Pharmacol 165 (6):1613–1616
6. Gao QB, Wang ZZ (2006) Classification of g-protein coupled receptors at four levels. Protein engineering, design & selection. PEDS 19 (11):511–516
7. Hu J, Xu Q, McTiernan C, Lai YC, Osei-Hwedieh D, Gladwin M (2015) Novel targets of drug treatment for pulmonary hypertension. Am J Cardiovasc Drugs 15(4):225–234
8. Wang G, Jacquet L, Karamariti E, Xu Q (2015) Origin and differentiation of vascular smooth muscle cells. J Physiol 593(14):3013–3030
9. Halayko AJ, Rector E, Stephens NL (1997) Characterization of molecular determinants of smooth muscle cell heterogeneity. Can J Physiol Pharmacol 75(7):917–929
10. Postolow F, Fediuk J, Nolette N, Hinton M, Dakshinamurti S (2011) Hypoxia and nitric oxide exposure promote apoptotic signaling in contractile pulmonary arterial smooth muscle but not in pulmonary epithelium. Pediatr Pulmonol 46(12):1194–1208
11. Fediuk J, Gutsol A, Nolette N, Dakshinamurti S (2012) Thromboxane-induced actin polymerization in hypoxic pulmonary artery is independent of rho. Am J Physiol Lung Cell Mol Physiol 302(1):L13–L26
12. Grynkiewicz G, Poenie M, Tsien RY (1985) A new generation of ca2+ indicators with greatly improved fluorescence properties. J Biol Chem 260(6):3440–3450
13. Spangler CM, Spangler C, Gottle M, Shen Y, Tang WJ, Seifert R, Schaferling M (2008) A fluorimetric assay for real-time monitoring of adenylyl cyclase activity based on terbium norfloxacin. Anal Biochem 381(1):86–93
14. Pfaltzgraff ER, Bader DM (2015) Heterogeneity in vascular smooth muscle cell embryonic origin in relation to adult structure, physiology, and disease. Dev Dyn 244(3):410–416

INDEX

A

Adeno-associated viruses (AAV) 363
Adenosine A_{2A} receptor (Adora2a) 58, 59, 61, 62, 66–71, 74, 75
Aggregation 34, 195, 284, 289, 297–299
Aggresome 289–291, 294, 295, 297, 299, 301
Agonist
  allosteric 105, 111
  biased 151
  orthosteric 9, 105, 127, 178, 179
Amido black 10B protein assay 38, 46, 48
Analgesia 352, 354
Antibody 86, 91, 94, 95, 99, 100, 207, 269–272, 277, 281, 282, 291, 295, 301, 307, 319, 320, 325, 326, 328, 330, 334, 335, 340, 341, 345, 346, 363–366, 373, 398
Apo-state 33, 34, 47
Autoradiography 66, 67, 69, 70, 72, 75

B

Bacterial transcription factor tTA 258
β-Arrestin 257–266, 323, 324
Bimolecular fluorescence complementation (BiFC) 199–214
Binding 4, 5, 7, 8, 10, 34–36, 49, 57, 68, 82, 95, 96, 101, 103, 105–107, 110, 111, 119, 127, 128, 130, 138, 151, 153, 156, 157, 159–167, 170, 183, 218, 245, 257, 318, 323, 339
Bioluminescence
  ex vivo 362, 364, 369–371
  imaging 366, 369
  in vitro 369
  in vivo 362, 366–369
Bioluminescence resonance energy transfer (BRET)
  acceptor:donor ratio (A:D) 184
  $BRET^2$ 200–202, 204, 207–210, 213, 214
  ΔBRET 178, 179
  efficiency ($BRET_{eff}$) 184, 185, 188, 190–197, 208, 214
  saturation assays 184, 213
  type-1 assays 184, 185, 187, 188, 193, 194
  type-2 assays 184, 185, 188, 193, 194
  type-3 assays 184, 185, 187, 188, 193, 195–197

C

cAMP 131, 217, 219, 220, 227–231, 258, 345, 362, 364, 367, 369, 371, 374, 389, 391
Cell-based assays 258
Cell culture
  transfection
    lipofectamine 113, 115, 173, 207, 276, 281, 306, 308, 340, 341
Cell-free protein synthesis
  amber suppression 61
  microsomes 58
  non-canonical amino acid 58, 69
  protein modification 58
Cerebral blood flow (CBF) 377–386
Chemical cross-linking 82, 83
Chemical ligation 32, 138
Chinese hamster ovary cells (CHO) 58, 61, 65, 73, 74, 118, 157, 159, 163, 219, 232, 233
$^{13}CH_3$-methionine 31, 43, 46, 47
Clozapine N-oxide (CNO) 105, 362, 364, 366, 372, 373
Coelenterazine-400A 186, 190, 196
Coelenterazine h 173, 176, 177, 180, 181, 186, 192, 196
Co-immunoprecipitation (co-IP) 200, 270, 274, 277, 280, 281, 283
Computational analysis 390
Confocal laser scanning microscopy
  image analysis 292
  image processing 292, 296, 297, 311
  Z-stack acquisition 311
Conformation-sensitive biosensors 170–172, 176
Constitutively active Rab5 (Rab5-Q79L) 304, 305, 308
CREB luciferase reporter 366, 368, 369, 371, 374
CRISPR/Cas9 .... 111, 169–182

Mario Tiberi (ed.), *G Protein-Coupled Receptor Signaling: Methods and Protocols*, Methods in Molecular Biology, vol. 1947, https://doi.org/10.1007/978-1-4939-9121-1,

D

Designer receptors exclusively activated by designer drugs (DREADDs) ...105, 361–374
D-luciferin ...260, 265, 364, 366, 371, 374
Dose–response curve ...129, 131, 224, 233, 234
Drug discovery ...82, 169, 170, 352

E

EC_{50} ...218–220, 222, 224, 226, 228–231, 233, 235
Electrophoresis
  Laemmli sample buffer ...274
  PVDF membrane ...89, 95, 364
  SDS-PAGE ...49, 85, 89, 94, 274
Electrophysiology
  G protein-coupled Inward Rectifying Potassium Channel (GIRK1) ...121
  whole cell patch clamp ...117–119
Endocytosis ...324
Endosomes ...303–305, 308, 309, 311–314, 319–321, 324
Enhanced orthogonal aminoacyl-tRNA-synthetase ...59, 60, 63, 64
Escherichia coli (*E. coli*) ...31–52, 58, 59, 63, 120, 205, 206

F

FlAsH ...170, 171, 173–181
Fluorescence resonance energy transfer (FRET)
  R_0 ...152, 219–222, 224, 228, 229, 231, 233
  R_{max} ...219–222, 224, 226, 228, 229, 231, 232
  R_{min} ...219, 222, 226, 228–231, 233, 234
  time-resolved (TR-FRET) ...137–145, 151–167
Fluorescent labeling
  bodipy-TMR-lysine ...65–67, 72
  in-gel fluorescence ...66, 67, 70–72
Fluorescent ligands ...152, 154–156, 161, 166, 319
Fluorescent probes ...137–145
Fluorophore characterization ...326, 330–333
Furan ...81–101
Fusion proteins ...33, 34, 43–45, 49, 50, 66, 213, 214, 258

G

Genome editing ...111, 171, 172, 178, 180
Glo reagent ...260, 262, 265
Glutathione Sepharose ...39, 44, 50, 245, 247
G protein-coupled receptors (GPCRs) ...v, 3–16, 21–29, 31, 57–75, 81–101, 103–132, 137–145, 151–167, 169–183, 199–214, 241, 249, 257, 259, 262, 264, 269–271, 282–284, 289, 297–299, 304, 323–346, 351, 361–374, 377–386, 389–401
Green fluorescent protein (GFP) ...49, 121, 155, 183, 200, 311, 365
Ground state depletion microscopy (GSDM) ...323–336
G_s ...131
GST-tagged HRV 3C protease ...39, 44, 50

H

HeLa cells ...99, 118
Hill coefficient ...218–220, 224, 226, 228–231, 233, 235
His_{10}-tag ...34–36, 43, 45
HTLA cells ...260, 265
HTRF® ...153, 156, 157
Human delta opioid receptor (DOPr) ...290, 298, 305
Human embryonic kidney 293 (HEK293) ...99, 157–159, 163, 233, 253, 265, 269–271, 276, 277, 290–294, 297, 304, 305, 308, 327, 334, 339, 340
Human embryonic kidney 293 T (HEK293T) ...260
Hydrazinopeptides ...137–145

I

Imidazole removal ...43, 44
Immobilized metal-affinity chromatography (IMAC) ...34, 38, 39, 41, 43–46, 49
Immunofluorescence ...320
Intraluminal vesicles (ILVs) ...303–305, 311
IRDye 680 Streptavidin ...86, 91
Isopropyl β-D-1-thiogalactopyranoside (IPTG) ...33, 37, 41, 59, 63

K

Knock-In ...361–374

L

Lanthanides ...139, 153–155, 163
Laser doppler flowmetry (LDF) ...377–386
Liquid chromatography ...84, 279
Localization microscopy ...337–346
Luciferase
  Renilla luciferase II (RlucII) ...170, 176–178

M

Mass spectrometry ...269, 271–274
MDA-MB-231 cells ...83, 89, 93, 99
Membrane proteins ...57, 58, 67, 69, 115, 137, 151, 289

Metabotropic glutamate (mGluRs) receptors
mGluR2 receptors 106, 108, 112, 117, 119–121, 123–127, 129
mGluR3 receptors 106, 108
$mGlu_5$ receptor 352
mGluR6 receptors 106, 108, 120, 209, 212, 213
Microplate
reader 63, 158, 164, 181
Misfolding 298
Molecular dynamics
communication networks 7
conformational ensemble 7, 10
Markov State Model 8, 15
multiscale modeling 5
root mean square deviations (RMSDs) 24, 25
water permeation 21
Mouse 112, 184, 219, 221, 230, 275, 281, 291, 297, 340, 353, 354, 356, 358, 362–366, 368–371, 374, 379, 381–384, 386
Multivesicular bodies (MVBs) 303
Myography 391, 393, 394, 396–398

N

NanoLuciferase (Nluc) 59, 61, 66
Neuromodulation 103
Neurotensin receptor 1 (NTS1) 31
Neurovascular coupling (NVC) 378, 383, 386
Nuclear magnetic resonance (NMR) 31

O

Oligomerization
hetero-oligomerization 157, 160
homo-oligomerization 151, 157, 160
Optogenetics 103
Optopharmacology 351, 352

P

Pain 4, 352, 381
Photoactivatable dyes (PD)
CAGE dyes 340, 341, 345
Photoactivated localization microscopy
PD-PALM 338–345
Photoswitchable orthogonal remotely tethered ligands (PORTLs) 103–132
Physiology .... v, 3, 5, 57, 98, 103, 104, 106, 112, 118, 126, 151, 169, 184, 213, 303, 338, 352, 354, 361, 373, 378, 380, 385, 389, 400
Poly-L-lysine (PLL) 113, 115, 259, 261, 262, 265, 291, 294, 300, 306, 309, 316, 334
Polymerase chain reaction (PCR) 60, 64, 65, 73, 170, 187, 206, 213
Post-endocytic sorting 303
Prokaryotic expression 63
Protein purification 37–39
Protein-tyrosine phosphatase (PTP)
colorimetric phosphatase assay 241, 249
in-gel phosphatase assay 241, 249
phosphatase activity 242, 243
Proteome v
PROTEOSTAT® 290, 291, 294, 297, 299
Pulmonary artery 389–401

R

Rat 33, 35, 219, 233, 272, 275, 277, 379
Reactive oxygen species (ROS) 89, 93, 100
Real time imaging 170, 217
Rhodopsin 3, 21–23, 25–27, 29, 32, 81, 105, 289
Rodent 352, 377–386

S

Scintillation counting 66, 67, 69, 72
Self-labeling enzyme 155, 160, 164, 165
Sequential BRET2-FRET (SRET2) 200, 301
Signaling .... v, 5, 8, 9, 35, 47, 57, 81, 103–105, 121, 131, 137, 241–244, 258, 303, 305, 323, 324, 361–374, 378, 389–391, 397
Silencing RNA (siRNA) 305, 306, 308, 309, 311, 313, 316, 318
Singlet oxygen 93, 100
Size exclusion chromatography (SEC) 34, 39, 43, 45, 46, 50
Smooth muscle cells 390, 391
SNAP tag 107, 110–113, 115, 116, 120, 121, 127, 130, 131, 155, 159, 164, 165
Solid phase peptide synthesis (SPPS) 83, 86, 92
Spatial assembly 337–346
Spectrophotometer 85, 98, 100, 113, 122, 123, 254
Stoichiometry 124, 183–197, 337–339
Super-resolution microscopy 323–336, 338, 339, 341, 344
Suppressor tRNA 58–61, 64–65, 68

T

Tagged receptors 127, 130, 131, 158, 166, 184
Tag-lite® screening 152, 158–161, 164
TALON® slurry 41, 44
Tamoxifen 363, 365

Tango 257, 259, 262, 264
Terbium 139, 153–155, 393, 395
Tethered ligands, *see* PORTLs
Tetracysteine tag 170, 171
Thermostable 31, 35
3C protease cleavage....34, 36, 43, 44
Tissue
  specific activation....361–374
  temporal activation....361–374
  tissue perfusion units (TPU)....383, 386
Tobacco etch virus protease cutting site (TEVcs)....258
Transactivator (tTA)-driven luciferase 260
Translocation 69, 258

V

Vasoconstriction 390, 391

W

Western blots....82, 85, 88, 89, 94–97, 99, 100, 249, 251, 254, 270, 271, 274–277, 280–283, 300, 319, 320, 364–366, 373
Whisker stimulation 383, 386

MIX
Papier aus verantwortungsvollen Quellen
Paper from responsible sources
FSC® C105338

If you have any concerns about our products,
you can contact us on
ProductSafety@springernature.com

In case Publisher is established outside the EU,
the EU authorized representative is:
Springer Nature Customer Service Center GmbH
Europaplatz 3, 69115 Heidelberg, Germany

Printed by Libri Plureos GmbH
in Hamburg, Germany